Current Topics in Behavioral Neurosciences

Volume 68

Current Topics in Behavioral Neurosciences provides critical and comprehensive discussions of the most significant areas of behavioral neuroscience research, written by leading international authorities. Each volume in the series represents the most informative and contemporary account of its subject available, making it an unrivalled reference source. Each volume will be made available in both print and electronic form.

With the development of new methodologies for brain imaging, genetic and genomic analyses, molecular engineering of mutant animals, novel routes for drug delivery, and sophisticated cross-species behavioral assessments, it is now possible to study behavior relevant to psychiatric and neurological diseases and disorders on the physiological level. The Behavioral Neurosciences series focuses on translational medicine and cutting-edge technologies. Preclinical and clinical trials for the development of new diagnostics and therapeutics as well as prevention efforts are covered whenever possible. Special attention is also drawn on epigenetical aspects, especially in psychiatric disorders.

CTBN series is indexed in PubMed and Scopus.

Tomáš Paus • Jeffrey R. Brook •
Katherine Keyes • Zdenka Pausova
Editors

Principles and Advances in Population Neuroscience

Springer

Editors
Tomáš Paus
Departments of Psychiatry and Neuroscience, Faculty of Medicine and Centre Hospitalier Universitaire
University of Montréal
Montreal, QC, Canada

Jeffrey R. Brook
Dalla Lana School of Public Health
University of Toronto
Toronto, ON, Canada

Katherine Keyes
Epidemiology
Columbia University
New York, NY, USA

Zdenka Pausova
Centre hospitalier universitaire Sainte-Justine and Department of Pediatrics
University of Montreal
Montreal, QC, Canada

ISSN 1866-3370 ISSN 1866-3389 (electronic)
Current Topics in Behavioral Neurosciences
ISBN 978-3-031-70136-8 ISBN 978-3-031-70137-5 (eBook)
https://doi.org/10.1007/978-3-031-70137-5

This Springer imprint is published by the registered company Springer Nature Switzerland AG
The registered company address is: Gewerbestrasse 11, 6330 Cham, Switzerland

Preface

Over 10 years ago, I sat down to write a small monograph on Population Neuroscience. It summarized basic concepts of a discipline that has emerged at the intersection of epidemiology, genetics and neuroscience, the latter enabled by advances in high-throughput brain imaging. By and large, the book was based on our experience with designing, and running, the Saguenay Youth Study. Together with my colleague and partner Zdenka Pausova, we combined our interests in brain & behaviour (TP) and cardio-metabolic health (ZP), and our expertise in brain imaging (TP) and genetics (ZP), to embark on studying the adolescent brain and body. This required moving outside of our respective disciplines, learning from each other and from our colleagues working in epidemiology and statistics, engineering, sociology, molecular biology and more... It was this journey—combined with our involvement in similar efforts—that led to the formation of an integrated multi-domain and multilevel approach of population neuroscience.

Since the initial studies, the field of population neuroscience has matured, new ever-larger cohorts were set up, and new technologies emerged. When approached by Springer to write another book on this topic, it was clear that help was needed. Over the past 10+ years, I have been fortunate to interact with many excellent scholars interested in various aspects of studying brain health across the human lifespan. For the new book, three colleagues helped with editing the following three sections. The Section on "Populations" was shaped by Kerry Keyes, an innovator in the field of epidemiology. It contains chapters on generalizability and transportability (Keyes and colleagues), data sharing and protection of privacy (Thorogood), diversity (Shaaban and Rosso) and sex & gender (Vosberg). The Section on "Genes and Molecules" was guided by Zdenka Pausova, an expert in molecular and population genetics and omics sciences. It contains chapters on genetic association studies (Sargurupremraj), transcriptomics (Liharska and Charney) and metabolomics (Pausova and Sliz). The Section on "Environment" was edited by Jeff Brook, a leader in geospatial science of urban health. It contains chapters on physical (Smith), urban (Balsa-Barreiro and colleagues) and social (Helbich and colleagues) environments, as well as one on open environmental-data platforms (Noaeen and colleagues). Finally, I have edited the Section on "Brains", containing chapters on development and maturation of the human brain (Paus), its aging (Fjell) and on large-scale studies of the human brain and mental health (Ching and colleagues).

Together with the three co-editors, I hope that exposing you—the reader—to this collection of ideas, approaches and knowledge, you will experience the same wonderment I have when working—over the past 20 years—with my colleagues and students! To whom I owe a debt of gratitude.

Montreal, Canada Tomáš Paus
July 4, 2024

Contents

Part I
Introduction

Population Neuroscience: Principles and Advances

Tomáš Paus

Contents

Abstract In population neuroscience, three disciplines come together to advance our knowledge of factors that shape the human brain: neuroscience, genetics, and epidemiology (Paus, Human Brain Mapping 31:891–903, 2010). Here, I will come back to some of the background material reviewed in more detail in our previous book (Paus, Population Neuroscience, 2013), followed by a brief overview of current advances and challenges faced by this integrative approach.

Keywords Aging · Development · Epidemiology · Genetics · History · Mental health · MRI · Neuroimaging

T. Paus (✉)
Department of Psychiatry and Neuroscience, Faculty of Medicine, University of Montreal, Montreal, QC, Canada

Curr Topics Behav Neurosci (2024) 68: 3–34
https://doi.org/10.1007/7854_2024_474
Published Online: 9 April 2024

1 Points of Departure

1.1 Neuroscience

Let us begin by looking back at some of the key approaches and discoveries with regard to the human brain and its functional and structural organization. The twentieth century generated an incredible amount of foundational knowledge about the primate brain, which provides an important stepping stone for today's work in population neuroscience.

In the case of the *human primate*, critical discoveries were made using two approaches: lesion studies (neuropsychology) and neuroimaging (cognitive neuroscience). The former approach (neuropsychology) is exemplified by the work carried out at the Montreal Neurological Institute by Brenda Milner and her colleagues, from the early 1950s onward; it revealed the role of the hippocampus in declarative memory (Scoville and Milner 1957; Milner 2005; Milner et al. 1997; Pigott and Milner 1993), prefrontal cortex in executive functions (Milner 1982), and hemispheric specialization in verbal and non-verbal domains (Frisk and Milner 1991; Smith and Milner 1989; Milner et al. 1968). The latter approach (cognitive neuroscience) started in earnest in the late 1980s and early 1990s with the use of positron emission tomography (PET) to measure regional changes in cerebral blood flow in healthy volunteers while performing a variety of sensory, motor, or cognitive tasks (Fox and Raichle 1984; Lueck et al. 1989; Talbot et al. 1991; Petersen et al. 1988). Since the late 1990s, PET has been largely replaced[1] by functional magnetic resonance imaging (MRI) that measures regional fluctuations in the blood-oxygen-level dependent (BOLD) signal, a hemodynamic correlate of brain "activity" acquired in a non-invasive manner (Belliveau et al. 1991). In addition to studying brain function, MRI offered another opportunity: mapping structural properties of the human brain, its gray and white matter. While T1-weighted images became the workhorse of brain "morphometrics" (e.g., hippocampal volume, cortical thickness), wealth of additional measures can be derived with multimodal imaging, which includes various MR sequences such as diffusion tensor imaging (DTI), T1 and T2 relaxometry, magnetization transfer (MT) imaging, and others (Lerch et al. 2017). But note the challenge of interpreting these metrics vis-à-vis their underlying neurobiology (see Sect. 2.2).

In the case of *non-human primates*, two domains stand out with regard to our knowledge of the structural and functional organization of the cerebral cortex. The first one is that of neuronal connectivity in general, and cortico-cortical connectivity in particular. From the early 1980s onward, a number of laboratories used tracers such as horseradish peroxidase to map in detail connections between small populations of neurons located in different brain regions (Mesulam 1982). This body of work represents foundational knowledge about the true neuronal connectivity of the primate brain, not available in humans. The so-called "wiring diagrams,"

[1] PET continues to be used for studies of specific neurotransmitters and other molecules of interest.

such as those of the visual system (Felleman and Van Essen 1991), are based on these experimental studies, which require injecting a tracer into a specific brain region in a live animal and studying (post mortem) the spread of the tracer (by axonal transport) under the microscope. The second domain is that of recording neuronal activity in behaving monkeys, an approach pioneered by Mountcastle and others in the late 1970s (Yin and Mountcastle 1977). Recording neuronal activity in various regions of the monkey cerebral cortex provided unprecedented insights into the functional variety of neurons found in different cortical regions and coding – on a millisecond scale – characteristics of various stimuli (Desimone and Ungerleider 1986; Gross et al. 1979), directions of an arm movement (Georgopoulos et al. 1982, 1989), or positions of targets after they disappeared, namely spatial working memory (Fuster and Alexander 1971; Kojima and Goldman-Rakic 1982), among many other features the external world, as well as its internal representations. In the early 2000s, the combination of neurophysiology with functional MRI revealed which neuronal events drive the hemodynamic response function (Logothetis et al. 2001), knowledge critical for the interpretation of the thousands of human studies using this indirect approach to measuring brain "activity." As we reviewed elsewhere, it is likely that regional variations in cerebral blood flow are driven by changes in (excitatory) synaptic activity, rather than neuronal firing (Paus 2003).

Collectively, the body of knowledge about the structural and functional organization of the primate brain in general, and cerebral cortex in particular, provides critical background for studies of factors shaping brain and behavior in health and diseases, and across the lifespan. For the latter perspective, foundational studies of brain development (Rakic 1988) and aging (Morrison and Hof 1997) are important sources of biological knowledge relevant for the interpretation of these studies.

1.2 Genetics

Needless to say, progress in genetics over the past 30+ years was nothing short of astounding. In 1990, the US National Institutes of Health (NIH) and Department of Energy (DOE) launched the Human Genome Project (HGP). Over the next 15 years, and with an annual budget of $ 200 million provided by the NIH and DOE, the HGP's plan was to map and sequence the 3.2-billion-nucleotide-long genome: creating genetic and physical maps of the human genome, identifying all genes, and sequencing the entire DNA (Watson and Jordan 1989). This has been achieved: libraries of 10 million+ single-base markers (single nucleotide polymorphisms, SNPs) exist, and the entire human genome has been sequenced (Nurk et al. 2022). Technologies for mapping SNPs (microarrays) and for DNA sequencing have become inexpensive, allowing for large-scale applications (see chapter on "Approaches to quantifying gene expression and their application to studying the human brain"). These advances enabled, in turn, large-scale genomic studies of complex traits, that is traits not following Mendelian (monogenic) inheritance (Lander and Schork 1994).

In Genome-Wide Association Studies (GWAS), one correlates – across the entire genome – allelic variations (captured by millions of SNPs) with a particular trait (e.g., hippocampal volume). Given the multigenic nature of complex traits, the fact that SNPs are located at genomic locations with various functions (e.g., coding, non-coding, regulatory), and due to other complexities of the genotype–phenotype relationships involved in non-Mendelian inheritance (Lander and Schork 1994), effect sizes revealed by GWASs at a given location are very small. Detecting effect sizes at this level with sufficient statistical power, especially given the large number of statistical tests, requires large number of individual genotype–phenotype datasets, often not available in a single study. For this reason, investigators have started pooling their data through meta-analyses, the so-called meta-GWAS. Furthermore, new tools have been developed for combining SNP-based variations across the genome, such as a software package for Genome-Wide Complex Trait Analysis (Yang et al. 2011) and PRSice software (https://choishingwan.github.io/PRSice) for calculating polygenic risk scores (Marees et al. 2018).

The following three international consortia brought together researchers interested in the human brain and behavior. Psychiatric Genomics Consortium started its work in 2007 (Sullivan 2010). It investigates genetics of psychiatric disorders, including attention deficit hyperactivity disorder (ADHD), autism, bipolar disorder, eating disorders, major depressive disorder, obsessive-compulsive disorder/Tourette syndrome, posttraumatic stress disorder, schizophrenia, and substance use disorders; at present, it includes over 400,000 cases (Sullivan et al. 2018). Enhancing Neuro-Imaging Genetics through Meta Analyses (ENIGMA) Consortium was initiated in the early 2010s (Thompson et al. 2014). It is focused on the relationship between genetic variations and various phenotypes derived from the brain MRI; at present, it includes over 30,000 datasets (Medland et al. 2022). The Cohorts for Heart and Aging Research in Genomic Epidemiology (CHARGE) consortium (Psaty et al. 2009), its NeuroCHARGE working group, came together in the early 2010s. Its efforts focus on genetics of cognitive abilities, all-cause dementia, and MRI-derived phenotypes; at present, it includes over 50,000 datasets (Davies et al. 2015; Bellenguez et al. 2022; Satizabal et al. 2019; Hofer et al. 2020; Shin et al. 2020; Duperron et al. 2023). Some of the recent outputs of these collaborations include reports on genetic variations associated with the tangential growth of cerebral cortex indexed by its surface area (Shin et al. 2020; Grasby et al. 2020), subcortical volumes (Satizabal et al. 2019), cerebral small vessel disease (Duperron et al. 2023), and attention deficit hyperactivity disorder (Demontis et al. 2023). Other international consortia are focusing on the genetics of specific neurological conditions, such as stroke (Malik et al. 2018), multiple sclerosis (International Multiple Sclerosis Genetics, and Multiple MS Consortium 2023), and brain cancers (Choi et al. 2023; Shete et al. 2011).

Interpretation of findings obtained in GWAS-based analyses is facilitated by the availability of tools allowing one to map the newly discovered genetic loci to functional genomics data, such as those provided by the ENCODE Consortium (ENCODE Project. Consortium et al. 2020). Furthermore, existing brain atlases of gene expression in the adult (Hawrylycz et al. 2012) and the developing (Li et al.

2018) cerebral cortex are indispensable for revealing neurobiological underpinnings of MRI-based phenotypes, such as cortical thickness or surface area, as well as for examining potential impact of genetic variations on gene expression, namely expression Quantitative Trait Loci, eQTLs. Finally, tools and atlases (Min et al. 2021) are available for deriving methylation QTLs (meQTLs) across the genome, thus facilitating work on the regulation of expression, in the human brain, of genes of interest. Some examples of the above are described in Sect. 2.3.

Altogether, genetics of brain and behavior reached a point where we are able to identify common genetic variants associated with multiple complex traits, but also begin to understand mechanistic pathways through which some of these variants shape these traits.

1.3 Epidemiology

Epidemics of communicable (infectious) diseases were the key raisons d'être of this field as it emerged in the mid-1800s (Morabia 2004). Major outbreaks of diseases such as cholera (e.g., London in 1854) or COVID-19 (global pandemic of 2020/2021) represent time-limited events whereby an external agent causes a disease. Note, however, that the causal agent may not be known at the time of the event. In case of the 1854 London epidemic of cholera, the distal cause of the disease, contaminated water, was identified through epidemiological investigation well before the proximal cause, the bacterium *Vibrio cholerae* (Morabia 2001).

In the 1960s, the interest of epidemiologists shifted to morbidity and mortality associated with non-communicable diseases, such as cardiometabolic diseases, psychiatric disorders, and cancer (Cohen et al. 2007). Unlike epidemics of communicable diseases mentioned above, many of the non-communicable diseases have a long non-symptomatic period during which relevant tissues (e.g., vessels, brain) undergo gradual deterioration of their function and structure. It is during this – often long - preclinical stage of the disease that a variety of external factors play a key role in shaping health-to-disease trajectories. Perhaps not surprisingly, epidemiology of non-communicable diseases turned its attention to "living conditions," with particular emphasis on social factors; Departments of "Social Medicine," "Community Medicine" or "Preventive Medicine" began opening their doors during this period (Vandenbroucke 1990). The complexity of identifying novel associations between possible "exposures" and "outcomes" at a population level stimulated the development of general concepts (e.g., prevalence, incidence, risk ratios) and designs (e.g., case–control, cohort) facilitating the work of epidemiologists (Morabia 2004).

In psychiatry, the emphasis on social factors was particularly strong. Between 1966 and 2004, the number of citations referencing psychiatric disorders together with subject headings indicating social factors, such as "residence characteristics," "social environment," or "social welfare," increased about eight fold (Cohen et al.

2007).[2] In the late 1980s, neurodevelopmental hypothesis of schizophrenia emerged, drawing attention to early developmental events affecting brain development and, in turn, increasing the probability that an individual may develop psychosis later in life (Lewis and Murray 1987; Weinberger 1987); recently, this idea received strong support in a systematic review and meta-analysis of the association between a variety of perinatal risk factors and psychosis (Davies et al. 2020). Preterm birth (Nosarti et al. 2012) and low birthweight even in newborns born at term (Abel et al. 2010) stand out in this respect. Needless to say, conceptualization of psychiatric disorders as disorders of brain development is consistent with the interest in social environment, especially during pregnancy. It is important to keep in mind, however, that the relationship between environment and genes is complex, often confounding each other (Havdahl et al. 2022; Pingault et al. 2023).

Assessment of our environment is challenging. As we described previously (Paus 2016), there are countless permutations of physical, built and social environments that surround us in space and time. We both "receive" and "create" our environments (Kendler et al. 2003), thus co-determining what air we breathe, how we go from one place to another, and who we see, hear and interact with when in school, at work or during our commutes.

A common way of collecting information about the participant's physical, built, and social environment is asking them about it. For example, the PhenX Toolkit contains protocols one can use when collecting information about social environment; using this term, a search under "Browse Protocols" yields 256 protocols classified into multiple domains, including Individual/Structural Determinant of Health, Social Environments, Psychosocial, Demographics, and many others (www.phenxtoolkit.org). Although valuable, there are three main disadvantages of such a survey-based approach: (1) participants' engagement (and consent); (2) their time; and (3) self-reported nature of the collected information. Given these limitations, survey-based assessment of the individual's environment is practical in studies targeting particular aspects of environment as experienced by the individual at a particular point in time.

Instead of asking a person to describe their environment in multiple domains, we can extract – from open sources – area-level characteristics of the physical, built, and social environment in which they live. This approach is described extensively in the "Environment" section of this book (Paus 2024a) and elsewhere (Paus 2024b). The key advantage of this approach is that it can be applied on a large scale, across countries and time periods. In turn, information about area-level environment (at a particular point in time) can be combined with an individual-level of information about the person's health (e.g., from administrative health databases) or individual-level information acquired in large research projects (see Sect. 2.4).

Briefly, geospatial science and related tools enable spatial analysis and visualization of external environments in which we spend considerable amount of our lives.

[2] This trend might have been attenuated by the concurrent emergence of psychopharmacology, which put an individual rather than society into the spotlight vis-à-vis treatment strategies.

Datasets can be created at different levels of spatial granularity matching the goals of a given study and availability of relevant data. In Canada, for example, geographic units including six-digit postal codes, Canadian Census geographic units such as dissemination areas (400 to 700 persons) and census tracts (2,500 to 8,000 persons), or larger areas such as city districts. The temporal dimension depends on the type of data; it may range from data sampled monthly (e.g., air quality), annually (e.g., public transportation) or up to every 5 years (e.g., the Canadian Census). The spatio-temporal datasets can be created using existing tools and databases provided by large GIS-based (*Geographic Information Systems*) organizations and companies, such as ESRI (www.esri.com), DMTI Spatial (www.dmtispatial.com), Google Earth Engine (https://earthengine.google.com), as well as open sources (e.g., www.openstreetmap.org), government sources (e.g., Statistics Canada), and academic organizations. One can also derive relevant metrics from new data streams such as high-resolution satellite and street-level imagery combined with machine-learning techniques (Weichenthal et al. 2019). The latter can exploit these image data to generate indirect indices of social environment (e.g., psychosocial stress) and physical environment (e.g. air or noise pollution) in a manner similar to that used by others to derive metrics such as living environment, health, and crime (Suel et al. 2019). Ultimately, one is often interested in linking aggregate-level "exposures" described above to the individual-level "outcomes." This can be achieved, for example, using administrative health databases or data acquired in large research cohorts. The former offers large population-base samples but coarse phenotypes (e.g., all hospital-based deliveries, birthweight) while the latter typically includes self-selected individuals who volunteered in a particular research study/cohort using deep phenotyping (e.g., the UKBB study, brain MRI).

Overall, sound epidemiological approaches to population-level studies, development of new tools for capturing area-level characteristics of the physical, built, and social environment, as well as access to (linked) individual-level data in large administrative databases and deep-phenotyped cohorts provide exciting opportunities for considering, simultaneously, the role of environment and genes vis-à-vis the development and aging of the human brain.

2 Advances and Challenges

2.1 Large Datasets and Sampling Biases

Over the past 10 years, we have seen remarkable growth in the size (number of participants) and breadth (number of phenotypes) of datasets suitable for answering questions about factors shaping the human brain in health and disease. Two open-science cohorts stand out. In 2014, the UK Biobank started an MRI arm of this cohort whereby a subset of the full sample of 500,000+ participants, recruited between 2006 and 2010 at 40 to 69 years of age (Sudlow et al. 2015), undergoes imaging of the brains, hearts, and bodies (Littlejohns et al. 2020); 10 years later,

~65,000 participants completed their first MRI visit, and ~5,000 returned for their second visit. In 2017, the first set of participants entered the ABCD Study of children and adolescents, with 11,875 children scanned during their first visit at 9–10 years of age (data made available in 2019; DOI: 10.15154/1503209), with subsequent visits planned in 2-year intervals for the next 10 years (Casey et al. 2018). Academic community has taken full advantage of the availability of these rich datasets and, in a span of 10 years, published 600+ brain-related reports.[3] In addition to such open resources, the majority of colleagues who have collected MRI datasets over the past 20+ years contribute to large-scale collaborative studies utilizing these to answer various questions. For example, a recent report on age-related variations in global and regional brain volumes (and other metrics) brought together MRI data collected in over 100,000 participants (Bethlehem et al. 2022).

But there are a number of challenges when using these and other research-based datasets. The first one concerns the heterogeneity across various samples related to ascertainment biases and the simple fact that all these studies rely on voluntary participation. As described in detail elsewhere in this book (chapter on "Population neuroscience: understanding concepts of generalizability and transportability and their application to improving the public's health"), it is useful to think about these issues by distinguishing between a *sample* in which all data are acquired, a *source* population from which this sample is recruited, and a *target* population for which one can make inferences based on findings obtained in the studied sample. *Generalizability* of the findings from the sample to the source population depends on the extent to which the two populations display similar characteristics (e.g., demographics, genetic makeup, neighborhood environment). Similarly, *transportability* of findings from the source to the target population depends on the similarity of the two populations at the time relevant for a given question/process. Let us use the UK Biobank as an example. All UK Biobank participants ($n = \sim 500{,}000$) were recruited, between 2006 and 2010, through offices of general practitioners located in the proximity of 22 assessment centers throughout England, Wales, and Scotland (Sudlow et al. 2015). A subset of this population, the "MRI" participants ($n = \sim 49{,}000$), volunteered to take part in an MRI session in one of the MRI hubs located in Bristol (South-West), Cheadle (Central), Reading (South-East), and Newcastle upon Tyne (North); these visits took place between 2014 and 2023 (Littlejohns et al. 2020). Table 1 shows basic demographics of the MRI participants (sample) and all UKBB participants (source population), with and without the sample. We can see that the MRI participants differ from non-MRI participants in all demographics considered here, including age, sex, income, and education; they also differ in the prevalence of certain health conditions, such as type 2 diabetes mellitus (MRI participants: 4.29%; non-MRI participants: 8.86%; all UKBB participants: 8.41%). As one would expect, both income and education are higher in the

[3] PubMed search (January 30, 2024) yielded a total of 472 results for the combination of "UK Biobank", "MRI", and "Brain" terms, and 163 results for the combination of "ABCD Study", "MRI", and "Brain" terms.

Table 1 Basic demographics of all UK Biobank participants (ALL), as well as those who did (MRI) or did not (NO-MRI) participate in the assessment with Magnetic Resonance Imaging (MRI). D, Cohen's d; SD, Standard Deviation; GBP, Great Britain Pound; A-level, Advanced level; AS-level, Advanced Subsidiary level; O-level, Ordinary Level; GCSE, General Certificate of Secondary Education; CSE, Certificate of Secondary Education; NVQ, National Vocational Qualification; HND, Higher National Diploma; HNC, Higher National Certificate

		All	No-MRI	MRI
Age (years)	Mean (years)	56.53	56.68	55.11
$d = 0.20$, $t = 43.6$	SD	8.09	8.14	7.57
$p = 2.2e\text{-}16$	Range	37–73	37–73	40–70
Sex (*n*)	Males	229,063	205,406	23,657
$\chi 2 = 126.16$	Females	273,293	247,656	25,637
$p < 2.2e\text{-}16$	Total	502,356	453,062	49,294
Sex (%)	Males	45.6%	45.3%	48.0%
	Females	54.4%	54.7%	52.0%
Income (*n*)	Less than 18,000 GBP	97,174	91,949	5,225
$\chi 2 = 5{,}266$	18,000 to 30,999	108,140	98,302	9,838
$p < 2.2e\text{-}16$	31,000 to 51,999	110,745	97,390	13,355
	52,000 to 100,000	86,240	73,698	12,542
	Greater than 100,000	22,923	19,461	3,462
	Total with income data	425,222	380,800	44,422
Income (%)	Less than 18,000	22.9%	24.1%	11.8%
	18,000 to 30,999	25.4%	25.8%	22.1%
	31,000 to 51,999	26.0%	25.6%	30.1%
	52,000 to 100,000	20.3%	19.4%	28.2%
	Greater than 100,000	5.4%	5.1%	7.8%
Education (*n*)	College or university degree	161,100	138,650	22,450
$\chi 2 = 1255.2$	A levels/AS levels or equivalent	130,170	112,578	17,592
$p < 2.2e\text{-}16$	O levels/GCSEs or equivalent	224,454	198,352	26,102
	CSEs or equivalent	64,490	57,837	6,653
	NVQ or HND or HNC or equivalent	90,976	81,569	9,407
	Other professional qualifications	139,233	121,444	17,789
	Total with education data	406,973	361,738	45,235
Education (%)	College or university degree	39.6%	38.3%	49.6%
	A levels/AS levels or equivalent	32.0%	31.1%	38.9%
	O levels/GCSEs or equivalent	55.2%	54.8%	57.7%
	CSEs or equivalent	15.8%	16.0%	14.7%
	NVQ or HND or HNC or equivalent	22.4%	22.5%	20.8%
	Other professional qualifications	34.2%	33.6%	39.3%

MRI (vs. non-MRI) participants. Next, we compared the MRI sample with all "non-MRI" UK Biobank participants with regard to common genetic variations (i.e., a GWAS where non-MRI participants are coded as "0" [$n = 339{,}861$] and those with MRI are coded "1" [$n = 38{,}428$]). This GWAS yielded differences in common genetic variations with suggestive significance only (684 SNPs [without clumping]

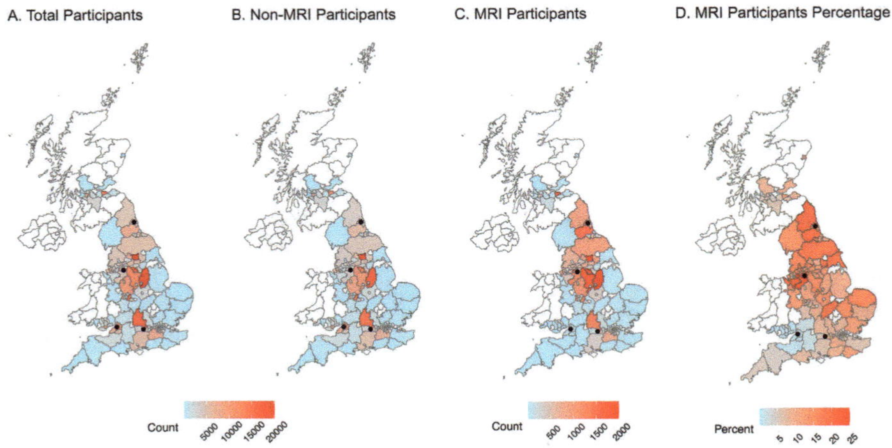

Fig. 1 Geospatial distribution of the UK Biobank participant. (**a**) All participants (102 to 20,091 participants per county); (**b**) Participants who did not participate in the MRI assessment (94 to 18,162 participants per county); (**c**) Participants with at least one MRI scan (1 to 1,929 participants per county); (**d**) Same as (**c**), in percent (0.07–22.74%). The four black dots indicate locations of the four MRI sites: Bristol (South-West), Cheadle (Central), Reading (South-East), and Newcastle upon Tyne (North). Counties with fewer than 100 UK Biobank participants (All) are shown in white. The maps were generated using R version 4.3.1 with the R package 'sf' version 1.0-15, using the December 2019 UK shapefile from the Office for National Statistics licensed under the Open Government Licence v.3.0

reached significance of $p < 1e\text{-}5$). Nevertheless, we observed a strong genetic correlation between this "MRI vs. no-MRI" GWAS and a published meta-GWAS of educational attainment (Lee et al. 2018); rg = 0.56, $p = 4.68e\text{-}72$ (Vosberg D, Paus T, unpublished observation, February 2, 2024). Thus, genetic variations associated with educational attainment explain 31% of variance in genetic differences associated with participating, in the UK Biobank, in an MRI session. Note that others addressed a similar issue with various approaches (Benonisdottir and Kong 2023; Schoeler et al. 2023; Cabrera-Mendoza et al. 2024). Finally, Fig. 1 shows the geospatial distribution, at the county level, of the residential locations of all UK Biobank participants and those who participated (or not) in the MRI assessment. Percentage of individuals who participated in the MRI study varies widely across counties (Fig. 1d); it is likely these variations bring along multiple differences in their physical, built, and social environment. Overall, perhaps not surprisingly, we see a number of (subtle) differences between the sample and source populations. Furthermore, if we compare the source population with the target population, defined here as the contemporary general population in England, Wales, and Scotland, we see – for example – differences between the source and target populations in basic demographics, such as education (UKBB: ~40% with college or university degree; England and Wales in 2011: 27% with Bachelor's degree or equivalent and higher)

(Office for National Statistics 2011), and in the prevalence of type 2 diabetes mellitus (UKBB: 8.4%; England and Wales: 9.0% [45–54 years of age], 12.7% [55–64 years] and 16.9% [65–74 years]) (Anonymous 2016). Thus, any statements about generalizability and transportability of findings obtained in the "MRI" sample should be qualified by these (and likely many other) differences between the sample, source, and target populations. Of course, any such differences may or may not be relevant for the question at hand.

2.2 *MRI and the Underlying Neurobiology*

In the past 10 years, we have seen great advancements in MRI technology, as well as further enhancements and developments of software tools for deriving MRI-based metrics. On the acquisition side, for example, the use of multichannel receive coil-arrays allows for the simultaneous acquisition of multiple slices; such multislice (multiband) acquisitions reduce the time needed for a scan and/or increase in the signal-to-noise ratio (SNR) per unit of time (Xu et al. 2013). Advances in the strength and rate of the spatial magnetic-field gradients (stronger and shorter pulses) facilitate diffusion-based imaging (Setsompop et al. 2013) and, in turn, estimations of distinct microstructural features (Panagiotaki et al. 2012). Going beyond the now standard 3-T scanners, both high-field (e.g., 7 T) and low-field (e.g., 0.06T) systems are likely to complement large-scale (3T-based) studies of brain structure and function in their own unique ways. The former (7 T) provides opportunities for high-resolution high-SNR imaging of structural and functional properties of the human brain in vivo (Marques and Norris 2018), albeit in relatively small samples of participants owing to the availability of these systems and technical challenges associated with their (consistent) use for multimodal imaging (Vachha and Huang 2021). The latter (0.06 T), such as the Swoop system by Hyperfine (https://hyperfine.io), facilitates accessibility and ease of data acquisition; it is particularly suitable for studies of early brain development (Deoni et al. 2021). On the image-analysis side, advances in multimodal imaging led to corresponding advancements and upgrades in the commonly used image-analysis tools, such as Freesurfer (Version 7.4.1) and FSL (Release 6.0). Various schemes for the parcellation of the human cerebral cortex have been added since the introduction of the Desikan-Killiani scheme (Desikan et al. 2006), such as those based on data-driven parcellations using multimodal (structural and functional) imaging datasets (Glasser et al. 2016) or fMRI data only (Schaefer et al. 2018). Other spatial maps and atlases have been collated at different academic sites (Van Essen et al. 2017), facilitating a multidimensional perspective on the structural and functional organization of the human cerebral cortex.

Without doubt, multimodal MRI provides a rich array of metrics, from basic structural measures, such as regional volumes and cortical thickness, to more complex ones, such as neurite orientation, dispersion, and density, as well as functional measures derived from correlational matrices of fMRI timeseries using, for example, the Brain Connectivity Toolbox (Rubinov and Sporns 2010). But whether basic or complex, the neurobiology underlying these measures is largely unknown. And yet, as we have discussed recently in the context of MRI studies in psychiatry: "For a biological psychiatrist, the ultimate goal is to understand biology underlying the course of a mental illness, from its development, presentation, treatment and remission, to its impact on everyday functioning. For this type of understanding, MRI cannot be treated as a black box, it needs to be rooted in neurobiology." (Paus 2023). The same can be said of any scientist interested in understanding factors that shape the human brain. For this reason, we have become interested in gaining insights into the cellular-level correlates of the various MRI-derived metrics and developed two general approaches, namely "virtual histology" and "virtual ontogeny."

In *virtual histology*, we take advantage of the regularities in the cellular-level organization of the cerebral cortex and its inter-regional variations, as first described by anatomists constructing maps of cytoarchitecture in the early 1900s (Zilles and Amunts 2010). As a proxy of the cellular composition of each of the cortical regions, we use inter-regional variations in the level of expression of genes specific for a given cell type. These cell-specific genes are identified using single-cell (or single-nucleus) RNA. Their expression in the human cerebral cortex is estimated from existing atlases and mapped to a set of cortical regions using a common parcellation scheme, such as the Desikan-Killiani parcellation (34 regions per hemisphere). Finally, we calculate a spatial correlation between the inter-regional profile of such cell-specific gene expression and that of an MRI-define phenotype (French and Paus 2015; Shin et al. 2018).

In a series of studies using this approach, we have identified cellular correlates of inter-regional (spatial) profiles of a number of MRI-derived metrics, including cortical thickness and its age-related variations during adolescence (Shin et al. 2018) and aging (Vidal-Pineiro et al. 2020), as well as those in magnetic transfer ratio and its changes during adolescence (Patel et al. 2019). Collectively, these studies pointed to the importance of inter-regional variations in a particular type of pyramidal cells, referred to as "CA1 pyramidal cells" in the original single-cell RNA paper by Zeisel and colleagues (Zeisel et al. 2015); expression of genes specific to this type of pyramidal cells is abundant in the human cerebral cortex, and it is characterized (Gene Ontology) by their involvement in the "regulation of dendrite extension" (Shin et al. 2018). As illustrated in Fig. 2, the distribution of (spatial) correlations between inter-regional variations in cell-specific gene panels and those of a given MRI metric suggests that this type of pyramidal cells correlate spatially with more than one MRI-derived characteristics (Patel et al. 2019, 2020). This observation is consistent with a "one-to-many" principle whereby one particular neurobiological entity underlies spatial variations in many MRI-derived metrics. Note that the opposite is also true: many types of cells appear to contribute to one particular MRI metrics, thus consistent with a "many-to-one" principle (Patel et al. 2020). Thus, perhaps not surprisingly, there is no one-to-one mapping between MRI-derived metrics and cellular-level neurobiology.

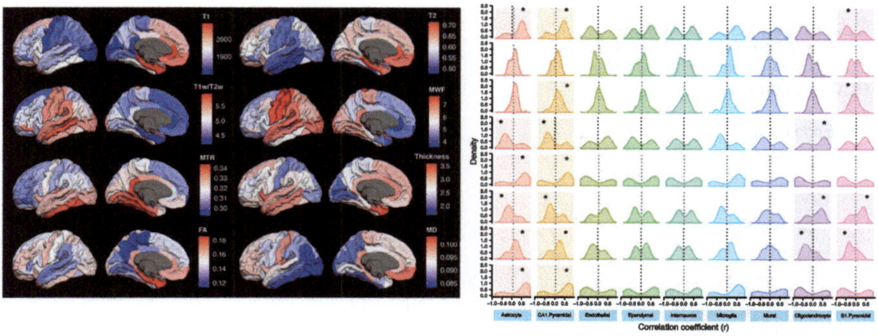

Fig. 2 Virtual histology of multimodal magnetic resonance imaging. The left panel shows maps (left hemisphere) of eight MRI-derived metrics. The right panel shows the results of virtual histology illustrated by the distribution of (spatial) correlations between inter-regional variations in cell-specific gene panels (names along the X axis) and those of a given MRI metric (names along the Y axis). Reprinted with permission from Patel et al. (Patel et al. 2020)

Using the same approach, we have also demonstrated a dominant role of the CA1 type of pyramidal cells in explaining inter-regional variations in group differences in cortical thickness between patients with psychiatric disorders and their healthy controls (Writing Committee for the Attention-Deficit/Hyperactivity, Autism Spectrum, Bipolar, Major Depressive, Obsessive-Compulsive, and Schizophrenia ENIGMA Working Groups et al. 2021). In this large study, based on the collaborative effort of the ENIGMA Consortium discussed more broadly elsewhere in this book (see chapter on "Large-scale neuroimaging of mental illness"), we showed that regions with high expression of these genes showed larger differences between the patients and controls (lower thickness in the patients); this was true about all six disorders. The inter-regional profile of these group differences common to the six disorders was also spatially correlated with that of the CA1 pyramidal cells (plus astrocytes and microglia) (Writing Committee for the Attention-Deficit/Hyperactivity, Autism Spectrum, Bipolar, Major Depressive, Obsessive-Compulsive, and Schizophrenia ENIGMA Working Groups et al. 2021). Altogether, the above studies with virtual histology point to the likely importance of dendritic arbor, and its modifications, as a neurobiological entity involved in the maturation and aging of the human cerebral cortex, as well in the pathophysiology of psychiatric disorders.

As pointed out in the background section (Sect. 1.3), many psychiatric disorders have their roots in prenatal and early post-natal development. For this reason, we have introduced "virtual ontogeny" as a tool for looking back in time into developmental events preceding the onset of mental illness.

In *virtual ontogeny*, we work with inter-regional profiles of gene expression assessed, across 11 regions of the human cerebral cortex, during fetal development (12 to 22 post-conception weeks). As shown in Fig. 3, we used single-cell RNA from the human cerebral cortex (Bhaduri et al. 2020) to identify genes specific to both

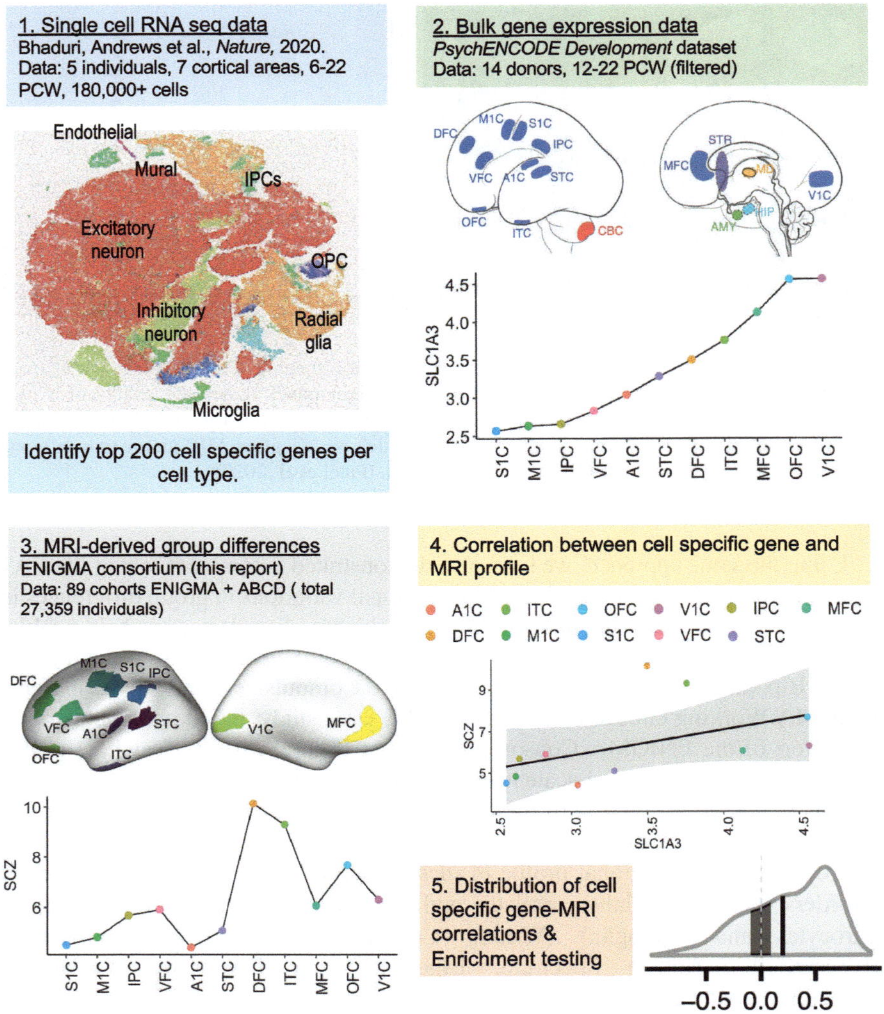

Fig. 3 Virtual ontogeny. In Step 1 (top left) – identify top 200 cell-specific genes from single-cell RNA seq data of the developing neocortex. Step 2 (top right) – quantify median gene expression across donors for each 11 cortical regions. For instance, the gene SLC1A3 was in the top 200 specific genes for the radial-glia panel. The expression of this gene is plotted in Step 2 (top right). Step 3 (bottom left) – quantifies meta-analytic group differences in surface area between cases and controls across the same 11 cortical regions. Step 4 (bottom right, top half) – correlation between cell-specific gene expression and an MRI-derived profile, in this case, case–control differences for SCZ and SLC1A3 expression. This is repeated for all 200 genes specific to a cell type (in this case radial glia) to create a distribution of correlation coefficients in Step 5 (bottom right, bottom half). OPC, oligodendrocyte precursor cell; IPCs, intermediate progenitor cells; DFC, dorsal frontal cortex; VFC, ventral frontal cortex; OFC, orbitofrontal cortex; MFC, medial frontal cortex; ITC, inferior temporal cortex; IPC, inferior parietal cortex; M1C, primary motor cortex; S1C, primary somatosensory cortex; A1C, primary auditory cortex; V1C, primary visual cortex. Reprinted with permission from Patel et al. (2022)

proliferative (radial glia, intermediate progenitor cells) and differentiated (e.g., excitatory neurons) cells during gestation (6 to 22 post-conception weeks). We then related these cell-specific inter-regional profiles to inter-regional profiles of group differences (patients vs. controls) in the tangential growth of the cerebral cortex, as indexed by surface area estimated from the MRI. Note that the cortical surface area increases dramatically during prenatal period and the first 2 years of post-natal life (Bethlehem et al. 2022), with relatively small additional changes later in life, thus providing a window into early brain development even when measured later in life (Patel et al. 2022). Here, we showed that cortical regions with high expression of genes specific to proliferative cells shows larger differences in surface area between patients and controls (lower surface area in patients). We also noted that these group differences appeared larger in multimodal (vs. unimodal) regions where the expression of the "proliferative" genes was higher during early fetal period. Finally, some of these genes intersected with those involved in mediating biological response to various perinatal adversities are known to be increasing the risk of psychosis, such as low birthweight (Patel et al. 2022).

Altogether, the above work – with virtual histology and virtual ontogeny – illustrates the usefulness of targeted transcriptomics approaches in facilitating the neurobiological interpretations of MRI-based (cortical) phenotypes. Owing to the relative (spatial) sparsity of gene-expression data available for non-cortical regions, however, it may be difficult to employ these approaches in gray-matter regions such as the hippocampus (6 subregions in the Allen Human Brain Atlas) or white-matter regions such as the corpus callosum (4 subregions in the Allen Human Brain Atlas). And yet, regional heterogeneity of white matter in its cellular composition, in particular axon diameter and the thickness of myelin sheath, is likely to influence which of the white-matter pathways may be more or less vulnerable to various perturbations, and when – during development and aging – this may happen (Paus 2023). We have attempted to circumvent this challenge in an investigation of the spatial relationship between fiber composition of the human corpus callosum, based on "true" (not virtual) histology, and MR-based metrics; we observed a strong spatial correlation between T1 and T2 relaxation times (and myelin water fraction) and fiber diameter; callosal subregions with small fibers had shorter T1 and T2 relaxation times and higher values of myelin water fraction (Bjornholm et al. 2017), likely related to higher relative content of myelin per unit of volume in these regions.

Overall, progress is being made with regard to our neurobiological understanding of various MRI-derived metrics. In addition to the approaches introduced above, studies in experimental animals are in an excellent position to relate MRI-based phenotypes to those measured ex vivo with multimodal microscopy (Woo et al. 2022; Munck et al. 2021).

2.3 Genomics

As pointed out in Sect. 1.2, an impressive growth of the available datasets and their pooling through meta-analyses have facilitated the identification of quantitative trait loci (QTLs) associated with a number of brain phenotypes, such as surface area of the cerebral cortex and volumes of subcortical structures. Combined with extensive genome-annotation resources and their current advancements (Miller et al. 2019), these and other studies are making inroads into uncovering molecular pathways underlying various brain phenotypes.

In addition to GWAS-based studies of common genetic variants, recent efforts are contributing knowledge about rare variants and their associations with the same brain phenotypes. Two types of Copy Number Variations (CNVs), namely deletions and duplications, have been studied extensively by the ENIGMA CNV group (Sonderby et al. 2022). The key advantage of studying such CNVs is that, unlike the majority of SNP-based QTLs revealed in GWAS studies, the deletion and duplication of a gene leads to a decrease and an increase in its expression, respectively (Handsaker et al. 2015; Han et al. 2020). To date, MRI-based brain phenotypes have been examined in a number of so-called recurrent CNVs, including 22q11.2, 16p11.2 distal, 15q11.2, and 1q21.1 (Kumar et al. 2023; Boen et al. 2024; Writing Committee for the ENIGMA-CNV Working Group, van der Meer et al. 2020). Furthermore, initial studies are underway to examine the genome-wide load of deletions and duplications in large community-based cohorts; such a genome-wide aggregation of all CNVs, regardless of their position in the genome, is necessitated by their very low frequency in general population. This approach has been used in case of cognitive abilities (Huguet et al. 2018, 2021), and it is in progress with regard to the tangential growth of the cerebral cortex (Liao et al. 2023a). Along similar lines, large efforts, such as the Trans-Omics for Precision Medicine (TOPMed) Program (https://topmed.nhlbi.nih.gov) and the Alzheimer's Disease Sequencing Project (Beecham et al. 2017), are on the way, taking advantage of whole-genome sequencing whereby Single Nucleotide Variants are identified in relation to various clinical phenotypes (Wang et al. 2023; Hu et al. 2022).

Let us close this section by illustrating briefly the usefulness of combining genomics with transcriptomics and epigenomics. In the same way GWAS is used to identify (common) genetic variations associated with system-level traits, such as MRI-derived brain phenotypes, one can use this approach to identify QTLs associated with molecular traits (Neumeyer et al. 2020), such as levels of mRNA (expression QTL [eQTL]) or levels of DNA methylation (methylation QTL [meQTL]). Typically, effect sizes of such genetic variations, located in regulatory (non-coding) regions, on the expression or methylation of nearby genes are larger compared with system-level traits (Min et al. 2021; GTEx Consortium 2020). Expression QTLs can be employed, for example, to strengthen causal inference with regard to an observed genotype–phenotype relationship; if the identified SNP is an eQTL changing expression of a nearby gene (causation), the inference about this gene's involvement in shaping the phenotype is strengthened (Porcu et al. 2021). A multivariable transcriptome-wide Mendelian randomization approach allows one to combine known eQTLs when testing for causality of GWAS-uncovered variants, associated with a given phenotype, by using "multiple SNPs as instruments and multiple gene expression traits as exposures, simultaneously" (Porcu et al. 2019). In our developmental studies, we are also using eQTLs to investigate the relationship between molecular processes at play in the fetal brain, such as neuronal proliferation, and their relationship with the tangential growth of the cerebral cortex (Liao et al. 2023b); in this case, we identified relevant eQTLs from the transcriptome obtained in 201 mid-gestational human brains (Walker et al. 2019). Finally, in case of meQTLs, we have used the robust genetic control of DNA methylation to model a methylation landscape, known to be associated with an exposure to famine, in a large cohort of individuals; we then examined moderating effects of such a virtual "famine exposure" on the relationship between birthweight genes and the tangential growth of the cerebral cortex (Vosberg et al. 2024).

Overall, advances in our knowledge of the genome and its regulatory elements, combined with whole-genome sequencing technology and rich -omics datasets, are facilitating our understanding of molecular pathways underlying complex traits. Large-scale international collaborations are essential for detecting small signals and replicating initial findings. Integration of knowledge derived through systems biology and systems neuroscience is important for generating new explanatory models of processes shaping the human brain.

2.4 Geospatial Mapping of Area-Level Environment

The detailed and systematic mapping of the human genome over the past 30+ years embodies an aspiration for an equally systematic mapping of the human "exposome" (Wild 2005; Munzel et al. 2023). Unlike genes, however, our (external) environment changes over time along multiple dimensions. Several approaches hold promise in addressing, at least in part, this challenge. First, high-throughput metabolomics may capture – directly or indirectly – some of the exposures experienced by an individual over time (Rattray et al. 2018). Second, along similar lines, Mendelian randomization may allow for modeling specific exposures using genetic variants known to predict levels of a particular molecule, such as folate levels during pregnancy (Davey Smith and Ebrahim 2005). Furthermore, as described above (Sect. 2.3), meQTLs can be used to model more complex exposures, such as famine (Vosberg et al. 2024).

Although the above two approaches are appealing due to their accessibility (blood sample), they are limited to a finite set of (specific) environmental exposures. And yet, the environment in which we live represents a complex interplay of multiple, often co-occurring, domains (see Sect. 1.3). Therefore, third, there is growing interest in assessing – simultaneously – the physical, built, and social environment at an area level. In this way, multiple exposures experienced by an individual living/working in a particular geospatial unit can be estimated from the characteristics of that unit, and compared with those of other units (at different levels of spatial granularity). Such characteristics can be extracted from publicly available data at multiple timepoints, often spanning several decades. For example, one can envisage extracting comparable characteristics for the majority of the UK counties shown in Fig. 1, going as far back as the birthyear of the oldest UK Biobank participant (1934) for some.

To develop conceptual and practical framework for the area-level assessment of the physical, built, and social environment, we convened an Ernst Strüngmann Forum focused on Digital Ethology. As described in the opening chapter of the ensuing book, "Conceptually, the Forum called for adopting an ethological approach whereby human behavior is observed, or inferred, in the 'wild'; that is, without influencing the observed individual (e.g., by asking them questions). Practically, it called for focusing on data sources that either exist or can be readily harnessed at an aggregate level, with different area-level (spatial) granularity (e.g., neighborhood, city, country). The ethological framework presented in Dumas et al. (Chapter 2) underpins the name to this Forum. By 'digital ethology,' we mean the observation of human behavior through its digital manifestations, such as the use of a search engine, a payment card, or through posting on social media. This behavior leaves 'digital footprints' that are particularly relevant for characterizing the social environment of a given area-level unit, as discussed by Weigle et al. (Chapter 4). Human behavior is also reflected and constrained by the surrounding physical and built environments, as outlined by Lovasi et al. (Chapter 3). Finally, variations in individual-level outcomes as a function of the multidimensional area-level environment can best be studied using large datasets; the practicalities, as well as legal and ethical considerations, are addressed by Bauzer Medeiros et al. (Chapter 5). Finally, Chapters 6 through 12 provide primers to many of the concepts and strategies that underpin digital ethology." (Paus 2024b). We encourage readers interested in this approach to consult this book.

2.5 Causality

In population neuroscience, we infer the role of various factors in shaping the human brain from data collected in observational studies. By their nature, these studies allow one to identify *correlations* between different sets of variables. But, of course, we are striving to reach causal understanding of the relationships discovered in these studies. We want to know whether exposure A leads to outcome X. This is difficult to achieve.

Let us start by a quote about causality, from our book on Digital Ethology (Paus 2024b). "Causality thus comes in different flavors and with different approaches to assessing it. In ethology we can at least describe four different types: ontogenetic (caused by the development), phylogenetic (caused by the evolution of species), proximate (caused by immediate physiological or environmental factors), and ultimate – associated with goals and function from an evolutionary point of view.

In a broader scientific context, three general classes of causality are usually defined: direct, structural, and logical. Observing direct and structural causality requires, most of the time, an experiment – an empirical intervention with planned perturbation of the system– or at least a quasi-experiment – the use of events that occur independently from the research planning but could nevertheless be exploited to infer how those events causally impact the system. While direct causality is associated with physical events and mechanisms in time (the earthquake destroyed the city), structural causality is more associated with physical objects and mechanisms in space (the transportation system constrains the growth of the city). Finally, logical causality is independent of time and space since it relies more on abstract propositions, reasoning, and implication." (Paus 2024b).

As stated elsewhere this book (chapter on "Population neuroscience: understanding concepts of generalizability and transportability and their application to improving the public's health"), "Often, we are collecting and analyzing observational data and we cannot make definitive claims about causation from the data that we have collected. We can identify associations, attempt to ensure that those associations are robust to known confounders, major selection and other biases, and design experimental and quasi-experimental studies that are well positioned to move towards identifying causation."

Thus, the first message is that of "validity" of the observational finding. Replication may be the answer here. As long as we can be confident that the "discovery" and "replication" samples do not share the same biases driving the finding (e.g., a particular stratification of the sample population or measurement error), we argue that replications is one of the best tests of validity of a new finding.

The second message is that of designing experimental or quasi-experimental studies. The former is possible when testing specific exposure-outcome relations. For example, cluster-randomized trials confirmed the positive relationship between breastfeeding and child's cognitive abilities (Kramer et al. 2008). The latter requires creative use of an external event, random with regard to who experienced it (or not), which can serve as an instrumental variable with regard to a particular outcome (Martens et al. 2006). For example, a policy instituted by the Dutch government led to a reduction in trans fatty acids (TFAs) in various food products; in temporal relation to the policy implementation, there was a decrease in plasma levels of TFAs in pregnant women participating in the Generation R study, and TFA-related variations in the head circumference of the fetus (Zou et al. 2022). It is therefore likely that, in this sample, exposure (TFAs) was causally related to the outcome (head circumference). Mendelian randomization represents a special case of the instrumental-variable approach whereby particular genetic variations serve as "proxies" of a particular exposure (Smith and Ebrahim 2003). For example, a genetic variant that "mimics" low levels of folate is associated with neural tube defects (Botto and Yang 2000), the latter is known to be associated with the actual folic acid levels and their change through food fortification (Atta et al. 2016). Furthermore, two-sample Mendelian randomization is used frequently to test directionality of relationships between two sets of variables, such as education and brain structure (Seyedsalehi et al. 2023).

Finally, one can also use negative controls and cross-contextual comparisons to strengthen causal inferences about an observed exposure-outcome association (Gage et al. 2016). A cross-country comparison of the association between breastfeeding and cognitive abilities in the UK and Brazil is an example of the latter; the very different confounding structure of the cohorts investigated in the two countries at the time of the exposure, namely breastfeeding associated with higher (UK) vs. lower (Brazil) maternal education, strengthens the causal interpretation of the same (positive) association between breastfeeding and cognitive abilities found in the two countries (Victora et al. 2015).

Overall, interpreting findings obtained in observational studies is challenging. We must pay close attention to potential biases in the design and ascertainment of our samples, as well as to potential confounders and measurement errors. When feasible, replications of the initial findings in independent – and ideally differently designed – cohorts increase confidence in these findings. Finally, causal inferences could be explored using experimental and quasi-experimental designs, instrumental variables, as well as negative controls and cross-contextual comparisons.

3 Conclusions

Population neuroscience builds on the past groundwork and achievements of the three constitutive disciplines, namely epidemiology, genetics, and neuroscience, over the past 50+ years. In the last 10+ years, we have made great progress in integrating our knowledge across these disciplines and, by doing so, generated new discoveries about factors shaping the human brain from conception onward. From a personal perspective, I believe further advancements will come from two directions: (1) systematic data-driven characterization of the physical, built, and social environment; and (2) modeling of MRI-based metrics at the cellular level. Furthermore, additional advancements in genetics and omics sciences will no doubt bring novel molecular phenotypes (e.g., those derived from tissue-specific extracellular vesicles), as well as new insights into the functioning of various molecular pathways and their interactions.

References

Abel KM, Wicks S, Susser ES, Dalman C, Pedersen MG, Mortensen PB, Webb RT (2010) Birth weight, schizophrenia, and adult mental disorder: is risk confined to the smallest babies? Arch Gen Psychiatry 67:923–930

Anonymous (2016) Diabetes prevalence model. Public Health England, London

Atta CA, Fiest KM, Frolkis AD, Jette N, Pringsheim T, St Germaine-Smith C, Rajapakse T, Kaplan GG, Metcalfe A (2016) Global birth prevalence of spina bifida by folic acid fortification status: a systematic review and meta-analysis. Am J Public Health 106:e24–e34

Beecham GW, Bis JC, Martin ER, Choi SH, DeStefano AL, van Duijn CM, Fornage M, Gabriel SB, Koboldt DC, Larson DE, Naj AC, Psaty BM, Salerno W, Bush WS, Foroud TM, Wijsman E, Farrer LA, Goate A, Haines JL, Pericak-Vance MA, Boerwinkle E, Mayeux R, Seshadri S, Schellenberg G (2017) The Alzheimer's disease sequencing project: study design and sample selection. Neurol Genet 3:e194

Bellenguez C, Kucukali F, Jansen IE, Kleineidam L, Moreno-Grau S, Amin N, Naj AC, Campos-Martin R, Grenier-Boley B, Andrade V, Holmans PA, Boland A, Damotte V, van der Lee SJ, Costa MR, Kuulasmaa T, Yang Q, de Rojas I, Bis JC, Yaqub A, Prokic I, Chapuis J, Ahmad S, Giedraitis V, Aarsland D, Garcia-Gonzalez P, Abdelnour C, Alarcon-Martin E, Alcolea D, Alegret M, Alvarez I, Alvarez V, Armstrong NJ, Tsolaki A, Antunez C, Appollonio I, Arcaro M, Archetti S, Pastor AA, Arosio B, Athanasiu L, Bailly H, Banaj N, Baquero M, Barral S, Beiser A, Pastor AB, Below JE, Benchek P, Benussi L, Berr C, Besse C, Bessi V, Binetti G, Bizarro A, Blesa R, Boada M, Boerwinkle E, Borroni B, Boschi S, Bossu P, Brathen G, Bressler J, Bresner C, Brodaty H, Brookes KJ, Brusco LI, Buiza-Rueda D, Burger K, Burholt V, Bush WS, Calero M, Cantwell LB, Chene G, Chung J, Cuccaro ML, Carracedo A, Cecchetti R, Cervera-Carles L, Charbonnier C, Chen HH, Chillotti C, Ciccone S, Claassen J, Clark C, Conti E, Corma-Gomez A, Costantini E, Custodero C, Daian D, Dalmasso MC, Daniele A, Dardiotis E, Dartigues JF, de Deyn PP, de Paiva Lopes K, de Witte LD, Debette S, Deckert J, Del Ser T et al (2022) New insights into the genetic etiology of Alzheimer's disease and related dementias. Nat Genet 54:412–436

Belliveau JW, Kennedy DN Jr, McKinstry RC, Buchbinder BR, Weisskoff RM, Cohen MS, Vevea JM, Brady TJ, Rosen BR (1991) Functional mapping of the human visual cortex by magnetic resonance imaging. Science 254:716–719

Benonisdottir S, Kong A (2023) Studying the genetics of participation using footprints left on the ascertained genotypes. Nat Genet 55:1413–1420

Bethlehem RAI, Seidlitz J, White SR, Vogel JW, Anderson KM, Adamson C, Adler S, Alexopoulos GS, Anagnostou E, Areces-Gonzalez A, Astle DE, Auyeung B, Ayub M, Bae J, Ball G, Baron-Cohen S, Beare R, Bedford SA, Benegal V, Beyer F, Blangero J, Blesa Cabez M, Boardman JP, Borzage M, Bosch-Bayard JF, Bourke N, Calhoun VD, Chakravarty MM, Chen C, Chertavian C, Chetelat G, Chong YS, Cole JH, Corvin A, Costantino M, Courchesne E, Crivello F, Cropley VL, Crosbie J, Crossley N, Delarue M, Delorme R, Desrivieres S, Devenyi GA, Di Biase MA, Dolan R, Donald KA, Donohoe G, Dunlop K, Edwards AD, Elison JT, Ellis CT, Elman JA, Eyler L, Fair DA, Feczko E, Fletcher PC, Fonagy P, Franz CE, Galan-Garcia L, Gholipour A, Giedd J, Gilmore JH, Glahn DC, Goodyer IM, Grant PE, Groenewold NA, Gunning FM, Gur RE, Gur RC, Hammill CF, Hansson O, Hedden T, Heinz A, Henson RN, Heuer K, Hoare J, Holla B, Holmes AJ, Holt R, Huang H, Im K, Ipser J, Jack CR Jr, Jackowski AP, Jia T, Johnson KA, Jones PB, Jones DT, Kahn RS, Karlsson H, Karlsson L, Kawashima R, Kelley EA, Kern S, Kim KW, Kitzbichler MG, Kremen WS, Lalonde F, Landeau B et al (2022) Brain charts for the human lifespan. Nature 604:525–533

Bhaduri A, Andrews MG, Mancia Leon W, Jung D, Shin D, Allen D, Jung D, Schmunk G, Haeussler M, Salma J, Pollen AA, Nowakowski TJ, Kriegstein AR (2020) Cell stress in cortical organoids impairs molecular subtype specification. Nature 578:142–148

Bjornholm L, Nikkinen J, Kiviniemi V, Nordstrom T, Niemela S, Drakesmith M, Evans JC, Pike GB, Veijola J, Paus T (2017) Structural properties of the human corpus callosum: multimodal assessment and sex differences. Neuroimage 152:108–118

Boen R, Kaufmann T, van der Meer D, Frei O, Agartz I, Ames D, Andersson M, Armstrong NJ, Artiges E, Atkins JR, Bauer J, Benedetti F, Boomsma DI, Brodaty H, Brosch K, Buckner RL, Cairns MJ, Calhoun V, Caspers S, Cichon S, Corvin AP, Crespo-Facorro B, Dannlowski U, David FS, de Geus EJC, de Zubicaray GI, Desrivieres S, Doherty JL, Donohoe G, Ehrlich S, Eising E, Espeseth T, Fisher SE, Forstner AJ, Fortaner-Uya L, Frouin V, Fukunaga M, Ge T, Glahn DC, Goltermann J, Grabe HJ, Green MJ, Groenewold NA, Grotegerd D, Grontvedt GR, Hahn T, Hashimoto R, Hehir-Kwa JY, Henskens FA, Holmes AJ, Haberg AK, Haavik J,

Jacquemont S, Jansen A, Jockwitz C, Jonsson EG, Kikuchi M, Kircher T, Kumar K, Le Hellard S, Leu C, Linden DE, Liu J, Loughnan R, Mather KA, McMahon KL, McRae AF, Medland SE, Meinert S, Moreau CA, Morris DW, Mowry BJ, Muhleisen TW, Nenadic I, Nothen MM, Nyberg L, Ophoff RA, Owen MJ, Pantelis C, Paolini M, Paus T, Pausova Z, Persson K, Quide Y, Marques TR, Sachdev PS, Sando SB, Schall U, Scott RJ, Selbaek G, Shumskaya E, Silva AI, Sisodiya SM, Stein F, Stein DJ, Straube B, Streit F, Strike LT, Teumer A, Teutenberg L et al (2024) Beyond the global brain differences: intraindividual variability differences in 1q21.1 distal and 15q11.2 BP1-BP2 deletion carriers. Biol Psychiatry 95:147–160

Botto LD, Yang Q (2000) 5,10-methylenetetrahydrofolate reductase gene variants and congenital anomalies: a HuGE review. Am J Epidemiol 151:862–877

Cabrera-Mendoza B, Wendt FR, Pathak GA, Yengo L, Polimanti R (2024) The impact of assortative mating, participation bias and socioeconomic status on the polygenic risk of behavioural and psychiatric traits. Nat Hum Behav

Casey BJ, Cannonier T, Conley MI, Cohen AO, Barch DM, Heitzeg MM, Soules ME, Teslovich T, Dellarco DV, Garavan H, Orr CA, Wager TD, Banich MT, Speer NK, Sutherland MT, Riedel MC, Dick AS, Bjork JM, Thomas KM, Chaarani B, Mejia MH, Hagler DJ Jr, Daniela Cornejo M, Sicat CS, Harms MP, Dosenbach NUF, Rosenberg M, Earl E, Bartsch H, Watts R, Polimeni JR, Kuperman JM, Fair DA, Dale AM, A.I.A. Workgroup (2018) The adolescent brain cognitive development (ABCD) study: Imaging acquisition across 21 sites. Dev Cogn Neurosci 32:43–54

Choi DJ, Armstrong G, Lozzi B, Vijayaraghavan P, Plon SE, Wong TC, Boerwinkle E, Muzny DM, Chen HC, Gibbs RA, Ostrom QT, Melin B, Deneen B, Bondy ML, Gliogene C, C. Genomics England Research, Bainbridge MN, Amos CI, Barnholtz-Sloan JS, Bernstein JL, Claus EB, Houlston RS, Il'yasova D, Jenkins RB, Johansen C, Lachance D, Lai R, Melin BS, Merrell RT, Olson SH, Sadetzki S, Schildkraut J, Shete S, Ambrose JC, Arumugam P, Bevers R, Bleda M, Boardman-Pretty F, Boustred CR, Brittain H, Brown MA, Caulfield MJ, Chan GC, Giess A, Griffin JN, Hamblin A, Henderson S, Hubbard TJP, Jackson R, Jones LJ, Kasperaviciute D, Kayikci M, Kousathanas A, Lahnstein L, Lakey A, Leigh SEA, Leong IUS, Lopez FJ, Maleady-Crowe F, McEntagart M, Minneci F, Mitchell J, Moutsianas L, Mueller M, Murugaesu N, Need AC, O'Donovan P, Odhams CA, Patch C, Perez-Gil D, Pereira MB, Pullinger J, Rahim T, Rendon A, Rogers T, Savage K, Sawant K, Scott RH, Siddiq A, Sieghart A, Smith SC, Sosinsky A, Stuckey A, Tanguy M, Tavares ALT, Thomas ERA, Thompson SR, Tucci A, Welland MJ, Williams E, Witkowska K, Wood SM, Zarowiecki M (2023) The genomic landscape of familial glioma. Sci Adv 9:eade2675

Cohen JM, Wilson ML, Aiello AE (2007) Analysis of social epidemiology research on infectious diseases: historical patterns and future opportunities. J Epidemiol Community Health 61:1021–1027

Davey Smith G, Ebrahim S (2005) What can mendelian randomisation tell us about modifiable behavioural and environmental exposures? BMJ 330:1076–1079

Davies G, Armstrong N, Bis JC, Bressler J, Chouraki V, Giddaluru S, Hofer E, Ibrahim-Verbaas CA, Kirin M, Lahti J, van der Lee SJ, Le Hellard S, Liu T, Marioni RE, Oldmeadow C, Postmus I, Smith AV, Smith JA, Thalamuthu A, Thomson R, Vitart V, Wang J, Yu L, Zgaga L, Zhao W, Boxall R, Harris SE, Hill WD, Liewald DC, Luciano M, Adams H, Ames D, Amin N, Amouyel P, Assareh AA, Au R, Becker JT, Beiser A, Berr C, Bertram L, Boerwinkle E, Buckley BM, Campbell H, Corley J, De Jager PL, Dufouil C, Eriksson JG, Espeseth T, Faul JD, Ford I, Generation S, Gottesman RF, Griswold ME, Gudnason V, Harris TB, Heiss G, Hofman A, Holliday EG, Huffman J, Kardia SL, Kochan N, Knopman DS, Kwok JB, Lambert JC, Lee T, Li G, Li SC, Loitfelder M, Lopez OL, Lundervold AJ, Lundqvist A, Mather KA, Mirza SS, Nyberg L, Oostra BA, Palotie A, Papenberg G, Pattie A, Petrovic K, Polasek O, Psaty BM, Redmond P, Reppermund S, Rotter JI, Schmidt H, Schuur M, Schofield PW, Scott RJ, Steen VM, Stott DJ, van Swieten JC, Taylor KD, Trollor J, Trompet S, Uitterlinden AG, Weinstein G, Widen E, Windham BG, Jukema JW, Wright AF et al (2015) Genetic contributions to variation in general cognitive function: a meta-analysis of genome-wide association studies in the CHARGE consortium (N=53949). Mol Psychiatry 20:183–192

Davies C, Segre G, Estrade A, Radua J, De Micheli A, Provenzani U, Oliver D, Salazar de Pablo G, Ramella-Cravaro V, Besozzi M, Dazzan P, Miele M, Caputo G, Spallarossa C, Crossland G, Ilyas A, Spada G, Politi P, Murray RM, McGuire P, Fusar-Poli P (2020) Prenatal and perinatal risk and protective factors for psychosis: a systematic review and meta-analysis. Lancet Psychiatry 7:399–410

Demontis D, Walters GB, Athanasiadis G, Walters R, Therrien K, Nielsen TT, Farajzadeh L, Voloudakis G, Bendl J, Zeng B, Zhang W, Grove J, Als TD, Duan J, Satterstrom FK, Bybjerg-Grauholm J, Baekved-Hansen M, Gudmundsson OO, Magnusson SH, Baldursson G, Davidsdottir K, Haraldsdottir GS, Agerbo E, Hoffman GE, Dalsgaard S, Martin J, Ribases M, Boomsma DI, Artigas MS, Mota NR, Howrigan D, Medland SE, Zayats T, Rajagopal VM, A. W.G.O.T.P.G. Consortium, iPSYCH-Broad Consortium, Nordentoft M, Mors O, Hougaard DM, Mortensen PB, Daly MJ, Faraone SV, Stefansson H, Roussos P, Franke B, Werge T, Neale BM, Stefansson K, Borglum AD (2023) Genome-wide analyses of ADHD identify 27 risk loci, refine the genetic architecture and implicate several cognitive domains. Nat Genet 55:198–208

Deoni SCL, Bruchhage MMK, Beauchemin J, Volpe A, D'Sa V, Huentelman M, Williams SCR (2021) Accessible pediatric neuroimaging using a low field strength MRI scanner. Neuroimage 238:118273

Desikan RS, Segonne F, Fischl B, Quinn BT, Dickerson BC, Blacker D, Buckner RL, Dale AM, Maguire RP, Hyman BT, Albert MS, Killiany RJ (2006) An automated labeling system for subdividing the human cerebral cortex on MRI scans into gyral based regions of interest. Neuroimage 31:968–980

Desimone R, Ungerleider LG (1986) Multiple visual areas in the caudal superior temporal sulcus of the macaque. J Comp Neurol 248:164–189

Duperron MG, Knol MJ, Le Grand Q, Evans TE, Mishra A, Tsuchida A, Roshchupkin G, Konuma T, Tregouet DA, Romero JR, Frenzel S, Luciano M, Hofer E, Bourgey M, Dueker ND, Delgado P, Hilal S, Tankard RM, Dubost F, Shin J, Saba Y, Armstrong NJ, Bordes C, Bastin ME, Beiser A, Brodaty H, Bulow R, Carrera C, Chen C, Cheng CY, Deary IJ, Gampawar PG, Himali JJ, Jiang J, Kawaguchi T, Li S, Macalli M, Marquis P, Morris Z, Munoz Maniega S, Miyamoto S, Okawa M, Paradise M, Parva P, Rundek T, Sargurupremraj M, Schilling S, Setoh K, Soukarieh O, Tabara Y, Teumer A, Thalamuthu A, Trollor JN, Valdes Hernandez MC, Vernooij MW, Volker U, Wittfeld K, Wong TY, Wright MJ, Zhang J, Zhao W, Zhu YC, Schmidt H, Sachdev PS, Wen W, Yoshida K, Joutel A, Satizabal CL, Sacco RL, Bourque G, Consortium C, Lathrop M, Paus T, Fernandez-Cadenas I, Yang Q, Mazoyer B, Boutinaud P, Okada Y, Grabe HJ, Mather KA, Schmidt R, Joliot M, Ikram MA, Matsuda F, Tzourio C, Wardlaw JM, Seshadri S, Adams HHH, Debette S (2023) Genomics of perivascular space burden unravels early mechanisms of cerebral small vessel disease. Nat Med 29:950–962

ENCODE Project Consortium, Moore JE, Purcaro MJ, Pratt HE, Epstein CB, Shoresh N, Adrian J, Kawli T, Davis CA, Dobin A, Kaul R, Halow J, Van Nostrand EL, Freese P, Gorkin DU, Shen Y, He Y, Mackiewicz M, Pauli-Behn F, Williams BA, Mortazavi A, Keller CA, Zhang XO, Elhajjajy SI, Huey J, Dickel DE, Snetkova V, Wei X, Wang X, Rivera-Mulia JC, Rozowsky J, Zhang J, Chhetri SB, Zhang J, Victorsen A, White KP, Visel A, Yeo GW, Burge CB, Lecuyer E, Gilbert DM, Dekker J, Rinn J, Mendenhall EM, Ecker JR, Kellis M, Klein RJ, Noble WS, Kundaje A, Guigo R, Farnham PJ, Cherry JM, Myers RM, Ren B, Graveley BR, Gerstein MB, Pennacchio LA, Snyder MP, Bernstein BE, Wold B, Hardison RC, Gingeras TR, Stamatoyannopoulos JA, Weng Z (2020) Expanded encyclopaedias of DNA elements in the human and mouse genomes. Nature 583:699–710

Felleman DJ, Van Essen DC (1991) Distributed hierarchical processing in the primate cerebral cortex. Cereb Cortex 1:1–47

Fox PT, Raichle ME (1984) Stimulus rate dependence of regional cerebral blood flow in human striate cortex, demonstrated by positron emission tomography. J Neurophysiol 51:1109–1120

French L, Paus T (2015) A FreeSurfer view of the cortical transcriptome generated from the Allen human brain atlas. Front Neurosci 9:323

Frisk V, Milner B (1991) Does left temporal lobectomy adversely affect the rate at which verbal material can be processed? Neuropsychologia 29:113–123

Fuster JM, Alexander GE (1971) Neuron activity related to short-term memory. Science 173:652–654

Gage SH, Munafo MR, Davey Smith G (2016) Causal inference in developmental origins of health and disease (DOHaD) research. Annu Rev Psychol 67:567–585

Georgopoulos AP, Kalaska JF, Caminiti R, Massey JT (1982) On the relations between the direction of two-dimensional arm movements and cell discharge in primate motor cortex. J Neurosci 2:1527–1537

Georgopoulos AP, Lurito JT, Petrides M, Schwartz AB, Massey JT (1989) Mental rotation of the neuronal population vector. Science 243:234–236

Glasser MF, Coalson TS, Robinson EC, Hacker CD, Harwell J, Yacoub E, Ugurbil K, Andersson J, Beckmann CF, Jenkinson M, Smith SM, Van Essen DC (2016) A multi-modal parcellation of human cerebral cortex. Nature 536:171–178

Grasby KL, Jahanshad N, Painter JN, Colodro-Conde L, Bralten J, Hibar DP, Lind PA, Pizzagalli F, Ching CRK, McMahon MAB, Shatokhina N, Zsembik LCP, Thomopoulos SI, Zhu AH, Strike LT, Agartz I, Alhusaini S, Almeida MAA, Alnaes D, Amlien IK, Andersson M, Ard T, Armstrong NJ, Ashley-Koch A, Atkins JR, Bernard M, Brouwer RM, Buimer EEL, Bulow R, Burger C, Cannon DM, Chakravarty M, Chen Q, Cheung JW, Couvy-Duchesne B, Dale AM, Dalvie S, de Araujo TK, de Zubicaray GI, de Zwarte SMC, den Braber A, Doan NT, Dohm K, Ehrlich S, Engelbrecht HR, Erk S, Fan CC, Fedko IO, Foley SF, Ford JM, Fukunaga M, Garrett ME, Ge T, Giddaluru S, Goldman AL, Green MJ, Groenewold NA, Grotegerd D, Gurholt TP, Gutman BA, Hansell NK, Harris MA, Harrison MB, Haswell CC, Hauser M, Herms S, Heslenfeld DJ, Ho NF, Hoehn D, Hoffmann P, Holleran L, Hoogman M, Hottenga JJ, Ikeda M, Janowitz D, Jansen IE, Jia T, Jockwitz C, Kanai R, Karama S, Kasperaviciute D, Kaufmann T, Kelly S, Kikuchi M, Klein M, Knapp M, Knodt AR, Kramer B, Lam M, Lancaster TM, Lee PH, Lett TA, Lewis LB, Lopes-Cendes I, Luciano M, Macciardi F, Marquand AF, Mathias SR, Melzer TR, Milaneschi Y et al (2020) The genetic architecture of the human cerebral cortex. Science 367

Gross CG, Bender DB, Gerstein GL (1979) Activity of inferior temporal neurons in behaving monkeys. Neuropsychologia 17:215–229

GTEs Consortium (2020) The GTEx Consortium atlas of genetic regulatory effects across human tissues. Science 369:1318–1330

Han L, Zhao X, Benton ML, Perumal T, Collins RL, Hoffman GE, Johnson JS, Sloofman L, Wang HZ, Stone MR, CommonMind C, Brennand KJ, Brand H, Sieberts SK, Marenco S, Peters MA, Lipska BK, Roussos P, Capra JA, Talkowski M, Ruderfer DM (2020) Functional annotation of rare structural variation in the human brain. Nat Commun 11:2990

Handsaker RE, Van Doren V, Berman JR, Genovese G, Kashin S, Boettger LM, McCarroll SA (2015) Large multiallelic copy number variations in humans. Nat Genet 47:296–303

Havdahl A, Wootton RE, Leppert B, Riglin L, Ask H, Tesli M, Bugge Askeland R, Hannigan LJ, Corfield E, Oyen AS, Andreassen OA, Tilling K, Davey Smith G, Thapar A, Reichborn-Kjennerud T, Stergiakouli E (2022) Associations between pregnancy-related predisposing factors for offspring neurodevelopmental conditions and parental genetic liability to attention-deficit/hyperactivity disorder, autism, and schizophrenia: the Norwegian mother, father and child cohort study (MoBa). JAMA Psychiatry 79:799–810

Hawrylycz MJ, Lein ES, Guillozet-Bongaarts AL, Shen EH, Ng L, Miller JA, van de Lagemaat LN, Smith KA, Ebbert A, Riley ZL, Abajian C, Beckmann CF, Bernard A, Bertagnolli D, Boe AF, Cartagena PM, Chakravarty MM, Chapin M, Chong J, Dalley RA, Daly BD, Dang C, Datta S, Dee N, Dolbeare TA, Faber V, Feng D, Fowler DR, Goldy J, Gregor BW, Haradon Z, Haynor DR, Hohmann JG, Horvath S, Howard RE, Jeromin A, Jochim JM, Kinnunen M, Lau C, Lazarz ET, Lee C, Lemon TA, Li L, Li Y, Morris JA, Overly CC, Parker PD, Parry SE, Reding M, Royall JJ, Schulkin J, Sequeira PA, Slaughterbeck CR, Smith SC, Sodt AJ, Sunkin SM, Swanson BE, Vawter MP, Williams D, Wohnoutka P, Zielke HR, Geschwind DH, Hof PR,

Smith SM, Koch C, Grant SG, Jones AR (2012) An anatomically comprehensive atlas of the adult human brain transcriptome. Nature 489:391–399

Hofer E, Roshchupkin GV, Adams HHH, Knol MJ, Lin H, Li S, Zare H, Ahmad S, Armstrong NJ, Satizabal CL, Bernard M, Bis JC, Gillespie NA, Luciano M, Mishra A, Scholz M, Teumer A, Xia R, Jian X, Mosley TH, Saba Y, Pirpamer L, Seiler S, Becker JT, Carmichael O, Rotter JI, Psaty BM, Lopez OL, Amin N, van der Lee SJ, Yang Q, Himali JJ, Maillard P, Beiser AS, DeCarli C, Karama S, Lewis L, Harris M, Bastin ME, Deary IJ, Witte AV, Beyer F, Loeffler M, Mather KA, Schofield PR, Thalamuthu A, Kwok JB, Wright MJ, Ames D, Trollor J, Jiang J, Brodaty H, Wen W, Vernooij MW, Hofman A, Uitterlinden AG, Niessen WJ, Wittfeld K, Bulow R, Volker U, Pausova Z, Pike GB, Maingault S, Crivello F, Tzourio C, Amouyel P, Mazoyer B, Neale MC, Franz CE, Lyons MJ, Panizzon MS, Andreassen OA, Dale AM, Logue M, Grasby KL, Jahanshad N, Painter JN, Colodro-Conde L, Bralten J, Hibar DP, Lind PA, Pizzagalli F, Stein JL, Thompson PM, Medland SE, Consortium E, Sachdev PS, Kremen WS, Wardlaw JM, Villringer A, van Duijn CM, Grabe HJ, Longstreth WT Jr, Fornage M, Paus T, Debette S, Ikram MA, Schmidt H, Schmidt R, Seshadri S (2020) Genetic correlations and genome-wide associations of cortical structure in general population samples of 22,824 adults. Nat Commun 11:4796

Hu Y, Haessler JW, Manansala R, Wiggins KL, Moscati A, Beiser A, Heard-Costa NL, Sarnowski C, Raffield LM, Chung J, Marini S, Anderson CD, Rosand J, Xu H, Sun X, Kelly TN, Wong Q, Lange LA, Rotter JI, Correa A, Vasan RS, Seshadri S, Rich SS, Do R, Loos RJF, Longstreth WT Jr, Bis JC, Psaty BM, Tirschwell DL, Assimes TL, Silver B, Liu S, Jackson R, Wassertheil-Smoller S, Mitchell BD, Fornage M, Auer PL, Reiner AP, Kooperberg C, t.N.T.-O. f.P.M.C. Trans-Omics for Precision Medicine Stroke Working Group (2022) Whole-genome sequencing association analyses of stroke and its subtypes in ancestrally diverse populations from trans-omics for precision medicine project. Stroke 53:875–885

Huguet G, Schramm C, Douard E, Jiang L, Labbe A, Tihy F, Mathonnet G, Nizard S, Lemyre E, Mathieu A, Poline JB, Loth E, Toro R, Schumann G, Conrod P, Pausova Z, Greenwood C, Paus T, Bourgeron T, Jacquemont S, I. Consortium (2018) Measuring and estimating the effect sizes of copy number variants on general intelligence in community-based samples. JAMA Psychiatry 75:447–457

Huguet G, Schramm C, Douard E, Tamer P, Main A, Monin P, England J, Jizi K, Renne T, Poirier M, Nowak S, Martin CO, Younis N, Knoth IS, Jean-Louis M, Saci Z, Auger M, Tihy F, Mathonnet G, Maftei C, Leveille F, Porteous D, Davies G, Redmond P, Harris SE, Hill WD, Lemyre E, Schumann G, Bourgeron T, Pausova Z, Paus T, Karama S, Lippe S, Deary IJ, Almasy L, Labbe A, Glahn D, Greenwood CMT, Jacquemont S (2021) Genome-wide analysis of gene dosage in 24,092 individuals estimates that 10,000 genes modulate cognitive ability. Mol Psychiatry 26:2663–2676

International Multiple Sclerosis Genetics Consortium, and MultipleMS Consortium (2023) Locus for severity implicates CNS resilience in progression of multiple sclerosis. Nature 619:323–331

Kendler KS, Gardner CO, Prescott CA (2003) Personality and the experience of environmental adversity. Psychol Med 33:1193–1202

Kojima S, Goldman-Rakic PS (1982) Delay-related activity of prefrontal neurons in rhesus monkeys performing delayed response. Brain Res 248:43–49

Kramer MS, Aboud F, Mironova E, Vanilovich I, Platt RW, Matush L, Igumnov S, Fombonne E, Bogdanovich N, Ducruet T, Collet JP, Chalmers B, Hodnett E, Davidovsky S, Skugarevsky O, Trofimovich O, Kozlova L, Shapiro S, G (2008) Promotion of breastfeeding intervention trial study, breastfeeding and child cognitive development: new evidence from a large randomized trial. Arch Gen Psychiatry 65:578–584

Kumar K, Modenato C, Moreau C, Ching CRK, Harvey A, Martin-Brevet S, Huguet G, Jean-Louis M, Douard E, Martin CO, Younis N, Tamer P, Maillard AM, Rodriguez-Herreros B, A. Pain, S.S.C. p11.2 European Consortium, Kushan L, Isaev D, Alpert K, Ragothaman A, Turner JA, Wang L, Ho TC, Schmaal L, Silva AI, van den Bree MBM, Linden DEJ, Owen MJ, Hall J, Lippe S, Dumas G, Draganski B, Gutman BA, Sonderby IE, Andreassen OA, Schultz

LM, Almasy L, Glahn DC, Bearden CE, Thompson PM, Jacquemont S (2023) Subcortical brain alterations in carriers of genomic copy number variants. Am J Psychiatry 180:685–698
Lander ES, Schork NJ (1994) Genetic dissection of complex traits. Science 265:2037–2048
Lee JJ, Wedow R, Okbay A, Kong E, Maghzian O, Zacher M, Nguyen-Viet TA, Bowers P, Sidorenko J, Linner RK, Fontana MA, Kundu T, Lee C, Li H, Li R, Royer R, Timshel PN, Walters RK, Willoughby EA, Yengo L, T. and Me Research, Cogent, C. Social Science Genetic Association, Alver M, Bao Y, Clark DW, Day FR, Furlotte NA, Joshi PK, Kemper KE, Kleinman A, Langenberg C, Magi R, Trampush JW, Verma SS, Wu Y, Lam M, Zhao JH, Zheng Z, Boardman JD, Campbell H, Freese J, Harris KM, Hayward C, Herd P, Kumari M, Lencz T, Luan J, Malhotra AK, Metspalu A, Milani L, Ong KK, Perry JRB, Porteous DJ, Ritchie MD, Smart MC, Smith BH, Tung JY, Wareham NJ, Wilson JF, Beauchamp JP, Conley DC, Esko T, Lehrer SF, Magnusson PKE, Oskarsson S, Pers TH, Robinson MR, Thom K, Watson C, Chabris CF, Meyer MN, Laibson DI, Yang J, Johannesson M, Koellinger PD, Turley P, Visscher PM, Benjamin DJ, Cesarini D (2018) Gene discovery and polygenic prediction from a genome-wide association study of educational attainment in 1.1 million individuals. Nat Genet 50:1112–1121
Lerch JP, van der Kouwe AJ, Raznahan A, Paus T, Johansen-Berg H, Miller KL, Smith SM, Fischl B, Sotiropoulos SN (2017) Studying neuroanatomy using MRI. Nat Neurosci 20:314–326
Lewis SW, Murray RM (1987) Obstetric complications, neurodevelopmental deviance, and risk of schizophrenia. J Psychiatr Res 21:413–421
Li M, Santpere G, Kawasawa YI, Evgrafov OV, Gulden FO, Pochareddy S, Sunkin SM, Li Z, Shin Y, Zhu Y, Sousa AMM, Werling DM, Kitchen RR, Kang HJ, Pletikos M, Choi J, Muchnik S, Xu X, Wang D, Lorente-Galdos B, Liu S, Giusti-Rodriguez P, Won H, de Leeuw CA, Pardinas AF, BrainSpan C, Psych EC, Psych EDS, Hu M, Jin F, Li Y, Owen MJ, O'Donovan MC, Walters JTR, Posthuma D, Reimers MA, Levitt P, Weinberger DR, Hyde TM, Kleinman JE, Geschwind DH, Hawrylycz MJ, State MW, Sanders SJ, Sullivan PF, Gerstein MB, Lein ES, Knowles JA, Sestan N (2018) Integrative functional genomic analysis of human brain development and neuropsychiatric risks. Science 362
Liao Z, Kopal J, Jacquemont S, Paus T (2023a) Copy number variants and the tangential growth of the cerebral cortex. Eur Neuropsychopharmacol 75:S138–S139
Liao Z, Kumar K, Kopal J, Jacquemont S, Paus T (2023b) Copy number variants and the tangential growth of the cerebral cortex. Eur Neuropsychopharmacol 75:S138–S139
Littlejohns TJ, Holliday J, Gibson LM, Garratt S, Oesingmann N, Alfaro-Almagro F, Bell JD, Boultwood C, Collins R, Conroy MC, Crabtree N, Doherty N, Frangi AF, Harvey NC, Leeson P, Miller KL, Neubauer S, Petersen SE, Sellors J, Sheard S, Smith SM, Sudlow CLM, Matthews PM, Allen NE (2020) The UK Biobank imaging enhancement of 100,000 participants: rationale, data collection, management and future directions. Nat Commun 11:2624
Logothetis NK, Pauls J, Augath M, Trinath T, Oeltermann A (2001) Neurophysiological investigation of the basis of the fMRI signal. Nature 412:150–157
Lueck CJ, Zeki S, Friston KJ, Deiber MP, Cope P, Cunningham VJ, Lammertsma AA, Kennard C, Frackowiak RS (1989) The colour centre in the cerebral cortex of man. Nature 340:386–389
Malik R, Chauhan G, Traylor M, Sargurupremraj M, Okada Y, Mishra A, Rutten-Jacobs L, Giese AK, van der Laan SW, Gretarsdottir S, Anderson CD, Chong M, Adams HHH, Ago T, Almgren P, Amouyel P, Ay H, Bartz TM, Benavente OR, Bevan S, Boncoraglio GB, Brown RD Jr, Butterworth AS, Carrera C, Carty CL, Chasman DI, Chen WM, Cole JW, Correa A, Cotlarciuc I, Cruchaga C, Danesh J, de Bakker PIW, DeStefano AL, den Hoed M, Duan Q, Engelter ST, Falcone GJ, Gottesman RF, Grewal RP, Gudnason V, Gustafsson S, Haessler J, Harris TB, Hassan A, Havulinna AS, Heckbert SR, Holliday EG, Howard G, Hsu FC, Hyacinth HI, Ikram MA, Ingelsson E, Irvin MR, Jian X, Jimenez-Conde J, Johnson JA, Jukema JW, Kanai M, Keene KL, Kissela BM, Kleindorfer DO, Kooperberg C, Kubo M, Lange LA, Langefeld CD, Langenberg C, Launer LJ, Lee JM, Lemmens R, Leys D, Lewis CM, Lin WY, Lindgren AG, Lorentzen E, Magnusson PK, Maguire J, Manichaikul A, McArdle PF,

Meschia JF, Mitchell BD, Mosley TH, Nalls MA, Ninomiya T, O'Donnell MJ, Psaty BM, Pulit SL, Rannikmae K, Reiner AP, Rexrode KM, Rice K, Rich SS, Ridker PM, Rost NS, Rothwell PM, Rotter JI, Rundek T, Sacco RL, Sakaue S, Sale MM et al (2018) Multiancestry genome-wide association study of 520,000 subjects identifies 32 loci associated with stroke and stroke subtypes. Nat Genet 50:524–537

Marees AT, de Kluiver H, Stringer S, Vorspan F, Curis E, Marie-Claire C, Derks EM (2018) A tutorial on conducting genome-wide association studies: quality control and statistical analysis. Int J Methods Psychiatr Res 27:e1608

Marques JP, Norris DG (2018) How to choose the right MR sequence for your research question at 7T and above? Neuroimage 168:119–140

Martens EP, Pestman WR, de Boer A, Belitser SV, Klungel OH (2006) Instrumental variables: application and limitations. Epidemiology 17:260–267

Medland SE, Grasby KL, Jahanshad N, Painter JN, Colodro-Conde L, Bralten J, Hibar DP, Lind PA, Pizzagalli F, Thomopoulos SI, Stein JL, Franke B, Martin NG, Thompson PM, E.G.W. Group (2022) Ten years of enhancing neuro-imaging genetics through meta-analysis: an overview from the ENIGMA Genetics Working Group. Hum Brain Mapp 43:292–299

Mesulam MM (1982) Tracing neural connections with horseradish peroxidase. Wiley, New York

Miller JE, Veturi Y, Ritchie MD (2019) Innovative strategies for annotating the "relationSNP" between variants and molecular phenotypes. BioData Min 12:10

Milner B (1982) Some cognitive effects of frontal-lobe lesions in man. Philos Trans R Soc Lond B Biol Sci 298:211–226

Milner B (2005) The medial temporal-lobe amnesic syndrome. Psychiatr Clin North Am 28:599–611. 609

Milner B, Taylor L, Sperry RW (1968) Lateralized suppression of dichotically presented digits after commissural section in man. Science 161:184–186

Milner B, Johnsrude I, Crane J (1997) Right medial temporal-lobe contribution to object-location memory. Philos Trans R Soc Lond B Biol Sci 352:1469–1474

Min JL, Hemani G, Hannon E, Dekkers KF, Castillo-Fernandez J, Luijk R, Carnero-Montoro E, Lawson DJ, Burrows K, Suderman M, Bretherick AD, Richardson TG, Klughammer J, Iotchkova V, Sharp G, Al Khleifat A, Shatunov A, Iacoangeli A, McArdle WL, Ho KM, Kumar A, Soderhall C, Soriano-Tarraga C, Giralt-Steinhauer E, Kazmi N, Mason D, McRae AF, Corcoran DL, Sugden K, Kasela S, Cardona A, Day FR, Cugliari G, Viberti C, Guarrera S, Lerro M, Gupta R, Bollepalli S, Mandaviya P, Zeng Y, Clarke TK, Walker RM, Schmoll V, Czamara D, Ruiz-Arenas C, Rezwan FI, Marioni RE, Lin T, Awaloff Y, Germain M, Aissi D, Zwamborn R, van Eijk K, Dekker A, van Dongen J, Hottenga JJ, Willemsen G, Xu CJ, Barturen G, Catala-Moll F, Kerick M, Wang C, Melton P, Elliott HR, Shin J, Bernard M, Yet I, Smart M, Gorrie-Stone T, Consortium B, Shaw C, Al Chalabi A, Ring SM, Pershagen G, Melen E, Jimenez-Conde J, Roquer J, Lawlor DA, Wright J, Martin NG, Montgomery GW, Moffitt TE, Poulton R, Esko T, Milani L, Metspalu A, Perry JRB, Ong KK, Wareham NJ, Matullo G, Sacerdote C, Panico S, Caspi A, Arseneault L, Gagnon F, Ollikainen M, Kaprio J, Felix JF, Rivadeneira F, Tiemeier H et al (2021) Genomic and phenotypic insights from an atlas of genetic effects on DNA methylation. Nat Genet 53:1311–1321

Morabia A (2001) Snow and Farr: a scientific duet. Soz Praventivmed 46:223–224

Morabia A (2004) History of epidemiologic methods and concepts. Birkhauser Verlag, Basel

Morrison JH, Hof PR (1997) Life and death of neurons in the aging brain. Science 278:412–419

Munck S, Swoger J, Coll-Llado M, Gritti N, Vande Velde G (2021) Maximizing content across scales: moving multimodal microscopy and mesoscopy toward molecular imaging. Curr Opin Chem Biol 63:188–199

Munzel T, Sorensen M, Hahad O, Nieuwenhuijsen M, Daiber A (2023) The contribution of the exposome to the burden of cardiovascular disease. Nat Rev Cardiol. https://doi.org/10.1038/s41569-023-00873-3

Neumeyer S, Hemani G, Zeggini E (2020) Strengthening causal inference for complex disease using molecular quantitative trait loci. Trends Mol Med 26:232–241

Nosarti C, Reichenberg A, Murray RM, Cnattingius S, Lambe MP, Yin L, MacCabe J, Rifkin L, Hultman CM (2012) Preterm birth and psychiatric disorders in young adult life. Arch Gen Psychiatry 69:E1–E8

Nurk S, Koren S, Rhie A, Rautiainen M, Bzikadze AV, Mikheenko A, Vollger MR, Altemose N, Uralsky L, Gershman A, Aganezov S, Hoyt SJ, Diekhans M, Logsdon GA, Alonge M, Antonarakis SE, Borchers M, Bouffard GG, Brooks SY, Caldas GV, Chen NC, Cheng H, Chin CS, Chow W, de Lima LG, Dishuck PC, Durbin R, Dvorkina T, Fiddes IT, Formenti G, Fulton RS, Fungtammasan A, Garrison E, Grady PGS, Graves-Lindsay TA, Hall IM, Hansen NF, Hartley GA, Haukness M, Howe K, Hunkapiller MW, Jain C, Jain M, Jarvis ED, Kerpedjiev P, Kirsche M, Kolmogorov M, Korlach J, Kremitzki M, Li H, Maduro VV, Marschall T, McCartney AM, McDaniel J, Miller DE, Mullikin JC, Myers EW, Olson ND, Paten B, Peluso P, Pevzner PA, Porubsky D, Potapova T, Rogaev EI, Rosenfeld JA, Salzberg SL, Schneider VA, Sedlazeck FJ, Shafin K, Shew CJ, Shumate A, Sims Y, Smit AFA, Soto DC, Sovic I, Storer JM, Streets A, Sullivan BA, Thibaud-Nissen F, Torrance J, Wagner J, Walenz BP, Wenger A, Wood JMD, Xiao C, Yan SM, Young AC, Zarate S, Surti U, McCoy RC, Dennis MY, Alexandrov IA, Gerton JL, O'Neill RJ, Timp W, Zook JM, Schatz MC, Eichler EE, Miga KH, Phillippy AM (2022) The complete sequence of a human genome. Science 376:44–53

Office for National Statistics (2011) Census: key statistics for England and Wales

Panagiotaki E, Schneider T, Siow B, Hall MG, Lythgoe MF, Alexander DC (2012) Compartment models of the diffusion MR signal in brain white matter: a taxonomy and comparison. Neuroimage 59:2241–2254

Patel Y, Shin J, Gowland PA, Pausova Z, Paus T, I. Consortium (2019) Maturation of the human cerebral cortex during adolescence: myelin or dendritic arbor? Cereb Cortex 29:3351–3362

Patel Y, Shin J, Drakesmith M, Evans J, Pausova Z, Paus T (2020) Virtual histology of multi-modal magnetic resonance imaging of cerebral cortex in young men. Neuroimage 218:116968

Patel Y, Shin J, Abe C, Agartz I, Alloza C, Alnaes D, Ambrogi S, Antonucci LA, Arango C, Arolt V, Auzias G, Ayesa-Arriola R, Banaj N, Banaschewski T, Bandeira C, Basgoze Z, Cupertino RB, Bau CHD, Bauer J, Baumeister S, Bernardoni F, Bertolino A, Bonnin CDM, Brandeis D, Brem S, Bruggemann J, Bulow R, Bustillo JR, Calderoni S, Calvo R, Canales-Rodriguez EJ, Cannon DM, Carmona S, Carr VJ, Catts SV, Chenji S, Chew QH, Coghill D, Connolly CG, Conzelmann A, Craven AR, Crespo-Facorro B, Cullen K, Dahl A, Dannlowski U, Davey CG, Deruelle C, Diaz-Caneja CM, Dohm K, Ehrlich S, Epstein J, Erwin-Grabner T, Eyler LT, Fedor J, Fitzgerald J, Foran W, Ford JM, Fortea L, Fuentes-Claramonte P, Fullerton J, Furlong L, Gallagher L, Gao B, Gao S, Goikolea JM, Gotlib I, Goya-Maldonado R, Grabe HJ, Green M, Grevet EH, Groenewold NA, Grotegerd D, Gruber O, Haavik J, Hahn T, Harrison BJ, Heindel W, Henskens F, Heslenfeld DJ, Hilland E, Hoekstra PJ, Hohmann S, Holz N, Howells FM, Ipser JC, Jahanshad N, Jakobi B, Jansen A, Janssen J, Jonassen R, Kaiser A, Kaleda V, Karantonis J, King JA, Kircher T, Kochunov P, Koopowitz SM, Landen M, Landro NI, Lawrie S et al (2022) Virtual ontogeny of cortical growth preceding mental illness. Biol Psychiatry 92:299–313

Paus T (2010) Population neuroscience: why and how. Hum Brain Mapp 31:891–903

Paus T (2013) Population neuroscience. Springer, Berlin

Paus T (2016) Population neuroscience. Handb Clin Neurol 138:17–37

Paus T (2023) Tracking development of connectivity in the human brain: axons and dendrites. Biol Psychiatry 93:455–463

Paus T, Brook J, Keyes KM and Pausova Z (Eds) (2024a) Principles and advances in population neuroscience. Springer, Cham

Paus T and Kum H-C (eds) (2024b) Digital ethology: human behavior in geospatial context. The MIT Press, Cambridge

Paus T (2003) Principles of functional neuroimaging. In: Schiffer RB, Fogel BS (eds) Neuropsychiatry. Lippincton Williams & Wilkins, Philadelphia, pp 63–90

Petersen SE, Fox PT, Posner MI, Mintun M, Raichle ME (1988) Positron emission tomographic studies of the cortical anatomy of single-word processing. Nature 331:585–589

Pigott S, Milner B (1993) Memory for different aspects of complex visual scenes after unilateral temporal- or frontal-lobe resection. Neuropsychologia 31:1–15

Pingault JB, Barkhuizen W, Wang B, Hannigan LJ, Eilertsen EM, Corfield E, Andreassen OA, Ask H, Tesli M, Askeland RB, Davey Smith G, Stoltenberg C, Davies NM, Reichborn-Kjennerud T, Ystrom E, Havdahl A (2023) Genetic nurture versus genetic transmission of risk for ADHD traits in the Norwegian mother, father and child cohort study. Mol Psychiatry 28: 1731–1738

Porcu E, Rueger S, Lepik K, Q.C. E, B. Consortium, Santoni FA, Reymond A, Kutalik Z (2019) Mendelian randomization integrating GWAS and eQTL data reveals genetic determinants of complex and clinical traits. Nat Commun 10:3300

Porcu E, Sjaarda J, Lepik K, Carmeli C, Darrous L, Sulc J, Mounier N, Kutalik Z (2021) Causal inference methods to integrate omics and complex traits. Cold Spring Harb Perspect Med 11

Psaty BM, O'Donnell CJ, Gudnason V, Lunetta KL, Folsom AR, Rotter JI, Uitterlinden AG, Harris TB, Witteman JC, Boerwinkle E, C. Consortium (2009) Cohorts for heart and aging research in genomic epidemiology (CHARGE) consortium: design of prospective meta-analyses of genome-wide association studies from 5 cohorts. Circ Cardiovasc Genet 2:73–80

Rakic P (1988) Specification of cerebral cortical areas. Science 241:170–176

Rattray NJW, Deziel NC, Wallach JD, Khan SA, Vasiliou V, Ioannidis JPA, Johnson CH (2018) Beyond genomics: understanding exposotypes through metabolomics. Hum Genomics 12:4

Rubinov M, Sporns O (2010) Complex network measures of brain connectivity: uses and interpretations. Neuroimage 52:1059–1069

Satizabal CL, Adams HHH, Hibar DP, White CC, Knol MJ, Stein JL, Scholz M, Sargurupremraj M, Jahanshad N, Roshchupkin GV, Smith AV, Bis JC, Jian X, Luciano M, Hofer E, Teumer A, van der Lee SJ, Yang J, Yanek LR, Lee TV, Li S, Hu Y, Koh JY, Eicher JD, Desrivieres S, Arias-Vasquez A, Chauhan G, Athanasiu L, Renteria ME, Kim S, Hoehn D, Armstrong NJ, Chen Q, Holmes AJ, den Braber A, Kloszewska I, Andersson M, Espeseth T, Grimm O, Abramovic L, Alhusaini S, Milaneschi Y, Papmeyer M, Axelsson T, Ehrlich S, Roiz-Santianez R, Kraemer B, Haberg AK, Jones HJ, Pike GB, Stein DJ, Stevens A, Bralten J, Vernooij MW, Harris TB, Filippi I, Witte AV, Guadalupe T, Wittfeld K, Mosley TH, Becker JT, Doan NT, Hagenaars SP, Saba Y, Cuellar-Partida G, Amin N, Hilal S, Nho K, Mirza-Schreiber N, Arfanakis K, Becker DM, Ames D, Goldman AL, Lee PH, Boomsma DI, Lovestone S, Giddaluru S, Le Hellard S, Mattheisen M, Bohlken MM, Kasperaviciute D, Schmaal L, Lawrie SM, Agartz I, Walton E, Tordesillas-Gutierrez D, Davies GE, Shin J, Ipser JC, Vinke LN, Hoogman M, Jia T, Burkhardt R, Klein M, Crivello F, Janowitz D, Carmichael O, Haukvik UK, Aribisala BS, Schmidt H et al (2019) Genetic architecture of subcortical brain structures in 38,851 individuals. Nat Genet 51:1624–1636

Schaefer A, Kong R, Gordon EM, Laumann TO, Zuo XN, Holmes AJ, Eickhoff SB, Yeo BTT (2018) Local-global parcellation of the human cerebral cortex from intrinsic functional connectivity MRI. Cereb Cortex 28:3095–3114

Schoeler T, Speed D, Porcu E, Pirastu N, Pingault JB, Kutalik Z (2023) Participation bias in the UK Biobank distorts genetic associations and downstream analyses. Nat Hum Behav 7:1216–1227

Scoville WB, Milner B (1957) Loss of recent memory after bilateral hippocampal lesions. J Neurol Neurosurg Psychiatry 20:11–21

Setsompop K, Kimmlingen R, Eberlein E, Witzel T, Cohen-Adad J, McNab JA, Keil B, Tisdall MD, Hoecht P, Dietz P, Cauley SF, Tountcheva V, Matschl V, Lenz VH, Heberlein K, Potthast A, Thein H, Van Horn J, Toga A, Schmitt F, Lehne D, Rosen BR, Wedeen V, Wald LL (2013) Pushing the limits of in vivo diffusion MRI for the human connectome project. Neuroimage 80:220–233

Seyedsalehi A, Warrier V, Bethlehem RAI, Perry BI, Burgess S, Murray GK (2023) Educational attainment, structural brain reserve and Alzheimer's disease: a Mendelian randomization analysis. Brain 146:2059–2074

Shete S, Lau CC, Houlston RS, Claus EB, Barnholtz-Sloan J, Lai R, Il'yasova D, Schildkraut J, Sadetzki S, Johansen C, Bernstein JL, Olson SH, Jenkins RB, Yang P, Vick NA, Wrensch M, Davis FG, McCarthy BJ, Leung EH, Davis C, Cheng R, Hosking FJ, Armstrong GN, Liu Y, Yu RK, Henriksson R, Gliogene C, Melin BS, Bondy ML (2011) Genome-wide high-density SNP linkage search for glioma susceptibility loci: results from the Gliogene Consortium. Cancer Res 71:7568–7575

Shin J, French L, Xu T, Leonard G, Perron M, Pike GB, Richer L, Veillette S, Pausova Z, Paus T (2018) Cell-specific gene-expression profiles and cortical thickness in the human brain. Cereb Cortex 28:3267–3277

Shin J, Ma S, Hofer E, Patel Y, Vosberg DE, Tilley S, Roshchupkin GV, Sousa AMM, Jian X, Gottesman R, Mosley TH, Fornage M, Saba Y, Pirpamer L, Schmidt R, Schmidt H, Carrion-Castillo A, Crivello F, Mazoyer B, Bis JC, Li S, Yang Q, Luciano M, Karama S, Lewis L, Bastin ME, Harris MA, Wardlaw JM, Deary IE, Scholz M, Loeffler M, Witte AV, Beyer F, Villringer A, Armstrong NJ, Mather KA, Ames D, Jiang J, Kwok JB, Schofield PR, Thalamuthu A, Trollor JN, Wright MJ, Brodaty H, Wen W, Sachdev PS, Terzikhan N, Evans TE, Adams H, Ikram MA, Frenzel S, Auwera-Palitschka SV, Wittfeld K, Bulow R, Grabe HJ, Tzourio C, Mishra A, Maingault S, Debette S, Gillespie NA, Franz CE, Kremen WS, Ding L, Jahanshad N, Consortium E, Sestan N, Pausova Z, Seshadri S, Paus T, Neuro CWG (2020) Global and regional development of the human cerebral cortex: molecular architecture and occupational aptitudes. Cereb Cortex 30:4121–4139

Smith GD, Ebrahim S (2003) 'Mendelian randomization': can genetic epidemiology contribute to understanding environmental determinants of disease? Int J Epidemiol 32:1–22

Smith ML, Milner B (1989) Right hippocampal impairment in the recall of spatial location: encoding deficit or rapid forgetting? Neuropsychologia 27:71–81

Sonderby IE, Ching CRK, Thomopoulos SI, van der Meer D, Sun D, Villalon-Reina JE, Agartz I, Amunts K, Arango C, Armstrong NJ, Ayesa-Arriola R, Bakker G, Bassett AS, Boomsma DI, Bulow R, Butcher NJ, Calhoun VD, Caspers S, Chow EWC, Cichon S, Ciufolini S, Craig MC, Crespo-Facorro B, Cunningham AC, Dale AM, Dazzan P, de Zubicaray GI, Djurovic S, Doherty JL, Donohoe G, Draganski B, Durdle CA, Ehrlich S, Emanuel BS, Espeseth T, Fisher SE, Ge T, Glahn DC, Grabe HJ, Gur RE, Gutman BA, Haavik J, Haberg AK, Hansen LA, Hashimoto R, Hibar DP, Holmes AJ, Hottenga JJ, Pol HEH, Jalbrzikowski M, Knowles EEM, Kushan L, Linden DEJ, Liu J, Lundervold AJ, Martin-Brevet S, Martinez K, Mather KA, Mathias SR, McDonald-McGinn DM, McRae AF, Medland SE, Moberget T, Modenato C, Monereo Sanchez J, Moreau CA, Muhleisen TW, Paus T, Pausova Z, Prieto C, Ragothaman A, Reinbold CS, Marques TR, Repetto GM, Reymond A, Roalf DR, Rodriguez-Herreros B, Rucker JJ, Sachdev PS, Schmitt JE, Schofield PR, Silva AI, Stefansson H, Stein DJ, Tamnes CK, Tordesillas-Gutierrez D, Ulfarsson MO, Vajdi A, van't Ent D, van den Bree MBM, Vassos E, Vazquez-Bourgon J, Vila-Rodriguez F, Walters GB, Wen W, Westlye LT, Wittfeld K, Zackai EH, Stefansson K, Jacquemont S et al (2022) Effects of copy number variations on brain structure and risk for psychiatric illness: large-scale studies from the ENIGMA working groups on CNVs. Hum Brain Mapp 43:300–328

Sudlow C, Gallacher J, Allen N, Beral V, Burton P, Danesh J, Downey P, Elliott P, Green J, Landray M, Liu B, Matthews P, Ong G, Pell J, Silman A, Young A, Sprosen T, Peakman T, Collins R (2015) UK biobank: an open access resource for identifying the causes of a wide range of complex diseases of middle and old age. PLoS Med 12:e1001779

Suel E, Polak JW, Bennett JE, Ezzati M (2019) Measuring social, environmental and health inequalities using deep learning and street imagery. Sci Rep 9:6229

Sullivan PF (2010) The psychiatric GWAS consortium: big science comes to psychiatry. Neuron 68:182–186

Sullivan PF, Agrawal A, Bulik CM, Andreassen OA, Borglum AD, Breen G, Cichon S, Edenberg HJ, Faraone SV, Gelernter J, Mathews CA, Nievergelt CM, Smoller JW, O'Donovan MC, C. (2018) Psychiatric genomics, psychiatric genomics: an update and an agenda. Am J Psychiatry 175:15–27

Talbot JD, Marrett S, Evans AC, Meyer E, Bushnell MC, Duncan GH (1991) Multiple representations of pain in human cerebral cortex. Science 251:1355–1358

Thompson PM, Stein JL, Medland SE, Hibar DP, Vasquez AA, Renteria ME, Toro R, Jahanshad N, Schumann G, Franke B, Wright MJ, Martin NG, Agartz I, Alda M, Alhusaini S, Almasy L, Almeida J, Alpert K, Andreasen NC, Andreassen OA, Apostolova LG, Appel K, Armstrong NJ, Aribisala B, Bastin ME, Bauer M, Bearden CE, Bergmann O, Binder EB, Blangero J, Bockholt HJ, Boen E, Bois C, Boomsma DI, Booth T, Bowman IJ, Bralten J, Brouwer RM, Brunner HG, Brohawn DG, Buckner RL, Buitelaar J, Bulayeva K, Bustillo JR, Calhoun VD, Cannon DM, Cantor RM, Carless MA, Caseras X, Cavalleri GL, Chakravarty MM, Chang KD, Ching CR, Christoforou A, Cichon S, Clark VP, Conrod P, Coppola G, Crespo-Facorro B, Curran JE, Czisch M, Deary IJ, de Geus EJ, den Braber A, Delvecchio G, Depondt C, de Haan L, de Zubicaray GI, Dima D, Dimitrova R, Djurovic S, Dong H, Donohoe G, Duggirala R, Dyer TD, Ehrlich S, Ekman CJ, Elvsashagen T, Emsell L, Erk S, Espeseth T, Fagerness J, Fears S, Fedko I, Fernandez G, Fisher SE, Foroud T, Fox PT, Francks C, Frangou S, Frey EM, Frodl T, Frouin V, Garavan H, Giddaluru S, Glahn DC, Godlewska B, Goldstein RZ, Gollub RL, Grabe HJ et al (2014) The ENIGMA Consortium: large-scale collaborative analyses of neuroimaging and genetic data. Brain Imaging Behav 8:153–182

Vachha B, Huang SY (2021) MRI with ultrahigh field strength and high-performance gradients: challenges and opportunities for clinical neuroimaging at 7 T and beyond. Eur Radiol Exp 5:35

Van Essen DC, Smith J, Glasser MF, Elam J, Donahue CJ, Dierker DL, Reid EK, Coalson T, Harwell J (2017) The brain analysis library of spatial maps and atlases (BALSA) database. Neuroimage 144:270–274

Vandenbroucke JP (1990) Epidemiology in transition: a historical hypothesis. Epidemiology 1: 164–167

Victora CG, Horta BL, Loret de Mola C, Quevedo L, Pinheiro RT, Gigante DP, Goncalves H, Barros FC (2015) Association between breastfeeding and intelligence, educational attainment, and income at 30 years of age: a prospective birth cohort study from Brazil. Lancet Glob Health 3:e199–e205

Vidal-Pineiro D, Parker N, Shin J, French L, Grydeland H, Jackowski AP, Mowinckel AM, Patel Y, Pausova Z, Salum G, Sorensen O, Walhovd KB, Paus T, Fjell AM, I. Alzheimer's Disease Neuroimaging, B. the Australian Imaging, and a (2020) Lifestyle flagship study of, cellular correlates of cortical thinning throughout the lifespan. Sci Rep 10:21803

Vosberg DE, Jurisica I, Pausova Z, Paus T (2024) Intrauterine growth and the tangential expansion of the human cerebral cortex in times of food scarcity and abundance. Nat Commun 15:1205

Walker RL, Ramaswami G, Hartl C, Mancuso N, Gandal MJ, de la Torre-Ubieta L, Pasaniuc B, Stein JL, Geschwind DH (2019) Genetic control of expression and splicing in developing human brain informs disease mechanisms. Cell 179:750–771 e22

Wang Y, Sarnowski C, Lin H, Pitsillides AN, Heard-Costa NL, Choi SH, Wang D, Bis JC, Blue EE, I. Alzheimer's Disease Neuroimaging, Boerwinkle E, De Jager PL, Fornage M, Wijsman EM, Seshadri S, Dupuis J, Peloso GM, DeStefano AL, P. Alzheimer's Disease Sequencing (2023) Key variants via Alzheimer's disease sequencing project whole genome sequence data. medRxiv

Watson JD, Jordan E (1989) The human genome program at the National Institutes of Health. Genomics 5:654–656

Weichenthal S, Hatzopoulou M, Brauer M (2019) A picture tells a thousand...exposures: opportunities and challenges of deep learning image analyses in exposure science and environmental epidemiology. Environ Int 122:3–10

Weinberger DR (1987) Implications of normal brain development for the pathogenesis of schizophrenia. Arch Gen Psychiatry 44:660–669

Wild CP (2005) Complementing the genome with an "exposome": the outstanding challenge of environmental exposure measurement in molecular epidemiology. Cancer Epidemiol Biomarkers Prev 14:1847–1850

Woo A, Botta A, Shi SSW, Paus T, Pausova Z (2022) Obesity-related neuroinflammation: magnetic resonance and microscopy Imaging of the brain. Int J Mol Sci 23

Writing Committee for the ENIGMA-CNV Working Group, D. van der Meer, Sonderby IE, Kaufmann T, Walters GB, Abdellaoui A, Ames D, Amunts K, Andersson M, Armstrong NJ, Bernard M, Blackburn NB, Blangero J, Boomsma DI, Brodaty H, Brouwer RM, Bulow R, Cahn W, Calhoun VD, Caspers S, Cavalleri GL, Ching CRK, Cichon S, Ciufolini S, Corvin A, Crespo-Facorro B, Curran JE, Dalvie S, Dazzan P, de Geus EJC, de Zubicaray GI, de Zwarte SMC, Delanty N, den Braber A, Desrivieres S, Di Forti M, Doherty JL, Donohoe G, Ehrlich S, Eising E, Espeseth T, Fisher SE, Fladby T, Frei O, Frouin V, Fukunaga M, Gareau T, Glahn DC, Grabe HJ, Groenewold NA, Gustafsson O, Haavik J, Haberg AK, Hashimoto R, Hehir-Kwa JY, Hibar DP, Hillegers MHJ, Hoffmann P, Holleran L, Hottenga JJ, Hulshoff Pol HE, Ikeda M, Jacquemont S, Jahanshad N, Jockwitz C, Johansson S, Jonsson EG, Kikuchi M, Knowles EEM, Kwok JB, Le Hellard S, Linden DEJ, Liu J, Lundervold A, Lundervold AJ, Martin NG, Mather KA, Mathias SR, McMahon KL, McRae AF, Medland SE, Moberget T, Moreau C, Morris DW, Muhleisen TW, Murray RM, Nordvik JE, Nyberg L, Olde Loohuis LM, Ophoff RA, Owen MJ, Paus T, Pausova Z, Peralta JM, Pike B, Prieto C, Quinlan EB, Reinbold CS, Reis Marques T, Rucker JJH et al (2020) Association of copy number variation of the 15q11.2 BP1-BP2 region with cortical and subcortical morphology and cognition. JAMA Psychiatry 77:420–430

Writing Committee for the Attention-Deficit/Hyperactivity, Autism Spectrum, Bipolar, Major Depressive, Obsessive-Compulsive, and Schizophrenia ENIGMA Writing Groups, Patel Y, Parker N, Shin J, Howard D, French L, Thomopoulos SI, Pozzi E, Abe Y, Abe C, Anticevic A, Alda M, Aleman A, Alloza C, Alonso-Lana S, Ameis SH, Anagnostou E, McIntosh AA, Arango C, Arnold PD, Asherson P, Assogna F, Auzias G, Ayesa-Arriola R, Bakker G, Banaj N, Banaschewski T, Bandeira CE, Baranov A, Bargallo N, Bau CHD, Baumeister S, Baune BT, Bellgrove MA, Benedetti F, Bertolino A, Boedhoe PSW, Boks M, Bollettini I, Bonnin CDM, Borgers T, Borgwardt S, Brandeis D, Brennan BP, Bruggemann JM, Bulow R, Busatto GF, Calderoni S, Calhoun VD, Calvo R, Canales-Rodriguez EJ, Cannon DM, Carr VJ, Cascella N, Cercignani M, Chaim-Avancini TM, Christakou A, Coghill D, Conzelmann A, Crespo-Facorro B, Cubillo AI, Cullen KR, Cupertino RB, Daly E, Dannlowski U, Davey CG, Denys D, Deruelle C, Di Giorgio A, Dickie EW, Dima D, Dohm K, Ehrlich S, Ely BA, Erwin-Grabner T, Ethofer T, Fair DA, Fallgatter AJ, Faraone SV, Fatjo-Vilas M, Fedor JM, Fitzgerald KD, Ford JM, Frodl T, Fu CHY, Fullerton JM, Gabel MC, Glahn DC, Roberts G, Gogberashvili T, Goikolea JM, Gotlib IH, Goya-Maldonado R, Grabe HJ, Green MJ et al (2021) Virtual histology of cortical thickness and shared neurobiology in 6 psychiatric disorders. JAMA Psychiatry 78:47–63

Xu J, Moeller S, Auerbach EJ, Strupp J, Smith SM, Feinberg DA, Yacoub E, Ugurbil K (2013) Evaluation of slice accelerations using multiband echo planar imaging at 3 T. Neuroimage 83: 991–1001

Yang J, Lee SH, Goddard ME, Visscher PM (2011) GCTA: a tool for genome-wide complex trait analysis. Am J Hum Genet 88:76–82

Yin TC, Mountcastle VB (1977) Visual input to the visuomotor mechanisms of the monkey's parietal lobe. Science 197:1381–1383

Zeisel A, Munoz-Manchado AB, Codeluppi S, Lonnerberg P, La Manno G, Jureus A, Marques S, Munguba H, He L, Betsholtz C, Rolny C, Castelo-Branco G, Hjerling-Leffler J, Linnarsson S (2015) Brain structure. Cell types in the mouse cortex and hippocampus revealed by single-cell RNA-seq. Science 347:1138–1142

Zilles K, Amunts K (2010) Centenary of Brodmann's map – conception and fate. Nat Rev Neurosci 11:139–145

Zou R, Labrecque JA, Swanson SA, Steegers EAP, White T, El Marroun H, Tiemeier H (2022) Prenatal exposure to trans fatty acids and head growth in fetal life and childhood: triangulating confounder-adjustment and instrumental variable approaches. Eur J Epidemiol 37:1171–1180

Part II
Populations

Population Neuroscience: Understanding Concepts of Generalizability and Transportability and Their Application to Improving the Public's Health

Katherine M. Keyes, Diana Pakserian, Kara E. Rudolph, Giovanni Salum, and Elizabeth A. Stuart

Contents

Abstract In population neuroscience, samples are not often selected with equal or known probability from an underlying population of interest; in other words, samples are not often formally representative of a specified underlying population. This chapter provides an overview of an epidemiological approach to considering the implications of selective participation on the value of our results for population health. We discuss definitions of generalizability and transportability, given the growing recognition that generalizability and transportability are central for interpreting data that are aiming to be population-based. We provide evidence that differences in the prevalence of effect measure modifiers between a study sample and a target population will lead to a lack of generalizability and transportability. We

K. M. Keyes (✉) and K. E. Rudolph
Department of Epidemiology, Columbia University Mailman School of Public Health, New York, NY, USA
e-mail: kmk2104@cumc.columbia.edu

D. Pakserian
National Institute of Mental Health, Bethesda, MD, USA

G. Salum
Child and Adolescent Mental Health Initiative, Child Mind Institute & Stavros Niarchos Foundation, New York, NY, USA

E. A. Stuart
Department of Biostatistics, Johns Hopkins Bloomberg School of Public Health, Baltimore, MD, USA

Curr Topics Behav Neurosci (2024) 68: 37–52
https://doi.org/10.1007/7854_2024_465
Published Online: 9 April 2024

provide an example of an association between a poly-genetic risk score and depression, showing how an internally valid association can differ based on the prevalence of effect measure modifiers. We show that when estimating associations, inferences from a study sample to a population can depend on clearly defining a target population. Given that representative sampling from explicitly defined target populations may not be feasible or realistic in many situations, especially given the sample sizes needed for statistical power for many exposures of interest (and especially when interactions are being tested), researchers should be well versed in tools available to enhance the interpretability of samples regarding target populations.

Keywords Effect measure modification · Gene by environment interaction · Generalizability · Transportability

In population neuroscience, samples are not often selected with equal or known probability from an underlying population of interest; in other words, samples are not often formally representative of a specified underlying population. Studies often are comprised of volunteers, individuals with clinical concerns, or those who meet stringent eligibility requirements. It is well recognized that these samples may thus not be "generalizable" to target populations of interest for intervention. Yet, less recognized and often under-discussed are the research questions for which samples do or do not need to be generalizable to a specific population, the reasons why study samples may not be generalizable, the conditions necessary for generalizability, and the implications for population science. This chapter provides an overview of an epidemiological approach to considering the implications of selective participation on the value of our results for population health.

It is important to define first some terms. We will provide an explanation for terms the way that they are most often used in epidemiology, and thus will be used in this chapter, but also acknowledge when readers from aligned disciplines may use them differently.

The first and perhaps most important is how we are going to use the term "causation." Of course, perhaps all scientists agree that causation occurs when the relationship between two variables is causal; that is, the action of one variable causes the action of another variable. Further, we all additionally agree that a fundamental goal of many scientific inquiries is to identify, ultimately, causes and causal processes. There is, however, often a difference between our ultimate goals and what we can endeavor to identify in any one study. Often, we are collecting and analyzing observational data and we cannot make definitive claims about causation from the data that we have collected. We can identify associations, attempt to ensure that those associations are robust to known confounders, major selection and other biases, and design experimental and quasi-experimental studies that are well positioned to move toward identifying causation. In this chapter, we will be explicit that the activity that we are ultimately interested in when we conduct a study is to identify causes – while this is a lofty goal in observational studies, it is nonetheless the ultimate goal of our work.

With the goal in mind that we engage in scientific studies to identify and understand causation, an additional set of terms that is important to clarify is: sample, source population, and target population. The *sample* is the group of participants among which data are collected. Each sample is recruited from an underlying *source population*, which may be defined by a geographic catchment area, time frame, and/or set of characteristics. A source population is the set of all people who would have been eligible to participate in the study if we were able to enumerated all eligible people.

We are oftentimes not inherently interested in the source population, however. We may instead be interested in whether the association we observe in our sample applies to a broader population beyond those who might have been eligible for the study. The population to which we want to make inferences about the associations observed in our study sample is the *target population*. The target population may overlap with the source population, wholly or partially, or may be a different population entirely. Critically, the target population is the group for which we are interested in estimating effects (Maldonado and Greenland 2002). For example, a pediatric neuroimaging sample of pre-adolescent children collected in the Bay Area may have as a source population all pre-adolescent children in the Bay Area who meet eligibility criteria; the target population may be pre-adolescent children in the USA (or beyond). Another way to consider the target population is the group to which the association observed in the study sample is intended to apply. In other words, the group that we are actually interested in applying the associations observed in the study sample. If we are interested in using the results of the study to understand the biology of or develop interventions for a specific population, that population is the target population.

From these definitions of samples and populations, we can then define an additional set of terms. First, *internal validity*: a study is internally valid when the association between an independent variable and a dependent variable reflects a causal association within the study sample; i.e., when the association of interest is causal for the sample (Stuart et al. 2018). We note that any one study and any one researcher cannot definitively prove that an observed association is causal and thus whether an observed study is internally valid, it is nonetheless the actual ultimate goal of why we do the work that we do. We can certainly confirm some elements of our study are valid, such as known confounders measured and controlled for, robust measurement used. Ultimately the study is not internally valid unless the observed association is causal, thus we can never know with certainty if internally validity is met. Nevertheless, we present internal validity in this way so that we can discuss and understand how internally validity can go awry.

Data collection and participation strategies in population neuroscience are often designed to maximize internal validity and statistical power, because the estimands of interest are typically etiological in nature. That is, often in this type of research we are interested in whether there are *proximal* causes of behavior that reside in the brain. Note that *distal* causes of behavior are in our genes and environment (and their interplay). Given those estimands of interest, internal validity has rightly been the focus of much of etiological research. While we cannot observe causal internal validity directly, we can conceptualize a study to be internally valid, in theory, if the association observed is a causal association among the participants in the study sample, regardless of the magnitude of the association in the source population or a target population. Ways that internal validity can go awry are well enumerated (e.g., unmeasured confounding, measurement error, random error), leading to false positive associations that are not causal. As such, sampling schemes are often developed to minimize bias and maximize statistical power. Randomized trials, and randomization generally, are design-based tools to help achieve internal validity, as they can provide an unbiased estimate, in expectation and in the absence of censoring and loss-to-follow-up, of the causal effect of interest for the participants in the trial. (Stuart et al. 2018) Furthermore, often internal validity is prioritized when biological functions under study area hypothesized to vary substantially across individuals, although such circumstances are increasingly understood to be rare. Thus, if we can observe a causal association in a study sample, it is assumed that we provide a window into the true biology or function of a wide variety of populations.

A final set of definitions important to clarify is generalizability and transportability. For the purpose of this chapter, we will define *generalizability* as the extent to which the association in the sample applies to the source population from which the sample is drawn. The extent to which the study sample reflects distributions in the source population of characteristics relevant to the observed associations will determine whether the sample association is generalizable to the source population. In other words, an association is generalizable to the source population if the observed association in the study sample is the same as the association we would have observed in the entire source population. Readers may have learned different definitions of generalizability during their disciplinary training. Sometimes, generalizability is considered more broadly than the specific source or target population to which we are interested in making inferences (e.g., if a study is conducted among mice, a researcher may state that the results may not be "generalizable" to humans). Within the field of epidemiology, however, a more narrow definition of generalizability has been adopted and is often clarifying when considering how to apply results within a study sample to beyond just the sample.

While the extent to which we are interested in estimating effects for the source population is referred to as generalizability, the extent to which we are interested in estimating effects for a target population is often referred to in epidemiology and biostatistics as *transportability*. Understanding variation in associations across populations is often framed in epidemiology as questions of external validity, of which transportability is often synonymous. Transportability refers to the extent to which the association in the study sample applies to populations that may, at least in part, be external from the source population from which the sample was drawn. For example, randomized controlled trials of new medications often have stringent inclusion and exclusion criteria, making the source population from which the sample for the trial is drawn often fairly narrow. Yet, there is abundant evidence in the clinical trial literature about differences between trial results and the population to which we want to make inference – the general patient population (He et al. 2020; Dahabreh et al. 2020). In this case, the target population is broader than either the study sample or the source population. Well-developed methods for understanding the extent to which findings from a randomized controlled trial apply beyond the trial have been articulated in the statistical literature for many years under the umbrella of understanding the transportability of a trial finding (Vose 2021; Rudolph and Laan 2017). As another example, in population genetics, it is increasingly understood that genetic findings in one group may not apply in a population of different ancestry or geographic area (Bigdeli et al. 2019) (although there are emergent examples where genetic findings in one group do apply across populations of different ancestry as well) thus the extent to which genetic findings are transportable across populations is one of increasing urgency to understand. Assumptions and methods for "transporting" a study result from one population to another are becoming more widely discussed and available.

Figure 1 provides a visual way to distinguish between the terms that we have introduced here. There is a sample, within which we will attempt to generate internally valid associations. That sample is situated within a source population; the extent to which the associations in the sample match the associations in the source population is the study finding's generalizability. The target population is often broader, and in the case of this figure includes the study sample (but in principle it may not); the extent to which the associations in the sample match the associations in the target population is the study finding's transportability.

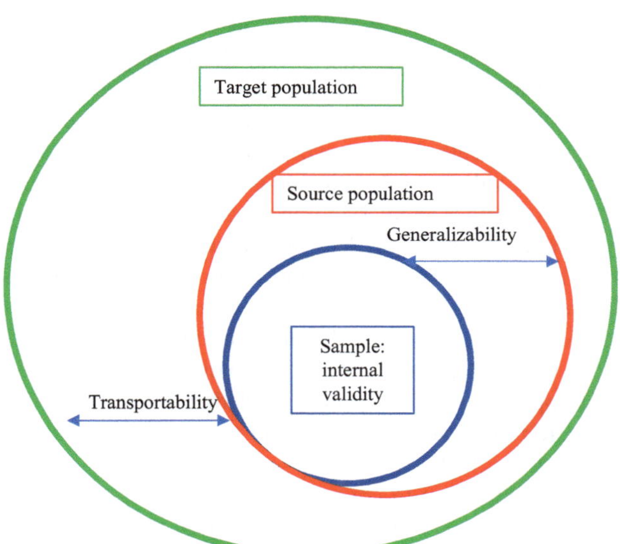

Fig. 1 Relationships among sample, score, and target population

In population neuroscience, while a well-conducted study focused on internal validity can be informative about effects of genes and environment on inter-individual variations in brain and behavior, studies often do not specify the source population from which the sample is drawn, or the target population to which inferences are going to be made, and therefore questions of whether sample association is generalizable or transportable remain opaque. This lack of consideration of generalizability and transportability may in some ways reflect underlying theories about neuroscientific, genetic, and environmental pathways. If it is hypothesized that pathways of interest largely do not vary across characteristics of populations, then generalizability and transportability may not be considered a central primary concern. Additionally, researchers may be interested in testing hypotheses about whether a certain exposure has an impact on outcomes in any context, a precursor to understanding its potential population level impact or an intervention impact (Rothman et al. 2013). Yet, there is increasing recognition that the strength and direction of many associations of interest may indeed vary across populations, prompting the need to specify target populations of interest, and catalyzing the need to also consider the factors that underlie variation across populations.

There is growing recognition that generalizability and transportability are central for interpreting data that are aiming to be population-based. There are increasing calls to diversify and conduct representative (e.g., probabilistic) sampling to increase the applicability of scientific findings to include broader populations (Peterson et al. 2019; MacDermod et al. 2022; Falk et al. 2013; Blanco et al. 2016; Charpentier et al. 2021; Garcini et al. 2022). Implicit in these calls for diversification of study-recruitment strategies is the recognition that in order for our scientific findings to be useful for intervention and prevention efforts at a population level, we need to consider formally defining target populations of interest and taking care to enroll study samples that are representative of those target populations. Using non-representative samples may be an essential first step of research when testing hypotheses about whether certain facets of genes and environment are distal causes of brain/behavior, specifically because we often know so little about what is and is not a threat to internal validity; this rationale is similar to that for doing small-scale efficacy trials to test whether an intervention works at all before testing on a broader or more representative population. But an aligned goal is in understanding the contexts in which those facets of genes and environment are associated with outcomes (i.e., brain/behavior), among whom, and what factors increase the strength of association, and what factors decrease the strength of association. In other words, we want to know to what extent the findings in a particular sample are relevant for a particular target population. Without this knowledge, we may develop intervention and prevention efforts that are not relevant to the populations for which they are targeted. (Bradley et al. 2021).

In summary, it is important to consider the sample, the source population from which a sample is drawn, and the target population of interest for estimating effects because, even when internal validity within the sample is high, differences between sample, source, and target populations can lead to results that are incomplete at best, and misleading or wrong at worst, with respect to generalizability and transportability (Westreich et al. 2019). Even for questions asked by population neuroscientists where there is an implicit (or explicit) assumption that observed effects do not vary across populations, we may want to think more carefully about generalizability from a sample to a source population, and about transportability from the source population to the target population of interest. While a central goal of all of health science is for scientific discovery in an effort to improve population health, a central goal of *population* neuroscience is to develop a scientific evidence base that directly assesses associations at an entire population level. Often, but not always, the source and target populations are implicit or undefined. There is no guarantee that even the most internally valid study in a selected or volunteer study sample will generalize to the source population, or be transportable to a target population, due to a multiple of factors which we will outline in this chapter.

Below we provide a basic overview of the epidemiological principles that guide study-recruitment decisions regarding the generalizability and transportability of study samples to target populations and beyond. We do this in three parts: first, we provide a conceptual overview of the concept of effect – measure modification and describe how it is a key consideration in generalizability and transportability. Second, we show a hypothetical example to illustrate how unmeasured effect-measure modifiers can lead study findings to be non-generalizable. Third, we discuss the implications for population neuroscience and provide recommendations for the field.

1 Conceptual Overview of Effect-Measure Modification with Application to Concepts of Generalizability

As described above, study samples are obtained from source populations, even if they are not explicitly identified. These populations may or may not reflect the actual target populations of research interest. For example, a nationally representative sample with probability sampling (whether the study design is cross-sectional or longitudinal, or more rarely, case control) is explicitly designed so that the results from the study sample will generalize formally to the source population of the nation from which the sample is drawn. We often do not, and cannot, design nationally representative studies in population neuroscience. Thus, the source population can become implicitly defined, hidden, or even undefinable. Even if the source population is defined, the target population may be only a part of the source population, broader than the source population, or wholly different than the source population. For example, suppose a large sample is collected in geographically diverse areas in the USA but with numerous inclusion and exclusion criteria. In that case, the target population of ultimate interest may implicitly be the entire USA, but the sample may not represent that target. Large studies across many countries, and those that use volunteers who contribute data, also reflect source and target populations that are difficult to define, given that there are many predictors (even genetic ones (Tyrrell et al. 2021)) of who volunteers for studies. Lastly, and perhaps most vague, are consortia studies, which combine samples from many sources, including across countries and sites. In this case, the source population is almost impossible to define – who are the underlying populations from which the many samples are drawn? And the target population may also remain opaque – who are the underlying population for which we are interested in estimating effects? Researchers often argue that results from cohort consortia studies are generalizable to target populations because of the large sample size or high degrees of variation in the study sample, but such arguments have no formal empirical basis (Bradley et al. 2021; Keyes and Westreich 2019) and may be erroneous. While replication across diverse study samples is a pathway toward making the argument that certain associations are generalizable, a counterargument is often that studies with similar designs may accumulate similar biases, thus replication, when it even occurs, may not be a safeguard against bias.

A key question is when a *scientific finding* from a study sample is or is not generalizable or transportable to a source or target population, and the implications of that for population neuroscience. While several factors determine the extent to which results from a study sample will apply to a target population (Hernán and Vanderweele 2011), the distributions of effect-measure modifiers in the sample and in the population are among the most central. An effect-measure modifier (also called an "effect modifier" or a "moderator") is a factor that increases/decreases the magnitude of an association. In some disciplines, such a situation (i.e., when the magnitude of association differs across levels of a third variable) is also termed an "interaction," although in other disciplines there is a conceptual contrast between effect-measure modification and interaction (VanderWeele and Robins 2007). In other words, an effect-measure modifier is a factor that changes the association between exposure or the independent variable of interest and the outcome.

For example, the association between polygenic risk scores (PRS) and depression is stronger among those with a particular set of environmental or social adversities – thus, those adversities are the effect-measure modifiers (Mitchell et al. 2021). The converse is also mathematically true – the association between adversities and depression will vary based on levels of the PRS. The designation of a variable as an exposure or moderator is a conceptual rather than mathematical decision. But in any case, for generalizability and transportability, it is the case that even if two populations have the same distribution of PRS, the PRS will have a different association with depression if the two populations have different prevalence of adversities.

Thus, if we measure social adversity and PRS in a sample, we will likely observe a statistical effect – measure modification; the relationship between PRS and depression will be stronger among those with social adversity than among those without it. But further, the relationship between PRS and depression will depend on the *prevalence* of social adversity in the sample. If there are low levels of social adversity in the sample, we will observe a weak, if non-existent, relationship between PRS and depression. Conversely, there will be a very strong relationship between PRS and depression in samples with high levels of social adversity.

Importantly, this is the case even if social adversity is unmeasured. Thus, the source populations from which we recruit for studies and the prevalence of social adversity in the resulting samples (measured or unmeasured) are critical for the signals that we detect for the PRS – and how well those signals might hold in other target populations.

It is important to note that while differences in the distribution of effect-measure modifiers is an important source of variation between a study sample and the source and target populations, any differences in the distribution of mediators or other intermediate variables on pathway between independent and dependent variables will also contribute to a lack of generalizability/external validity.

2 Hypothetical Example to Illustrate How Unmeasured Effect-Measure Modifiers Can Lead Study Findings to be Non-Generalizable

As described above, the prevalence of measured or unmeasured effect modifiers in our sample determines whether we observe strong or weak associations in our sample data. In addition, the difference in distribution of these factors (as well as mediators or other intermediate variables) between the sample and a source population, or the sample and another population entirely, relates directly to the question of to whom the result from a study sample is generalizable, and the extent to which study findings are transportable. We provide the following example, which quantifies the principles that were described conceptually in Part 1, to describe the relationship between target populations and samples, and where generalizability can go awry.

Figure 2 shows a hypothetical example of a source population and study sample from that source population, in which the association observed is not generalizable to the source population due to differing prevalence of an unmeasured effect-measure modifier. In Fig. 2, the source population includes 100 individuals, 10% of whom are high on the polygenic risk score. Among those high on the PRS, 50% develop depression, and among those who are low on the PRS, 4.4% develop depression, for a risk ratio of 11.25. A non-representative sample is then taken from the source population, so that 20.0% of the sample are high on PRS. Among those high on PRS, 28.6% develop depression, compared with 14.3% among those low on PRS, for a sample-estimated risk ratio of 2.0.

Why is the risk ratio so much higher in the source population than in the sample? Because: (1) social adversity is an effect – measure modifier, and (2) the prevalence of social adversity is very different in the source population than in the sample. In the source population, 50% are exposed to social adversity, compared with just 22.9% in the sample.

It is important to note that while the example in Fig. 2 is about the relationship between a sample and a source population, the same concepts apply when considering the relationship between a sample and a target population. To the extent that the distribution of an effect-measure modifier (or mediator) differs between the

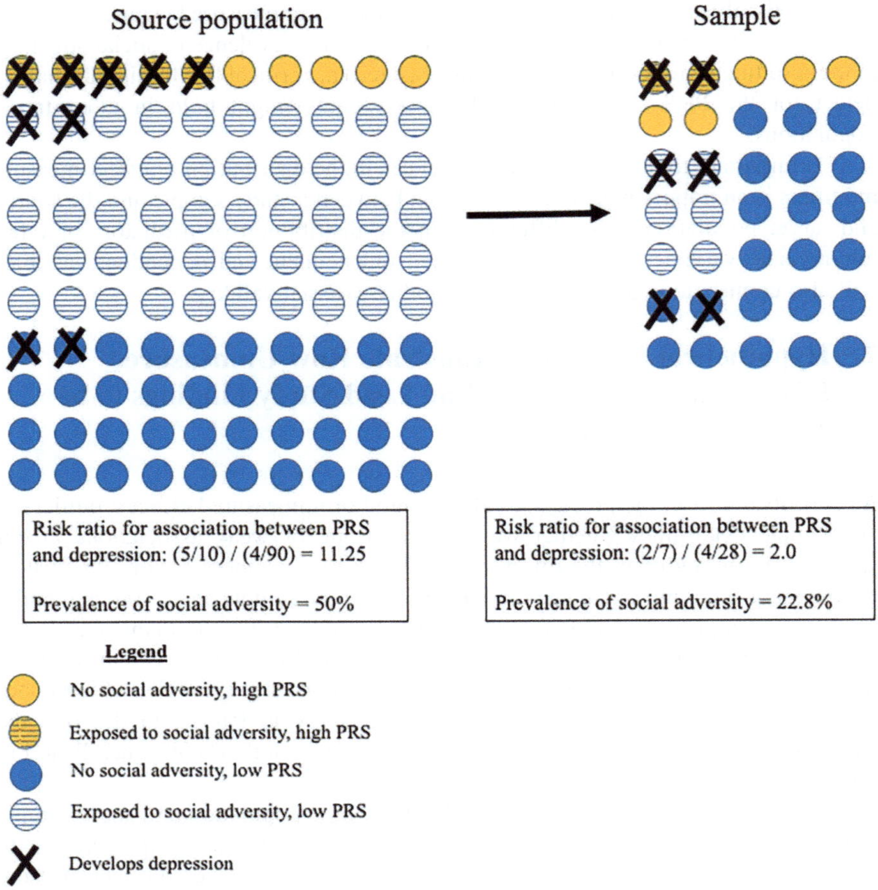

Fig. 2 Hypothetical example showing how an association in a study sample can differ from the same association in its source population

sample and the target population, the associations observed in the sample will be different than the associations in the target population.

This example illustrates the concept described in Part 1 that the magnitude of the measure of association – whether that measure of association is a risk ratio, regression coefficient, heritability estimate, or any other measure – is dependent on the prevalence of the effect modifiers. In the best case scenario, we know (or at least can hypothesize) important effect-measure modifiers, and know what their distribution is in within the study sample and perhaps in the source population, target population of interest, or another target population entirely. Indeed, when we create post-stratification weights in data to anchor our study samples to a target population, we are adjusting for the distribution of potential effect-measure modifiers. We may not know (or have no hypotheses), however, about all of the relevant effect-measure modifiers; and yet, their prevalence will be deterministic about the sample association magnitude.

Regardless of our knowledge of the distribution of effect-measure modifiers in the study sample and target population, this example makes clear that explicitly defining source and target populations and hypothesizing about (or having data on) the effect-measure modifiers and their distribution in the source and target population is key to producing science that is not only generalizable but interpretable and intervenable. Further, this concept can be applied even more generally beyond the source and target population. Suppose we conduct studies in several samples but do not measure social adversity; we only measure the PRS and depression. Within these samples, there may be a strong signal for a relationship between PRS and depression in some, and in others, it may be weak. Why would this be the case? These different samples may have a different prevalence (unmeasured) of social adversity, or any other potential effect modifier. Indeed, we would *expect* the signal for the relationship between PRS and depression would be strongest in the samples most exposed to environmental or social adversity and weakest in the samples least exposed.

3 Implications for Population Neuroscience and Recommendations for the Field

We have exemplified in this chapter that while estimating associations, for the ultimate goal of understanding causal effects, within study samples requires a number of criteria well known in the sciences such as assessing confounding and selection bias, inferences from a study sample to a population can depend on clearly defining a target population. Differences in the prevalence of measured or unmeasured effect-measure modifiers (and differences in the prevalence of mediators and other intermediate variables) between the source and target populations will lead to findings that do not generalize between source and target populations. This is true even when an effect estimated from a sample has perfect internal validity. Further, samples that do not have a defined target population are largely uninterpretable for population-health goals. This has implications for population neuroscience, as conclusions about biological underpinnings and social consequences may or may not apply to the target populations for which intervention and prevention efforts aim to be targeted.

Even with regard to molecular/genetic and environmental exposures, effect sizes are not necessarily immutable or constant across samples and populations. This is well recognized in many areas of science, such as in the variability of heritability estimates depending on sample characteristics (Turkheimer et al. 2009; Smith 2011), and in many meta-regression analyses that demonstrate variation in effect-size magnitude based on composition and design of the selected samples. And yet, recognition of this point is not universal, with many studies including policy and programmatic recommendations based on non-representative samples without regard to target populations, and an increasing number of consortia studies that conceptualize findings without regard to potential heterogeneity across studies and samples. Acknowledging the important role of sample selection in interpreting findings from population neuroscience studies would increase the rigor of our science, provide a platform for more evidence-based recommendations that draw from our work, and elucidate a framework for better sampling practices that acknowledge target populations of interest.

It should be noted, of course, that just because associations and effect sizes can vary across samples does not mean they must vary across samples. Some exposures or independent variables of interest may have few if any effect-measure modifiers, or if they do, the effect-measure modifiers may be equally distributed across all relevant target populations for which we would like to generalize or transport findings from a study sample. In other cases, while there may be variation across study samples, the actual magnitude of the variation is not meaningful to inform intervention or policy. For example, pack years of cigarette smoking is associated in a relatively dose-response fashion across many different samples collected in many different target populations (Lee et al. 2012); while the magnitude may vary to some extent based on the distribution of factors that modify the effect of smoking such as other environmental respiratory irritants, interventions to reduce cigarette smoking are likely to be of high public health impact in most if not all populations where cigarette smoking is prevalent.

Given that representative sampling from explicitly defined target populations may not be feasible or realistic in many situations, especially given the sample sizes needed for statistical power for many exposures of interest (and especially when interactions are being tested), researchers should be well versed in tools available to enhance the interpretability of samples regarding target populations. Three action items are worth consideration. First, one immediate action item is being explicit about the target population to whom the investigators is interested in inferring results from the study sample. Second, in addition to stating explicitly the target population, investigators should be clear about the source populations from which samples were obtained, including what the sampling strategy was, how people were recruited and consented, including clearly stated participation rates (number approached for participation, number initially agreed, number eventually consented and included), which may give some sense for the potential unmeasured factors that could relate to the selection process. Third, making an effort to collect demographic and socioeconomic data and other likely effect-measure modifiers would allow researchers much more information on the generalizability and transportability of samples that what is currently typically included in neuroscientific studies. For example, Keyes and colleagues recently attempted a meta-regression of the association between hippocampal volume and depression to see if the associations varied by socioeconomic status, but were unable to even run a regression because so few studies report basic education or income data (Keyes et al. 2023).

These issues become even more complicated when study data are combined and harmonized from multiple sites. Interpreting data harmonized across study sites is not only made difficult when measures need to be calibrated, anchored together and equated, but meeting the assumptions for transportability of associations from one site to another becomes much more challenging. That is, even if measures are perfectly aligned across sites, assuming that associations found in one site can be transported to another site requires testing before, during, and after data are combined. Yet, such practices are rarely done, and commonly accepted practices may do more harm than good. For example, when conducting analyses, it is a common practice to control statistically for study site, which removes any confounding related to fixed characteristics that differ across study site from the resulting harmonized sample estimates. But if these characteristics modify the effects of the exposure under investigation (which could be likely), controlling for study site provides a weighted average of the effect sizes across site. This weighted average does not correspond directly to any effect size of interest in the source or any target population and may not be useful in determining the existence, strength, or applicability of the exposures effect in a population of actual interest for intervention and prevention. A better approach in this situation may be to define an underlying target population of interest, and design and implement estimation approaches, including weights or other estimators (Rudolph and Laan 2017; Rudolph et al. 2018), that allow the sample to approximate distributions of key characteristics in the target population (Westreich et al. 2017).

We provided an example where associations vary across levels of social adversity. Hypotheses about effect-measure modification between social exposures are common in psychiatric studies (e.g., stressful life events and social adversity with biologically relevant population variation such as brain structure and function or molecular and environmental factors) and have formed the basis of thousands of study designs and statistical analyses. Nonetheless, social exposures are but one source of potential effect modification that may cause sample estimates to differ from those in the target population. There are many potentially relevant effect-measure modifiers for processes such as age, environmental exposures, healthcare access and history, historical legacies of subjugation that lead to racially motivated categorizations, and many others. Yet, the fact that these factors interact with our exposures of interest is not nuisance or noise; they form the relevant contexts in which biology unfurls and are rife with opportunities for enhancing scientific discovery and designing interventions. Human variation in biological vulnerability through genetic or environmental factors explains individual differences within a specific population, but the implications of this variation for health status depend on the social context and social structures that encode risk at high levels of organization. By recognizing and harnessing this structural variation and hypothesizing about effect-measure modifiers at these higher levels, we can improve our ability to use population neuroscience to improve public health.

Acknowledgments This work was funded by a seed grant from the Center for Science and Society at Columbia University. We thank Megan Finsaas, PhD, for comments and suggestions on earlier versions of this chapter.

References

Bigdeli TB, Genovese G, Georgakopoulos P et al (2019) Contributions of common genetic variants to risk of schizophrenia among individuals of African and Latino ancestry. Mol Psychiatry 25(10):2455–2467. https://doi.org/10.1038/S41380-019-0517-Y

Blanco C, Wall MM, Lindquist MA et al (2016) Generalizability of neuroimaging studies in 5 common psychiatric disorders based on the National Epidemiologic Survey on Alcohol and Related Conditions (NESARC). J Clin Psychiatry 77(12):7699

Bradley VC, Kuriwaki S, Isakov M, Sejdinovic D, Meng X-L, Flaxman S (2021) Unrepresentative big surveys significantly overestimated US vaccine uptake. Nature 2021:1–6. https://doi.org/10.1038/S41586-021-04198-4

Charpentier CJ, Faulkner P, Pool ER et al (2021) How representative are neuroimaging samples? Large-scale evidence for trait anxiety differences between fMRI and behaviour-only research participants. Soc Cogn Affect Neurosci 16(10):1057–1070

Dahabreh IJ, Robertson SE, Steingrimsson JA, Stuart EA, Hernán MA (2020) Extending inferences from a randomized trial to a new target population. Stat Med 39(14):1999–2014. https://doi.org/10.1002/SIM.8426

Falk EB, Hyde LW, Mitchell C et al (2013) What is a representative brain? Neuroscience meets population science. Proc Natl Acad Sci 110(44):17615–17622

Garcini LM, Arredondo MM, Berry O et al (2022) Increasing diversity in developmental cognitive neuroscience: a roadmap for increasing representation in pediatric neuroimaging research. Dev Cogn Neurosci:101167

He Z, Tang X, Yang X et al (2020) Clinical trial generalizability assessment in the big data era: a review. Clin Transl Sci 13(4):675. https://doi.org/10.1111/CTS.12764

Hernán MA, Vanderweele TJ (2011) Compound treatments and transportability of causal inference. Epidemiology 22(3):368. https://doi.org/10.1097/EDE.0B013E3182109296

Keyes KM, Westreich D (2019) UK biobank, big data, and the consequences of non-representativeness. Lancet. in press

Keyes KM, Kreski NT, Joseph VA et al (2023) What is not measured cannot be counted: sample characteristics reported in studies of hippocampal volume and depression in neuroimaging studies. Biol Psychiatry Cogn Neurosci Neuroimaging 8(5):492–494

Lee PN, Forey BA, Coombs KJ (2012) Systematic review with meta-analysis of the epidemiological evidence in the 1900s relating smoking to lung cancer. BMC Cancer 12:1–90

MacDermod C, Pettie MA, Carrino EA et al (2022) Recommendations to encourage participation of individuals from diverse backgrounds in psychiatric genetic studies. Am J Med Genet Part B Neuropsychiatr Genet 189(5):163–173

Maldonado G, Greenland S (2002) Estimating causal effects. Int J Epidemiol 31(2):422–429. https://doi.org/10.1093/intjepid/31.2.422

Mitchell BL, Thorp JG, Wu Y et al (2021) Polygenic risk scores derived from varying definitions of depression and risk of depression. JAMA Psychiatry 78(10):1152–1160. https://doi.org/10.1001/JAMAPSYCHIATRY.2021.1988

Peterson RE, Kuchenbaecker K, Walters RK et al (2019) Genome-wide association studies in ancestrally diverse populations: opportunities, methods, pitfalls, and recommendations. Cell 179(3):589–603

Rothman KJ, Gallacher JEJ, Hatch EE (2013) Why representativeness should be avoided. Int J Epidemiol 42(4):1012–1014. https://doi.org/10.1093/ije/dys223

Rudolph KE, Laan MJ (2017) Robust estimation of encouragement design intervention effects transported across sites. J R Stat Soc Ser B Stat Methodol 79(5):1509–1525

Rudolph KE, Schmidt NM, Glymour MM et al (2018) Composition or context: using transportability to understand drivers of site differences in a large-scale housing experiment. Epidemiology 29(2):199

Smith GD (2011) Epidemiology, epigenetics and the 'gloomy Prospect': embracing randomness in population health research and practice. Int J Epidemiol 40(3):537–562. https://doi.org/10.1093/ije/dyr117

Stuart EA, Ackerman B, Westreich D (2018) Generalizability of randomized trial results to target populations: design and analysis possibilities. Res Soc Work Pract 28(5):532–537

Turkheimer E, Harden KP, D'Onofrio B, Gottesman II (2009) The Scarr-Rowe interaction between measured socioeconomic status and the heritability of cognitive ability. Exp Dev A festschrift Honor Sandra Wood Scarr:81–97

Tyrrell J, Zheng J, Beaumont R et al (2021) Genetic predictors of participation in optional components of UK Biobank. Nat Commun 12(1):1–13. https://doi.org/10.1038/S41467-021-21073-Y

VanderWeele TJ, Robins JM (2007) The identification of synergism in the sufficient-component-cause framework. Epidemiology 18(3):329–339

Vose J (2021) Minority enrollment to clinical trials: road to increased access. Oncology (willist Park NY) 35(3):107

Westreich D, Edwards JK, Lesko CR, Stuart E, Cole SR (2017) Transportability of trial results using inverse odds of sampling weights. Am J Epidemiol 186(8):1010–1014

Westreich D, Edwards JK, Lesko CR, Cole SR, Stuart EA (2019) Target validity and the hierarchy of study designs. Am J Epidemiol 188(2):438–443. https://doi.org/10.1093/AJE/KWY228

Population Neuroscience: Strategies to Promote Data Sharing While Protecting Privacy

Adrian Thorogood

Contents

Abstract Population neuroscience aims to advance our understanding of how genetic and environmental factors influence brain development and brain health over the life span, by integrating genomics, epidemiology, and neuroscience at population scale. This big data approach depends on data sharing strategies at both the micro- and macro-level, as well as attention to effective data management and protection of participant privacy. At the micro-level, researchers participate in international consortia that support collaboration, standards, and data sharing. They also seek to link together cohort studies, administrative health databases, and measures of the physical, built, and social environment in creative ways. Large-scale, longitudinal, and multi-modal cohorts are being designed to support explorations of genetic and environmental impacts on the brain. At a macro-level, funding agency policies now require data across health research domains to be managed according to the FAIR (findable, accessible, interoperable, and re-useable) Data

A. Thorogood (✉)
The Terry Fox Research Institute, Vancouver, BC, Canada
e-mail: athorogood@tfri.ca

Curr Topics Behav Neurosci (2024) 68: 53–66
https://doi.org/10.1007/7854_2024_467
Published Online: 21 March 2024

principles and made available to the research community in a timely manner to support reproducibility and re-use. Data repositories provide technical infrastructure for storing, accessing, and increasingly also analyzing rich population-level data. Federated and cloud-based approaches are being leveraged to improve the security, remote accessibility, and performance of repositories. Finally, legal frameworks are being developed to facilitate secure health data access, integration, and analysis, providing new opportunities for the field.

Keywords Data linkage · Data protection · Data sharing · FAIR · Population neuroscience · Privacy

1 Introduction

The biology of the brain, including its structure and function, is shaped by complex interactions between genes and the environment, across the lifespan (Paus 2010). Population neuroscience aims to advance our understanding of genetic and environmental influences on the brain and behavior from conception, through development and aging, by studying large cohorts of participants representative of the general population, using a combination of state-of-the-art approaches from epidemiology, genomics, and neuroimaging (Paus 2016; Paus et al. 2022b). Falk et al. describe population neuroscience as an interdisciplinary convergence of fields, improving the generalizability of neuroscience through sampling approaches that strengthen the link with target populations, and encouraging population studies to consider the biology of the brain (Falk et al. 2013). Indeed, population studies are increasingly conceived of as large-scale, multi-modal, longitudinal cohorts that integrate genetic and biomarkers as well as neuroimaging and cognition measures so as to better understand the mechanisms by which environmental and genetic factors affect brain biology (Nees et al. 2021). Challenges for population neuroscience include the complexity of understanding complex gene–environment interactions and how they impact on the complex structure and function of the brain over the human life span (Paus et al. 2022b). Data sharing is essential for advancing population neuroscience by bringing together rich and diverse data at population scale and supporting scientific collaboration and reproducibility. Policies, infrastructure, and legal frameworks are being established to facilitate data sharing, with increasing emphasis on data stewardship and management principles like the FAIR (findable, accessible, interoperable, and re-useable) Data principles (Wilkinson et al. 2016). Sharing rich and sensitive data at population scale also presents important privacy risks for research participants, which must be addressed using a range of complementary safeguards.

2 Big Data Approaches: Opportunities and Risks

Population neuroscience aims to bring together information about genetics, environment, and the brain in a Big Data approach. Big Data is characterized by the 4Vs: data volume, variety, velocity, and (uncertain) veracity (Schadt 2012). It is worth taking a moment to consider the evolving data richness of each of these respective research areas, which is what fuels both the interests of bringing them together for population neuroscience and the potential risks of doing so. Data sharing policies and practices should aim to promote the opportunities of integrating these data types and methods to advance science and medicine, while also mitigating risks to individual privacy and welfare.

2.1 Genetics and Genomics

Genome-wide association studies (GWAS) have been the hallmark of population-based genomics research, including for studies of mental health and illness. These studies seek to identify the genetic correlates of human health and disease by conducting genome-wide genotyping of many individuals and examining differences between those with and those without the phenotype of interest (Uffelmann et al. 2021). Reliability concerns about GWAS have faded as studies or meta-analyses have attained large sample sizes, and as methods and replicability standards have improved (Uffelmann et al. 2021). Part of the evolution of population neuroscience is to seek deeper brain phenotyping of participants to start to develop mechanistic insights into how genetic variants influence neurobiology. Increasingly, building on studies of constitutional genetic variation, researchers are employing a spectrum of tools such as transcriptomics and proteomics to understand the complexities of genetic regulation and its interplay with environmental factors. Acknowledging the strength and importance of GWAS, these studies have also come under scrutiny for a lack of diversity, given the tendency to focus on individuals of European descent, bringing into question the generalizability of findings across groups with different biogeographical origins (de Hemptinne and Posthuma 2023; Popejoy and Fullerton 2016). Genetic information, particularly where it is genome-wide, can also present privacy and welfare issues for sequenced individuals and their biological family members. The genome can contain a range of socially relevant information, including predictive information about various health risks and physical and cognitive traits (Hallinan 2021). The disclosure of such information presents risks of stigma within families and communities, as well as risks of genetic discrimination (e.g., in insurance or employment), though laws have been adopted in many countries to prevent such applications (Joly and Dalpe 2022). There is public concern over with the sale of human tissue and genomic data to private companies, as well as about commercial access and use more generally (Critchley et al. 2021). Constitutional genomic data can also serve as a near-perfect biometric identifier, raising the question of whether or not such data can ever be effectively anonymized. As genomic databases reach population scale, they risk being re-purposed by law enforcement as forensic databases for identifying suspects directly or through familial linkage; as is already a reality with direct-to-consumer genetic testing companies and recreational genetic ancestry databases (Wan et al. 2022). Additionally, the use of genome-wide assays or whole-genome sequencing raises ethical challenges for researchers to manage the range of potential findings of potential health relevance to participants and their families, not to mention how to inform appropriately participants about the attendant implications, limits, and psychosocial risks (Thorogood et al. 2019).

2.2 Environmental and Health Research

Interdisciplinary approaches are now being used to explore how environmental exposures (e.g., chemicals, food, residence, and behavior) perturb the biological functions of the brain (Eva et al. 2022). Understanding this "exposome" is challenging (Wild 2005; Vermeulen et al. 2020). It requires new techniques to measure virtually limitless aspects of our physical, built, and social environments including the fact that individual behavior co-creates environment (T. Paus 2016). Surveys are a common methodology for collecting these data but can be burdensome on participants and affected by self-reporting bias. Environment and geography researchers are therefore developing new approaches, such as obtaining geospatial data from satellites and other sensors at different levels of spatial and temporal resolution that can be linked to health and genetic databases (Paus et al. 2022a). As environmental measures can increasingly be established with greater spatial and temporal resolution, researchers are also seeking to generate more personalized exposure measures at the individual level, potentially at the level of real-time geo-localization data generated by mobile devices. Increasingly, much of our online behavior leaves a geo-coded, digital footprint. The field of digital ethology aims to measure such behavior to assess aspects of our social environment (Paus et al. 2022a). A challenge with this diversity of data is ensuring reliability and validity. Privacy challenges presented by these research approaches are twofold. First, geocoding at high spatial and temporal resolution can support singling out and re-identification of participants in combination with other sources, making de-identification difficult. Second, capturing information about individual behavior (such as unhealthy or socially undesirable ones) can lead to new forms of informational risks. There is also an undeniable link between the study of environmental effects on health and the politics of distributive and environmental justice. On the one hand, population neuroscience can allow us to understand and address the sources of these inequalities. On the other hand, there are risks that political interests distort or co-opt scientific agendas or communication to support particular political ends.

3 Cohort Studies, Data Sharing, and Data Linkage

On a practical level, no single researcher or project alone can achieve the scale and richness of data needed for effective population neuroscience (Paus et al. 2022a). Big Data approaches require a range of data sharing strategies at both the micro- and macro-level. The most obvious strategy for conducting population neuroscience research is through prospective cohort studies. International research consortia and open science initiatives present opportunities for researchers to share data to validate findings or conduct new studies. Finally, data linkage approaches present another strategy to enrich existing health data resources with measures of the physical, built, and social environment. These activities also present privacy risks that must be addressed through increased attention to data governance.

3.1 Cohort Studies

Cohort studies are a popular strategy relevant to epidemiology generally and aim to be sufficiently large-scale as well as demographically and geographically diverse to produce results generalizable to target populations. A number of cohort studies focus on longitudinal study of healthy adult populations, often enriched with genotyping, cognitive assessments, and neuroimaging to learn about the factors that cause some to stay healthy and others to develop mental or other types of illnesses as they age. Examples of resources relevant to population neuroscience include the UK Biobank Study (over 500,000 participants https://www.ukbiobank.ac.uk/), the US NIH All of Us Initiative (over one million participants with an emphasis on diversity https://allofus.nih.gov/), and Canada's CanPATH network of provincial population cohorts (+330,000 participants https://canpath.ca/). As brain development is influenced from the earliest stages of development, population neuroscience has a particular interest in studying development using birth, child, and adolescent cohorts. Birth cohort studies are an important subcategory, a number of these have been started around the world, driven by "increasing awareness of the long-term health effects of intrauterine and early life combined with a renewed attention to the life-course approach in epidemiology." (Birthcohorts.net). Birth cohorts often involve a large number of participants with longitudinal, intensive data and biospecimen collection and can be a relatively unbiased sample of the population. The BirthCohorts inventory helpfully describes the range of data categories collected by different cohorts, including childhood exposures (e.g., air pollution), health and development measures (including neurological or cognitive), as well as rich biological data (e.g., whole-genome sequence data or neuroimaging data). Examples include: Avon Longitudinal Study of Parents and Children (www.bristol.ac.uk/alspac); Northern Finland Birth Cohorts (www.oulu.fi/nfbc); Generation R (https://generationr.nl); the HEALthy Brain and Child Development (HBCD) study (https://heal.nih.gov/research/infants-and-children/healthy-brain), and the Dunedin Multidisciplinary Health and Development Study (https://dunedinstudy.otago.ac.nz).

Large-scale, multi-modal, longitudinal cohort studies present privacy risks for the studied individuals. Establishing and managing these resources typically involves the collection and storage of rich personal information about health and genetics generally considered by law and ethics as sensitive data meriting additional protection. Information related to genetics (inherited conditions) and/or mental health status (potentially stigmatizing) may be perceived or treated as having an even greater level of sensitivity. A threshold ethical and compliance issue is whether access is being provided to identifiable data (as opposed to non-identifiable data), which is typically the threshold that triggers applicable data privacy/protection laws as well as human subjects research protections. Where data are identifiable, there are general requirements to obtain individual consent to research; to safeguard privacy, confidentiality, and security; and to obtain a Research Ethics Committee (REC) approval. Exceptional waivers of consent requirements may be available from an REC under certain conditions. A variety of safeguards are commonly adopted to mitigate privacy risks while supporting cohort research, including informed consent (though not typically practicable in the context of administrative health database); data privacy, confidentiality, and security controls; and access oversight by RECs or Data Access Committees. (Paprica et al. 2023).

3.2 International Research Consortia

Data sharing is the practice of making data collected for a research study or other purpose available to the broader research community for validation and secondary research purposes. Population neuroscience has been supported by the participation of researchers in international research consortia who share data and support collaboration. This community has established several prominent international research consortia to bring together like-minded researchers, support training and knowledge exchange, promote collaboration, and share resources (including data, results, and algorithms). These consortia may support collaboration on both prospective efforts to scan the brains of larger cohorts, as well as retrospective efforts to piece together data from existing cohorts to support re-use and meta-analysis (Falk et al. 2013). The ENIGMA Consortium (Enhancing Neuro Imaging Genetics through Meta-Analysis) is a prominent example organized around genomics and brain imaging technologies as well as numerous working groups focus on different mental illnesses, as well as development and aging (https://enigma.ini.usc.edu/) (Palk et al. 2020). Such consortia promote team science as well as voluntary data sharing to foster collaboration, pool resources, and allow replication of promising findings. Open science is also popular in genomics and neuroscience and involves a range of practices to accelerate scientific progress through transparency and collaboration, including open access to publications as well as open data and open source software (Harding et al. 2023).

Establishing rich data repositories for broad sharing and re-use can also strain privacy protection. In Big Data contexts it is no longer feasible to rely solely on anonymization to protect individual privacy. Anonymization involves the irreversible transformation of data to ensure there is no reasonable likelihood of directly or indirectly identifying an individual. Anonymization will often present an unacceptable trade-off with preserving the scientific utility of data. There are also both scientific and ethical reasons for preserving a secure link back to an individual's identity, which enables re-contact for future data collection or recruitment, or for reporting results of individual health significance. A compromise is to ensure data are coded or pseudonymized before sharing, which typically means separating direct identifiers and replacing them with a random code allowing for secure and authorized re-identification of participants. The aim of making rich individual-level data available to researchers across institutions, sectors, and countries, for not fully specified future research uses, is also in tension with legal and ethical principles of informed consent to research participation and personal data processing. Broad consent approaches have been introduced to inform individuals of the potential scope of data sharing, while also placing emphasis on downstream oversight and transparency (Longstaff et al. 2022). Data sharing across research teams, institutions, and countries can also raise concerns about a loss of control and accountability. Such pooling is therefore often pursued under the rubric of common project or consortia governance approaches that emphasize the need for common and coordinated governance of management and access to the resource (Bernier et al. 2022). Access governance involves transparent processes for submitting an access request, review of requests by a data access committee, and execution of a data use agreement providing for privacy, confidentiality, and security controls. There is also growing interest in providing approved users with remote data access in secure and monitored research environments, to avoid the loss of control inherent in distributing copies of data.

3.3 *Data Linkage*

To be of great utility for population neuroscience, environment and health researchers increasingly seek to link existing administrative health databases and/or population cohort studies together, or with data resources describing the physical, built, and social environment. Administrative health databases are an important resource for population health research generally. These are often national or provincial/territorial databases of data generated as a by-product of governments delivering health and/or social services, and subsequently made available to researchers for research studies. A prominent example is Finland's FinData (https://findata.fi/en/), which supports data integration and researcher access to health and social service data at national scale. Such databases contain valuable data about health and social service use and outcomes, with longitudinal data collection and population-wide coverage. The weaknesses often have to do with data quality for secondary research purposes. For mental health research, administrative data typically only captures the most serious cases of mental illness. Legal and administrative barriers also hinder researcher access. Administrative health researchers have also established networks to facilitate access to administrative databases, as well as to support multi-jurisdictional studies or collaborations (e.g., Health Data Research Network Canada https://www.hdrn.ca/en/). Cohort study data can potentially be linked to administrative data to study longitudinal health outcomes.

Cohort study or administrative databases can also be linked to environmental data to support population neuroscience. There are different approaches that may be adopted to link these health databases to environmental data resources. One common approach is to map measures of environmental exposure according to a certain standard geospatial areas, such as census tracts, or postal or zip codes (Eva et al. 2022). There are also a number of researcher Networks established to advance environment and health research (e.g., Canadian Urban Environmental Health Research Consortium (CANUE) https://canue.ca). Researchers are also seeking ways to link data at higher levels of resolution, e.g., by linking environmental data from satellites and other sources at the individual-subject level (postal code or address of home and work).

Data linkage also raises practical and privacy challenges. One of the challenges with analyzing the ethics of data linkage generally is the diversity of connections that may be made; researchers constantly dream up new opportunities to link existing data resources together. Many forms of linkage are only practically possible for data being stored in coded/pseudonymized form. Certain identifiers must be maintained to be able to link different data sources at an individual level, so each data source must collect and retain those identifiers (e.g., name, data of birth, healthcare number). Direct identifiers are separated from the research data, but mechanisms are established to link back securely to an individual's identity for certain purposes. This presents a privacy risk of unauthorized access to the re-identification mechanism (e.g., key code) resulting in re-identification of participants and disclosure of sensitive information. Even where key codes are securely protected, and where linkage is conducted in a privacy-preserving manner, linking datasets together may increase the overall risk of re-identification in the resulting dataset. "Individual-level" data linkage produces a richer dataset, which may increase the residual privacy risk of re-identification. It is important to keep a proportionate view of the dangers of linkage, however; not all linkages are at the individual-level and not all individual-level linkages necessarily increase privacy risk. As with data access, a balance needs to be struck between promoting the benefits of linkage while minimizing the individual privacy risks. As a result of these privacy concerns, linkage may be frustrated by legal, ethical, or administrative barriers designed to protect confidentiality and individual privacy. Indeed, data access policies and agreement often include a blanket prohibition on data linkage (at least without the consent of the custodian).

One solution available to cohort studies is to obtain explicit consent from participants to conduct data linkage with an administrative database. For example, guidance for population health researchers in Canada calls for consents to address specific information about the types of data to be linked; consent to the collection and retention of the particular identifiers needed; information on which databases/linkage units this information will be disclosed to; how security will be ensured; and potential risks (Health Data Research Network Canada 2021). Another solution is to ensure high levels of security during the linkage process, i.e., to ensure that the people and organizations with access to data to conduct the linkage are limited. Identifiers can be removed from the linked database before being made available for the research. Some jurisdictions have introduced secure-linkage units authorized to link together data sources and then remove the identifiers before sharing or providing secure access to the dataset. Risks of increased re-identification in the resulting dataset can be addressed through privacy impact assessments or ethics review of the linkage proposal. Researchers can also reduce privacy risks by conducting linkage in an ephemeral way for single study, reflecting the principle of data minimization. Unfortunately, this may hinder reproducibility of the study or re-use of the now enriched dataset. Another strategy is to seek to enrich or "pre-link" the administrative health database or cohort study with the linked environmental data, which can then be shared in the future with other researchers through the standard access procedures. A creative technology solution would be to create interfaces or scripts allowing authorized researchers to connect administrative health data resources with the latest environmental data in real time.

4 Macro-Level Strategies to Support Data Sharing and Linkage

It is clear from the challenges outlined above that effective and privacy-preserving population neuroscience requires supporting policies, infrastructure, and even new legislative frameworks to promote secure access to health data.

4.1 Data Management and Sharing Policies

Data sharing policies by journals and funding agencies have increasingly required data sharing as a condition of study publication or funding, both to improve the transparency and reproducibility of research, as well as to promote data re-use by the research community to accelerate scientific progress. While in the past, funder policies focused on certain types of research (e.g., large-scale genomics projects), recent policies aim to apply broadly but flexibly across health research domains. To cover a broad scope, these policies emphasize flexibility. They highlight general principles of scientific data management and sharing, leaving details to be worked out by specific disciplines and projects. The central elements of these policies include requirements for researchers to 1) Establish data management and sharing plans, 2) Deposit study data in a suitable digital repository as early as practicable; 3) Maximize the openness of research data; and 4) Anticipate and address ethical and legal requirements. Examples include the US NIH's *Policy for Data Management and Sharing* (2022) (effective 2023), Canada's *Tri-Agency Statement of Principles on Digital Data Management* (2021) (effective 2023), and the European Commission (EC)'s Horizon 2020 policy and model grant agreement.

Data availability is necessary but not sufficient to realize the opportunities of Big Data. Funder policies are putting greater emphasis on scientific data management across the data lifecycle. This highlights that data cannot be meaningfully made available and re-used unless it is managed from the outset of each research project toward that end. The FAIR Data principles have become a talisman for appropriate data stewardship and management (Wilkinson et al. 2016). Funders in the USA, Canada, and Europe now require research projects to establish data management and sharing plans that outline what data will be collected, according to what formats and standards, and where and when data will be made available. These plans may be required to be submitted as part of grant applications or before projects start and are expected to be periodically updated as projects evolve.

The second element is the timely deposit of research data in an accessible, secure, and curated repository for preservation, discovery, and re-use by the research community. Definitions of research data in these policies are general and inclusive, covering data, metadata, and code. Not all data need to be preserved or made available; rather, only data that directly support published research findings. Particular repositories may already be specified for certain domains, otherwise the repository selected should meet general data management standards.

Data sharing policies covering broad areas of science generally avoid being too prescriptive. When it comes to the timing of data sharing, the general principle is that data should be shared as early as possible, and no later than the time of publication (or project completion). In terms of openness, the general principle is that research data should be made "as open as possible, as close as necessary," in recognition of ethical and legal requirements. The NIH Policy calls on researchers to *anticipate* these requirements and to ensure that steps are taken from project outset to maximize the ability to share in a compliant manner. In particular, the NIH recommends that data can still be broadly shared through a combination of safeguards, including informed consent, de-identification, and/or controlled access.

4.2 Data Sharing Infrastructure

Large-scale repositories have been established to support the integration and sharing of data from diverse projects. Pooling population neuroscience data in a central repository, however, is not always possible because of technical, privacy, or sovereignty barriers. Federated approaches offer an alternative where independent data providers maintain their own secure databases, while coordinating on data and technical standards to enable researchers to distribute queries or analytics across the network and receive an integrated result (Thorogood et al. 2021). Indeed, a goal of the ENIGMA consortium has been to "send protocols to distributed sites and analyze summaries of the resulting data" in order to reduce the privacy risks of data sharing (Christoforou et al. 2014). Data and technical standards are also key intangible infrastructure for meaningful data sharing. The International Neuroinformatics Coordinating Facility (INCF), for example, supports the coordination of standards and best practices for the neuroscience community (Abrams et al. 2022). As data resources grow in scale and sensitivity, however, copying and distributing data becomes legally and technically more difficult. This has prompted greater interest in cloud computing approaches. The cloud is characterized by remote access to data, and scalable storage and computing resources, which can allow dispersed researchers to collaborate in a single computing environment with common data and tools (Bain et al. 2020). The cloud presents opportunities to support collaborative research at large-scale, but by doing so may exacerbate existing concerns about privacy, consent, and credit sharing. Introducing commercial cloud service providers into an increasingly complex data sharing ecosystem can also raise new challenges relating to control over data (e.g., access by the provider for its own purposes or compelled access to data stored in the cloud by foreign intelligence agencies), the division of responsibility between users and providers over cybersecurity, and the long-term sustainability of research infrastructure. If implemented appropriately, the cloud has the potential to facilitate collaborative research and data sharing, while preserving local data control and providing high data security standards.

4.3 Legal Frameworks

There are also various governmental and legislative initiatives under development to enhance access to administrative and other health data for health research. Building on a strong data protection framework, Europe is now proposing to support and facilitate secure, innovative integration and re-use of data through initiatives like the European Health Data Space (EHDS). The EHDS takes its inspiration from successful initiatives to support access to and linkage between administrative health and social service data sources in countries like Finland (FinData) and France (Health Data Hub) (TEHDAS 2023). Outlined in a proposed Regulation, the EHDS would provide a comprehensive governance framework to promote the interoperability of routinely collected health data. One of its key aims is to establish a common framework for compelling holders of electronic health data to make these data available for legitimate secondary uses, including research and innovation (European Commission 2020). National access bodies would be responsible for reviewing access requests, integrating data from different sources, and providing requestors with remote access to data in secure processing environments. The promise of the EHDS is that it offers a clear legal foundation for data integration and access, allowing studies of large-scale, representative datasets. This has, however, raised competing concerns about legally overriding individual control over data (consent is not required) and undermining the confidentiality underpinning the doctor–patient relationship (Shabani 2022). It also remains unclear how the EHDS will interface across countries or with health research data resources containing genomic or neuroimaging data.

5 Conclusion

Population neuroscience promises to realize the opportunities of Big Data approaches to unravel the mysteries of gene–environment interactions on brain development, aging, and illness. Initial advances have been unlocked by multi-modal, longitudinal population studies, data sharing initiatives, and creative efforts to link together new measures of the physical, built, and social environment with rich genotype and brain phenotype data. Data access and integration at population scale is increasingly supported by advances in data sharing policy, infrastructure, and supporting legal frameworks that given due attention to data management and privacy protection.

References

Abrams MB, Bjaalie JG, Das S, Egan GF, Ghosh SS, Goscinski WJ, Grethe JS et al (2022) A standards organization for open and FAIR neuroscience: the international neuroinformatics coordinating facility. Neuroinformatics 20(1):25–36. https://doi.org/10.1007/s12021-020-09509-0

Bain L, Gee AW, Stroud C (eds) (2020) Neuroscience data in the cloud: opportunities and challenges: proceedings of a workshop. National Academies Press, Washington, DC. https://doi.org/10.17226/25653

Bernier A, Molnár-Gábor F, Knoppers BM (2022) The international data governance landscape. J Law Biosci 9(1):lsac005. https://doi.org/10.1093/jlb/lsac005

Christoforou A, Le Hellard S, Giddaluru S, Steen VM, The ENIGMA Consortium (2014) The ENIGMA consortium: large-scale collaborative analyses of neuroimaging and genetic data. Brain Imaging Behav 8:153., June. https://doi.org/10.1007/s11682-013-9269-5

Critchley CR, Fleming J, Nicol D, Marlton P, Ellis M, Devereux L, Bruce G, Kerridge I (2021) Identifying the nature and extent of public and donor concern about the commercialisation of biobanks for genomic research. Eur J Hum Genet 29(3):503–511. https://doi.org/10.1038/s41431-020-00746-0

de Hemptinne MC, Posthuma D (2023) Addressing the ethical and societal challenges posed by genome-wide association studies of Behavioral and brain-related traits. Nat Neurosci 26(6): 932–941. https://doi.org/10.1038/s41593-023-01333-4

European Commission, Vincent (2020) "European health data space." Text. Public Health - European Commission. September 18, 2020. https://ec.europa.eu/health/ehealth/dataspace_en

Eva G, Liese G, Stephanie B, Petr H, Leslie M, Roel V, Martine V et al (2022) Position paper on management of personal data in environment and health research in Europe. Environ Int 165 (July):107334. https://doi.org/10.1016/j.envint.2022.107334

Falk EB, Hyde LW, Mitchell C, Faul J, Gonzalez R, Heitzeg MM, Keating DP et al (2013) What is a representative brain? Neuroscience meets population science. Proc Natl Acad Sci 110(44): 17615–17622. https://doi.org/10.1073/pnas.1310134110

Hallinan D (2021) Protecting genetic privacy in biobanking through data protection law. Oxford University Press, Oxford

Harding RJ, Bermudez P, Bernier A, Beauvais M, Bellec P, Hill S, Karakuzu A et al (2023) The Canadian open neuroscience platform—an open science framework for the neuroscience community. PLoS Comput Biol 19(7):e1011230. https://doi.org/10.1371/journal.pcbi.1011230

Health Data Research Network Canada (2021) Guidelines: informed consent wording for administrative data linking. https://www.hdrn.ca/sites/default/files/2021-05/Administrative%20Data%20Linking%20Consent%20Wording%20Tool%20V1.0_20210507.pdf

Joly Y, Dalpe G (2022) Genetic discrimination still casts a large shadow in 2022. Eur J Hum Genet 30(12):1320–1322. https://doi.org/10.1038/s41431-022-01194-8

Longstaff H, Flamenbaum J, Richer E, Egar J, McMaster CR, Zawati M'n H (2022) Core elements of participant consent documents for Canadian Human Genomics Research and the National Human Genome Library: guidance for policy. CMAJ 194(44):E1500–E1508. https://doi.org/10.1503/cmaj.212063

Nees F, Deserno L, Holz NE, Romanos M, Banaschewski T (2021) Prediction along a developmental perspective in psychiatry: how far might we go? Front Syst Neurosci 15:670404. https://doi.org/10.3389/fnsys.2021.670404

Palk A, Illes J, Thompson PM, Stein DJ (2020) Ethical issues in global neuroimaging genetics collaborations. NeuroImage 221(November):117208. https://doi.org/10.1016/j.neuroimage.2020.117208

Paprica PA, Crichlow M, Maillet DC, Kesselring S, Pow C, Scarnecchia TP, Schull MJ et al (2023) Essential requirements for the governance and management of data trusts, data repositories, and other data collaborations. Int J Popul Data Sci 8(4):1. https://doi.org/10.23889/ijpds.v8i4.2142

Paus T (2010) Population neuroscience: why and how. Hum Brain Mapp 31(6):891–903. https://doi.org/10.1002/hbm.21069

Paus T (2016) Chapter 2 - Population neuroscience. In: Aminoff MJ, Boller F, Swaab DF (eds) Handbook of clinical neurology, Neuroepidemiology, vol 138. Elsevier, pp 17–37. https://doi.org/10.1016/B978-0-12-802973-2.00002-1

Paus T, Brook J, Doiron D (2022a) Mapping inequalities in the physical, built and social environment in population-based studies of brain health. Front Neuroimaging 1:897943. https://doi.org/10.3389/fnimg.2022.884191

Paus T, Debette S, Seshadri S (2022b) Editorial: population neuroscience of development and aging. Front Syst Neurosci 16:897943. https://doi.org/10.3389/fnsys.2022.897943

Popejoy AB, Fullerton SM (2016) Genomics is failing on diversity. Nature 538(7624):161–164. https://doi.org/10.1038/538161a

Schadt EE (2012) The changing privacy landscape in the era of big data. Mol Syst Biol 8(1):612

Shabani M (2022) Will the European health data space change data sharing rules? Science 375(6587):1357–1359. https://doi.org/10.1126/science.abn4874

TEHDAS (2023) TEHDAS identifies needs for recommendations on cross-border collaboration between health data access bodies. https://tehdas.eu/results/tehdas-identifies-needs-for-recommendations-on-cross-border-collaboration-between-health-data-access-bodies/

Thorogood A, Dalpé G, Knoppers BM (2019) Return of individual genomic research results: are laws and policies keeping step? Eur J Hum Genet., January, 1 27:535. https://doi.org/10.1038/s41431-018-0311-3

Thorogood A, Rehm HL, Goodhand P, Page AJH, Joly Y, Baudis M, Rambla J et al (2021) International Federation of Genomic Medicine Databases using GA4GH standards. Cell 1(2): 100032. https://doi.org/10.1016/j.xgen.2021.100032

Uffelmann E, Huang QQ, Munung NS, de Vries J, Okada Y, Martin AR, Martin HC, Lappalainen T, Posthuma D (2021) Genome-wide association studies. Nat Rev Methods Primers 1(1):1–21. https://doi.org/10.1038/s43586-021-00056-9

Vermeulen R, Schymanski EL, Barabási A-L, Miller GW (2020) The exposome and health: where chemistry meets biology. Science 367(6476):392–396. https://doi.org/10.1126/science.aay3164

Wan Z, Hazel JW, Clayton EW, Vorobeychik Y, Kantarcioglu M, Malin BA (2022) Sociotechnical safeguards for genomic data privacy. Nat Rev Genet 23(7):429–445. https://doi.org/10.1038/s41576-022-00455-y

Wild CP (2005) Complementing the genome with an 'exposome': the outstanding challenge of environmental exposure measurement in molecular epidemiology. Cancer Epidemiol Biomarkers Prev 14(8):1847–1850. https://doi.org/10.1158/1055-9965.EPI-05-0456

Wilkinson MD, Dumontier M, Aalbersberg IJJ, Appleton G, Axton M, Baak A, Blomberg N, Boiten J-W, da Silva Santos LB, Bourne PE (2016) The FAIR guiding principles for scientific data management and stewardship. Sci Data 3:160018

Racial, Ethnic, and Geographic Diversity in Population Neuroscience

C. Elizabeth Shaaban and Andrea L. Rosso

Contents

Abstract In this chapter, we consider lack of racial, ethnic, and geographic diversity in research studies from a public health perspective in which representation of a target population is critical. We review the state of the research field with respect to racial, ethnic, and geographic diversity in study participants. We next focus on key factors which can arise from the lack of diversity and can negatively impact external validity. Finally, we argue that the public's health, and future research, will ultimately be served by approaches from both recruitment and representation science and population neuroscience, and we close with recommendations from these two fields to improve diversity in studies.

Keywords Alzheimer's disease and related dementias · External validity · Public health · Recruitment and representation science · Representativeness

---

C. E. Shaaban (✉)
Department of Epidemiology, School of Public Health, University of Pittsburgh, Pittsburgh, PA, USA

Alzheimer's Disease Research Center, University of Pittsburgh, Pittsburgh, PA, USA
e-mail: Beth.Shaaban@pitt.edu

A. L. Rosso
Department of Epidemiology, School of Public Health, University of Pittsburgh, Pittsburgh, PA, USA

Curr Topics Behav Neurosci (2024) 68: 67–86
https://doi.org/10.1007/7854_2024_475
Published Online: 7 June 2024

1 Introduction

Inclusion of diverse samples within research studies is a critical goal for ensuring justice, trust from communities, and validity of research findings. Justice is a key principle of human-subjects research as outlined in the Belmont report and requires that the risks and benefits of research are fairly distributed (US Department of Health, Education, and Welfare 1979; Gilmore-Bykovskyi et al. 2021). Further, trust in research findings and willingness to adhere to recommendations is eroded when members of the public do not see themselves reflected in research samples (Schwartz et al. 2023; Alsan et al. 2022). Importantly, validity of findings can also be compromised when samples lack diversity that reflects the target population for potential intervention (Gilmore-Bykovskyi et al. 2021, 2022; Schwartz et al. 2023; Denny et al. 2020). Here, we focus on the impact of lack of diversity on external validity, or the ability to apply research findings to populations of interest. We consider diversity across racial/ethnic identity, geographic location (e.g., rural vs urban), and country of residence (e.g., high-income countries (HICs) vs low- and middle-income countries (LMICs)), though many more dimensions of diversity may be relevant to consider.

Research in population neuroscience fuses the knowledge and methods of brain, behavior, and population sciences to make inferences about exposure-outcome relations. Its goal is to understand better variation in development, aging, and disease related to the brain at the population level. These inferences are only useful for public health and clinical implementation if the findings are relevant to a target population of interest. Often, the target population for a research study is not explicitly defined or is presumed to be universal, regardless of how selected the study sample may be. If the study sample does not represent the intended target population, then results may have limited usefulness for informing implementation strategies. This is true even when the internal validity of the study is high. Therefore, ensuring diversity across a number of characteristics relevant to the target population is critical as is thorough characterization of included samples to understand to which population(s) inferences can be made. Without representative samples, we risk producing scientific findings with limited usefulness in the real world. We refer the reader to another Chapter "Population Neuroscience: Understanding Concepts of Generalizability and Transportability and Their Application to Improving the Public's Health" in this book for a detailed introduction to issues of internal and external validity.

Definitions of Key Terms

- Differential item functioning (DIF) – differences in how assessment items are interpreted across subgroups, which indicate that the construct is not being measured similarly across samples;
- Effect-measure modification – difference in association between an exposure and outcome based on another variable; may be a difference (indicated by a significant interaction term) in direction of effects (positive in one group, negative in another), significance (stratified results that are significant in one group but not in the other), or in magnitude of effect (associations are significant and in the same direction in both groups but larger in one compared to the other); see also Chapter "Population Neuroscience: Understanding Concepts of Generalizability and Transportability and Their Application to Improving the Public's Health" in this book;
- External validity – extent to which study results represent the truth in the source population (generalizability) or the truth in the target population (transportability); see also Chapter "Population Neuroscience: Understanding Concepts of Generalizability and Transportability and Their Application to Improving the Public's Health" in this book;
- High-income countries (HICs)—countries with a per capita gross national income ≥$13,846 (World Bank 2024).
- Internal validity – extent to which study results represent the truth within the study sample; see also Chapter "Population Neuroscience: Understanding Concepts of Generalizability and Transportability and Their Application to Improving the Public's Health" in this book;
- Low- and middle-income countries (LMICs)—countries with a per capita gross national income ≤$13,845 (World Bank 2024);
- Population attributable fraction (PAF) – proportion of disease cases within a population that can be attributed to the exposure, or risk factor, of interest;
- Psychometric properties – the extent to which a measurement instrument is valid (accurately measures what is intended to be measured), reliable (performs consistently across time and individuals), and responsive (has the ability to detect change);
- Study sample – the participants included in a study; should be representative of the target population to which inferences are intended to be applied; see also Chapter "Population Neuroscience: Understanding Concepts of Generalizability and Transportability and Their Application to Improving the Public's Health" in this book;
- Target population – the population to which sample estimates are intended to be applied and for which inferences are made; see also Chapter "Population Neuroscience: Understanding Concepts of Generalizability and Transportability and Their Application to Improving the Public's Health" in this book.

2 Where Is the Field Currently?

For much of its history, research in brain and cognitive science has focused on convenience samples, often without specifying a target population or determining representativeness. Both brain and behavioral research studies have focused on samples from countries that are Western, Educated, Industrialized, Rich, and Democratic (WEIRD) (Chiao and Cheon 2010; Henrich et al. 2010). For example, a review of cognitive psychology literature that was completed from 2003 to 2007 demonstrated that 96% of study participants came from Western industrialized countries and that the majority were from the United States (Henrich et al. 2010). Within those Western industrialized samples, the majority of samples were composed of undergraduate students. While such results would appropriately apply if the researchers wished to understand something specific to a country's undergraduates (e.g., how undergraduates' brains function under the influence of alcohol), these samples were not representative of the target populations of interest, which included other age groups (Henrich et al. 2010). In addition, as of 2009, 90% of neuroimaging studies were conducted in Western countries (Chiao 2009).

With the growing awareness of the need for broader representation in research, several groups have recently set out to quantify the extent to which lack of diversity has been a problem to date. Below, we provide examples of this work to demonstrate the current state of research. Most of this work has focused on racial/ethnic representation, largely in the United States, with some studies assessing representation across global geographies. Reviews have not assessed rural representativeness but given that most research centers are located in more urban settings, it is likely that individuals living in urban and suburban settings are overrepresented in research studies. In addition, most studies of representativeness have been done within the field of Alzheimer's disease and related dementias (ADRD), as represented below. There is likely a similar state of lack of representation across other sources of diversity (for example, socioeconomic status, sexual orientation) and for other neurological conditions, but the evidence is not yet documented.

Several reviews have documented diversity of samples included in research on dementia (Mooldijk et al. 2021; Mielke et al. 2022; Fiest et al. 2016), including ADRD neuroimaging studies (Lim et al. 2023) and drug trials (Franzen et al. 2022). All have demonstrated that samples are largely unrepresentative of target populations. Contemporary dementia research is almost exclusively conducted within North America and Europe (89% of reviewed studies) and the samples overwhelmingly included White (i.e., individuals of European ancestry) participants (median of 89% White participants) (Mooldijk et al. 2021). Many studies of dementia incidence and prevalence have not conducted analyses by sex or gender (see Chapter on "Sex and Gender in Population Neuroscience") for definitions and more information), and most studies that have evaluated this have been conducted in HICs (Mielke et al. 2022; Fiest et al. 2016). Neuroimaging studies of Alzheimer's disease in the United States have also included predominantly non-Hispanic White participants. While there has been a notable increase in representation by race and ethnicity in recent years, particularly for Black/African American participants, the numbers still do not reflect population demographics (Lim et al. 2023). Lack of racial/ethnic representation in Alzheimer's drug trials is even lower, with the median percentage of White participants being 95% and with no significant change over time (Franzen et al. 2022). The authors of this review note that eligibility criteria, including exclusions based on comorbid psychiatric or cardiometabolic conditions and requirements for a caregiver to attend study visits, were a key source of racial/ethnic underrepresentation. In addition, global diversity for drug trials was low with 79% of studies including sites in North America, 60% including sites in Europe, and few studies including sites in South America (15%) or Africa (7%) (does not total 100% because some studies included sites in more than one part of the world).

These reviews note the lack of reporting on race/ethnicity with only 22% of dementia studies (Mooldijk et al. 2021) and 50% of Alzheimer's drug trials (Franzen et al. 2022) reporting the race/ethnicity of participants. The review of Alzheimer's neuroimaging studies only included papers that reported race/ethnicity and thus excluded a large number of studies (n = 1,160) for not reporting race/ethnicity (Lim et al. 2023). Among the studies which could be included, a large number required the authors to indirectly derive race/ethnicity estimates (n = 1,745/2,464 (71%)). The authors were able to do this because the papers came from large, well-published cohorts that have previously reported these data. This is in comparison with the n = 719/2,464 (29%) studies that directly reported race/ethnicity data within the review. Another notable finding from these reviews is that several large cohorts served as the primary source of data for numerous published studies. For example, 3 cohorts, all based in the United States, represented 21% of study samples in overall dementia research (Mooldijk et al. 2021). Among the AD neuroimaging studies which indirectly reported race, 10 cohorts were the source of 94% of study samples (Lim et al. 2023).

Additional reviews have assessed diversity for other neurologic and psychiatric outcomes and have demonstrated a lack of diversity similar to that seen for ADRD research. Cognitive neuroscience studies underreport race/ethnicity and socioeconomic status, with only 14% and 18% of studies, respectively, providing descriptive data on these demographic characteristics (Dotson and Duarte 2020). Based on the few studies reporting on race/ethnicity, the authors demonstrate that cognitive neuroscience studies are much less diverse than, and not representative of, the target population. Two reviews assessed global diversity in studies of Parkinson's genetics in individuals of non-European descent (Schumacher-Schuh et al. 2022) and in pediatric psychiatric disorders in LMICs (Cirillo et al. 2020). Even among studies conducted in samples of individuals with non-European ancestry, representation may be limited. The review of Parkinson's genetic studies demonstrated that the majority of studies in those without European ancestry were conducted with Asian participants from Greater China (57%) with far fewer conducted in sub-Saharan Africa (4%), Southeast Asia (3%), or Central Asia (0.5%). Similarly, a review of studies of psychiatric disorders in children and adolescents in LMICs identified only 6 studies from 4 countries, including 3 from Brazil, 1 from China (Hong Kong), 1 from South Africa, and 1 from Mauritius (Cirillo et al. 2020). Clearly, research

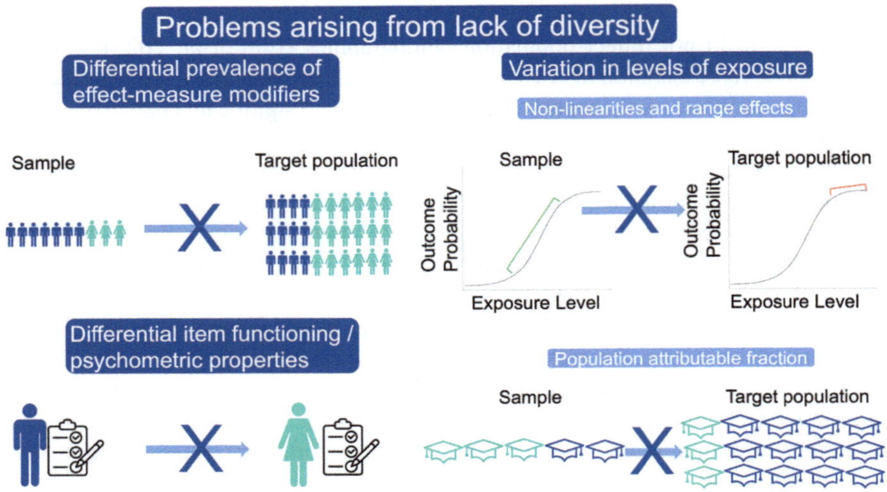

Fig. 1 Examples of problems arising from lack of diversity. Top left: The exposure-outcome relationship varies by sex assigned at birth, and the sample has a greater proportion of male participants than the target population. Bottom left: An activities of daily living measure works differently in men vs. women. Top right: The exposure-outcome association is strong in the sample, but not in the target population due to differing ranges of exposure values in the sample vs. target population. Bottom right: The sample has a greater proportion of participants with high education (aqua) than the target population

across multiple disorders is not representative of the global population and even in countries such as the United States that are well-represented in research, there are large portions of the population that are underrepresented.

3 What Problems Arise from Lack of Diversity?

There are several common circumstances in which inferences from a non-diverse, non-representative sample could be invalid when applied to the target population, including differential prevalence of effect-measure modifiers across populations, differing ranges or prevalence of exposures, and differential item functioning (DIF) of key measures across groups (Fig. 1).

3.1 Differential Prevalence of Effect-Measure Modifiers

Presence of unaccounted for effect-measure modification of an exposure-outcome relation coupled with differences in prevalence of the effect-measure modifier by group can lead to non-comparability of findings across populations. See Chapter "Population Neuroscience: Understanding Concepts of Generalizability and Transportability and Their Application to Improving the Public's Health" for more detail about effect-measure modification. Briefly, effect-measure modification occurs when the relation between an exposure and outcome differs by level of a third variable – the effect-measure modifier. If effect-measure modification is not accounted for in analyses, through methods such as stratification or inclusion of interaction terms, the overall estimated effect size will be an average of the effect at all levels of the effect-measure modifier, weighted by the distribution of that effect-measure modifier within the sample. Therefore, differences in prevalence or distribution of the effect-measure modifier across populations would lead to different overall effect estimates by population.

An example of this can be seen in a comparison of strengths of exposure-outcome associations in a highly selected, clinic-based sample, the Alzheimer's Disease Neuroimaging Initiative (ADNI), and a community-based sample, Atherosclerosis Risk in Communities (ARIC) (Gianattasio et al. 2021). The two samples differed in distribution of a number of important factors that could act as effect-measure modifiers in analyses of associations; for example, educational attainment, sex, and race. At least ¼ of tested associations varied significantly between the two studies and approximately ½ of the point estimates varied by >50%. Some of these differences were highly consequential. For example, the exposure-outcome association of *APOE4* genotype with amyloid beta (Aβ) positivity was an odds ratio (OR) of 2.8 in the community-based study but 8.6 in the clinic-based study. This suggests that a variable that modifies the effect of *APOE4* may have had a higher prevalence in the clinic-based study than in the community-based study. This type of difference matters for prioritization of intervention targets and choices about where to place public health resources. For example, if the OR is truly 8.6 in the target population, interventionists might prioritize developing new AD interventions that target *APOE* (e.g., via modulation of APOE protein levels or gene therapy approaches), but they might make different decisions if the OR in the community is truly 2.6 while other risk factors (e.g., physical health) have a stronger relationship with Aβ positivity. These concerns extend beyond clinic-based vs. community-based studies and are also relevant for convenience samples vs. population-representative community-based studies. A relevant example is the UK Biobank, a volunteer-based study with an impressive and well-characterized sample of 500,000 individuals from the United Kingdom. The study was highlighted in a correspondence about non-representativeness (Keyes and Westreich 2019). The authors of the correspondence point out that associations can differ due to different prevalence of effect modifiers in the study sample vs. the target population (i.e., the entire UK), and

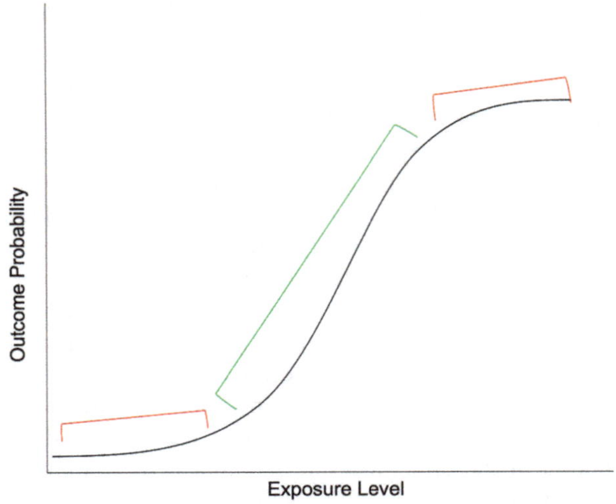

Fig. 2 Non-linearities and range effects. Samples drawn from populations in the red areas (top and bottom of the sigmoid) would have no association between exposure and outcome whereas samples drawn from populations in the green area (middle of the sigmoid) would have a strong positive correlation between exposure and outcome

that the large sample size does nothing to address this particular problem (Keyes and Westreich 2019).

3.2 Variation in Levels of Exposure

3.2.1 Non-linearities and Range Effects

It is not unusual for levels of an exposure to vary across populations based on demographics, geography, or other characteristics. A difference in prevalence of exposure across populations does not necessarily imply a difference in the strength of association with outcomes of interest. But inferences about exposure-outcome relationships in one population may not apply to another population when there are non-linear associations across levels of exposure and levels of exposure differ substantially between those populations (Fig. 2). For example, if a particular threshold must be reached before an exposure causes disease, associations will not be observed in populations with overall low exposure levels. Further, this can cause inference problems if the range of exposure level is too narrow within a single population, even if the relation is linear. In this case, the narrow range of exposure will limit the ability to detect associations due to the lack of variance. This type of relationship is highly relevant in AD research. For example, there is a tipping point at which tau accumulation rapidly accelerates – when Aβ levels are ≥40 centiloids on Positron Emission Tomography (PET) imaging (Doré et al. 2021). Because Aβ accumulation plateaus in the clinical stages of the disease, it is often found to be only weakly or not associated with cognition in that stage, while tau is strongly associated with cognition (Jack et al. 2013; Ossenkoppele et al. 2019). Researchers should report the mean and range or variance of their exposure variables within their samples to allow for comparison with other samples or with known distributions in the target population, when available. In addition, non-linearities of exposure-outcome relations should be assessed.

3.2.2 Population Attributable Fraction (PAF)

Another way in which differences in exposure prevalence can affect interpretation of results across populations is in calculation of the PAF for a given exposure or set of exposures. PAF is the proportion of disease cases within a population that can be attributed to the exposure or risk factor of interest and is calculated as

$$\mathrm{PAF} = \frac{P_e \times (\mathrm{RR} - 1)}{P_e \times (\mathrm{RR} - 1) + 1}$$

where P_e is the prevalence of the exposure or risk factor and RR is the relative risk for the risk factor on the outcome of interest.

We discuss two examples of this – one using dementia-related PAF estimates in LMICs vs. HICs, and one using dementia-related PAF estimates across ethnoracial groups in the United States. The 2020 Lancet report on dementia prevention, intervention, and care calculated PAFs for established risk factors of dementia and suggested that at least 40% of risk for dementia is due to 12 modifiable dementia risk factors (this is the PAF summed across risk factors) while the remaining 60% is either non-modifiable or not yet identified (Livingston et al. 2020). This built upon the group's 2017 work, which included 9 modifiable dementia risk factors and found that they accounted for an estimated 35% of dementia risk (Livingston et al. 2017). These estimates were based on meta-analyses to define both P_e and RR.

Although international studies were used to represent global estimates, it is important to point out that the studies used to obtain the P_e and RR estimates have generally been conducted in relatively homogeneous samples from HICs. Because the PAF is a function of both the P_e and RR, variations in their estimates will cause the PAF to vary. A summary of estimates of P_e, RR, and PAFs from various studies and populations may be found in Table 1. PAFs were weighted to account for correlations between risk factors and try to give an estimate of the unique contribution of each exposure to overall dementia risk. All included studies used the same RR estimates, while estimates of P_e were study specific. RRs represent a weighted average RR over relevant population subgroups. Because risk factor-dementia associations vary based on these subgroups defined by different prevalence of effect-measure modifiers, RRs will differ in populations with different distributions of subgroups (Keyes and Westreich 2019; Cole and Stuart 2010). The RRs drawn from the meta-analyses included in the 2017 and 2020 Lancet dementia reports will be relevant for global estimates to the extent that the prevalence of unmeasured effect-measure modifiers is similar across those populations not represented. This

Table 1 Population attributable fractions (PAFs) of modifiable dementia risk factors across various study populations

Study	Measure	↓ Education	↓ Hearing	tbi	htn	↑ Alcohol	Obesity	Smoking	Depression	Social isolation	Physical inactivity	Diabetes	Air pollution	Total weighted PAF
RR's for all studies are from: Livingston et al. (2017, 2020)	RR	1.6 (1.3–2.0)	1.9 (1.4–2.7)	1.8 (1.5–2.2)	1.6 (1.2–2.2)	1.2 (1.1–1.3)	1.6 (1.3–1.9)	1.6 (1.2–2.2)	1.9 (1.6–2.3)	1.6 (1.3–1.9)	1.4 (1.2–1.7)	1.5 (1.3–1.8)	1.1 (1.1–1.1)	
Lancet Commission Report (Livingston et al. 2020) 12 exposures	P_e	40.0%	31.7%	12.1%	8.9%	11.8%	3.4%	27.4%	13.2%	11.0%	17.7%	6.4%	75.0%	
	PAF	7.1%	8.2%	3.4%	1.9%	0.8%	0.7%	5.2%	3.9%	3.5%	1.6%	1.1%	2.3%	39.7%
Lancet Commission Report (Livingston et al. 2017) 9 exposures	P_e	40.0%	31.7%		8.9%		3.4%	27.4%	13.2%	11.0%	17.7%	6.4%		
	PAF	7.5%	9.1%		2.0%		0.8%	5.5%	4.0%	2.3%	2.6%	1.2%		35.0%
Borelli et al. (2022) Brazil 10 exposures	P_e	51.1%	29.4%		52.0%	4.3%	29.2%	17.1%	18.1%	5.1%	47.2%	15.7%		
	PAF	9.5%	14.2%		10.4%	0.2%	6.3%	2.9%	6.9%	1.5%	11.3%	3.6%		50.5%
Mukadam et al. (2019). 6 Latin American LMICs 9 exposures	P_e	68.8%	28.8%		55.6%		44.8%	30.0%	23.9%	0.5%	34.2%	18.5%		
	PAF	10.9%	7.7%		9.3%		7.9%	5.7%	6.6%	0.1%	4.5%	3.2%		55.8%
Mukadam et al. (2019) India 9 exposures	P_e	92.2%	22.3%		19.3%		13.7%	39.9%	5.2%	10.4%	15.3%	9.3%		
	PAF	13.6%	6.4%		4.0%		2.9%	6.4%	1.7%	2.3%	2.2%	1.7%		41.2%
Mukadam et al. (2019) China 9 exposures	P_e	75.9%	14.3%		38.1%		32.0%	23.0%	1.5%	3.4%	50.7%	9.4%		
	PAF	10.8%	3.9%		6.4%		5.6%	4.2%	0.5%	0.7%	5.8%	1.6%		39.5%

Lee et al. (2022)	P_e	10.7%	10.8%	17.1%	42.2%	3.6%	44.0%	8.5%	7.4%	11.9%	62.8%	28.6%	22.8%	
US PAFs overall	PAF	2.0%	3.0%	4.1%	6.8%	0.2%	7.1%	1.6%	2.1%	2.3%	6.8%	4.2%	0.8%	41.0%
Lee et al. (2022).	P_e	27.1%	13.1%	10.3%	38.5%	2.0%	48.3%	6.9%	10.7%	24.0%	68.6%	41.0%	44.4%	
US PAFs, Hispanic 12 exposures	PAF[a]	4.6%	3.5%	2.5%	6.2%	0.1%	7.4%	1.3%	2.9%	4.1%	7.1%	5.6%	1.4%	46.7%
Lee et al. (2022).	P_e	6.4%	6.9%	6.0%	38.5%	0.7%	14.6%	4.9%	4.3%	8.0%	56.6%	44.1%	55.2%	
US PAFs, Non-Hispanic Asian 12 exposures	PAF[a]	1.4%	2.2%	1.7%	7.1%	0.1%	3.1%	1.1%	1.4%	1.7%	7.0%	6.9%	2.0%	35.6%
Lee et al. (2022).	P_e	10.6%	6.5%	9.2%	61.0%	2.7%	54.3%	11.7%	6.6%	12.1%	73.2%	37.2%	41.3%	
US PAFs, Non-Hispanic Black 12 exposures	PAF[a]	2.1%	1.9%	2.4%	9.3%	0.2%	8.5%	2.3%	1.9%	2.3%	7.8%	5.4%	1.4%	45.6%
Lee et al. (2022).	P_e	5.5%	10.6%	20.1%	39.8%	4.2%	43.5%	8.4%	7.2%	10.8%	61.3%	25.4%	17.2%	
US PAFs, Non-Hispanic White 12 exposures	PAF[a]	1.1%	3.0%	4.7%	6.5%	0.3%	7.0%	1.6%	2.1%	2.1%	6.7%	3.8%	0.6%	39.4%

Note: tbi = traumatic brain injury; htn = hypertension. All PAFs are weighted

[a] Weighted PAFs obtained via direct correspondence with the authors

assumption may be met for many variables but would probably not be met for several important variables such as age, education, or gender. Even if the strength of the association between the exposure and outcome does not differ between populations, the relative importance of the exposure as a target for public health interventions could differ substantially based on the P_e and the PAF, as demonstrated in Table 1.

LMIC vs HIC PAFs

Based on data from the ELSI-Brazil (Lima-Costa et al. 2018), the Brazilian member of the global Health and Retirement Studies, which was designed to be population-representative of individuals 50+ years of age in Brazil, the PAF for dementia was as 50.5%, higher than that from the 2020 Lancet report (39.7%) despite being calculated from fewer risk factors (10 vs 12) (Borelli et al. 2022). The PAF estimates for the 9 dementia risk factors included in the 2017 Lancet report (35.0%) in several LMIC locations were also higher as follows: across 6 Latin American countries (Cuba, Dominican Republic, Mexico, Peru, Puerto Rico, and Venezuela), 55.8%; in India, 41.2%, and in China, 39.5% (Mukadam et al. 2019). Not only are the RRs likely to have been different in these populations as discussed above, but the P_e estimates also varied substantially. For example, the prevalence of low education was 10.7% in the United States, 40.0% globally, relying on mostly HIC data, and 92.2% in India.

PAFs in Various US Ethnoracial Groups

In the United States, there are notable variations in dementia risk and risk factor prevalence between ethnoracial groups. Thus, national PAF estimates (41.0% based on 12 risk factors) do not necessarily apply to the entire US population, nor do PAF estimates for one ethnoracial group apply to others within the United States (Lee et al. 2022). For example, based on a study using data from ARIC, low education was most common for Hispanic individuals in the United States and least common for non-Hispanic Asian and White individuals leading to a greater PAF due to low education and in sum for those identifying as Hispanic vs. those identifying as non-Hispanic Asian or White (Table 1) (Lee et al. 2022). Hypertension was present in 61.0% of non-Hispanic Black individuals but <40% in the other ethnoracial groups. The consequences of this difference are a greater PAF due to hypertension and in sum for those identifying as non-Hispanic Black vs. those identifying as non-Hispanic Asian or White (Table 1) (Lee et al. 2022). In addition, non-Hispanic White individuals had lower exposure to air pollution with a P_e of 17.2%, while the remaining US ethnoracial groups had P_es that were more than double that. This leads to a lower PAF due to air pollution and in sum for those identifying as non-Hispanic White vs. those identifying as Hispanic or non-Hispanic Black.

3.3 *Differential Item Functioning (DIF) and Different Psychometric Properties of Instruments Across Groups*

DIF indicates that an item does not measure the same construct across different groups. DIF can also impact sum-scored instruments when items within the instrument function differently; this is important for population-based health research, which often uses sum scores from established scales (Jones 2019). DIF and differences in psychometric properties of instruments across groups can influence validity of findings between convenience samples and the broader population. This can arise due to differences in interpretation of wording in questionnaires or in reaction to the content of an instrument. For example, the Center for Epidemiologic Studies Scale-Depression (CES-D), a commonly used questionnaire that generates a sum score for depressive symptomology, has been shown to have differential functioning by race (Kato 2021; Assari and Moazen-Zadeh 2016), ethnicity (Macintosh and Strickland 2010), immigration status (Van Lieshout et al. 2011), country of study (Bergenfeld et al. 2023), and sexual orientation (Gomez and Mclaren 2017), though it should be noted that many other studies have demonstrated measurement invariance of the CES-D across different groups, including those listed above. Additional examples of potentially different measurement properties by populations include autism spectrum characteristics by gender or country of origin (Belcher et al. 2023; Chee et al. 2023) and suicidality by ethnicity (Mcclure et al. 2023). DIF is beginning to be assessed for the tasks used in task-related functional MRI (Demidenko et al. 2024). Differences in psychometric properties (i.e. validity, reliability, responsiveness) of an assessment across groups can also lead to differences in sensitivity/specificity to an underlying construct and as a result, differences in ability to detect significant associations within certain groups.

4 Recommendations to Improve Diversity in Research Studies

Clearly there is a lot of room to enhance diversity and external validity in research, and we now move to a discussion of recommended strategies for improvement. Two main categories of strategies can be used to enhance diversity and external validity in studies: (1) study-design approaches implemented before data collection, including specific sampling and recruitment strategies, and (2) analytic approaches that can enhance external validity in some instances when data have already been collected. Ultimately, we see enhancing diversity and external validity as requiring a merging of recruitment and representation science with population neuroscience. A number of study-design problems have limited diversification of samples, and recruitment and representation science offers strategies to address these barriers. Recruitment and representation science is a growing field focused on improving recruitment strategies that enhance inclusion and retention of individuals from underrepresented backgrounds (Dilworth-Anderson 2011; Gilmore-Bykovskyi et al. 2019). This field is identifying best practices that should be consulted to improve representation of samples.

A major barrier to diversity in research has been a lack of understanding by researchers of the needs of diverse communities that would allow them to participate in research. Often, researchers have attempted to diversify their samples by continuing to utilize recruitment strategies and data collection approaches that they had been using and which have consistently resulted in the non-generalizable/non-transportable samples in prior research. This includes largely passive recruitment methods that rely on volunteers to see and respond to advertisements to participate. Location and method of recruitment should be carefully considered and sufficient resources devoted to recruitment efforts. Often, researchers do not have familiarity with the communities that have been underrepresented in research, which leads to a lack of knowledge about the needs and desires of those communities with regard to their ability, willingness, and motivation to participate in research studies. Overall, research efforts that reduce participant burden, build trust between the participant and research team, and improve participant knowledge of the research study will enhance research participation (Gabel et al. 2022). Having greater representation at both the staff and investigator level of members of these communities will greatly enhance the ability to recruit and retain diverse samples (Denny et al. 2020; Alsan et al. 2023).

The location and timing of research visits can be a barrier to participation, particularly for those from lower socioeconomic conditions, rural communities, and communities of color. Consideration should be given to reducing participant burden by providing mobile clinic sites, providing in-home visits, and/or locating clinic sites within target communities. Accessibility for those with disabilities should also be considered. Researchers should make efforts to provide flexibility in timing of visits and options outside of traditional working hours. Imaging modalities that are more mobile include electroencephalography (EEG), functional near-infrared spectroscopy (fNIRS) and, most recently, mobile low-field magnetic resonance imaging, MRI (see Chapter on "Population Neuroscience: Principles and Advances"). For studies using non-portable imaging modalities (e.g. 3T MRI), researchers should consider paying for or providing transportation to the study site and offering scan times outside of normal business hours. Analytic approaches (discussed below) can help with addressing the selection that often occurs in imaging studies due to the inconvenience of visits, discomfort with procedures, and safety-related exclusion criteria. Blood biomarkers of brain pathologies such as Aβ, tau, neurodegeneration, and neuroinflammation are also a promising approach to increasing inclusion in future population neuroscience research.

Unnecessarily strict exclusion criteria can also impact the diversity of individuals included in research samples (Franzen et al. 2022). All inclusion and exclusion criteria should be carefully considered with regard to their impact on external validity and their necessity for maintaining participant safety and internal validity. Communities that are underrepresented in research may be hesitant to participate in studies, lack trust in researchers, and/or lack knowledge about what it means to participate in research. In all of these cases, the burden should be on researchers to work with communities to overcome these barriers. Randomized controlled trial evidence shows that participant compensation increases recruitment (Abdelazeem et al. 2022; Halpern et al. 2021). Best practices in the Alzheimer's Disease Research Centers' Network have recently been developed and are available via the National Alzheimer's Coordinating Center website (https://files.alz.washington.edu/documentation/remuneration-guidelines.pdf). Finally, giving to the community (at least) twice over is key for engagement, recruitment, and retention: first, before asking for research participation to help meet community needs (e.g., health fairs or educational talks), and second, after participation to return group and/or individual research results to participants and community members (Denny et al. 2020). This return of research results and sharing educational information with participants has recently been conceptualized as return of value (ROV) (Wilkins et al. 2019).

To draw relevant inferences about the target population of interest, the study must include diverse participants with all of the characteristics observed in the target population. External validity will be enhanced if the study sample is population representative. There are circumstances, however, where this is neither feasible nor desirable. For example, getting valid estimates of prevalence or of exposure-outcome associations in smaller subgroups may require oversampling of that group relative to their population proportion to achieve sufficient statistical power. We discuss analytic strategies that may be used in such circumstances below. Variables to describe relevant aspects of diversity must be collected by design; this includes race/ethnicity and geographic factors as well as factors which may vary across diverse groups such as health conditions, resilience factors and coping strategies, and sociocultural, political, economic, and environmental factors. Classification of race/ethnicity and geographic factors should provide the appropriate level of specificity to capture the heterogeneity in researchers' outcomes of interest.

Analytic approaches to enhance external validity also warrant discussion for cases where data have already been collected. A key goal of some studies will be to understand how multiple factors jointly contribute to risk and resilience, and such analyses should test for additive and multiplicate interactions. Knowledge of diverse populations is enhanced when contexts (e.g., social, environmental) are considered; such studies may need to employ multilevel analytic approaches, which account for correlated data. Analytic approaches can also be used to enhance both internal and external validity in studies. Approaches such as inverse probability of selection weighting, propensity score matching, and the like can be used to weight subsamples, for example a neuroimaging sub-study, to the full parent study, reducing selection bias and thereby enhancing internal validity (Ganguli et al. 2014; Babulal et al. 2023). Joint modeling can enhance internal validity of longitudinal studies by addressing differential attrition due to causes such as drop out and death (Griswold et al. 2021). External validity can be improved analytically by using transportability estimators (Hayes-Larson et al. 2022). This can be thought of as an extension of survey weighting methods to weight the study sample to the population. This approach can be useful when taking a population representative sample was not feasible or when oversampling of smaller subgroups was required.

It is critical to train the next generation of researchers for community engagement and participant recruitment as well as fluent study design and analysis to advance diverse and representative research that can serve the population. Unique training approaches such as training individuals in population neuroscience and recruitment and representation science can help to bring knowledge of representative and diverse population characteristics to research (Denny et al. 2020; Rosano 2022; Falk et al. 2013; Butler 3rd et al. 2018). Additional funding that specifically requires representative samples can improve diversity in studies. This may be accomplished, for example, by policies such as the recent announcement that the US National Institute on Aging (NIA) will prioritize new grant submissions that represent the diversity of populations experiencing the disease of interest and that are designed to include participants from groups experiencing health disparities (National Institute on Aging 2023). This is a particularly difficult barrier in LMICs due to limited funding to conduct health research. Greater funding from organizations which do not limit investigator location can help fill the gap. Funding researchers who are themselves from groups historically underrepresented in research can improve relationships with the community and inform scientific hypotheses to test and interpretation of results.

5 Summary and Conclusions

There is a growing awareness of the need to discuss target populations, define them explicitly, and to sample systematically to create representativeness of research samples to those target populations. This awareness is evidenced by the development and adoption of population-neuroscience approaches (Ganguli et al. 2014, 2018; Jorgensen et al. 2018; Paus 2010) and a push from funding agencies to diversify research participants to create more representative samples. Our review of the state of the field shows that diversification across multiple domains of identity and geography has been slow, however. We discussed three key factors which can negatively impact external validity including differential prevalence of effect modifiers across populations, variations in levels of exposures, and differential item functioning and differing psychometric properties of instruments across groups. We discussed approaches to enhance diversity in studies and suggest that a merging of perspectives from recruitment and representation science and population neuroscience can best increase diversity and improve external validity. These methodological improvements will most serve the public's health.

Acknowledgments This work was supported by the National Institute on Aging at the National Institutes of Health by grant number K01AG071849 to CES and grant number T32AG055381. The content is solely the responsibility of the authors and does not necessarily represent the official views of the National Institutes of Health.

We thank Andrea M. Weinstein, PhD; Beth E. Snitz, PhD; Caterina Rosano, MD, MPH; Jennifer H. Lingler, PhD, MA, CRNP, FAAN; Mary Ganguli, MD, MPH; and Michelle M. Mielke, PhD for feedback on earlier versions of this chapter.

References

Abdelazeem B, Abbas KS, Amin MA et al (2022) The effectiveness of incentives for research participation: a systematic review and meta-analysis of randomized controlled trials. PLoS One 17(4):E0267534

Alsan M, Durvasula M, Gupta H, Schwartzstein J, Williams H (2022) Representation and extrapolation: evidence from clinical trials. National Bureau of Economic Research, Inc

Alsan M, Campbell RA, Leister L, Ojo A (2023) Investigator racial diversity and clinical trial participation. National Bureau of Economic Research

Assari S, Moazen-Zadeh E (2016) Confirmatory factor analysis of the 12-item center for epidemiologic studies depression scale among blacks and whites. Front Psych 7:178

Babulal GM, Zhu Y, Trani JF (2023) Racial and ethnic differences in neuropsychiatric symptoms and progression to incident cognitive impairment among community-dwelling participants. Alzheimers Dement 19(8):3635–3643

Belcher H, Uglik-Marucha N, Vitoratou S, Ford RM, Morein-Zamir C (2023) Gender bias in autism screening: measurement invariance of different model frameworks of the autism spectrum quotient. Bjpsych Open 9(5):E173

Bergenfeld I, Kaslow NJ, Yount KM, Cheong YF, Johnson ER, Clark CJ (2023) Measurement invariance of the center for epidemiologic studies scale-depression within and across six diverse intervention trials. Psychol Assess 35(10):805–820

Borelli WV, Leotti VB, Strelow MZ, Chaves MLF, Castilhos RM (2022) Preventable risk factors of dementia: population attributable fractions in a Brazilian population-based study. Lancet Reg Health Am 11:100256

Butler J 3rd, Fryer CS, Garza MA, Quinn SC, Thomas SB (2018) Commentary: critical race theory training to eliminate racial and ethnic health disparities: the public health critical race praxis institute. Ethn Dis 28(Suppl 1):279–284

Chee ZJ, Scheeren AM, De Vries M (2023) The factor structure and measurement invariance of the autism spectrum quotient-28: a cross-cultural comparison between Malaysia and The Netherlands. Autism:13623613221147395

Chiao JY (2009) Cultural neuroscience: a once and future discipline. Prog Brain Res 178:287–304

Chiao JY, Cheon BK (2010) The weirdest brains in the world. Behav Brain Sci 33(2–3):88–90

Cirillo A, Diniz E, Gadelha A et al (2020) Population neuroscience: challenges and opportunities for psychiatric research in low- and middle-income countries. Braz J Psychiatry 42(4):442–448

Cole SR, Stuart EA (2010) Generalizing evidence from randomized clinical trials to target populations: the Actg 320 trial. Am J Epidemiol 172(1):107–115

Demidenko MI, Mumford JA, Ram N, Poldrack RA (2024) A multi-sample evaluation of the measurement structure and function of The modified monetary incentive delay task in adolescents. Dev Cogn Neurosci 65:101337

Denny A, Streitz M, Stock K et al (2020) Perspective on the "African American Participation In Alzheimer Disease Research: Effective Strategies" workshop, 2018. Alzheimers Dement 16(12):1734–1744

Dilworth-Anderson P (2011) Introduction to the science of recruitment and retention among ethnically diverse populations. Gerontologist 51(Suppl 1):S1–S4

Doré V, Krishnadas N, Bourgeat P et al (2021) Relationship between amyloid and Tau levels and its impact on Tau spreading. Eur J Nucl Med Mol Imaging 48(7):2225–2232

Dotson VM, Duarte A (2020) The importance of diversity in cognitive neuroscience. Ann N Y Acad Sci 1464(1):181–191

Falk EB, Hyde LW, Mitchell C et al (2013) What is a representative brain? Neuroscience meets population science. Proc Natl Acad Sci U S A 110(44):17615–17622

Fiest KM, Roberts JI, Maxwell CJ et al (2016) The prevalence and incidence of dementia due to Alzheimer's disease: a systematic review and meta-analysis. Can J Neurol Sci 43(Suppl 1):S51–S82

Franzen S, Smith JE, Van Den Berg E et al (2022) Diversity in Alzheimer's disease drug trials: the importance of eligibility criteria. Alzheimers Dement 18(4):810–823
Gabel M, Bollinger RM, Knox M et al (2022) Perceptions of research burden and retention among participants in Adrc cohorts. Alzheimer Dis Assoc Disord 36(4):281–287
Ganguli M, Lee CW, Hughes T et al (2014) Who wants a free brain scan? Assessing and correcting for recruitment biases in a population-based Smri pilot study. Brain Imaging Behav
Ganguli M, Albanese E, Seshadri S et al (2018) Population neuroscience: dementia epidemiology serving precision medicine and population health. Alzheimer Dis Assoc Disord 32(1):1–9
Gianattasio KZ, Bennett EE, Wei J et al (2021) Generalizability of findings from a clinical sample to a community-based sample: a comparison of Adni and Aric. Alzheimers Dement
Gilmore-Bykovskyi AL, Jin Y, Gleason C et al (2019) Recruitment and retention of underrepresented populations. In: Alzheimer's disease research: a systematic review. Alzheimer's & Dementia (New York, N Y) 5:751–770
Gilmore-Bykovskyi A, Jackson JD, Wilkins CH (2021) The urgency of justice in research: beyond Covid-19. Trends Mol Med 27(2):97–100
Gilmore-Bykovskyi A, Croff R, Glover CM et al (2022) Traversing the aging research and health equity divide: toward intersectional frameworks of research justice and participation. Gerontologist 62(5):711–720
Gomez R, Mclaren S (2017) The center for epidemiologic studies depression scale: invariance across heterosexual men, heterosexual women, gay men, and lesbians. Psychol Assess 29(4): 361–371
Griswold ME, Talluri R, Zhu X et al (2021) Reflection on modern methods: shared-parameter models for longitudinal studies with missing data. Int J Epidemiol 50(4):1384–1393
Halpern S, Chowdhury M, Bayes B et al (2021) Effectiveness and ethics of incentives for research participation: 2 randomized clinical trials. JAMA Intern Med 181(11):1479–1488
Hayes-Larson E, Mobley TM, Mungas D et al (2022) Accounting for lack of representation in dementia research: generalizing Khandle study findings on the prevalence of cognitive impairment to the California older population. Alzheimers Dement 18(11):2209–2217
Henrich J, Heine SJ, Norenzayan A (2010) The Weirdest people in the world? Behav Brain Sci 33(2–3):61–83. Discussion 83–135
Jack CR Jr, Knopman DS, Jagust WJ et al (2013) Tracking pathophysiological processes in Alzheimer's disease: an updated hypothetical model of dynamic biomarkers. Lancet Neurol 12(2):207–216
Jones RN (2019) Differential item functioning and its relevance to epidemiology. Curr Epidemiol Rep 6:174–183
Jorgensen DR, Shaaban CE, Wiley CA, Gianaros PJ, Mettenburg J, Rosano C (2018) A population neuroscience approach to the study of cerebral small vessel disease in midlife and late life: an invited review. Am J Physiol Heart Circ Physiol 314(6):H1117–H1136
Kato T (2021) Measurement invariance in The center for epidemiologic studies-depression (Ces-D) scale among English-speaking whites and Asians. Int J Environ Res Public Health 18(10)
Keyes KM, Westreich D (2019) UK Biobank, Big Data, and the consequences of non-representativeness. Lancet (London, England) 393(10178):1297
Lee M, Whitsel E, Avery C et al (2022) Variation in population attributable fraction of dementia associated with potentially modifiable risk factors by race and ethnicity in the us. JAMA Netw Open 5(7):E2219672
Lim AC, Barnes L, Weissberger GH et al (2023) Quantification of race/ethnicity representation in Alzheimer's disease neuroimaging research in the USA: a systematic review. Commun Med (Lond) 3(1):101
Lima-Costa MF, De Andrade FB, De Souza PRB Jr et al (2018) The Brazilian longitudinal study of aging (Elsi-Brazil): objectives and design. Am J Epidemiol 187(7):1345–1353
Livingston G, Sommerlad A, Orgeta V et al (2017) Dementia prevention, intervention, and care. Lancet (London, England) 390(10113):2673–2734

Livingston G, Huntley J, Sommerlad A et al (2020) Dementia prevention, intervention, and care: 2020 report of the lancet commission. Lancet 396(10248):413–446

Macintosh RC, Strickland OJ (2010) Differential item responses on Ces-D inventory: a comparison of elderly hispanics and non-hispanic whites in the United States and item usage by elderly hispanics across time. Aging Ment Health 14(5):556–564

Mcclure K, Bell KA, Jacobucci R, Ammerman BA (2023) Measurement invariance and response consistency of single-item assessments for suicidal thoughts and behaviors. Psychol Assess 35(10):830–841

Mielke MM, Aggarwal NT, Vila-Castelar C et al (2022) Consideration of sex and gender in Alzheimer's disease and related disorders from a global perspective. Alzheimers Dement 18(12):2707–2724

Mooldijk SS, Licher S, Wolters FJ (2021) Characterizing demographic, racial, and geographic diversity in dementia research: a systematic review. JAMA Neurol 78(10):1255–1261

Mukadam N, Sommerlad A, Huntley J, Livingston G (2019) Population attributable fractions for risk factors for dementia in low-income and middle-income countries: an analysis using cross-sectional survey data. Lancet Glob Health 7(5):E596–E603

National Institute on Aging. Nia's Commitment to Inclusion https://www.Nia.Nih.Gov/Research/Blog/2023/11/Nias-Commitment-Inclusion. Published 2023. Accessed 10 Dec 2023

Ossenkoppele R, Smith R, Ohlsson T et al (2019) Associations between Tau, Aβ, and cortical thickness with cognition in Alzheimer disease. Neurology 92(6):E601–E612

Paus T (2010) Population neuroscience: why and how. Hum Brain Mapp 31(6):891–903

Rosano C (2022) A training program for researchers in population neuroimaging: early experiences. Front Neuroimaging 1:896350

Schumacher-Schuh AF, Bieger A, Okunoye O et al (2022) Underrepresented populations in Parkinson's genetics research: current landscape and future directions. Mov Disord 37(8): 1593–1604

Schwartz AL, Alsan M, Morris AA, Halpern SD (2023) Why diverse clinical trial participation matters. N Engl J Med 388(14):1252–1254

US Department of Health, Education, and Welfare (1979) The Belmont report: ethical principles and guidelines for the protection of human subjects of research. National Institute of Health; Bethesda.

Van Lieshout RJ, Cleverley K, Jenkins JM, Georgiades K (2011) Assessing the measurement invariance of the center for epidemiologic studies depression scale across immigrant and non-immigrant women in the postpartum period. Arch Womens Ment Health 14(5):413–423

Wilkins CH, Mapes BM, Jerome RN, Villalta-Gil V, Pulley JM, Harris PA (2019) Understanding what information is valued by research participants, and why. Health Aff 38(3):399–407

World Bank. World Bank country and lending groups. https://datahelpdesk.worldbank.org/knowledgebase/articles/906519-world-bank-country-and-lending-groups. Accessed 17 Feb 2024

Sex and Gender in Population Neuroscience

Daniel E. Vosberg

Contents

Abstract To understand psychiatric and neurological disorders and the structural and functional properties of the human brain, it is essential to consider the roles of sex and gender. In this chapter, I first define sex and gender and describe studies of sex differences in non-human animals. In humans, I describe the sex differences in behavioral and clinical phenotypes and neuroimaging-derived phenotypes, including whole-brain measures, regional subcortical and cortical measures, and structural and functional connectivity. Although structural whole-brain sex differences are large, regional effects (adjusting for whole-brain volumes) are typically much smaller and often fail to replicate. Nevertheless, while an individual neuroimaging feature may

D. E. Vosberg (✉)
Centre Hospitalier Universitaire Sainte-Justine, University of Montreal, Montreal, QC, Canada

Department of Neuroscience, Faculty of Medicine, University of Montreal, Montreal, QC, Canada
e-mail: daniel.vosberg@umontreal.ca

Curr Topics Behav Neurosci (2024) 68: 87–106
https://doi.org/10.1007/7854_2024_468
Published Online: 21 March 2024

have a small effect size, aggregating them in a "maleness/femaleness" score or machine learning multivariate paradigm may prove to be predictive and informative of sex- and gender-related traits. Finally, I conclude by summarizing emerging investigations of gender norms and gender identity and provide methodological recommendations to incorporate sex and gender in population neuroscience research.

Keywords Sex · Gender · Population neuroscience · Neuroimaging · Psychiatric disorders · "Maleness/femaleness" · Gender norms · Gender identity

1 Introduction

The consideration of sex and gender is critical to the advancement of population neuroscience, an emerging field combining genetics, epidemiology, and neuroscience (Paus 2013, 2010a). Here, I begin with fundamental definitions of sex and gender and introduce the approaches of non-human animal research on the neuroscience of sex differences. Following this, I summarize known sex differences in behaviors, psychiatric disorders, and neurological diseases in humans, as well as the current state of knowledge and methodological considerations of human neuroimaging studies of sex differences. Furthermore, I evaluate studies carried out with structural magnetic resonance imaging (MRI) examining sex differences in measures of global, regional, and connectivity-based neuroanatomy. In addition to considering average sex differences, I consider also sex differences in the population variability, as well as in continuous measures of "maleness/femaleness," the latter capturing interindividual variability within the sexes. Next, I review neuroimaging investigations of gender norms and gender identity. Finally, based on the synthesis of the literature, I provide recommendations for appropriate research practices concerning sex and gender in population neuroscience.

2 Biological Sex

Sex is nearly universal among multicellular eukaryotic organisms (Goodenough and Heitman 2014). While the evolutionary origins and adaptive significance of sex remain elusive, biologists have speculated that sexual reproduction is adaptive due to its ability to increase genetic variation in a population, producing more varied phenotypes upon which natural selection may act (Barton 2009). Others have argued, however, that this hypothesis is oversimplistic (Barton 2009) and, in some instances, sexual reproduction actually decreases major (often deleterious) genetic variation, compared with asexual reproduction (Gorelick and Heng 2011). Across species, females and males have been defined by the size and morphology of the gametes, such that larger eggs are female while smaller sperm are male (anisogamy)

(Monro and Marshall 2016). In prokaryotic species that reproduce sexually, the gametes are indistinguishable (isogamy), and thus, there are no females or males (Lehtonen et al. 2016). Moreover, although sex is binary and genetically determined in most sexually reproducing species, hermaphroditism (i.e., organisms that produce male and female gametes) is observed in some invertebrate species, and there are species of fish who transition between sexes under specific environmental conditions (Avise and Mank 2009; Pla et al. 2021). It is also vital to recognize intersex individuals, among whom there is a divergence of the typical congruence in chromosomes, gonads, genitalia, or hormones.

Sex chromosomes have evolved from autosomal chromosomes multiple times across evolutionary history, resulting in, for instance, heterogametic males (XY) and homogametic females (XX) among mammals, and heterogametic females (ZW) and homogametic males (ZZ) among birds (Fridolfsson et al. 1998). In humans, the X chromosome is much larger (~150 million base pairs, ~800 genes) than the Y chromosome (~23 million base pairs, 80 genes). It is estimated that the X and Y chromosomes of our ancestors were the same size ~160 million years ago, but the Y chromosome has gradually degraded. While two X chromosomes recombine with each other, the Y chromosome does not undergo recombination with the X chromosome, except in the few homologous genetic loci between the chromosomes, resulting in the accumulation of deleterious mutations and genetic loss (Sun and Heitman 2012). Although this lack of recombination has profoundly degraded the Y chromosome, it serves to preserve the sex-determining gene, *SRY*.

The presence of *SRY* initiates the differentiation of gonads into testes (Berta et al. 1990) whereas its absence produces ovary differentiation (Kashimada and Koopman 2010). Consequently, levels of sex hormone differ greatly between the sexes, whereby testes in males secrete testosterone while ovaries in females secrete estradiol and progesterone (Paus 2010b). Whereas testosterone has both organizational (irreversible) and activational (reversible) effects, ovarian hormones primarily exert activational effects, beginning in puberty (Paus 2010b). Phenotypically, sexual dimorphisms may arise due to sexual selection, first proposed by Charles Darwin, which selects for traits that provide a reproductive advantage by enhancing intrasexual or intersexual competition (Székely et al. 2000; Servedio and Boughman 2017). Mechanistically, sexual dimorphisms were historically attributed uniquely to differing sex hormone levels, but it is now recognized that sexual dimorphisms may also arise from non-hormonal effects of sex chromosomal genes (Arnold 2017). For instance, in females only, X-inactivation silences one of the two X chromosomes in each cell, producing a mosaic of cells with either a paternal or maternal X chromosome (Migeon 2007a). In contrast, cells in males all contain a maternal X chromosome only. Furthermore, some genes escape X-inactivation, doubling their dosage in females; these genes may be overexpressed in females relative to males or, if the X gene has a homologous Y gene, parallel the expression in males (Berletch et al. 2011). While there are no sex differences in autosomal DNA sequences, there are robust sex differences in the expression, epigenetics, and regulation of autosomal genes (McCarthy et al. 2009; Trabzuni et al. 2013).

3 Gender

The Canadian Institutes of Health Research (CIHR) defines gender as "the socially constructed roles, behaviors, expressions and identities of girls, women, boys, men, and gender diverse people" (https://cihr-irsc.gc.ca/e/48642.html). Similarly, the National Institutes of Health (NIH) defines gender as "a multidimensional construct that encompasses gender identity and expression, as well as social and cultural expectations about status, characteristics, and behavior as they are associated with certain sex traits" (Bates et al. 2022). Whereas sex applies to all sexually reproducing animals, gender is only relevant for humans (Bates et al. 2022). While biological sex and gender identity are evidently correlated, they are distinct, given that 0.33% of Canadians identify as transgender or non-binary, according to the first country-wide census data including gender diverse people (Statistics Canada 2022). Moreover, younger generations self-identify as gender diverse at higher rates than older generations (Gen Z: 0.79%; Millennials: 0.51%; Generation X: 0.19%; Baby Boomers: 0.15%; Interwar and Greatest Generations: 0.12%) (Statistics Canada 2022). Nevertheless, some authors have proposed that sex and gender are inextricably linked, preferring the term "sex/gender," given that supposedly "hard-wired" sex differences in the brain may partially or fully be explained by environmental, rather than biological, factors (Jordan-Young and Rumiati 2012). On the other hand, some behaviors and characteristics that may fit the definition of gender (e.g., higher average empathy in females) may have a genetic component, as suggested by sex differences in the genetic architecture of empathy (Warrier et al. 2018). Additionally, studies of the heritability of gender identity, gender dysphoria, and related phenomena reported that these traits were mostly in the range of 30–60% heritable, with little evidence of a shared environmental factor, and some evidence of a non-shared environmental factor (Coolidge et al. 2002; Polderman et al. 2018). Despite these limitations, for clarity in this chapter, I use "sex" whenever referring to males and females and, where applicable, specify whether I am referring to "gender identity" or "gender norms." As addressed in the recommendations below, it is ideal for researchers to ask participants for both their biological sex at birth and gender identity, as well as measuring their genetic sex, when possible.

4 The Neuroscience of Sex Differences in Non-Human Animals

Rodent studies have determined that prenatal testosterone secreted by the testes is aromatized into estradiol and is responsible for the masculinization of the brain, as assessed by the development of male reproductive anatomy and behaviors (Schwarz and McCarthy 2008). In rodents, the appropriately named sexually dimorphic nucleus of the preoptic area is 3–5 times larger in males, compared with females, and is necessary for male sexual behaviors (Schwarz and McCarthy 2008; Gorski et al. 1980). A structure necessary for female sexual behaviors, the ventromedial nucleus of the hypothalamus, is larger in females (Matsumoto and Arai 1983). There are also sexually dimorphic structures in the brains of song birds, related to the preference of female song birds for male song birds capable of producing more complex songs (Garamszegi et al. 2005). In comparison, such clear regional neuroanatomical dimorphisms are exceedingly rare in humans, as described below.

In order to investigate whether sex differences in the brain are attributable to chromosomes (i.e., XX vs XY) or gonadal hormone secretion (i.e., ovaries vs. testes), the "four core genotypes" (FCG) mouse model may be considered (Arnold 2014). The four mouse models are as follows: typical XX females, typical XY males, XY females with ovaries (*SRY* inactivation), XX males with testes (*SRY* addition). These studies have examined many facets of the nervous system, ranging from defects in the neural tube closure to various measures of brain morphology assessed with magnetic resonance imaging (MRI) in mice (Arnold 2020). In the latter studies, among 27 regions exhibiting volumetric sex differences, 16 were attributable to the gonads while 11 were attributable to the sex chromosomes (Arnold 2020).

Historically, sex differences in the brains of non-human animals were presumed to be minimal, and concerns over the variability introduced by the estrus cycles of females led to the exclusion of females in research studies. It is now clear that one cannot extrapolate findings from males to those in females. While preclinical discoveries in males have accumulated over decades, there is a profound gap in knowledge regarding the neurobiology of females (Beery and Zucker 2011). As a probable consequence of the exclusion of females in preclinical research, side effects in clinical trials are reported approximately twice as frequently by females, compared with males (Zucker and Prendergast 2020). While females should certainly not be excluded from preclinical or clinical research, some investigators advocate for the consideration of estrus cycles in their studies, increasing inclusivity while enhancing the validity of their findings (Rocks et al. 2022).

5 Sex Differences in Personality Traits, Psychiatric Disorders, and Neurological Diseases

There are numerous sex differences across behavioral phenotypes in humans. Meta-analyses have identified a male advantage for mental rotation (Maeda and Yoon 2013; Lauer et al. 2019) and a female advantage for verbal working memory (Voyer et al. 2021). Moreover, in a large study of the Big Five Personality Traits in 320,128 participants aged 19–69 years old, it was reported that females scored higher in agreeableness (Cohen's $d = 0.58$), neuroticism ($d = 0.40$), openness ($d = 0.19$), and conscientiousness ($d = 0.12$), with a negligible effect size for extraversion ($d = 0.01$) (Kajonius and Johnson 2018). Among the extraversion facets, males scored higher in excitement-seeking ($d = 0.29$) and assertiveness ($d = 0.16$), while females scored higher in activity ($d = 0.25$), cheerfulness ($d = 0.11$), friendliness ($d = 0.07$), and gregariousness ($d = 0.05$) (Kajonius and Johnson 2018).

Moreover, there are robust sex differences across psychiatric disorders. For autism spectrum disorder (ASD), the male-to-female prevalence ratio is 3:1 according to a meta-analysis ($n = 13.8$ million), after adjusting for diagnostic sex bias (Loomes et al. 2017). Furthermore, in a nation-wide cohort study of Denmark ($n = 1.3$ million), diagnoses before 18 years of age were more common among males for oppositional defiant/conduct disorder (2.8:1), ASD (2.4:1), and attention deficit hyperactivity disorder (ADHD; 1.9:1), while diagnoses were more common among females for eating disorders (6.4:1), personality disorders (3.5:1), depression (2.6:1), anxiety disorders (1.7:1), schizophrenia-spectrum disorders (1.6:1), and obsessive compulsive disorder (OCD; 1.5:1) (Dalsgaard et al. 2020).

Several theoretical frameworks have been proposed to explain these sex differences. First, there are reports of higher internalizing traits in females and higher externalizing traits in males, according to confirmatory factor analyses (Caspi et al. 2015). It has been posited that higher childhood-onset externalizing disorders in males are explained by greater prenatal testosterone exposure, modulating the dopamine system and related behavioral traits (Martel 2013). Moreover, this theory posits that higher adolescent-emerging internalizing disorders in females are driven by pubertal estradiol, modulating the serotonin system and related traits (e.g., neuroticism, rumination) (Martel 2013). Second, using self-report measures, ASD is characterized by lower empathizing and higher systemizing, compared with controls (discovery $n = 671{,}606$; validation $n = 14{,}354$), representing "extreme male" scores on these two traits among autistic individuals of both sexes (Greenberg et al. 2018). Notably, empathizing and systemizing account for 19 times more of the variance in autistic traits than demographic traits, including biological sex (Greenberg et al. 2018). Third, based on sex chromosomal aneuploidy studies, the overall chromosome number and the Y chromosome in particular are associated with higher ASD risk (Green et al. 2019). In comparison, preliminary studies indicate that additional X chromosomes are associated with depression and anxiety, while the Y chromosome number is not (Green et al. 2019). The mechanisms by which varying chromosomal numbers modulate the risk of developing psychiatric disorders are unclear but may involve differences in the levels of sex steroid hormones at various developmental stages. Importantly, socio-cultural gender norms also moderate sex differences in mental health, as described below.

Regarding neurological diseases, studies have estimated that the prevalence of Parkinson's disease is twice as high among males, compared with females, and this may be explained (partially) by sex differences in the nigrostriatal dopamine system (Gillies et al. 2014). Moreover, the prevalence of Alzheimer's disease is twice as high among females, compared with males, and following diagnosis, cognitive decline and brain atrophy are faster among females (Ferretti et al. 2018). Similarly, multiple sclerosis is twice as prevalent among females, compared with males (Kingwell et al. 2013); autoimmune diseases are broadly more common among females, perhaps relating to sex differences in antibody composition and levels (Kronzer et al. 2020). Finally, migraine is 2–3 times more prevalent among females, relative to males, with females exhibiting greater attack duration and severity, potentially related to levels of female sex hormones (Vetvik and MacGregor 2017).

Despite the apparent sex differences in the prevalence and expression of personality traits, psychiatric disorders, and neurological diseases, investigators have noted that these behavioral differences do not necessarily correspond to a particular sex difference observed at the level of the brain (Rippon et al. 2014). Thus, upon discovering specific sex differences in the brain, in the following sections, researchers must be cautious before considering these as causal explanations for sex differences in particular behavioral phenotypes, even if they are correlated.

6 Sex Differences in Structural MRI Measures

6.1 Global Structural Effects

The most robust neuroanatomical sex difference in humans is that, on average, the brains of males are 8–14% larger than those of females, typically operationalized as total brain volume (TBV) or intracranial volume (ICV) (Paus 2010b; Ruigrok et al. 2014). This difference exists across the lifespan, from birth (8% larger in males, $d = 0.65$) to pre-menopausal adulthood (14% larger in males, $d = 1.6$) (Paus 2010b). After adjusting for body size (e.g., height squared), the sex difference in total brain volume drops but remains large (e.g., $d = 1.66$ to $d = 0.76$ (Burgaleta et al. 2012)). Males have greater volumes of both gray matter (GM) and white matter (WM), but while the GM differences are relatively stable across the lifespan, the differences in WM are greatest in adolescence and adulthood (~10–22%), relative to birth and childhood (~6–12%) (Paus 2010b). This apparent developmental sex difference may be explained by the male-specific increase in WM volume during adolescence, mediated by testosterone (Paus 2010b; Paus and Toro 2009; Perrin et al. 2008). Moreover, although there is a weak, positive correlation between brain volume and general cognitive function ($r = 0.24$, $r^2 = 0.06$, $n = \sim 8{,}000$), the sex differences in brain volume are unrelated to cognitive function; studies have not identified a significant mean difference in general cognitive function between the sexes (Burgaleta et al. 2012; Pietschnig et al. 2015). While surface area of the cerebral cortex is also larger among males, mean cortical thickness is reported to be greater among females in some reports (Williams et al. 2021a; Ritchie et al. 2018), but not all (Vijayakumar et al. 2016). As a potential explanation for this, our group has identified that, during adolescence, cortical thinning (i.e., the negative relationship between cortical thickness and age) is steeper among males ($r^2 = 0.28$), compared with females ($r^2 = 0.1$), and related to levels of testosterone (Paus et al. 2010, 2017). Moreover, in a study of 17,075 (adult) participants from the Enhancing NeuroImaging Genetics through Meta-Analysis (ENIGMA) Consortium, cortical thinning was steeper among males ($r = -0.39$ to -0.38), compared with females ($r = -0.27$), aged 30–59 years old (Frangou et al. 2022).

6.2 Regional Structural Effects

In an extensive survey of neuroimaging studies of sex differences, the authors asserted that besides the large sex difference in global brain volume, the alleged sex differences in other neuroimaging measures (e.g., white matter/gray matter ratio, structural connectivity, regional cortical and subcortical volumes, and functional magnetic resonance imaging) are "trivial" and "population-specific" (Eliot et al. 2021). Across these studies, following adjustments for global brain volume, the findings frequently failed to replicate, varied based on the adjustment method for global brain size, and among the findings that did replicate, the effect sizes were small (Eliot et al. 2021). In their summary of 33 highly-cited neuroimaging studies of sex differences in subcortical volumes (n range = 20–5,216), the findings were largely not statistically significant or the reported directions of effects were inconsistent across studies (Eliot et al. 2021). While the reports of larger putamen and amygdala volumes in males did replicate, the effect sizes were small, representing volumetric differences of 1–3% between the sexes (Eliot et al. 2021). Moreover, in their survey of 25 studies of sex differences in cortical regions (n range = 35–5,216), while there was moderate agreement among the higher-powered studies that used Freesurfer, the findings differed with those that used other MRI processing approaches, namely voxel-based morphometry (VBM) or Diffeomorphic Anatomical Registration Through Exponentiated Lie Algebra (DARTEL) (Eliot et al. 2021).

The conclusion that the regional effects are "trivial" has been challenged, however, by the authors of a study that used a much larger sample size, and adjusted not only for the sex difference in global brain volume, but also brain allometry, defined as the non-linear and non-proportional relationship between regional and global brain measures (Williams et al. 2021a,b; de Jong et al. 2017). In other words, as the volume of the whole brain increases, the volume of a particular region does not increase in a proportional or linear manner. In this UK Biobank neuroimaging study ($n = 40{,}028$, ages 40–69), the authors evaluated sex differences in structural MRI measures, using Freesurfer and FSL pipelines to extract 629 imaging-derived phenotypes, including regional cortical measures (volumes, surface areas, thicknesses) and subcortical volumes and segmentations (Williams et al. 2021a). Across these phenotypes, and adjusting for global brain size and allometry, 37% were larger in males, 30% were larger in females, and the remaining 33% were not statistically significant (Williams et al. 2021a). Specifically, after the adjustments, there were significant sex differences in cortical regions for 66% of volumes (42% M > F; 24% F > M), 57% of surface areas (34% M > F; 23% F > M), and 66% of thicknesses (34% M > F; 32% F > M). Among the subcortical structures, the hippocampus and left nucleus accumbens were larger in females (standardized β range: 0.06–0.10), while the thalamus, putamen, left amygdala, and left pallidum were larger in males (β range: 0.08–0.18) (Williams et al. 2021a).

Other investigators, also challenging the notion that regional effects are "trivial," have argued that these regional sex differences are particularly meaningful when considered in the aggregate, such that machine learning algorithms using a multivariate approach can predict the sex of a brain with 60–70% accuracy, even after adjusting for global brain size (Hirnstein and Hausmann 2021), citing (Chekroud et al. 2016; Sanchis-Segura et al. 2020). From the population neuroscience perspective, this may be thought of as analogous to the effect sizes between a single nucleotide polymorphism (SNPs) and a phenotype, which almost always explain very little variance in a trait (Manolio et al. 2009). But considering SNPs in the aggregate as polygenic scores, by summing up the number risk or effect alleles, weighted by the respective effect size per SNP, greatly increases the percentage of variance explained in a phenotype, in independent samples (Plomin and von Stumm 2022). Thus, small regional effects should not be dismissed, if they are statistically robust and replicable. As described in the section below, "Individual variability: 'maleness/femaleness' and mosaics," our group has proposed an approach to the study of sex differences which, rather than focusing on individual traits, sums up values on quantitative traits, weighted by their respective sex difference effect sizes, generating "sex-scores" (Vosberg et al. 2021).

7 Sex Differences in Structural and Functional Connectivity

While early studies indicated that the volume of the corpus callosum, the largest white-matter structure interconnecting the hemispheres, differed between the sexes, the preponderance of the evidence indicates that this is not the case, following an appropriate adjustment for whole-brain volume (Luders and Kurth 2020; Bishop and Wahlsten 1997). Another claim made in the literature was that male brains exhibit greater hemispheric asymmetry (often assessed behaviorally with sensory tasks) and that this may explain the sex differences in cognitive traits (e.g., spatial and verbal functions) (Hirnstein et al. 2019). A systematic review indicates, however, that these differences in hemispheric asymmetry are robust but small and highly unlikely to account for the sex differences in cognitive domains (Hirnstein et al. 2019). Following these earlier investigations, there has been a focus on the potential sex differences in structural connectivity and functional connectivity, across the brain.

Structural connectivity is assessed in vivo using diffusion tensor imaging (DTI), which measures the diffusion of water molecules to estimate white-matter tracts; an individual's whole-brain white- matter network constitutes their structural connectome (Sporns 2013). Functional connectivity, on the other hand, is assessed based on the degree to which blood oxygenation level dependent (BOLD) responses are correlated – over the acquisition time – between disparate regions, typically evaluated using "resting state" functional magnetic resonance imaging (rs-fMRI), during which the participant is not instructed to perform any particular task (Fingelkurts et al. 2005). As a caveat for the studies described below, a correlation between a neuroimaging and behavioral phenotype does not necessarily indicate a causal relationship.

In a study of children and adolescents ($n = 949$; 8 to 22 years of age), greater interhemispheric structural connectivity was observed in females while greater intrahemispheric structural connectivity was observed in males (Ingalhalikar et al. 2014). In a follow-up study of the same sample, the authors considered subnetworks, defined as a network composed of nodes (regions of interest) and edges (white-matter tracts), associated with a particular function/behavior (Tunç et al. 2016). Here, the investigators reported greater structural connectivity in memory, attention, and social motivation subnetworks in females, and greater in executive, sensory, and motor subnetworks in males (Tunç et al. 2016). Following this, the authors trained a pattern classifier to predict participant sex according to both their structural connectivity and behavioral phenotypes (cognitive and psychiatric measures), with greater accuracy observed with increasing age; the structural connectivity and behavioral measures were significantly correlated, albeit weakly (Tunç et al. 2016). Most recently, in a UK Biobank study ($n = 28{,}821$) evaluating associations between brain connectivity and cognition, it was reported that the sex differences in verbal memory (higher in females; $d = 0.36$) and processing speed (faster in females; $d = 0.17$) were mediated by "global efficiency" (higher in males; $d = 0.30$), defined as a structural network with shorter paths between nodes (Yang et al. 2023).

There is also evidence of sex differences in functional connectivity, beginning in prenatal development, as demonstrated by a study of 95 fetuses (19 to 40 gestational weeks) in which the authors used machine learning to classify sex based on functional connectivity with an accuracy of 73% (Cook et al. 2023). In an rs-fMRI study of youths and young adults ($n = 674$; 9–22 years), machine learning was used to classify participants' sex with 71% accuracy based on their functional-connectivity profile, which was greater than the accuracy based on their sex-related cognitive profile (63%) (Satterthwaite et al. 2015). Notably, the degree of "maleness/femaleness" of the cognitive and functional-connectivity profiles were correlated with one another (Satterthwaite et al. 2015). In middle-aged adults, a UK Biobank study ($n = 5{,}216$) reported that females exhibited greater functional connectivity in the default mode network, while males exhibited greater functional connectivity in motor, unimodal, and prefrontal cortical regions (Ritchie et al. 2018). In a more recent UK Biobank study ($n = 37{,}543$), it was reported that while functional connectivity differed between the sexes for younger middle-aged participants (lower average connectivity and greater negative connections in males), this sex difference diminished with increasing age, in this cross-sectional dataset (Mijalkov et al. 2023).

8 Sex Differences in Variability

While most neuroscience and biomedical studies focus on average sex differences, sex differences in variability are of increasing interest. For instance, an ENIGMA (Enhancing NeuroImaging Genetics Through Meta-Analysis) Consortium mega-analysis ($n = 16{,}683$; 1–90 years of age) identified greater variability among males, compared with females, for the volumes of seven subcortical structures (i.e., amygdala, hippocampus, accumbens, caudate, putamen, pallidum, and thalamus), the surface areas of all cortical regions, and the thicknesses of 60% of cortical regions (68 Desikan-Killiany ROIs), effects which were stable across the lifespan, using a cross-sectional design (Wierenga et al. 2022). In a study of three cohorts (the Philadelphia Neurodevelopmental Cohort [PNC], the Human Connectome Project [HCP], and the Open Access Series of Imaging Studies [OASIS-3]; $n = 3{,}069$; 8–95 years of age), the authors replicated findings regarding the subcortical volumetric and cortical surface area, but did not identify differences in variability between the sexes for cortical thickness for most regions (Forde et al. 2020). Similar patterns of findings were observed in these studies after adjusting for total brain volume, albeit with smaller effect sizes (Wierenga et al. 2022; Forde et al. 2020). In accordance with this, in our phenotypic "sex-score" study, ranging from "maleness" to "femaleness" based on 48 quantitative traits of the brain and body, we identified significantly greater variability among males (Vosberg et al. 2021). Several hypotheses have been put forth regarding the observation of greater male variability for many, but not all, traits (Lehre et al. 2009). Firstly, organisms that are heterogametic (e.g., XY in male mammals, ZW in female birds) have greater variability than those that are homogametic (e.g., XX in female mammals, ZZ in male birds), according to a meta-analysis spanning 292 studies (Reinhold and Engqvist 2013). A non-mutually exclusive hypothesis stems from the phenomenon of X-chromosomal mosaicism (Migeon 2007b), introduced above. Since females contain a mosaic of X-chromosome variants across cells, the effects of differing variants across cells may cancel each other out and result in more average phenotypes (Lehre et al. 2009). In contrast, since males possess the same X-chromosomal variants across cells, there may be a summation of effects across the body, generating more extreme phenotypes (Lehre et al. 2009). It must be emphasized, however, that this "greater male variability" hypothesis is by no means universal and, when observed, is not necessarily explained by genetic or biological factors. For instance, a meta-analysis of 204 studies found no evidence of greater male variability in behavioral traits in non-human animals (e.g., aggression, exploration, and sociality) (Harrison et al. 2022).

8.1 Individual Variability: "Maleness/Femaleness" and Mosaics

While there are undoubtedly average sex differences across traits of the body and brain, the distributions for a given trait for males and females almost always overlap, even for the most robust and replicable sex differences with large effect sizes. Thus, as discussed above, we developed a polyphenotypic quantitative sex-score that captures interindividual variability in maleness/femaleness in the brain and body, and was found to correlate with testosterone in males, as well as personality and psychiatric traits within males and females, separately (Vosberg et al. 2021). Additionally, by conducting genome-wide association studies (GWAS), we identified that the genes underlying sex-scores in the UK Biobank (females: $n = 161{,}906$; males: $n = 141{,}980$) were enriched across multiple brain tissues and the cervix, particularly in females (Vosberg et al. 2023). Finally, we identified that the genetics underlying testosterone are highly distinct between the sexes and that females with the autosomal genetics of higher testosterone identified in males had a more male-typical pattern of associations with cardiometabolic traits; males with the genetics of testosterone identified in females had a more female-typical pattern of associations with cardiometabolic traits (Vosberg et al. 2022). Beyond simply disputing that the human brain is sexually dimorphic, Joel et al. have argued that the human brain is a mosaic, given that most individuals do not have a consistently "male brain" or consistently "female brain," based on sex differences across regions (Joel et al. 2015, 2018). While they have argued against a single maleness/femaleness continuum, we believe that individuals with a mixture of female-typical and male-typical neuroanatomical traits would simply fall toward the middle of this continuum (Vosberg et al. 2021), while fully acknowledging its limitations and inability to capture all interindividual variability. The utility of this approach has been reviewed and advocated for elsewhere (Rauch and Eliot 2022), and while in its infancy, assessing maleness/femaleness or masculinity/femininity, using machine learning or simpler score-based approaches, may uncover individual variability within the sexes and genders with potential applications for sex-biased disorders.

9 Gender Norms: Mental Health and Neuroimaging

Gender norms vary across time and space and have the potential to modify disparities in mental health and neuroanatomy on a timescale that is much more rapid than the gradual evolution of genetic variations related to sex. For instance, while there are gender gaps in depression (higher in women) and substance-use disorders (higher in men), a cross-national study ($n = 72{,}933$) indicates that these gaps are narrowing, concomitant with reductions in "female gender role traditionality," which can be operationalized by aggregating measures of female employment, education, marital timing, and birth control (McHugh et al. 2018; Seedat et al. 2009). Moreover, two US-based studies ($n = 7{,}789$; $n = 34{,}653$) identified that women living in states with greater gender equality, captured by reproductive rights, financial independence, income, and political participation, reported fewer symptoms of depression and post-traumatic stress disorder (Chen et al. 2005; McLaughlin et al. 2011). Finally, in a study across European countries ($n = 116{,}783$), the largest sex differences for depression were found among older individuals in societies with lower gender equality; young men and women in societies with higher gender equality had the most similar rates of depression (Bracke et al. 2020). In alignment with the mental health studies, in an international neuroimaging study spanning 29 countries ($n = 7{,}876$), differences between the sexes in patterns of cortical thickness and surface area varied as a function of the degree of societal gender equality (Zugman et al. 2023).

10 Gender Identity

Given the relatively small population of individuals who identity as transgender or non-binary, and the paucity of studies that have asked participants to report their gender identities, little is known about the neuroanatomical basis of gender identity. Nevertheless, in a mega-analysis study from the ENIGMA Transgender Persons Working Group ($n = 172$ transgender women [assigned male at birth], $n = 214$ transgender men [assigned female at birth], $n = 196$ cisgender women, and $n = 221$ cisgender men), investigators assessed sex and gender differences in cortical measures (i.e., volume, thickness, surface area) and subcortical volumes (Mueller et al. 2021). For cortical and subcortical volumes, the findings were highly heterogeneous across regions, with the emergence of five distinct patterns. To illustrate the complexity of the patterns of effects, across different regions, the patterns were as follows: (1) cismen > ciswomen & transwomen, (2) cismen > ciswomen & transwomen & transmen, (3) cismen > transwomen > ciswomen and transmen, (4) cismen > ciswomen > transwomen, (5) cismen and transwomen different from ciswomen and transmen (Mueller et al. 2021). The cortical surface area findings were similarly complex, grouped into four patterns, and no significant group differences were identified for cortical thickness (Mueller et al. 2021). Therefore, the neuroanatomy of gender identity appears far more complex than a simple similarity to one's gender identity compared with one's sex assigned at birth. Moreover, there is some evidence that machine learning algorithms using neuroimaging data can classify the sex (assigned at birth) of cisgender, compared with transgender, individuals with greater accuracy (Baldinger-Melich et al. 2020), but other studies indicate similar prediction accuracy of sex for both cisgender and transgender participants (Wiersch et al. 2023). Ultimately, the study sample sizes of transgender or non-binary individuals are very small and our understanding of the neuroanatomy of gender identity is poor, so we must exert great caution and provide appropriate justification for studies of the etiology of gender dysphoria, which should promote inclusivity and avoid stigmatization (Levin et al. 2023).

11 Recommendations

Unfortunately, poor research practices and poor science communication have led to the perpetuation of sexist stereotypes and the exaggerations of phenotypic sex differences in the brain (Jordan-Young and Rumiati 2012). In order to combat this misinformation and improve our understanding of sex- and gender-related factors related to the human brain, improving our research practices is imperative. Firstly, increasing sample sizes is essential, given the replication crisis in biomedical and psychological research (Open Science Collaboration 2015) and following the example of genome-wide association studies (e.g., $n = 146$ in 2005 (Klein et al. 2005); $n = 5.4$ million in 2022 (Yengo et al. 2022)), leading to robust, replicable findings.

Moreover, while the sex difference in global brain volume is large, most of the regional effects with global adjustments are markedly smaller, requiring greater power with larger sample sizes to detect. With the promotion of open science and the formation of large cohorts and international consortia (e.g., Generation R, Avon Longitudinal Study of Parents and Children [ALSPAC], Northern Finland Birth Cohort [NFBC], Adolescent Brain Cognitive Development [ABCD], Saguenay Youth Study [SYS], Imaging Genetics for Mental Disorders [IMAGEN], UK Biobank, Enhancing NeuroImaging Genetics through Meta-Analysis [ENIGMA]), neuroimaging sample sizes will continue to grow at a rapid pace. In the recruitment and assessment of participants, investigators should ask participants to report their sex assigned at birth, their gender identities, and measure their genetic sex when it is feasible.

It is vital also to report the method of computing the global brain measure, how adjustment for the global brain measure was done, and how brain allometry was considered (DeCasien 2022). Whenever possible, investigators should perform analyses in males and females separately, as well as evaluate the interaction terms between sex and the independent variables of interest, before drawing conclusions regarding sex differences (Peters and Woodward 2023). As with all scientific studies, it is necessary to adjust for multiple p-values and report which method was applied in reports and any custom code should be made accessible to other researchers (Poldrack et al. 2017).

In addition to the consideration of sex differences in averages and variabilities, researchers should consider potential environmental mediators of neuroimaging-derived phenotypes. As discussed above, greater attention should be placed on the aggregate effects of phenotypes with known sex differences, which may have cumulative effects across the lifespan, and inform a precision-medicine approach to the treatments of sex-biased or gender-biased psychiatric and neurological disorders. Finally, researchers should be careful to ensure accurate and clear science communication in their scientific publications and in their public discourse, in order to minimize the risk that research findings are distorted by the media or the public. Altogether, if the population neuroscience community continues to engage ever-improving methodologically rigorous and ethical research practices, vastly more will be learned about the roles of sex and gender in the human brain.

Acknowledgments I would like to thank Drs. Tomáš Paus and Katherine M. Keyes for their helpful feedback on this work.

Competing Interests The author declares that he has no competing interests.

Funding Canadian Institutes of Health Research (CIHR) Postdoctoral Fellowship (MFE – 187903).

References

Arnold AP (2014) Conceptual frameworks and mouse models for studying sex differences in physiology and disease: why compensation changes the game. Exp Neurol 259:2–9
Arnold AP (2017) A general theory of sexual differentiation. J Neurosci Res 95:291–300
Arnold AP (2020) Four Core genotypes and XY* mouse models: update on impact on SABV research. Neurosci Biobehav Rev 119:1–8
Avise JC, Mank JE (2009) Evolutionary perspectives on hermaphroditism in fishes. Sex Dev 3: 152–163
Baldinger-Melich P et al (2020) Sex matters: a multivariate pattern analysis of sex- and gender-related neuroanatomical differences in cis- and transgender individuals using structural magnetic resonance imaging. Cereb Cortex 30:1345–1356
Barton NH (2009) Why sex and recombination? Cold Spring Harb Symp Quant Biol 74:187–195
Bates N, Chin M, Becker T (eds) (2022) Measuring sex, gender identity, and sexual orientation. National Academies Press, Washington, DC. https://doi.org/10.17226/26424
Beery AK, Zucker I (2011) Sex bias in neuroscience and biomedical research. Neurosci Biobehav Rev 35:565–572
Berletch JB, Yang F, Xu J, Carrel L, Disteche CM (2011) Genes that escape from X inactivation. Hum Genet 130:237–245
Berta P et al (1990) Genetic evidence equating SRY and the testis-determining factor. Nature 348: 448–450
Bishop KM, Wahlsten D (1997) Sex differences in the human corpus callosum: myth or reality? Neurosci Biobehav Rev 21:581–601
Bracke P, Delaruelle K, Dereuddre R, Van de Velde S (2020) Depression in women and men, cumulative disadvantage and gender inequality in 29 European countries. Soc Sci Med 267: 113354
Burgaleta M et al (2012) Sex differences in brain volume are related to specific skills, not to general intelligence. Intelligence 40:60–68
Caspi A, Houts RM, Belsky DW, Goldman-mellor SJ (2015) The p factor : one general psychopathology factor in the. Clin Psychol Sci 2:119–137
Chekroud AM, Ward EJ, Rosenberg MD, Holmes AJ (2016) Patterns in the human brain mosaic discriminate males from females. Proc Natl Acad Sci 113:E1968–E1968
Chen Y-Y, Subramanian S, Acevedo-Garcia D, Kawachi I (2005) Women's status and depressive symptoms: a multilevel analysis. Soc Sci Med 60:49–60
Cook KM et al (2023) Robust sex differences in functional brain connectivity are present in utero. Cereb Cortex 33:2441–2454
Coolidge FL, Thede LL, Young SE (2002) The heritability of gender identity disorder in a child and adolescent twin sample. Behav Genet 32:251–257
Dalsgaard S et al (2020) Incidence rates and cumulative incidences of the full Spectrum of diagnosed mental disorders in childhood and adolescence. JAMA Psychiatry 77:155–164
de Jong LW et al (2017) Allometric scaling of brain regions to intra-cranial volume: an epidemiological MRI study. Hum Brain Mapp 38:151–164
DeCasien AR (2022) Sex differences in the human brain: a roadmap for more careful analysis and interpretation of a biological reality. Biol Sex Differ 13:43
Eliot L, Ahmed A, Khan H, Patel J (2021) Dump the "dimorphism": comprehensive synthesis of human brain studies reveals few male-female differences beyond size. Neurosci Biobehav Rev 125:667–697
Ferretti MT et al (2018) Sex differences in Alzheimer disease — the gateway to precision medicine. Nat Rev Neurol 14:457–469
Fingelkurts AA, Fingelkurts AA, Kähkönen S (2005) Functional connectivity in the brain—is it an elusive concept? Neurosci Biobehav Rev 28:827–836

Forde NJ et al (2020) Sex differences in variability of brain structure across the lifespan. Cereb Cortex 30:5420–5430

Frangou S et al (2022) Cortical thickness across the lifespan: data from 17,075 healthy individuals aged 3–90 years. Hum Brain Mapp 43:431–451

Fridolfsson A-K, Cheng H, Copeland NG, Jenkins NA, Liu H-C, Raudsepp T, Woodage T, Chowdhary B, Halverson J, Ellegren H (1998) Evolution of the avian sex chromosomes from an ancestral pair of autosomes. Proc Natl Acad Sci 95(14):8147–8152

Garamszegi LZ, Eens M, Erritzøe J, Møller AP (2005) Sexually size dimorphic brains and song complexity in passerine birds. Behav Ecol 16:335–345

Gillies GE, Pienaar IS, Vohra S, Qamhawi Z (2014) Sex differences in Parkinson's disease. Front Neuroendocrinol 35:370–384

Goodenough U, Heitman J (2014) Origins of eukaryotic sexual reproduction. Cold Spring Harb Perspect Biol 6

Gorelick R, Heng HHQ (2011) Sex reduces genetic variation: a multidisciplinary review. Evolution 65:1088–1098

Gorski RA, Harlan RE, Jacobson CD, Shryne JE, Southam AM (1980) Evidence for the existence of a sexually dimorphic nucleus in the preoptic area of the rat. J Comp Neurol 193:529–539

Green T, Flash S, Reiss AL (2019) Sex differences in psychiatric disorders: what we can learn from sex chromosome aneuploidies. Neuropsychopharmacology 44:9–21

Greenberg DM, Warrier V, Allison C, Baron-Cohen S (2018) Testing the empathizing–systemizing theory of sex differences and the extreme male brain theory of autism in half a million people. Proc Natl Acad Sci USA 115:12152–12157

Harrison LM, Noble DW, Jennions MD (2022) A meta-analysis of sex differences in animal personality: no evidence for the greater male variability hypothesis. Biol Rev 97:679–707

Hirnstein M, Hausmann M (2021) Sex/gender differences in the brain are not trivial—a commentary on Eliot et al. (2021). Neurosci Biobehav Rev 130:408–409

Hirnstein M, Hugdahl K, Hausmann M (2019) Cognitive sex differences and hemispheric asymmetry: a critical review of 40 years of research. Laterality 24:204–252

Ingalhalikar M et al (2014) Sex differences in the structural connectome of the human brain. Proc Natl Acad Sci 111:823–828

Joel D et al (2015) Sex beyond the genitalia: the human brain mosaic. Proc Natl Acad Sci 112: 15468–15473

Joel D et al (2018) Analysis of human brain structure reveals that the brain "types" typical of males are also typical of females, and vice versa. Front Hum Neurosci 12:399

Jordan-Young R, Rumiati RI (2012) Hardwired for sexism? Approaches to sex/gender in neuroscience. Neuroethics 5:305–315

Kajonius PJ, Johnson J (2018) Sex differences in 30 facets of the five factor model of personality in the large public (N = 320,128). Personal Individ Differ 129:126–130

Kashimada K, Koopman P (2010) Sry: the master switch in mammalian sex determination. Development 137:3921–3930

Kingwell E et al (2013) Incidence and prevalence of multiple sclerosis in Europe: a systematic review. BMC Neurol 13:128

Klein RJ et al (2005) Complement factor H polymorphism in age-related macular degeneration. Science 308:385–389

Kronzer VL, Bridges SL, Davis JM (2020) Why women have more autoimmune diseases than men: an evolutionary perspective. Evol Appl 14:629–633

Lauer JE, Yhang E, Lourenco SF (2019) The development of gender differences in spatial reasoning: a meta-analytic review. Psychol Bull 145:537–565

Lehre A-C, Lehre KP, Laake P, Danbolt NC (2009) Greater intrasex phenotype variability in males than in females is a fundamental aspect of the gender differences in humans. Dev Psychobiol 51: 198–206

Lehtonen J, Kokko H, Parker GA (2016) What do isogamous organisms teach us about sex and the two sexes? Philos Trans R Soc B Biol Sci 371:20150532

Levin RN, Erickson-Schroth L, Mak K, Edmiston EK (2023) Biological studies of transgender identity: a critical review. J Gay Lesbian Ment Health 27:254–283
Loomes R, Hull L, Mandy WPL (2017) What is the male-to-female ratio in autism Spectrum disorder? A systematic review and meta-analysis. J Am Acad Child Adolesc Psychiatry 56:466–474
Luders E, Kurth F (2020) Chapter 1 - Structural differences between male and female brains. In: Lanzenberger R, Kranz GS, Savic I (eds) Handbook of clinical neurology, vol 175. Elsevier, pp 3–11
Maeda Y, Yoon SY (2013) A meta-analysis on gender differences in mental rotation ability measured by the Purdue spatial visualization tests: visualization of rotations (PSVT:R). Educ Psychol Rev 25:69–94
Manolio TA et al (2009) Finding the missing heritability of complex diseases. Nature 461:747–753
Martel MM (2013) Sexual selection and sex differences in the prevalence of childhood externalizing and adolescent internalizing disorders. Psychol Bull 139:1221–1259
Matsumoto A, Arai Y (1983) Sex difference in volume of the ventromedial nucleus of the hypothalamus in the rat. Endocrinol Jpn 30:277–280
McCarthy MM et al (2009) The epigenetics of sex differences in the brain. J Neurosci 29:12815–12823
McHugh RK, Votaw VR, Sugarman DE, Greenfield SF (2018) Sex and gender differences in substance use disorders. Clin Psychol Rev 66:12–23
McLaughlin KA, Xuan Z, Subramanian S, Koenen KC (2011) State-level women's status and psychiatric disorders among US women. Soc Psychiatry Psychiatr Epidemiol 46:1161–1171
Migeon BR (2007a) Why females are mosaics, x-chromosome inactivation, and sex differences in disease. Gend Med 4:97–105
Migeon BR (2007b) Why females are mosaics, x-chromosome inactivation, and sex differences in disease. Gend Med 4:97–105
Mijalkov M et al (2023) Sex differences in multilayer functional network topology over the course of aging in 37543 UK biobank participants. Netw Neurosci 7:351–376
Monro K, Marshall DJ (2016) Unravelling anisogamy: egg size and ejaculate size mediate selection on morphology in free-swimming sperm. Proc R Soc B Biol Sci 283:20160671
Mueller SC et al (2021) The neuroanatomy of transgender identity: mega-analytic findings from the ENIGMA transgender persons working group. J Sex Med 18:1122–1129
Open Science Collaboration (2015) Estimating the reproducibility of psychological science. Science 349:aac4716
Paus T (2010a) Population neuroscience: why and how. Hum Brain Mapp 31:891–903
Paus T (2010b) Sex differences in the human brain. A developmental perspective. Prog Brain Res 186:13–28
Paus T (2013) Population neuroscience. Springer Science & Business Media
Paus T, Toro R (2009) Could sex differences in white matter be explained by g ratio? Front Neuroanat 3:14
Paus T et al (2010) Sexual dimorphism in the adolescent brain: role of testosterone and androgen receptor in global and local volumes of grey and white matter. Horm Behav 57:63–75
Paus T, Wong AP-Y, Syme C, Pausova Z (2017) Sex differences in the adolescent brain and body: findings from the Saguenay youth study. J Neurosci Res 95:362–370
Perrin JS et al (2008) Growth of white matter in the adolescent brain: role of testosterone and androgen receptor. J Neurosci 28:9519–9524
Peters SAE, Woodward M (2023) A roadmap for sex- and gender-disaggregated health research. BMC Med 21:354
Pietschnig J, Penke L, Wicherts JM, Zeiler M, Voracek M (2015) Meta-analysis of associations between human brain volume and intelligence differences: how strong are they and what do they mean? Neurosci Biobehav Rev 57:411–432
Pla S, Maynou F, Piferrer F (2021) Hermaphroditism in fish: incidence, distribution and associations with abiotic environmental factors. Rev Fish Biol Fisheries 31:935–955

Plomin R, von Stumm S (2022) Polygenic scores: prediction versus explanation. Mol Psychiatry 27:49–52
Polderman TJC et al (2018) The biological contributions to gender identity and gender diversity: bringing data to the table. Behav Genet 48:95–108
Poldrack RA et al (2017) Scanning the horizon: towards transparent and reproducible neuroimaging research. Nat Rev Neurosci 18:115–126
Rauch JM, Eliot L (2022) Breaking the binary: gender versus sex analysis in human brain imaging. NeuroImage 264:119732
Reinhold K, Engqvist L (2013) The variability is in the sex chromosomes. Evolution 67:3662–3668
Rippon G, Jordan-Young R, Kaiser A, Fine C (2014) Recommendations for sex/gender neuroimaging research: key principles and implications for research design, analysis, and interpretation. Front Hum Neurosci 8:650
Ritchie SJ et al (2018) Sex differences in the adult human brain: evidence from 5216 UK biobank participants. Cereb Cortex 28:2959–2975
Rocks D, Cham H, Kundakovic M (2022) Why the estrous cycle matters for neuroscience. Biol Sex Differ 13:62
Ruigrok ANV et al (2014) A meta-analysis of sex differences in human brain structure. Neurosci Biobehav Rev 39:34–50
Sanchis-Segura C, Ibañez-Gual MV, Aguirre N, Cruz-Gómez ÁJ, Forn C (2020) Effects of different intracranial volume correction methods on univariate sex differences in grey matter volume and multivariate sex prediction. Sci Rep 10:12953
Satterthwaite TD et al (2015) Linked sex differences in cognition and functional connectivity in youth. Cereb Cortex 25:2383–2394
Schwarz JM, McCarthy MM (2008) Cellular mechanisms of estradiol-mediated masculinization of the brain. J Steroid Biochem Mol Biol 109:300–306
Seedat S et al (2009) Cross-national associations between gender and mental disorders in the World Health Organization world mental health surveys. Arch Gen Psychiatry 66:785–795
Servedio MR, Boughman JW (2017) The role of sexual selection in local adaptation and speciation. Annu Rev Ecol Evol Syst 48:85–109
Sporns O (2013) The human connectome: origins and challenges. NeuroImage 80:53–61
Statistics Canada (2022) Canada is the first country to provide census data on transgender and non-binary people. The Daily
Sun S, Heitman J (2012) Should Y stay or should Y go: the evolution of non-recombining sex chromosomes. BioEssays 34:938–942
Székely T, Reynolds JD, Figuerola J (2000) Sexual size dimorphism in shorebirds, gulls, and alcids: the influence of sexual and natural selection. Evolution 54:1404–1413
Trabzuni D et al (2013) Widespread sex differences in gene expression and splicing in the adult human brain. Nat Commun 4:2771
Tunç B et al (2016) Establishing a link between sex-related differences in the structural connectome and behaviour. Philos Trans R Soc B Biol Sci 371:20150111
Vetvik KG, MacGregor EA (2017) Sex differences in the epidemiology, clinical features, and pathophysiology of migraine. Lancet Neurol 16:76–87
Vijayakumar N et al (2016) Brain development during adolescence: a mixed-longitudinal investigation of cortical thickness, surface area, and volume. Hum Brain Mapp 37:2027–2038
Vosberg DE et al (2021) Sex continuum in the brain and body during adolescence and psychological traits. Nat Hum Behav 5:265–272
Vosberg DE, Parker N, Shin J, Pausova Z, Paus T (2022) The genetics of testosterone contributes to "femaleness/maleness" of cardiometabolic traits and type 2 diabetes. Int J Obes 46:235–237
Vosberg DE, Pausova Z, Paus T (2023) The genetics of a "femaleness/maleness" score in cardiometabolic traits in the UK biobank. Sci Rep 13:9109
Voyer D, Saint Aubin J, Altman K, Gallant G (2021) Sex differences in verbal working memory: a systematic review and meta-analysis. Psychol Bull 147:352–398

Warrier V et al (2018) Genome-wide meta-analysis of cognitive empathy: heritability, and correlates with sex, neuropsychiatric conditions and cognition. Mol Psychiatry 23:1402–1409

Wierenga LM et al (2022) Greater male than female variability in regional brain structure across the lifespan. Hum Brain Mapp 43:470–499

Wiersch L et al (2023) Accurate sex prediction of cisgender and transgender individuals without brain size bias. Sci Rep 13:13868

Williams CM, Peyre H, Toro R, Ramus F (2021a) Neuroanatomical norms in the UK biobank: the impact of allometric scaling, sex, and age. Hum Brain Mapp 42:4623–4642

Williams CM, Peyre H, Toro R, Ramus F (2021b) Sex differences in the brain are not reduced to differences in body size. Neurosci Biobehav Rev 130:509–511

Yang CC, Totzek JF, Lepage M, Lavigne KM (2023) Sex differences in cognition and structural covariance-based morphometric connectivity: evidence from 28,000+ UK biobank participants. Cereb Cortex 33:10341–10354

Yengo L et al (2022) A saturated map of common genetic variants associated with human height. Nature 610:704–712

Zucker I, Prendergast BJ (2020) Sex differences in pharmacokinetics predict adverse drug reactions in women. Biol Sex Differ 11:32

Zugman A et al (2023) Country-level gender inequality is associated with structural differences in the brains of women and men. Proc Natl Acad Sci 120:e2218782120

Part III
Genes and Molecules

Genetic Architecture of Neurological Disorders and Their Endophenotypes: Insights from Genetic Association Studies

Muralidharan Sargurupremraj

Contents

Abstract Population-scale genetic association studies of complex neurologic diseases have identified the underlying genetic architecture as multifactorial. Despite the study sample sizes reaching the millions, the identified disease-related genes explain only a small fraction of the phenotypic variance. Notable advancements in statistical methods now enable researchers to gain insights even from genomic regions where genotype–phenotype associations do not reach statistical significance. Such studies confirm a highly interconnected molecular network comprising a core group of genes directly involved in the disease process, alongside an expanded peripheral network, each contributing a small but potentially important (modulatory) effect. Additionally, causal inference methods, utilizing genetic instruments, have shed light on putative causal links between risk factors and clinical endpoints. In light of the pervasive genetic overlap or pleiotropy, however, caution is warranted in interpreting causal relationships inferred from these analyses. In this chapter, I will introduce the genetic association model, provide insights into the current state of genetic association studies, and discuss potential future directions.

Keywords GWAS · Mendelian randomization · Omnigenic · Pleiotropy

M. Sargurupremraj (✉)
Glenn Biggs Institute for Alzheimer's and Neurodegenerative Diseases, University of Texas Health Sciences Center, San Antonio, TX, USA
e-mail: sargurupremr@uthscsa.edu

Curr Topics Behav Neurosci (2024) 68: 109–128
https://doi.org/10.1007/7854_2024_513
Published Online: 14 August 2024

1 Introduction

In his groundbreaking 1918 paper, R.A. Fisher extended our understanding of the variability observed in human traits (Fisher 1919). He proposed that the inheritance patterns of quantitative traits, such as human height, could be elucidated through genetic units within the Mendelian framework of inheritance. This laid the foundation for what we now know as "quantitative genetics." Moving forward to the 2000s, advancements in genomic technologies revolutionized our ability to profile millions of these genetic units, known as single-nucleotide polymorphisms (SNPs), and test their associations with phenotypes or disease-trait information across hundreds of thousands of individuals. This paved the way for genome-wide association studies (GWAS), a research design that examines the statistical links between individual SNPs and various complex traits.

The NHGRI-EBI (National Human Genome Research Institute-European Bioinformatics Institute) GWAS catalogue currently boasts nearly 6,500 publications dedicated to exploring SNP associations with diverse complex traits (Sollis et al. 2023). These studies have identified over half-a-million SNPs exhibiting strong statistical evidence of such associations. In the context of neurological and neurodegenerative diseases, GWAS has revealed more than 5,000 genetic loci, collectively highlighting the polygenic architecture of complex traits. This genetic architecture encompasses multiple genetic loci, each making modest (infinitesimal) genetic contributions (Uffelmann et al. 2021). The infinitesimal genetic architecture, in conjunction with the population frequency of the second most common of two alleles at a genetic locus (or minor allele frequency, MAF), dictates the requisite sample size for detecting a statistically significant link between a given SNP and a trait. Moreover, given that the true causal genetic variant remains elusive, the capability of a GWAS-identified hit to represent potentially a causal variant scales linearly with the sample size (Visscher et al. 2017; Wray 2005). Recent investigations suggest that achieving an attribution of 50% of the phenotypic variability of complex disease traits to the underlying genetic variability ("heritability") may necessitate sample sizes ranging from several hundred thousand for conditions such as coronary artery disease to millions for Alzheimer's disease (AD) (O'Connor 2021).

In addition to the disease-specific consortia, with information on the genotype and phenotype data collected prospectively, the availability of large biobanks (e.g., Global Biobank Meta-analysis Initiative) (Zhou et al. 2022) offers a valuable resource to meet the methodological prerequisites. These biobanks represent millions of individuals with genetic data linked to electronic health records (EHRs), including those from underrepresented populations (e.g., Biobank Japan, Qatar Biobank, Taiwan Biobank) (Cho et al. 2012; Feng et al. 2022; Nagai et al. 2017; Al Kuwari et al. 2015). Large-scale meta-analysis initiatives leveraging such resources continue to increase the statistical power to uncover novel genetic variants for human diseases. They also contribute to a more nuanced understanding of the shared genetic susceptibility observed in closely related traits (Bellenguez et al.

2022; Mishra et al. 2022). In particular, biobanks that are linked to EHRs provide a global overview of the phenome-wide association of disease-susceptible genetic variants across a wide range of traits and disorders (Denny et al. 2013). Furthermore, the release of GWAS association statistics to the broader research community has facilitated the rapid expansion of analyses employing genetic instruments—an approach referred to as "Mendelian randomization"—that untangles the causal relationships between diverse risk factors and clinical endpoints. Lastly, the integration of GWAS association statistics with molecular and cellular phenotypes, encompassing transcriptomics and epigenomics, has proven to be a valuable strategy for fine-mapping of the putative disease-causing genes and identifying disease-relevant cell types. In the forthcoming sections, I aim to provide a comprehensive overview of genome-wide association methods, with a specific focus on cerebrovascular disorders, such as stroke, and their preclinical markers or endophenotypes.

2 GWAS Design

A typical GWAS workflow involves three major steps.

(i) *Data collection* step involves gathering DNA and phenotypic information from a representative population (see chapter on "Population neuroscience: understanding concepts of generalizability and transportability and their application to improving the public's health" in this book), and then genotyping each individual using microarrays for SNPs that commonly segregate in the population, or using next-generation sequencing methods, such as whole-exome or whole-genome sequencing, which also cover rare genetic variants. The tendency of the genotyped SNPs to tag blocks of proximal SNPs (due to the non-random association of genetic alleles at different loci in a population, i.e., "linkage disequilibrium," LD) enables capturing the genetic variation across the genome in a high-throughput manner.

(ii) *Data processing* step enhances the genotyping coverage by implementing variant-level and sample-level quality control measures. This ensures that only common, bi-allelic variants that adhere to an idealized state of non-random genetic variations across generations ("Hardy-Weinberg equilibrium") and are well-represented in the sample set are considered. Following the quality control, variants are phased to the paternal or maternal allele, determining the combination of alleles inherited together on the same chromosome (haplotyping). This process allows comparing the phased genomic data to the reference haplotypes from deeply sequenced populations (e.g., 1,000 Genomes project or TOPMed) (Taliun et al. 2021; The 1000 Genomes Project Consortium 2015), and imputing the missing genotype across the genome with high confidence ("imputation"). Finally, high-quality common SNPs are used to account for the underlying population substructure due to ancestral differences, where pair-wise similarity ("genetic relationship matrix") metrics between pairs of individuals are constructed from the genotype, and their representative components ("principal components," PCs) are adjusted for in the data analysis step.

(iii) *Data analysis* step employs either a linear or logistic regression model that tests for associations with the genotype, depending on whether the phenotype is continuous (e.g., hippocampal volume) or binary (e.g., AD case vs. control status), respectively. The linear regression (1) in a GWAS can be modeled as:

$$Y \sim W\alpha + \beta_S G_S + e + g \tag{1}$$

where:

- Y is a vector of quantitative measures, representing each individual, predicted from a set of independent variables (genotype and covariates).
- W represents a matrix of covariates along with the corresponding effect estimates α and the intercept term.
- G_S is a vector of genotype values for a given SNP S, and β_S is the effect estimate of interest in the GWAS model.
- e is an error term describing the deviations of the fitted value from the observed response value.

In the case of case–control phenotypes, logistic regression (2) is modeled using a function (logit link) that tests odds of the outcome from binomially distributed binary measures (Zhou et al. 2018).

$$\text{logit}(\mu) \sim W\alpha + \beta_S G_S + e + g \tag{2}$$

Recent advances in statistical approaches allow modeling family relatedness by accounting for the genetic relationship matrix as random effects g in a mixed model setting ("linear mixed models"), in addition to the fixed or predictable contribution from known covariates (Loh et al. 2015, 2018). Mixed model approaches are also effective in studying the individual effects of genetic variants while adjusting for the LD structure, which helps in identifying causal variants (Mbatchou et al. 2021).

The association test with a phenotype of interest involves millions of SNPs, necessitating stringent multiple-testing corrections to minimize false positives. The standard "Bonferroni correction" strategy, correcting for the number of genetic markers tested ($0.05/10^6 = 5 \times 10^{-8}$), has been commonly used in GWASs (Uffelmann et al. 2021), despite its conservative nature due to the lack of accounting for LD between genetic markers. To balance false positives and false negatives more effectively, alternative multiple-testing correction methods that consider the dependent nature of genetic variants (LD blocks) have been proposed (Gao et al. 2010; Johnson et al. 2010). A systematic comparison of different multiple-testing thresholds suggests a more relaxed threshold (5×10^{-7}) for well-powered GWASs with sample sizes greater than 20,000, while the standard threshold (5×10^{-8}) is recommended for modest-sized GWASs (Chen et al. 2021). Consortia of well-characterized cohorts are instrumental in conducting large-scale GWASs, where the availability of harmonized phenotype and genotype information and the use of various meta-analysis strategies enable effective ways of combining the individual study-wise GWAS results. The meta-analysis strategy allows for testing whether the observed genetic effect aligns in all included studies and adjusts for any deviations by weighting study-specific genetic-marker association results by the study sample size or certainty of the genetic effects (known as "inverse variance weighting"), thereby increasing statistical power (Willer et al. 2010).

The GWAS hits satisfying the statistical evidence for the genotype–phenotype association are still considered a putative signal, requiring detailed downstream analyses to define genetic effects and to assign functional consequences and biological relevance. Implementing an independent replication set in addition to the discovery set in a two-stage GWAS design is an effective downstream strategy for validating putative signals and for controlling biases inherent in large-scale GWASs. One such bias is the winner's curse bias, where the effect sizes of lead variants that pass the genome-wide (GW) significance threshold are often inflated. The inflation can be attributed to the replication efforts concentrating solely on these lead variants while overlooking the broader context of the locus harboring these variants. Calibration efforts testing the phenotype association of the genetic loci in its entirety, accounting for the population-specific LD patterns, have been shown to minimize this bias between the observed and the expected effect estimates (Palmer and Pe'er 2017; Li et al. 2012). The classical two-stage GWAS design can also aid in the fine-mapping efforts, as the disease-causal variants identified in one subset of a population are expected to replicate in a different subset of the same population due to the high LD between the variants (Palmer and Pe'er 2017). Moreover, replicating association signals in independent ancestry groups can implicate common genetic variants across different populations, enhancing the transportability of findings from well-represented (e.g., European) to underrepresented (e.g., African Americans, East Asians) populations (Marigorta and Navarro 2013; Atkinson 2023). It is also common for GWASs to meta-analyze jointly the association statistics from discovery and replication datasets to increase the statistical power and implicate additional genetic loci. Recent large-scale GWASs for AD (Bellenguez et al. 2022) and stroke (and its subtypes) (Mishra et al. 2022) adopted this strategy and identified multiple genetic loci across all the autosomes. Increasing sample size by adding cohorts and biobanks provides a statistical advantage, doubling the number of disease-associated genetic loci compared with earlier studies (Kunkle et al. 2021; Schwartzentruber et al. 2021; Malik et al. 2018a). Joint-analysis strategies require, however, additional fine-mapping approaches that take into account the sparse genetic effects dispersed across the multiple linked variants (Hormozdiari et al. 2014; Wang et al. 2020) and the ancestry-specific linkage patterns (Lu et al. 2022) to identify probable sets of causal variants ("credible sets"). For instance, credible-set construction for stroke-risk genetic loci prioritizes a smaller set of putative causal SNPs from thousands of associated SNPs, potentially shared across ancestries (Mishra et al. 2022).

3 Exploring Genetic Architecture: Lessons from Current and Future GWASs

GWAS-identified "sentinel" variants, which are GW significant, LD-independent, and representative of disease-susceptible loci, explain only a small fraction of phenotypic variability or heritability. For example, sentinel variants associated with stroke risk individually explain between 0.6% and 1.8% of the disease variance on the liability scale and, in total, only 0.8% (Malik et al. 2018a). The discrepancy between the individual variant-level estimates and the aggregate measure could be attributed to the methodological challenges in accurately estimating the former measure (Shi et al. 2016). While the largest-effect variants are concentrated in genic regions with specific disease-function roles, the bulk of the heritability comes from genetic variants with small effects spread across the genome (Boyle et al. 2017). This underscores the polygenic nature of complex diseases. Given that much of the heritability for complex traits is distributed across multiple other genetic loci that do not reach the statistical significance threshold, it is a common procedure to estimate the heritability accounting for the genome-wide LD structure (Finucane et al. 2015). LD score regression, a commonly used post-GWAS method (Bulik-Sullivan et al. 2015), involves regressing variant-level phenotype association statistic on a metric that reflects the variant's ability to tag putative causal variants due to LD (LD score) and estimate heritability explained by the SNPs across the genome. Genome-wide SNP-based heritability from LD score regression for stroke (all stroke) and AD remains modest, however, with recent GWASs explaining only 3% and 2% of the disease variance, respectively (Mishra et al. 2022; Escott-Price and Hardy 2022).

Heritability, in a broader sense, indicates the phenotypic variance in a trait explained by (i) the additive effects of the genetic markers, (ii) non-additive effects due to inter-marker interactions (e.g., dominance, epistasis), and (iii) interactions with the environment. Since the sample-size requirement for studying non-additive effects is much larger than the currently available biobank-scale resources (O'Connor 2021), most of the GWASs and family-based association studies focus on studying the additive genetic effects. Family-based studies, such as twin studies leveraging the controlled environment, estimate the additive or narrow-sense heritability for AD and stroke to be about 70% and 40%, respectively (Karlsson et al. 2022; Bak et al. 2002; Bevan et al. 2012). This is in contrast to the much smaller heritability estimates obtained from GWASs of unrelated individuals, and this disparity is termed the "missing heritability." In addition to the differences in the study design, the observed differences can be attributed to the types of genetic markers used, as most GWASs focus on common (MAF $>$ 5%) and low-frequency (MAF 1–5%) variations, which are presumed to have small effects, unlike the large-effect rare (MAF $<$ 1%) or ultrarare (MAF $<$ 0.1%) variants (Weiner et al. 2023). Phenotype heterogeneity in GWASs also plays a crucial role in heritability estimations. Inaccuracies in disease diagnosis and the meta-analysis of clinically defined cases with those defined based on proxy status, such as the parental history of a given disease, can cause a downward bias in the heritability estimates (De La Fuente et al. 2022). Interestingly, GWASs focusing on preclinical markers (e.g., white matter hyperintensities [WMH]), which are continuous in nature, yield consistently larger estimates for the phenotypic variance explained by genetic variations (10 to 29%) (Armstrong et al. 2020; Sargurupremraj et al. 2020; Duperron et al. 2018). Moreover, decomposing the genome-wide heritability estimates to the constituent individual genetic loci (Shi et al. 2016) helped prioritize specific loci (*SH3PXD2A*) with a notable contribution to the genome-wide heritability of preclinical markers of stroke (WMH burden) (Sargurupremraj et al. 2020). Interestingly, *SH3PXD2A*, involved in the extracellular matrix degradation and tissue repair process, was later prioritized as one of the disease-causal genes for ischemic stroke (Mishra et al. 2022). Earlier research has revealed genetic variations in the promoter sequence of *SH3PXD2A* influencing expression levels of the gene in macrophages, thereby regulating inflammatory responses—a crucial mechanism implicated in stroke pathobiology (Ma et al. 2017; Var et al. 2021). This example highlights the importance of post-GWAS analyses in strengthening the evidence for involvement of putative risk loci in a phenotype of interest that may not reach genome-wide significance solely due to statistical power and sample-size constraints.

One can study the distribution of the genetic effect sizes from common genetic variants across the genome to estimate the sample size required to identify disease-susceptibility loci. Such power calculations indicate that a substantial sample size—ranging from 10^6 to 10^8 individuals—is required to identify risk variants explaining the phenotypic variance satisfactorily (Visscher et al. 2017). While the more in-depth whole-genome sequencing method studying low (MAF 1–5%) and rare frequency (MAF $<$ 1%) variants may theoretically perform better with smaller sample sizes, this is only achievable under the unrealistic assumption of genetic effects being close to 1. Modeling based on expected genetic effect sizes ranging from 0.1 to 0.01 reveals the need for an unfeasibly large sample size of 10 (Nagai et al. 2017). Recent advancements in statistical methods, accounting for the underlying polygenic architecture, such as Fourier transform regression, offer more accurate estimations of case numbers required for future GWASs, ranging from 100,000 for vascular risk factors to nearly a million for AD (O'Connor 2021). Understandably, conducting such studies is challenging, with the added complexity of heterogeneity in data acquisition and disease ascertainment. But the availability of large-scale GWASs from consortia, which meta-analyze harmonized cohorts, and the application of variance-components methods (Bulik-Sullivan et al. 2015) have streamlined the post-GWAS efforts, already yielding a comprehensive understanding of the genetic architecture of complex diseases. Notably, SNP-based heritability analysis examining independent LD blocks has shown that the heritability distribution is also proportional to the physical length of chromosomes (Shi et al. 2016), with a relatively smaller contribution from the shorter chromosomes (chromosomes 13–22). Spatial proximity maps of the human genome mapping of the inter-chromosomal contact probabilities show that the shorter chromosomes centered in the nucleus preferentially interact with each other (Lieberman-Aiden et al. 2009). This underscores the significant influence of additional factors, such as epigenetic

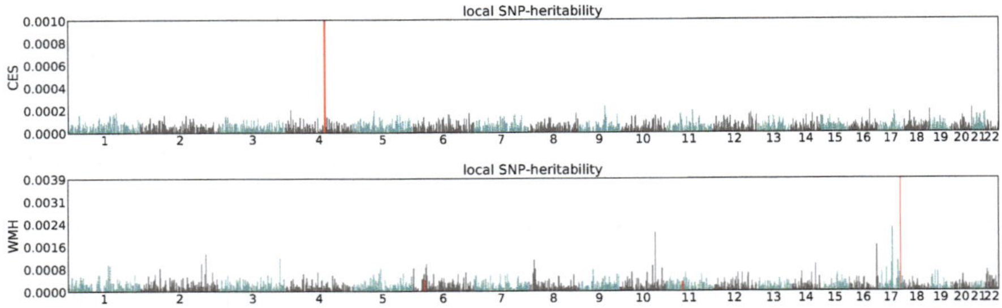

Fig. 1 Local SNP-based heritability for stroke and white matter hyperintensities. Only showing traits harboring significant local SNP heritability (red bars). (CES: 4q25 region harboring *PITX2*; WMH: 17q25.1 harboring *TRIM65/TRIM47*). CES: cardioembolic stroke, WMH: white matter hyperintensities

variations and gene expression modulations, on phenotypic variability and heritability.

Indeed, estimating SNP-based heritability using HESS (Shi et al. 2016) (method described elsewhere (Sargurupremraj et al. 2020)) for stroke and WMH in each of the independent LD blocks (Berisa and Pickrell 2016) confirmed that the genetic variance is uniformly distributed across the genome, with the notable yet modest contribution by regions hosting "core genes" that are implicated directly in the disease process (4q25:*PITX2*-CES, 17q25.1:*TRIM65/TRIM47*-WMH) (Fig. 1). Additionally, a systematic examination of the fraction of variance contributed by these "core genes" to the overall heritability for cerebral small vessel disease (cSVD) traits (Fig. 2) revealed that, except for a few loci (*PITX2* and *SH3PXD2A*), most established genetic loci made only minor contributions. *PITX2*—a gene encoding β-catenin-regulated transcription factor—consistently emerges as a risk factor for ischemic and cardioembolic stroke (Traylor et al. 2019; von Berg et al. 2020; Malik and Dichgans 2018), potentially sharing underlying mechanisms with early risk factors like atrial fibrillation (Roselli et al. 2018). *SH3PXD2A*—encoding an adaptor protein involved in extracellular matrix degradation—exhibits widespread associations with various brain-imaging endophenotypes (Sargurupremraj et al. 2020; Fornage et al. 2011; Verhaaren et al. 2015; Persyn et al. 2020) and stroke (Traylor et al. 2019). Genetic variants from this locus interact in a network fashion with nearby genes mediating the neurotoxic effect of amyloid-β peptide, possibly contributing to the vascular aspects of neurodegenerative disorders (Laumet et al. 2010).

Collectively, the above observations lend support to the "omnigenic disease model" (Liu et al. 2019), a special entity within the polygenic spectrum where the distant peripheral genes possibly implicated in multiple correlated traits greatly influence the disease process compared with the "core genes." In their "omnigenic model hypothesis" (Stephens 2017), Boyle et al. (2017) studied the effect-size distribution postulating that genes harboring putative causal variants are interconnected in a vast network termed the "peripheral" set. Perturbation of this network in a cell-type-specific manner could impact the "core" set of genes, which play a biologically direct and interpretable role in the disease process. Furthermore,

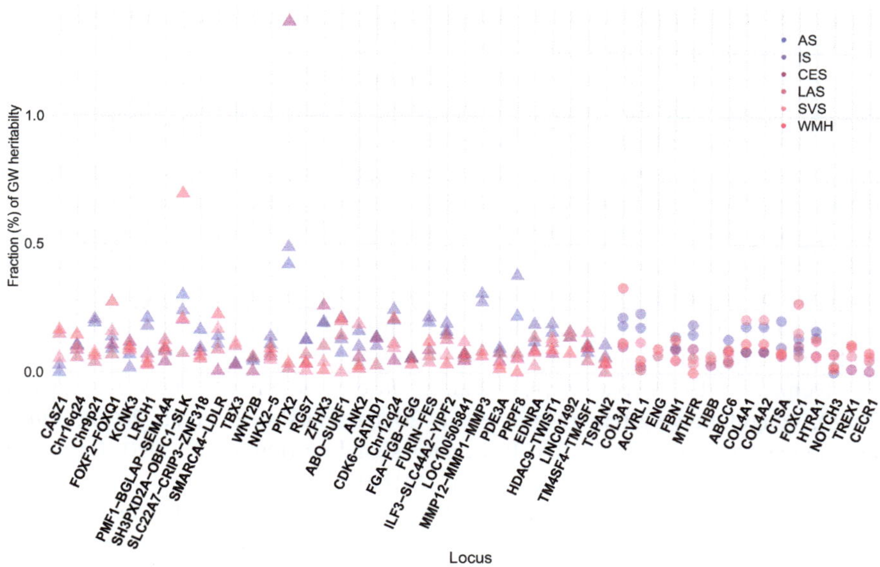

Fig. 2 Contribution of the stroke-risk loci (estimated using HESS, Shi H. et al.) to the genome-wide heritability of cSVD traits. Genes implicated in sporadic (triangle) or Mendelian (circle) disease forms. AS: any stroke, SVS: small vessel stroke, LAS: large artery stroke, CES: cardioembolic stroke, IS: Ischemic stroke, WMH: white matter hyperintensities

the study revealed a notable concentration of non-zero effect variants or putative causal variants within gene regulatory regions. Building upon this insight, Liu X. et al. (2019) introduced a quantitative model to the omnigenic framework to gene expression measures and delineated the "cis-direct" effect from disease-relevant core genes and the "trans-indirect" effect from numerous peripheral genes onto the core genes in a network fashion. A significant insight from the omnigenic model is the widespread or pervasive pleiotropy of the peripheral genes implicated for a disease trait with multiple other risk factors, which in turn could consequentially impact the causal inference models.

4 Comparative Look at Pleiotropy and Causal Inference: Related Yet Distinct Concepts

Pleiotropy is an umbrella term referring to the putative mechanisms whereby (i) a single genetic variant or a group of LD-linked variants spanning a definite region (locus) influences multiple traits (locus-level pleiotropy), (ii) the causal link observed between an exposure and outcome in the causal inference models is the result of indirect association mediated by unmeasured confounders (horizontal pleiotropy) (Morrison et al. 2020), or (iii) cell-type-specific gene expression signatures exhibiting association with multiple closely related traits (cell type-mediated pleiotropy) (Xue et al. 2020).

It is increasingly clear that locus-level pleiotropy, where multiple genetic variants impact various traits, is more prevalent than pleiotropy at the single-unit level (Watanabe et al. 2019). This prevalence is largely attributed to the positional proximity of trait-associated variants to gene regulatory regions, which influences the expression of multiple genes and affects diverse traits in a network fashion. Investigating locus-level pleiotropy of stroke-risk loci alongside other preclinical risk factors could offer valuable insights into the covert and distant biological mechanisms contributing to the disease process. Supporting this notion, pathway-enrichment analyses, utilizing gene sets constructed based on the positional proximity of stroke-risk variants to gene regulatory regions, have already revealed the involvement of diverse pathways, including nitric oxide release (Malik et al. 2018a, 2018), coagulation system (Malik et al. 2018a; Harshfield et al. 2020), innate immune system, such as the T-cell receptor signaling (Qin et al. 2020; Barr et al. 2010), and cardiac pathways (Malik et al. 2018a).

Conversely, examining pleiotropy at the single-unit level enables the identification of risk factors directly influencing the disease risk, often due to the strategic positioning of variants within sites disrupting specific gene functions. While such relationships are readily discernible from the replication efforts of disease-risk genes in relation to closely related risk factors (Fig. 3, for stroke loci), employing relevant methods within a causal inference framework to address potential confounders is essential for validating any putative causal relationships.

Interactive Sankey Plot URL: Download
GWAS also offers a valuable tool for exploring causal relationships between exposures and outcomes by leveraging genetic variants strongly associated with a phenotype as unconfounded proxies in an instrumental variable (IV) analysis framework. Mendelian randomization (MR), an IV method, examines whether a phenotype (exposure) in the causal pathway influences an outcome solely through that exposure, avoiding potential confounding factors associated with both exposure and outcome (Davey Smith and Ebrahim 2003). Conceptually akin to a randomized trial, MR utilizes the random "allocation" of genetic alleles as instrumental variables to determine exposure levels or risks (Smith and Ebrahim 2005). One of the key assumptions of MR is that all of the genetic variants in the analysis are valid instrumental variables and not associated *independently* with the outcome, and with any unmeasured confounders. The most common MR approach, known as two-sample MR (2SMR), functions akin to a meta-analysis, aggregating causal estimates from multiple genetic IVs into a weighted average. Similar to observational studies, MR using genetic instruments with imprecise standard errors may suffer from regression dilution bias toward the null, particularly when the genetic instruments lack strength. This bias tends to diminish as the strength of the genetic instruments increases (Bowden and Holmes 2019). Construction of genetic IVs from well-powered GWAS with a large number of genetic variants is a common strategy

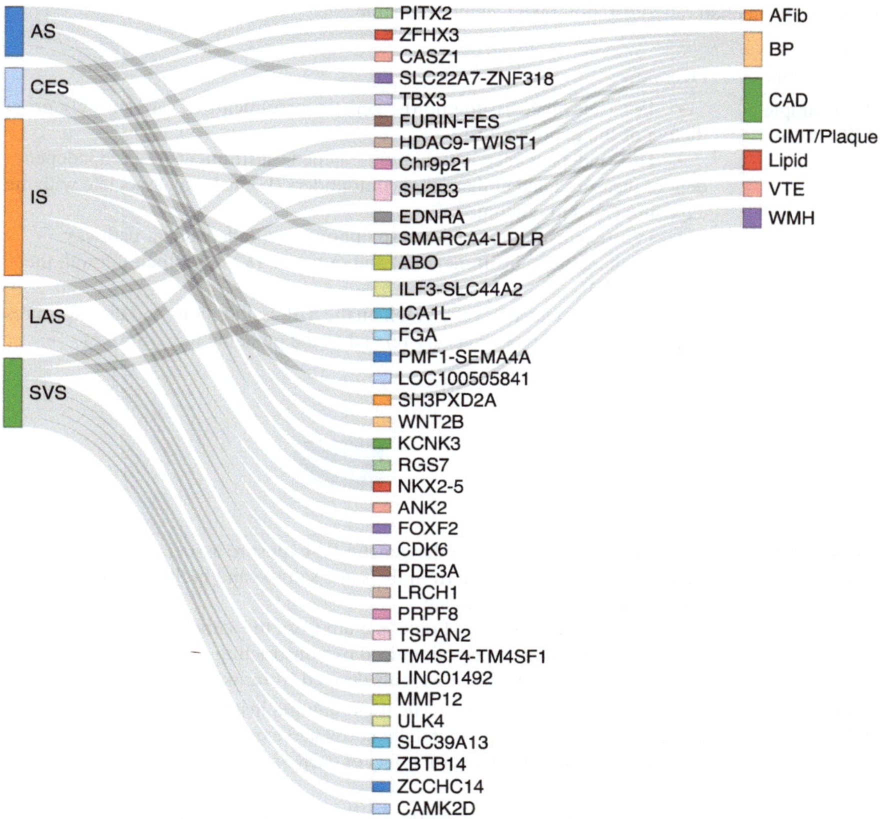

Fig. 3 Sankey plot showing the gene-level (middle) overlap of stroke-risk loci (left) with vascular risk factors (right). AS: any stroke, CES: cardioembolic stroke, IS: ischemic stroke, LAS: large artery stroke, SVS: small vessel stroke, AFib: atrial fibrillation, BP: blood pressure CAD: coronary artery disease, CIMT: carotid intima-media thickness; VTE: venous thromboembolism; WMH: white matter hyperintensity burden

for increasing the instrument's strength (Burgess and Thompson 2011), which could potentially also increase the likelihood of adding pleiotropic variants, mandating the application of various pleiotropy robust methods (MR-Egger, MR-RAPS, weighted-median, mode-based MR) (Hemani et al. 2018). Table 1 presents the MR analysis scheme and highlights key MR assumptions.

A considerable body of literature has delved into the potential causal relationship between various risk factors and stroke risk, as evidenced by the close to 170 publications yielded from a PubMed search using the terms "[stroke AND Mendelian randomization]." Fig. 4 provides a summarized overview of findings from studies employing the hypothesis-driven 2SMR approach to investigate the causal associations of vascular risk factors, circulating biomarkers, and drug targets as exposures with stroke (any stroke), ischemic stroke (IS), and its subtypes (cardioembolic; CES,

Table 1 Mendelian randomization (MR) analysis scheme

MR components	Overview
Two-sample MR	• IV1 (relevance assumption)—genetic instruments (Z) are strongly predictive of the exposure (X) • IV2 (independence assumption)—genetic instruments (Z) are independent of measured and unmeasured confounders (U) that may associate with the exposure (X) and outcome (Y) • IV3 (exclusion restriction assumption)—no independent pathway between the genetic instrument (Z) and outcome (Y) other than through the exposure (X) • Figure: Directed acyclic graph on the MR assumptions and the exposures (e.g., WMH) putatively causally related to the outcome (e.g., Stroke) 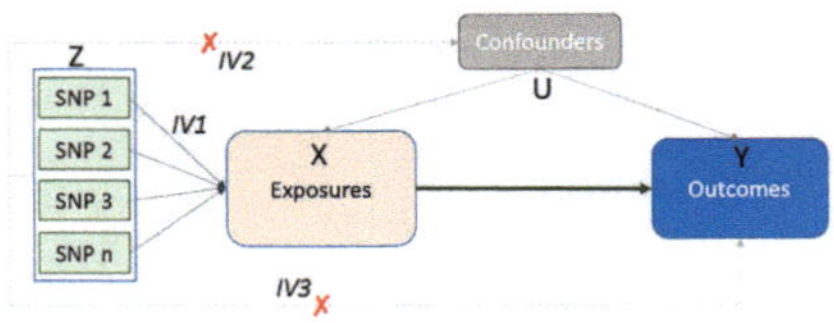
One-sample MR	• MR analyses in which the genetic IV-exposure association and genetic IV-outcome association are from the same sample and individual-level data is used to derive the MR estimate • Since the exposure and outcome are from the same individuals, any statistical overfitting can be minimized by constructing a weighted genetic risk score for the exposure using weights derived from independent GWASs
Multivariable MR	• MVMR—multivariable MR uses multiple instruments (Z1, . . ., Zn) associated with multiple, potentially correlated exposures (X,. . .,Xn) to jointly estimate the independent causal effect of each of the exposures on the outcome (Y) • Instrument selection criteria: SNPs associated with at least one of the exposures
Sensitivity analyses	• Violation from the IV1 assumption: Weak instruments (F statistic less than 10) can be excluded (Burgess and Thompson 2011; Burgess et al. 2017) • Violation from the IV2 assumption: Confounding variables (e.g., smoking status, diabetes) exhibiting association with the genetic risk score for the main exposures (BP, lipid traits) can be additionally adjusted • Violation from the IV3 assumption: – Strategy 1: Use of multivariable MR, which accounts for potential pleiotropic effects on the outcome that is not through the exposure of interest – Strategy 2: Filtering pleiotropic outliers using cutting-edge statistical tools (MR-GENIUS) (Tchetgen et al. 2017) that identifies genetic instruments with differences in the variance explaining the exposure

large artery; LAS, small vessel; SVS). Most of these studies focus on the European population, employing multiple and single genetic instruments to predict exposure levels or risks. Certain associations depicted at the periphery of Fig. 4 exhibit stroke subtype specific associations pointing to specific disease mechanisms. The majority of the exposures (positioned at the center) are associated with all-stroke and all

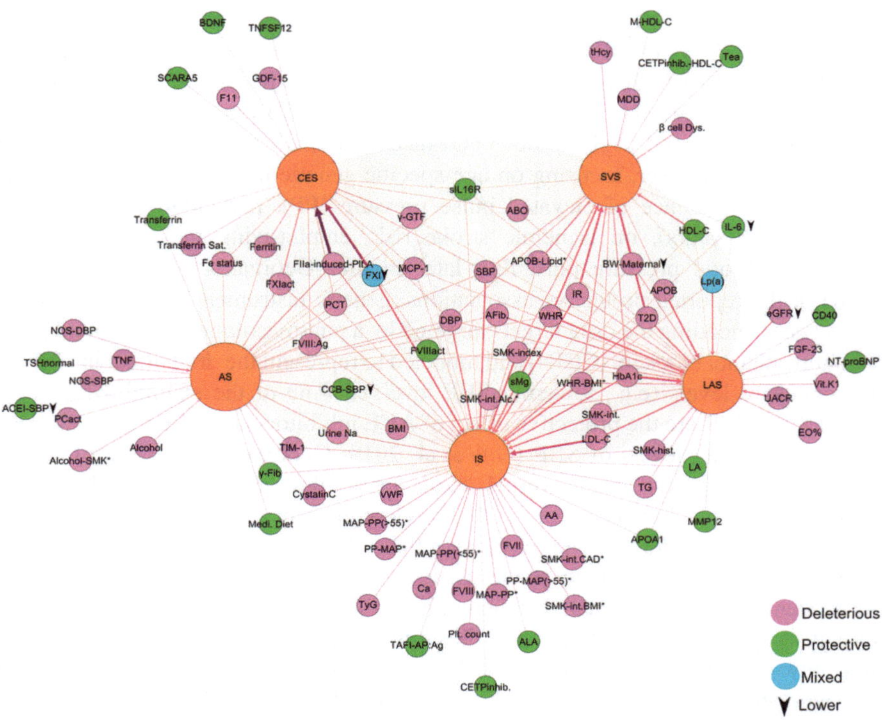

Fig. 4 Summary of 2SMR results with stroke as the outcome. Genetically predicted higher levels of the exposure are associated with either higher risk (purple circle) or lower risk (green circle) of stroke, backed by multiple literature sources. Results with mixed findings are depicted in the blue circle. Arrow marks facing down: genetic instruments defined as exposure decreasing

ischemic subtypes. Notable associations include the strong effect of higher genetically predicted levels of blood pressure traits, obesity-related measures, lipid traits, and lifestyle factors, such as smoking, with higher stroke risk.

Certain associations highlight notable physiological links. For example, thrombin-induced platelet aggregation (FIIa-induced-Plt.A) and factor XI (FXI) showing clear links to cardioembolic stroke, vitamin K1 involved in blood coagulation, and fibroblast growth factor 23 (FGF23) known as an early marker of chronic kidney disease (CKD) exhibit distinct associations with large artery stroke (LAS). The association of FGF23 with LAS underscores a potential causal link with kidney function, as indicated by the association of low estimated glomerular filtration rate (eGFR) and urinary albumin to creatinine ratio (UACR) with LAS (Marini et al. 2020a). Considering the prominent role of CKD and its markers in the causal pathway for stroke risk (Longenecker et al. 2002), such exposures can function as latent variables in future risk-stratification strategies. Multivariable MR analyses provide additional nuanced insights, such as highlighting the detrimental impact of smoking even after accounting for BMI, coronary artery disease (CAD) status, and alcohol intake (Fig. 4: SMK-int. BMI, SMK-int. CAD, SMK-int. Alc.) (Larsson et al. 2020). They unveil BMI-independent associations, such as waist–hip ratio (WHR), linked to both LAS and small vessel stroke (Marini et al. 2020b), and underscore the crucial role of apolipoprotein B (ApoB) in the pathogenesis of LDL-cholesterol- and triglycerides (TG)-mediated small vessel stroke risk (Yuan et al. 2020). Moreover, focusing on age-specific genetic instruments highlights the independent influence of elevated pulse pressure (PP), particularly in older age groups (>55 years), on ischemic stroke risk, emphasizing the significance of addressing blood pressure pulsatility later in life (Georgakis et al. 2020a). For some of the reported protective associations, genetic instruments were constructed from genes encoding the drug targets targeted by a pharmacological compound [e.g., calcium channel blockers (CCB)] and modeled to represent the drug effect in altering the exposure levels (e.g., lowering SBP levels). Some of the examples for drug-target-based MR are the protective effect of ACE inhibitors and CCBs on stroke risk through lowering levels of SBP (Gill et al. 2019; Georgakis et al. 2020b) and of CETP inhibitors on small vessel stroke risk through increasing levels of HDL-cholesterol (Johannsen et al. 2012; Georgakis et al. 2020c). Furthermore, focusing on exposures (proteome) that reflect closely underlying biological processes can mitigate effectively issues associated with biological pleiotropy, such as those stemming from alternative splicing, micro-RNA regulation, or protein-folding effects. A systematic investigation of the circulating proteome using MR techniques not only validated established targets like apolipoprotein A, but it also unearthed novel targets specific to cardioembolic stroke (SCARA5, TNSF12) and LAS (CD40, NT-proBNP) mechanisms (Chong et al. 2019). With the availability of a genome-wide map of the protein quantitative trait locus (pQTLs) (Sun et al. 2018) from multiple blood protein analytes, it is now possible to extend such exploration to a near hypothesis-free strategy identifying novel drug targets and assess their safety implications.

Recently, a global study examining the enrichment of disease-risk SNPs across 24 trait domains, including neurological traits, revealed significant enrichment of the disease variants in active chromatin regions (Watanabe et al. 2019). This underscores the regulatory activity of disease-risk SNPs in a cell-type-specific manner (Roadmap Epigenomics et al. 2015), suggesting potential pleiotropy also at the cell-type level (cell-type-mediated pleiotropy). Single-cell RNA sequencing (scRNA-seq) data offer a robust platform for testing this hypothesis. Analyses focused on cell-type specificity across multiple scRNA-seq datasets showed a notable enrichment of the expression profile of cardiovascular risk (including stroke) loci in endothelial and glial cells (Watanabe et al. 2019). Similarly, WMH and stroke, which share over 60% of their risk loci with cardiovascular risk factors, exhibit a similar enrichment profile for endothelial cells (Sargurupremraj et al. 2020). Thus, the shared cell and tissue types among closely related disease mechanisms have major implications for investigating causal inference using genetic determinants of cellular phenotypes (e.g., gene expression, protein levels).

5 Conclusions

In summary, while GWASs have identified numerous disease-risk loci, challenges reminiscent of the linkage-analysis era persist when it comes to identifying the causal variant or gene. But recent advances in statistical methods have enabled the identification of putative causal variants, even from regions not reaching statistical significance (due to sample-size limitations). Moreover, profiling cellular phenotypes at the single-cell level has expanded the application of GWAS findings, elucidating how alterations in gene expression and protein levels mediate putative causal relationships between risk factors and disease outcomes. As the promise of rare variants to explain missing heritability faces scrutiny (Weiner et al. 2023), and with the wealth of data from biobank-scale studies, it is imperative to focus on other genetic variations like structural (e.g., splice) variants with regulatory potential. These variants might play a crucial role in enhancing our efforts in fine-mapping of genetic loci associated with complex traits. Lastly, comprehensive perspectives, including corroborating evidence from well-designed observational studies, are essential in causal inference studies to mitigate better the influence of confounders (Sargurupremraj et al. 2024).

References

Al Kuwari H, Al Thani A, Al Marri A et al (2015) The Qatar Biobank: background and methods. BMC Public Health 15(1). https://doi.org/10.1186/s12889-015-2522-7

Armstrong NJ, Mather KA, Sargurupremraj M et al (2020) Common genetic variation indicates separate causes for periventricular and deep white matter hyperintensities. Stroke 51(7): 2111–2121. https://doi.org/10.1161/strokeaha.119.027544

Atkinson EG (2023) Estimation of cross-ancestry genetic correlations within ancestry tracts of admixed samples. Nat Genet 55(4):527–529. https://doi.org/10.1038/s41588-023-01325-x

Bak S, Gaist D, Sindrup SH, Skytthe A, Christensen K (2002) Genetic liability in stroke. Stroke 33(3):769–774. https://doi.org/10.1161/hs0302.103619

Barr TL, Conley Y, Ding J et al (2010) Genomic biomarkers and cellular pathways of ischemic stroke by RNA gene expression profiling. Neurology 75(11):1009–1014. https://doi.org/10.1212/WNL.0b013e3181f2b37f

Bellenguez C, Küçükali F, Jansen IE et al (2022) New insights into the genetic etiology of Alzheimer's disease and related dementias. Nat Genet. https://doi.org/10.1038/s41588-022-01024-z

Berisa T, Pickrell JK (2016) Approximately independent linkage disequilibrium blocks in human populations. Bioinformatics 32(2):283–285. https://doi.org/10.1093/bioinformatics/btv546

Bevan S, Traylor M, Adib-Samii P et al (2012) Genetic heritability of ischemic stroke and the contribution of previously reported candidate gene and Genomewide associations. Stroke 43(12):3161–3167. https://doi.org/10.1161/strokeaha.112.665760

Bowden J, Holmes MV (2019) Meta-analysis and Mendelian randomization: a review. Res Synth Methods 10(4):486–496. https://doi.org/10.1002/jrsm.1346

Boyle AE, Li IY, Pritchard KJ (2017) An expanded view of complex traits: from polygenic to Omnigenic. Cell 169(7):1177–1186. https://doi.org/10.1016/j.cell.2017.05.038

Bulik-Sullivan BK, Loh PR, Finucane HK et al (2015) LD score regression distinguishes confounding from polygenicity in genome-wide association studies. Nat Genet 47(3):291–295. https://doi.org/10.1038/ng.3211

Burgess S, Thompson SG (2011) Avoiding bias from weak instruments in Mendelian randomization studies. Int J Epidemiol 40(3):755–764. https://doi.org/10.1093/ije/dyr036

Burgess S, Small DS, Thompson SG (2017) A review of instrumental variable estimators for Mendelian randomization. Stat Methods Med Res 26(5):2333–2355. https://doi.org/10.1177/0962280215597579

Chen Z, Boehnke M, Wen X, Mukherjee B (2021) Revisiting the genome-wide significance threshold for common variant GWAS. G3 (Bethesda) 11(2). https://doi.org/10.1093/g3journal/jkaa056

Cho SY, Hong EJ, Nam JM, Han B, Chu C, Park O (2012) Opening of the national Biobank of Korea as the infrastructure of future biomedical science in Korea. Osong Public Health Res Perspect 3(3):177–184. https://doi.org/10.1016/j.phrp.2012.07.004

Chong M, Sjaarda J, Pigeyre M et al (2019) Novel drug targets for ischemic stroke identified through Mendelian randomization analysis of the blood proteome. Circulation 140(10):819–830. https://doi.org/10.1161/CIRCULATIONAHA.119.040180

Davey Smith G, Ebrahim S (2003) 'Mendelian randomization': can genetic epidemiology contribute to understanding environmental determinants of disease?*. Int J Epidemiol 32(1):1–22. https://doi.org/10.1093/ije/dyg070

De La Fuente J, Grotzinger AD, Marioni RE, Nivard MG, Tucker-Drob EM (2022) Integrated analysis of direct and proxy genome wide association studies highlights polygenicity of Alzheimer's disease outside of the APOE region. PLoS Genet 18(6):e1010208. https://doi.org/10.1371/journal.pgen.1010208

Denny JC, Bastarache L, Ritchie MD et al (2013) Systematic comparison of phenome-wide association study of electronic medical record data and genome-wide association study data. Nat Biotechnol 31(12):1102–1111. https://doi.org/10.1038/nbt.2749

Duperron MG, Tzourio C, Sargurupremraj M et al (2018) Burden of dilated perivascular spaces, an emerging marker of cerebral small vessel disease, is highly heritable. Stroke 49(2):282–287. https://doi.org/10.1161/strokeaha.117.019309

Escott-Price V, Hardy J (2022) Genome-wide association studies for Alzheimer's disease: bigger is not always better. Brain Commun 4(3):fcac125. https://doi.org/10.1093/braincomms/fcac125

Feng YA, Chen CY, Chen TT et al (2022) Taiwan biobank: a rich biomedical research database of the Taiwanese population. Cell Genom 2(11):100197. https://doi.org/10.1016/j.xgen.2022.100197

Finucane HK, Bulik-Sullivan B, Gusev A et al (2015) Partitioning heritability by functional annotation using genome-wide association summary statistics. Nat Genet 47(11):1228–1235. https://doi.org/10.1038/ng.3404

Fisher RA (1919) XV. The correlation between relatives on the supposition of Mendelian inheritance. Trans R Soc Edinburgh 52(2):399–433. https://doi.org/10.1017/s0080456800012163

Fornage M, Debette S, Bis JC et al (2011) Genome-wide association studies of cerebral white matter lesion burden: the CHARGE consortium. Ann Neurol 69(6):928–939. https://doi.org/10.1002/ana.22403

Gao X, Becker LC, Becker DM, Starmer JD, Province MA (2010) Avoiding the high Bonferroni penalty in genome-wide association studies. Genet Epidemiol 34(1):100–105. https://doi.org/10.1002/gepi.20430

Georgakis MK, Gill D, Malik R, Protogerou AD, Webb AJS, Dichgans M (2020a) Genetically predicted blood pressure across the lifespan: differential effects of mean and pulse pressure on stroke risk. Hypertension 76(3):953–961. https://doi.org/10.1161/HYPERTENSIONAHA.120.15136

Georgakis MK, Gill D, Webb AJS et al (2020b) Genetically determined blood pressure, antihypertensive drug classes, and risk of stroke subtypes. Neurology 95(4):e353–e361. https://doi.org/10.1212/WNL.0000000000009814

Georgakis MK, Malik R, Anderson CD, Parhofer KG, Hopewell JC, Dichgans M (2020c) Genetic determinants of blood lipids and cerebral small vessel disease: role of high-density lipoprotein cholesterol. Brain 143(2):597–610. https://doi.org/10.1093/brain/awz413

Gill D, Georgakis MK, Koskeridis F et al (2019) Use of genetic variants related to antihypertensive drugs to inform on efficacy and side effects. Circulation 140(4):270–279. https://doi.org/10.1161/CIRCULATIONAHA.118.038814

Harshfield EL, Sims MC, Traylor M, Ouwehand WH, Markus HS (2020) The role of haematological traits in risk of ischaemic stroke and its subtypes. Brain 143(1):210–221. https://doi.org/10.1093/brain/awz362

Hemani G, Bowden J, Davey SG (2018) Evaluating the potential role of pleiotropy in Mendelian randomization studies. Hum Mol Genet 27(R2):R195–R208. https://doi.org/10.1093/hmg/ddy163

Hormozdiari F, Kostem E, Kang EY, Pasaniuc B, Eskin E (2014) Identifying causal variants at loci with multiple signals of association. Genetics 198(2):497–508. https://doi.org/10.1534/genetics.114.167908

Johannsen TH, Frikke-Schmidt R, Schou J, Nordestgaard BG, Tybjaerg-Hansen A (2012) Genetic inhibition of CETP, ischemic vascular disease and mortality, and possible adverse effects. J Am Coll Cardiol 60(20):2041–2048. https://doi.org/10.1016/j.jacc.2012.07.045

Johnson RC, Nelson GW, Troyer JL et al (2010) Accounting for multiple comparisons in a genome-wide association study (GWAS). BMC Genomics 11:724. https://doi.org/10.1186/1471-2164-11-724

Karlsson IK, Escott-Price V, Gatz M et al (2022) Measuring heritable contributions to Alzheimer's disease: polygenic risk score analysis with twins. Brain Commun 4(1):fcab308. https://doi.org/10.1093/braincomms/fcab308

Kunkle BW, Schmidt M, Klein HU et al (2021) Novel Alzheimer disease risk loci and pathways in African American individuals using the African genome resources panel: a meta-analysis. JAMA Neurol 78(1):102–113. https://doi.org/10.1001/jamaneurol.2020.3536

Larsson SC, Mason AM, Back M et al (2020) Genetic predisposition to smoking in relation to 14 cardiovascular diseases. Eur Heart J 41(35):3304–3310. https://doi.org/10.1093/eurheartj/ehaa193

Laumet G, Petitprez V, Sillaire A et al (2010) A study of the association between the ADAM12 and SH3PXD2A (SH3MD1) genes and Alzheimer's disease. Neurosci Lett 468(1):1–2. https://doi.org/10.1016/j.neulet.2009.10.040

Li M-X, Yeung JMY, Cherny SS, Sham PC (2012) Evaluating the effective numbers of independent tests and significant p-value thresholds in commercial genotyping arrays and public imputation reference datasets. Hum Genet 131(5):747–756. https://doi.org/10.1007/s00439-011-1118-2

Lieberman-Aiden E, Van Berkum NL, Williams L et al (2009) Comprehensive mapping of long-range interactions reveals folding principles of the human genome. Science 326(5950):289–293. https://doi.org/10.1126/science.1181369

Liu X, Li YI, Pritchard JK (2019) Trans effects on gene expression can drive omnigenic inheritance. Cell 177(4):1022–1034.e6. https://doi.org/10.1016/j.cell.2019.04.014

Loh PR, Tucker G, Bulik-Sullivan BK et al (2015) Efficient Bayesian mixed-model analysis increases association power in large cohorts. Nat Genet 47(3):284–290. https://doi.org/10.1038/ng.3190

Loh PR, Kichaev G, Gazal S, Schoech AP, Price AL (2018) Mixed-model association for Biobank-scale datasets. Nat Genet 50(7):906–908. https://doi.org/10.1038/s41588-018-0144-6

Longenecker JC, Coresh J, Powe NR et al (2002) Traditional cardiovascular disease risk factors in dialysis patients compared with the general population: the CHOICE study. J Am Soc Nephrol 13(7):1918–1927. https://doi.org/10.1097/01.asn.0000019641.41496.1e

Lu Z, Gopalan S, Yuan D et al (2022) Multi-ancestry fine-mapping improves precision to identify causal genes in transcriptome-wide association studies. Am J Hum Genet 109(8):1388–1404. https://doi.org/10.1016/j.ajhg.2022.07.002

Ma D, Zheng B, Suzuki T et al (2017) Inhibition of KLF5-Myo9b-RhoA pathway-mediated podosome formation in macrophages ameliorates abdominal aortic aneurysm. Circ Res 120(5):799–815. https://doi.org/10.1161/CIRCRESAHA.116.310367

Malik R, Dichgans M (2018) Challenges and opportunities in stroke genetics. Oxford University Press, pp 1226–1240

Malik R, Chauhan G, Traylor M et al (2018a) Multiancestry genome-wide association study of 520,000 subjects identifies 32 loci associated with stroke and stroke subtypes. Nat Genet 50(4): 524–537. https://doi.org/10.1038/s41588-018-0058-3

Malik R, Rannikmäe K, Traylor M et al (2018b) Genome-wide meta-analysis identifies 3 novel loci associated with stroke. Ann Neurol 84(6):934–939. https://doi.org/10.1002/ana.25369

Marigorta UM, Navarro A (2013) High trans-ethnic replicability of GWAS results implies common causal variants. PLoS Genet 9(6):e1003566. https://doi.org/10.1371/journal.pgen.1003566

Marini S, Georgakis MK, Chung J et al (2020a) Genetic overlap and causal inferences between kidney function and cerebrovascular disease. Neurology 94(24):e2581–e2591. https://doi.org/10.1212/WNL.0000000000009642

Marini S, Merino J, Montgomery BE et al (2020b) Mendelian randomization study of obesity and cerebrovascular disease. Ann Neurol 87(4):516–524. https://doi.org/10.1002/ana.25686

Mbatchou J, Barnard L, Backman J et al (2021) Computationally efficient whole-genome regression for quantitative and binary traits. Nat Genet 53(7):1097–1103. https://doi.org/10.1038/s41588-021-00870-7

Mishra A, Malik R, Hachiya T et al (2022) Stroke genetics informs drug discovery and risk prediction across ancestries. Nature 611(7934):115–123. https://doi.org/10.1038/s41586-022-05165-3

Morrison J, Knoblauch N, Marcus JH, Stephens M, He X (2020) Mendelian randomization accounting for correlated and uncorrelated pleiotropic effects using genome-wide summary statistics. Nat Genet 52(7):740–747. https://doi.org/10.1038/s41588-020-0631-4

Nagai A, Hirata M, Kamatani Y et al (2017) Overview of the BioBank Japan project: study design and profile. J Epidemiol 27(3S):S2–S8. https://doi.org/10.1016/j.je.2016.12.005

O'Connor LJ (2021) The distribution of common-variant effect sizes. Nat Genet 53(8):1243–1249. https://doi.org/10.1038/s41588-021-00901-3

Palmer C, Pe'er I (2017) Statistical correction of the winner's curse explains replication variability in quantitative trait genome-wide association studies. PLoS Genet 13(7):e1006916. https://doi.org/10.1371/journal.pgen.1006916

Persyn E, Hanscombe KB, Howson JMM, Lewis CM, Traylor M, Markus HS (2020) Genome-wide association study of MRI markers of cerebral small vessel disease in 42,310 participants. Nat Commun 11(1):1–12. https://doi.org/10.1038/s41467-020-15932-3

Qin X, Akter F, Qin L et al (2020) Adaptive immunity regulation and cerebral ischemia. Front Immunol 11:689. https://doi.org/10.3389/fimmu.2020.00689

Roadmap Epigenomics C, Kundaje A, Meuleman W et al (2015) Integrative analysis of 111 reference human epigenomes. Nature 518(7539):317–330. https://doi.org/10.1038/nature14248

Roselli C, Chaffin MD, Weng LC et al (2018) Multi-ethnic genome-wide association study for atrial fibrillation. Nat Genet 50(9):1225–1233. https://doi.org/10.1038/s41588-018-0133-9

Sargurupremraj M, Suzuki H, Jian X et al (2020) Cerebral small vessel disease genomics and its implications across the lifespan. Nat Commun 11(1). https://doi.org/10.1038/s41467-020-19111-2

Sargurupremraj M, Soumaré A, Bis JC et al (2024) Genetic complexities of cerebral small vessel disease, blood pressure, and dementia. JAMA Netw Open 7(5):e2412824. https://doi.org/10.1001/jamanetworkopen.2024.12824

Schwartzentruber J, Cooper S, Liu JZ et al (2021) Genome-wide meta-analysis, fine-mapping and integrative prioritization implicate new Alzheimer's disease risk genes. Nat Genet 53(3): 392–402. https://doi.org/10.1038/s41588-020-00776-w

Shi H, Kichaev G, Pasaniuc B (2016) Contrasting the genetic architecture of 30 complex traits from summary association data. Am J Hum Genet 99(1):139–153. https://doi.org/10.1016/j.ajhg.2016.05.013

Smith GD, Ebrahim S (2005) What can mendelian randomisation tell us about modifiable behavioural and environmental exposures? BMJ Publishing Group, pp 1076–1079

Sollis E, Mosaku A, Abid A et al (2023) The NHGRI-EBI GWAS catalog: knowledgebase and deposition resource. Nucleic Acids Res 51(D1):D977–D985. https://doi.org/10.1093/nar/gkac1010

Stephens M (2017) False discovery rates: a new deal. Biostatistics 18(2):275–294. https://doi.org/10.1093/biostatistics/kxw041

Sun BB, Maranville JC, Peters JE et al (2018) Genomic atlas of the human plasma proteome. Nature 558(7708):73–79. https://doi.org/10.1038/s41586-018-0175-2

Taliun D, Harris DN, Kessler MD et al (2021) Sequencing of 53,831 diverse genomes from the NHLBI TOPMed program. Nature 590(7845):290–299. https://doi.org/10.1038/s41586-021-03205-y

Tchetgen EJT, Sun B, Walter S (2017) The GENIUS approach to robust Mendelian randomization inference. Cold Spring Harbor Laboratory

The 1000 Genomes Project Consortium (2015) A global reference for human genetic variation. Nature 526(7571):68–74. https://doi.org/10.1038/nature15393

Traylor M, Anderson CD, Rutten-Jacobs LCA et al (2019) Subtype specificity of genetic loci associated with stroke in 16 664 cases and 32 792 controls. Circ Genom Precis Med 12(7): e002338. https://doi.org/10.1161/CIRCGEN.118.002338

Uffelmann E, Huang QQ, Munung NS et al (2021) Genome-wide association studies. Nat Rev Methods Primers 1(1). https://doi.org/10.1038/s43586-021-00056-9

Var SR, Shetty AV, Grande AW, Low WC, Cheeran MC (2021) Microglia and macrophages in neuroprotection, neurogenesis, and emerging therapies for stroke. Cells 10(12). https://doi.org/10.3390/cells10123555

Verhaaren BFJ, Debette S, Bis JC et al (2015) Multiethnic genome-wide association study of cerebral white matter hyperintensities on MRI. Circ Cardiovasc Genet 8(2):398–409. https://doi.org/10.1161/CIRCGENETICS.114.000858

Visscher PM, Wray NR, Zhang Q et al (2017) 10 years of GWAS discovery: biology, function, and translation. Am J Hum Genet 101(1):5–22. https://doi.org/10.1016/j.ajhg.2017.06.005

von Berg J, van der Laan SW, McArdle PF et al (2020) Alternate approach to stroke phenotyping identifies a genetic risk locus for small vessel stroke. Eur J Hum Genet 28(7):963–972. https://doi.org/10.1038/s41431-020-0580-5

Wang G, Sarkar A, Carbonetto P, Stephens M (2020) A simple new approach to variable selection in regression, with application to genetic fine mapping. J R Stat Soc Ser B Stat Methodol 82(5): 1273–1300. https://doi.org/10.1111/rssb.12388

Watanabe K, Stringer S, Frei O et al (2019) A global overview of pleiotropy and genetic architecture in complex traits. Nat Genet 51(9):1339–1348. https://doi.org/10.1038/s41588-019-0481-0

Weiner DJ, Nadig A, Jagadeesh KA et al (2023) Polygenic architecture of rare coding variation across 394,783 exomes. Nature 614(7948):492–499. https://doi.org/10.1038/s41586-022-05684-z

Willer CJ, Li Y, Abecasis GR (2010) METAL: fast and efficient meta-analysis of genomewide association scans. Bioinformatics 26(17):2190–2191. https://doi.org/10.1093/bioinformatics/btq340

Wray NR (2005) Allele frequencies and the r2 measure of linkage disequilibrium: impact on design and interpretation of association studies. Twin Res Hum Genet 8(2):87–94. https://doi.org/10.1375/1832427053738827

Xue C, Jiang L, Long Q, Chen Y, Li X, Li M (2020) A global overview of single-cell type selectivity and pleiotropy in complex diseases and traits. bioRxiv

Yuan S, Tang B, Zheng J, Larsson SC (2020) Circulating lipoprotein lipids, apolipoproteins and ischemic stroke. Ann Neurol 88(6):1229–1236. https://doi.org/10.1002/ana.25916

Zhou W, Nielsen JB, Fritsche LG et al (2018) Efficiently controlling for case-control imbalance and sample relatedness in large-scale genetic association studies. Nat Genet 50(9):1335–1341. https://doi.org/10.1038/s41588-018-0184-y

Zhou W, Kanai M, Wu KH et al (2022) Global Biobank meta-analysis initiative: powering genetic discovery across human disease. Cell Genom 2(10):100192. https://doi.org/10.1016/j.xgen.2022.100192

Transcriptomics

Approaches to Quantifying Gene Expression and Their Application to Studying the Human Brain

Lora Liharska and Alexander Charney

Contents

Abstract To date, the field of transcriptomics has been characterized by rapid methods development and technological advancement, with new technologies continuously rendering older ones obsolete.

L. Liharska (✉) and A. Charney (✉)
Icahn School of Medicine at Mount Sinai, New York, NY, USA
e-mail: lora.liharska@icahn.mssm.edu; alexander.charney@mssm.edu

Curr Topics Behav Neurosci (2024) 68: 129–176
https://doi.org/10.1007/7854_2024_466
Published Online: 8 July 2024

This chapter traces the evolution of approaches to quantifying gene expression and provides an overall view of the current state of the field of transcriptomics, its applications to the study of the human brain, and its place in the broader emerging multiomics landscape.

Keywords Data analysis · Gene expression · Multiomics · NGS · RNA · RNA-seq · Transcriptomics

Alternative splicing	Cellular process occurring during or after transcription by which different combinations of exons are included or excluded from a final processed (i.e., mature) RNA transcript, allowing for the production of multiple different (alternative) RNA transcripts from the same DNA sequence, which can be translated into multiple protein isoforms
Messenger RNA (mRNA)	Single-stranded RNA molecule created from a template protein-coding DNA sequence during transcription (i.e., a transcript) that corresponds to the genetic sequence of the DNA template and is in turn used by a ribosome as the template for synthesizing a protein
Microarray	Hybridization-based laboratory tool used to detect gene expression or genetic variation in many genes at the same time
Multiomics	Field of systems biology that integrates multiple biological data types into analyses to enable more comprehensive understanding of the molecular basis of health and disease
Ribonucleic acid (RNA)	Polymer molecule composed of a chain of (ribo)nucleotides that serves a multitude of diverse cellular functions required for survival
RNA-seq (i.e., RNA sequencing)	Refers to the next-generation sequencing technologies and associated computational methods for capturing and quantifying RNA transcripts present in a sample
RNA transcript (i.e., transcript)	Single-stranded RNA molecule synthesized by transcription of a DNA sequence; through the process of alternative splicing, multiple different RNA transcripts can be produced from the same DNA sequence
Transcriptome	The set of all RNA molecules (i.e., transcripts: protein-coding (mRNA) and non-coding RNA, including rRNA, tRNA, lncRNA, pri-miRNA, and others)

	expressed in an entire organism or in a specific tissue type or cell type at a given time
Transcriptomics	The study of the transcriptome; encompasses all aspects of RNA synthesis, structure, function, regulation, localization, trafficking, splicing, post-transcriptional modification, and degradation
Transcript expression	The expression of a specific RNA transcript of a DNA sequence (multiple different (i.e., alternative) RNA transcripts can be produced from the same DNA sequence through the process of alternative splicing)
Read	In DNA and RNA sequencing, a portion of a DNA/cDNA fragment that has been sequenced
Read count	In DNA and RNA sequencing, the number of reads that map (i.e., align) to a given genomic feature such as a gene using computational methods
Read length	In DNA and RNA sequencing, the number of bases of a DNA/cDNA fragment that are sequenced at a time
Sequencing depth (i.e., read depth; depth of coverage)	In DNA and RNA sequencing, the average number of times that each base is sequenced during a sequencing run; i.e., the average number of times a given position in the genome is sequenced
Sequencing coverage (i.e., coverage; read coverage)	In DNA and RNA sequencing, the proportion or percentage of a genome or genomic region that has been sequenced at a certain depth; calculated based on the number of unique reads mapped/aligned to a region in a reference genome
RNA splicing (i.e., splicing)	The process of removing introns and ligating together exons of a newly synthesized eukaryotic precursor RNA transcript in order to form the mature RNA transcript

1 Introduction to Transcriptomics

The transcriptome is the set of all ribonucleic acid (RNA) molecules – including protein-coding RNA (messenger RNA [mRNA]) and non-coding RNA (ribosomal RNA [rRNA], transfer RNA [tRNA], long non-coding RNA [lncRNA], microRNA [miRNA], and others) – expressed in an entire organism or in a specific tissue type or cell type at a given time (Thompson et al. 2016). The transcriptome is a link between the information contained in an organism's genome and the functioning of the organism. Transcriptomics is the study of the transcriptome and encompasses all aspects of RNA synthesis, structure, function, regulation, localization, trafficking, splicing, post-transcriptional modification, and degradation (Thompson et al. 2016).

Since the first (partial) human transcriptome was published in 1991 (Adams et al. 1991), the field of transcriptomics has been characterized by rapid methods development and technological advancement, resulting in new techniques continuously rendering previous technologies obsolete. Over the decades, different types of high-throughput technologies have been developed for in-depth characterization and quantification of RNA transcript expression. The study of the whole transcriptome has not been feasible until more recently, as the complexity of capturing the full range of transcripts expressed by an organism or tissue necessitates the use of sophisticated technology. The introduction of next-generation sequencing (NGS) technology in 2000 (Barba et al. 2014), which allows for comprehensive sequencing of the transcriptome with deep coverage and base-level resolution, has made it possible to detect and identify nearly every transcribed RNA molecule within a sample (Gupta and Gupta 2014).

This chapter discusses the role of transcriptomics in the study of the human brain, with a particular focus on evolving methods for quantification and characterization of gene expression, and how these have been applied to human brain research. Terms used throughout the text are defined in context in the Glossary section at the end of the chapter.

Genome Biology

In brief, information stored in DNA is utilized via two processes: transcription and translation. Transcription is the process by which a segment of DNA is copied to RNA. Translation is the process by which RNA is used to produce protein. Transcription and translation are complex multi-step cellular processes subject to many regulatory mechanisms (Clark et al. 2019).

In transcription, an RNA polymerase enzyme (with assistance from other transcription factor proteins) unwinds a portion of double-stranded DNA and uses base-pairing complementarity to create a complementary, anti-parallel, single-stranded RNA copy of one of the unwound DNA strands (i.e., the template strand). The complementary RNA molecule incorporates the nucleotide uracil (U) instead of thymine (T) when base-pairing with the alanine (A) present in the DNA sequence of the template strand.

After transcription, some RNA serves a functional role within the cell (i.e., non-coding RNA [ncRNA]), while other RNA undergoes translation (i.e., messenger RNA [mRNA]), ultimately resulting in the production of protein.

A segment of DNA encoding a protein typically includes (1) the coding sequence that is transcribed to mRNA and then translated into the protein and (2) regulatory sequences (that direct and regulate the synthesis of the protein) that are transcribed to ncRNA.

2 History of Assay Development

The discovery of the double helix structure of DNA in 1953 led to a paradigm shift in biological science. This ground-breaking finding gave rise to modern molecular biology, a new field focusing on the molecular basis of biological activity as the foundation for life. In the 1970s and 1980s, continuing advances in knowledge of the structural basis of DNA led to the development of novel scientific techniques that relied on modification of DNA to achieve desired results (e.g., recombinant DNA, genetic engineering, rapid gene sequencing, monoclonal antibodies). These techniques formed the basis for modern biotechnology, including the technology for sequencing the nucleic acids of the genome and transcriptome. Major advances in sequencing technology since the beginning of the twenty-first century have established nucleic acid sequencing as an integral tool for biomedical research and the basis of transcriptomics.

2.1 Beginnings

"... [K]nowledge of sequences could contribute much to our understanding of living matter." – Frederick Sanger (Odelberg and Nobel Foundation 1981).

The central dogma of molecular biology states that all of the biological information necessary for an organism to exist is encoded in the organism's DNA sequence and transferred to RNA sequences that the organism is able to use. The transcription of regions of DNA into complementary RNA molecules defines cellular identity and regulates cellular activity. As the sequential order of nucleotides in DNA and RNA polynucleotide chains ultimately encodes the hereditary and biochemical information that directs the development and functioning of the organism, the ability to extract this order is a necessary prerequisite to progress in the biological sciences.

In 1953, James Watson and Francis Crick solved the structure of DNA using data generated by Rosalind Franklin and Maurice Wilkins. Watson and Crick described DNA as two long helical strands wound together in a double helix, with each strand composed of individual bases that match the bases on the other strand. In addition to the structure of DNA, Watson and Crick introduced a conceptual framework for the mechanism of DNA replication (Crick and Watson 1954) and proposed that genes are determined by the specific sequence arrangement of bases in DNA (Watson and Crick 1953). Nevertheless, it took over a decade following the introduction of Watson and Crick's DNA model for the first sequence of bases in a polynucleotide chain to be decoded (Holley 1965), and nearly a quarter century for the first robust method for sequencing of DNA and RNA to be developed (Sanger et al. 1977a, b).

By the time Watson and Crick's DNA model was introduced, strategies for inferring the sequence of amino acids in protein chains were already well underway. By 1955, Frederick Sanger and colleagues (Sanger et al. 1955) had sequenced all 51 amino acids comprising the insulin molecule using the "degradation procedure," a method involving breaking up the insulin molecule into small fragments, identifying where the amino acids of the fragments overlap, and reconstructing the polypeptide chains based on the overlap (Sanger et al. 1955; Heather and Chain 2016).

One of the earliest nucleotide sequencing methods was an adaptation of Sanger's degradation procedure for sequencing of amino acids. Developed by biochemist Robert Holley and his colleagues and published in 1965, this method leveraged two newly discovered ribonucleases (i.e., RNase enzymes able to cut RNA molecules at specific locations) – pancreatic ribonuclease and takadiastase ribonuclease T1. Using this method, a 77-nucleotide-long purified yeast alanine tRNA molecule was split by the ribonucleases into 16 fragments at locations corresponding to specific nucleotides, and the tRNA sequence was then manually assembled from the overlapping fragments (Holley 1965).

In 1965, Sanger and colleagues sequenced another short RNA (a bacterial 5S rRNA consisting of 120 nucleotides) using a radioactive phosphorous isotope (32P) to label nucleotides and a two-dimensional fractionation procedure to separate nucleotides. In the following years, additional tRNAs were sequenced in rapid succession using various other fractionation methods developed by Sanger and colleagues (Brownlee et al. 1967).

In 1968, the first piece of DNA was sequenced by Ray Wu and A Dale Kaiser, which consisted of 12 nucleotides from the end of the linear genome of the recently purified lambda bacteriophage (Wu and Kaiser 1968). Wu and Kaiser's multi-step approach to deducing the DNA sequence included many techniques similar to those developed for RNA by Sanger and colleagues, and a novel technique that utilized the DNA polymerase enzyme discovered in 1955 to fill in the 5′ overhanging "cohesive" end of the DNA strand with radioactive nucleotides, one at a time, and measured their incorporation (Wu and Kaiser 1968; Heather and Chain 2016).

In 1973, after developing an eight-nucleotide-long primer for DNA polymerase, Sanger and colleagues announced the successful sequencing of 50 nucleotides of the DNA of the filamentous bacteriophage (f1 phage) (Sanger et al. 1973). The f1 DNA sequenced by Sanger was the longest piece of DNA sequenced up to that point, and its success demonstrated the feasibility of sequencing DNA by (1) using primer and DNA polymerase to substitute specific deoxynucleotides in a DNA chain with specific ribonucleotides (i.e., ribosubstitution), (2) using ribonuclease to cut DNA subject to ribosubstitution into fragments ending in specific ribonucleotides, and (3) manually piecing together the overlapping fragments.

In 1975, after years of testing different polymerases and refining fractionation techniques for separating polynucleotides by length using electrophoresis through multi-lane polyacrylamide gels, Sanger and Alan Coulson introduced the plus/minus technique. This novel method, which utilized DNA polymerase and radiolabeled nucleotides, did not require the manual piecing together of a sequence based on the overlaps in DNA fragments (Sanger and Coulson 1975). Rather, the technique produced a series of DNA molecules of varying lengths that could be separated using polyacrylamide gel electrophoresis (Sanger and Coulson 1975; Sanger et al. 1977a, b). Despite the plus/minus system's improvements over previous methods, technical problems with the system led Sanger and colleagues to continue developing other sequencing methods.

2.2 First-Generation Sequencing

The major breakthrough that forever changed the course of DNA sequencing technology was based on the work of Arthur Kornberg and colleagues, who in 1969 showed that it is possible to synthesize DNA chains that end in the nucleotide base thymine by adding the chain-terminating dideoxythymidine triphosphate (ddTTP), a chemical analogue of the thymine nucleotide that lacks the 3′ hydroxyl group required for the nucleotide to form a phosphodiester bond with the 5′ phosphate of the next nucleotide in the DNA (dNTP) (Atkinson et al. 1969). In 1975, Sanger's application of dideoxynucleotides (ddNTPs) gave rise to the chain-termination method (i.e., the Sanger dideoxy method), subsequently referred to as "Sanger sequencing" (Sanger et al. 1977a, b).

Sanger sequencing is considered a "first-generation" nucleic acid sequencing method, representing the birth of genome sequencing technology. In Sanger sequencing, the addition of radiolabeled chain-terminating ddNTPs along with the four standard dNTPs (i.e., dATP, dGTP, dCTP, and dTTP) to a DNA extension reaction results in the generation of DNA fragments of every possible length, because as a strand extends, an available chain-terminating ddNTP is randomly incorporated into the strand, terminating further elongation at that point in the strand. Applying this method in separate parallel reactions with each of the four ddNTPs (i.e., ddATP, ddGTP, ddCTP, or ddTTP), separating the generated DNA fragments by size on a four-lane polyacrylamide gel, and visualizing them by autoradiography or UV light on X-ray film or gel image allows for the inference of the correct nucleotide in each position of the sequence (Zhou and Li 2015).

In 1977, Allan Maxam and Walter Gilbert introduced another first-generation sequencing method developed in parallel to Sanger sequencing (Maxam and Gilbert 1977). In this method, termed "Maxam–Gilbert sequencing," DNA fragments were labeled using chemical treatments that introduced chemical modifications to DNA bases. Labeled DNA fragments were then separated by gel electrophoresis and visualized with autoradiography on X-ray film. As Maxam–Gilbert sequencing did not require the generation of single-stranded DNA, this method was initially more popular than Sanger sequencing. But as the method was technically complex, necessitated extensive use of toxic chemicals, and was difficult to scale up (Saccone and Pesole 2005), it was ultimately made obsolete following additional improvements to and automation of Sanger sequencing.

In the following decades, Sanger sequencing, which was more rapid, accurate, safe, and efficient than all previous methods, became the most widely used DNA sequencing technology. Ultimately, fully automated implementations of Sanger sequencing were used by the Human Genome Project to complete the original sequencing of the human genome over the course of 10 years at a cost of approximately \$3 billion (National Human Genome Research Institute 2022a; Schadt et al. 2010). Due to its ultra-high accuracy rate (approximately 99.99%) and long read length (average of 800 DNA bases sequenced at a time) (Venter et al. 2001; Shendure and Ji 2008), Sanger sequencing remains the gold standard in sequencing accuracy and is used for validation of results from second- and third-generation sequencing technologies (Heather and Chain 2016). Nevertheless, the major limitations of this method – its high cost and low throughput (i.e., inability to process large amounts of DNA per unit of time) – led to the development of the subsequent generation of sequencing technology (i.e., next-generation sequencing [NGS] methods).

2.3 *Hybridization-Based Transcriptome Profiling*

2.3.1 Southern Blot

In 1975, in parallel to the sequencing efforts of Sanger and others, Edwin Southern introduced the Southern blot, a method used to identify and quantify specific sequences of interest in DNA samples after separating DNA fragments by size (Southern 1975). In Southern blotting, DNA is digested into fragments using sequence-specific restriction enzymes, separated by size via gel electrophoresis, denatured in the gel, and transferred to a membrane (i.e., "blotted") (Merker and Arber 2010). The DNA is fixed to the membrane and incubated with radioactively labeled single-stranded probes, which hybridize to the complementary DNA sequences of interest on the membrane (Belmont 2008). The unhybridized probes are washed off and the membrane is analyzed by autoradiography and X-ray film to identify the bands of radioactively labeled DNA (Belmont 2008), which represent particular DNA sequences. Variations in experimental conditions (i.e., hybridization temperature and washing solutions) can allow probes to hybridize to DNA fragments that are only partially complementary to the probe sequences (Nickle and Barrette-Ng 2016). Since the introduction of the Southern blot, nonradioactive fluorescent labeling methods have been developed. One example of an application of Southern blotting is a genetic test for sickle cell anemia, which identifies the presence of the mutated gene that causes the disease (Chang and Kan 1982).

2.3.2 Northern Blot

In 1977, two years following the introduction of the Southern blot, James Alwine, David Kemp, and George Stark introduced the Northern blot, an analogous technique for the detection and quantification of specific RNA sequences in an RNA sample using radioactively labeled oligonucleotide probes complementary to all or part of the sequences of interest (Alwine et al. 1977). Northern blotting allows for determination of the size and relative abundance of RNA transcripts (including alternatively spliced transcripts), which provides insights into expression levels of individual genes (Arabameri and Tseng 2023). Although advances in transcriptomic technologies have replaced Northern blotting as a method of studying gene expression, this technique is still used as a gold standard for validation of data obtained from high-throughput whole-genome transcriptomic methods (Yang et al. 2022; Arabameri and Tseng 2023).

2.3.3 Polymerase Chain Reaction (PCR)

Developed in 1984 by Kary Mullis, the polymerase chain reaction (PCR) has become an essential tool in biomedical research. PCR is a technique for rapid production of millions or billions of copies of small DNA segments through multiple cycles of DNA synthesis (Mullis et al. 1986). Once copied (i.e., "amplified"), the DNA produced by PCR is used for many downstream molecular and genetic assays and analyses (National Human Genome Research Institute 2022b).

In the 1990s, modifications to the PCR technique led to the development of real-time PCR, a tool for detecting and quantifying expression profiles of individual genes in real time (Deepak et al. 2007). Real-time PCR uses fluorescent technology to contemporaneously monitor the accumulation of amplified DNA as the reaction progresses, allowing for the quantification of DNA abundance at the end of each cycle, rather than just at the end of the reaction, as in regular PCR (Adams 2020). This allows for the accurate and sensitive determination of the initial number of copies of the target DNA template being amplified. Results of real-time PCR can detect the presence or absence of a DNA sequence (i.e., qualitative real-time PCR) or the copy number of the DNA sequence (i.e., quantitative real-time PCR; qPCR) (Bio-Rad n.d.). Reverse transcription PCR (RT-PCR) approaches adapt PCR and real-time PCR for the quantification of gene expression by first reverse transcribing RNA into cDNA, which is used as the template for PCR and real-time PCR amplification (ThermoFisher Scientific n.d.).

Due to its easier automation, higher-throughput, lower required amount of starting DNA, and much higher sensitivity, real-time PCR techniques have largely replaced Southern and Northern blotting for addressing the same experimental questions (Hoebeeck et al. 2007; Fehr et al. 2000).

2.3.4 Colony Hybridization

In 1975, the same year as the introduction of the Southern blot technique, Michael Grunstein and David Hogness introduced the colony hybridization method for isolating specific DNA sequences from bacterial colonies containing hybrid DNA (Grunstein and Hogness 1975; Bumgarner 2013). In this method, DNA of interest is randomly inserted into *E. coli* plasmids through recombination, and these bacterial plasmids are cultured to form bacterial colonies, some of which continue to contain the DNA of interest. Bacterial colonies are lysed to access their plasmids, and the

DNA is denatured and fixed to a nitrocellulose filter (Grunstein and Hogness 1975). Radioactively labeled nucleic acid probes complementary to the sequences of interest are then hybridized to the respective DNA clusters on the filter, and the clusters are identified by autoradiography, thus allowing for rapid screening of thousands of bacterial colonies to identify clones containing the DNA of interest (Grunstein and Hogness 1975). The colony hybridization method, an early example of a labeled probe being used to identify sequences of interest, can be considered as the original DNA array (Bumgarner 2013). In 1979, Grunstein and Hogness's method was adapted to create ordered arrays of colonies in 144-well microplates, and these protocols were further refined over the next decade (Gergen et al. 1979; Bumgarner 2013).

2.3.5 Microarray

The late 1980s and early 1990s saw significant advances in methods development for DNA arrays, including the introduction of robotic technology for automating and parallelizing the processes for creating arrays of DNA clones in defined patterns (Bumgarner 2013). In 1995, the first miniaturized microarray was introduced, which utilized high-speed robotic printing of cDNA samples onto a standard glass microscope slide (Schena et al. 1995). A microarray is a device used to detect the simultaneous expression of thousands of genes in an experimental sample (Sturla et al. 2003). A DNA microarray is a solid surface (usually a glass slide; "called a 'gene chip' or 'DNA chip'") containing thousands to millions of known DNA sequences in defined positions acting as probes that detect gene expression in an experimental sample by hybridizing to fluorescently tagged complementary sequences within the sample (Govindarajan et al. 2012). DNA microarrays can be of various types, including oligonucleotide DNA microarrays, BAC (bacterial artificial chromosome) clone chips, and cDNA microarrays (consisting of cDNA fragments reverse transcribed from mRNA) (Brown et al. 2024). Microarray probe-target hybridization is detected and quantified using various techniques and detection devices.

Unlike older techniques like the Northern blot and RT-PCR, which allow testing of a few genes per experiment, microarray technology enables the simultaneous assaying of thousands of cDNA-converted transcripts on the same surface. In this way, microarrays represent a relatively fast, cost-effective, and high-throughput method for studying coordinated gene expression on an organismal scale. The first genome-wide brain gene expression studies using microarray began in 1999, and two of these pioneering studies reported differentially expressed genes in chromosomal regions with known quantitative trait loci (QTLs) (Hitzemann et al. 2014).

2.4 *Sequencing-Based Transcriptome Profiling*

DNA microarray approaches, which are hybridization-based, high-throughput, and relatively cost-effective, pioneered gene expression profiling on a genomic scale. Nevertheless, the fundamental limitations inherent to hybridization-based technology constrain the use of DNA microarrays in important ways. As a result, microarray-based technologies:

1. Detect only known genomic sequences corresponding to pre-defined probes on an individual array and thus cannot detect novel transcripts, novel sequence variations (e.g., single-nucleotide polymorphisms [SNPs]), gene fusions, and indels (i.e., small insertions and deletions) (Illumina n.d.; Shendure 2008; Wang et al. 2009)
2. Demonstrate limited specificity (i.e., limited capacity for unambiguous differentiation between transcripts with significant sequence homology or multiple splicing variants) due to high background levels (i.e., high ambient, non-specific fluorescent signal due to binding of transcripts to non-target probes, processing effects, and optical noise from scanner) (Okoniewski and Miller 2006; Koltai and Weingarten-Baror 2008; Ritchie et al. 2007)
3. Exhibit narrow dynamic range (i.e., limited sensitivity for simultaneous detection of genes expressed at very low levels and genes expressed at very high levels) due to high background levels at low concentrations of a hybridizing species, and saturation of the array at high concentrations (Illumina n.d.; Shendure 2008)
4. Require complex data normalization and analysis processes for comparing expression levels across different experiments (Wang et al. 2009).

These limitations have led to the eventual substitution of microarray-based measurements of global gene expression with the fundamentally different "next-generation" massively parallel DNA sequencing technology arising out of Sanger sequencing.

2.4.1 Tag-Based Approaches

In contrast to hybridization-based approaches, sequencing-based methods allow for the direct determination of cDNA sequences. In the 1980s and 1990s, improvements in automation of the low-throughput Sanger sequencing led to the development of tag-based approaches (i.e., approaches utilizing expressed sequence tags [ESTs]). ESTs are short, unedited, randomly selected single-pass sequence reads from mRNA (cDNA) that are usually produced in large batches (BLAST 2017). ESTs represent portions of expressed genes and thus provide a snapshot of the transcriptional activity of a given tissue at a given time.

Tag-based approaches – including serial analysis of gene expression (SAGE), cap analysis of gene expression (CAGE), and massively parallel signature sequencing (MPSS) (Wang et al. 2009) – were early sequencing-based transcriptomic methods developed to increase throughput of generated ESTs and provide precise quantification of gene expression levels. These methods, however, were subject to serious inherent limitations (e.g., significant portion of the short tags cannot be uniquely mapped to the reference genome; only a portion of a transcript is analyzed, and isoforms are generally indistinguishable from each other; sample preparation and data analysis are labor-intensive) (Lowe et al. 2017).

2.4.2 Next-Generation Sequencing

Next-generation sequencing (NGS) (i.e., massively parallel sequencing; deep sequencing) technologies refer to various related large-scale, high-throughput sequencing technologies that perform sequencing of millions of small DNA fragments in parallel (Behjati and Tarpey 2013). In the early 2000s, NGS sequencing methods and technologies started to become commercially available as higher-throughput alternatives to the "first-generation" automated Sanger-sequencing-based dideoxy sequencing machines used to sequence the first human genome (Barba et al. 2014). In 2004, the first NGS platform, the Roche 454, was introduced, and this was quickly followed by the release of other platforms by other manufacturers (Barba et al. 2014). Although these instruments varied in their underlying chemistry and technical methodology, all had the capability to sequence large numbers of nucleotides simultaneously, quickly, accurately, and at reduced cost (Richards 2015; Behjati and Tarpey 2013). For reference, sequencing of the human genome using Sanger sequencing technology took the Human Genome Project over a decade to complete (completed in 2003), while current NGS technologies can sequence an entire human genome in a single day (Behjati and Tarpey 2013). By the early 2010s, the NGS technology for RNA sequencing (i.e., sequencing of the transcriptome), known as RNA-seq, had largely replaced microarray-based assessments of global gene expression (Stark et al. 2019).

2.4.3 RNA-seq

RNA-seq refers to the NGS technologies and associated computational methods for capturing and quantifying transcripts present in an RNA extract (Lowe et al. 2017). Some of the key advantages of RNA-seq over existing hybridization-based microarray technologies are that RNA-seq:

1. Directly determines cDNA sequences, including novel sequences and sequence variations (e.g., SNPs) in transcribed regions
2. Allows for unambiguous mapping of cDNA sequences to unique regions of the reference genome, resulting in low background signal
3. Exhibits wide dynamic range (i.e., high sensitivity for simultaneous detection of genes expressed at very low levels and genes expressed at very high levels)

4. Accurately and reproducibly quantifies expression levels of biological and technical replicates (Wang et al. 2009; Shendure 2008).

Various RNA-seq methods have been developed for studying the transcriptome at different resolution.

2.4.4 Short-Read Sequencing

NGS technologies utilize short-read sequencing methods, which sequence large stretches of DNA (or cDNA in the case of RNA-seq) by breaking the genome into small fragments, amplifying the fragments before sequencing, sequencing the fragments, and using computational methods to assemble the random duplicated sequenced reads into a continuous sequence (Deshpande et al. 2023).

2.4.5 Bulk RNA-seq

Bulk RNA-seq is used to capture and quantify pooled RNA from cells or tissues. Bulk RNA-seq measures the average gene expression of individual genes across the hundreds to millions of cells in each sample to provide a global view of gene expression within samples and differences in gene expression between samples (Stark et al. 2019).

2.4.6 Short-Read Bulk RNA-seq

In a typical short-read bulk RNA-seq experiment, (1) a sequencing library (i.e., a pool of cDNA fragments with sequencing adapters attached) is prepared from RNA that has been extracted and converted to cDNA, and (2) the library is sequenced on an NGS platform (Wang et al. 2009).

In brief, first, RNA is isolated from the biological sample. Next, the sequencing library is constructed by isolating the desired population of RNA species (e.g., mRNA via selection of polyadenylated (poly-A) RNA species, selective depletion of rRNA species), reverse-transcribing the RNA to cDNA, and adding sequencing adaptors to one or both ends of each cDNA molecule. Next, the library of cDNA fragments is amplified via PCR and each cDNA fragment is sequenced using a high-throughput NGS sequencing platform to obtain short sequences of 30–400 base pairs (i.e., reads), depending on the sequencing technology used. cDNA fragments can be sequenced from one end only (i.e., single-end sequencing) or from both ends (i.e., paired-end sequencing). Sequencing depth (i.e., the total number of reads sequenced in an RNA sequencing experiment) is one of the most important considerations when designing an RNA sequencing experiment and depends on the goals of the study (e.g., studies aiming to identify genes with low expression levels require higher sequencing depth, while studies measuring differential expression between genes in two groups may not) (Wang et al. 2009). Following completion of sequencing, the sequence reads are computationally aligned to a reference genome or transcriptome or assembled de novo.

2.4.7 Single-Cell RNA-seq (scRNA-seq)

Bulk RNA-seq requires homogenization of each tissue sample, which masks the differences in gene expression between individual cells and cell types within the sample. Therefore, for a given gene, bulk RNA-seq measures the gene's average expression across a heterogenous population of cells (Yu et al. 2021). As the human brain is comprised of many different types of cells situated in close proximity, understanding brain function necessitates determining the gene expression profiles of individual cells and the differences in expression across diverse cell types. Transcriptome sequencing of individual cells can facilitate categorization of cell types based on their distinct molecular profiles, identification of new cell types, characterization of cell-type-specific gene expression, and understanding of the organization and function of cell types within tissues.

In 2009, the first protocol for single-cell transcriptome analysis using NGS technology was introduced, with which single mouse blastomeres and oocytes were manually picked under the microscope and subjected to mRNA sequencing (Piwecka et al. 2023). Single-cell RNA sequencing methods and technologies (scRNA-seq) have rapidly evolved since then, including improvements in optimization of workflows, scaling up of the technology, introduction of unique molecular identifiers and cell barcoding for accurate annotation of cells and counting of RNA molecules, and advances in computational analysis (Piwecka et al. 2023). In 2015, improvements in adaptation of microfluidics for capture of cells led to the introduction of droplet-based microfluidic methods for scRNA-seq, which significantly increase reaction throughput (Zhang et al. 2019).

scRNA-seq measures both cytoplasmic and nuclear transcripts. But as most cellular mRNA is mature cytoplasmic mRNA that has been exported out of the nucleus, scRNA-seq typically focuses on exon reads and provides limited insight into the pre-mRNA status of genes (Kang et al. 2023). The main process steps of a scRNA-seq experiment are analogous to those of a bulk RNA-seq experiment. In a typical scRNA-seq experiment, (1) the sample is dissociated into single cells floating in suspension, (2) live single cells are isolated – using a variety of methods depending on the contents of the sample and the processing steps to be performed following isolation (e.g., fluorescence-activated cell sorting [FACS], magnetic-activated cell sorting [MACS], laser capture microdissection [LCM], and micromanipulation) – and then lysed (Gross et al. 2015; Valihrach et al. 2018), (3) for each isolated cell, RNA transcripts are captured and reverse-transcribed to single-stranded cDNA, (4) cDNA is amplified and cell-specific barcodes and specific adapter sequences are added to the amplified cDNA fragments at one or both ends, (5) all material from all cells in the sample is multiplexed (i.e., pooled into a single sequencing library based on cell-specific barcodes) and a sample-specific barcode is added to all material from the sample, (6) the library is sequenced on an NGS platform, and (7) the raw sequencing data is aligned to the reference transcriptome, processed, and analyzed (Vaga 2022).

Table 1 outlines some of the most common applications of bulk RNA-seq and scRNA-seq data.

Table 1 Common applications of bulk RNA-seq and scRNA-seq data

Common application of RNA-seq data	Bulk RNA-seq	scRNA-seq
Differential gene expression analysis: Identification of differentially expressed genes and transcripts between tissues, cell types, and experimental conditions (e.g., disease state, drug treatment, time course)	✔	✔
Transcriptome annotation Identification of novel transcripts, isoforms, and non-coding RNAs, and refinement of known annotations	✔	✔
Alternative splicing analysis: Understanding of alternative splicing events, regulation of alternative splicing, and the functional consequences of alternative splicing	✔	✔
Identification of gene fusion events and fusion genes: Fusion genes are often associated with specific cancer types	✔	✔
Co-expression analysis: Identification of gene expression patterns (i.e., modules) that are coordinated across tissues, cell types, regions, experimental conditions, etc.	✔	✔
Enrichment analysis: Elucidation of molecular pathways and biological processes associated with differentially expressed genes and co-expression modules	✔	✔
Understanding of developmental biology and mechanisms for determination of cellular fate: Including cellular differentiation, lineage tracing, and developmental trajectories of cells and cell types	✔	✔
Gene regulatory network analysis: Inference of gene regulatory networks and identification of key regulators of cellular processes from gene expression patterns	✔	✔
Analyses integrating other data types: Including clinical data, imaging data, other omics data (e.g., proteomics, metabolomics, genomics)	✔	✔
Identification and characterization of cell types: Comparison of gene expression profiles of individual cells to identify and characterize novel and rare cell types and further refine knowledge of identified cell types	×	✔
Understanding of cellular heterogeneity within tissues: Identification of similar and distinct cell states and populations, which may be masked in bulk RNA-seq data (e.g., tumor microenvironment, immune cell populations)	×	✔

2.4.8 Single-Nucleus RNA-seq (snRNA-seq)

Despite the advantages of scRNA-seq over bulk RNA-seq in terms of the increased resolution to study the transcriptomes of individual cells within the same tissue sample, limitations to scRNA-seq prevent its use in all cases. Intact single cells of certain cell types and conditions (e.g., adipocytes, neurons, multinucleated cells, cells with damaged membranes in archived frozen tissue) are difficult to isolate due to their tight interconnections, large size, fragility, and other factors (Nadelmann et al. 2021). Neurons are particularly affected by these limitations – isolation of intact single neurons is complicated by the interconnectedness of neurons in the brain (Lake et al. 2017) and the damaging effects of the enzymatic dissociation processes required to isolate the cells from tissue, which can also result in stress-induced transcriptional artifacts (Habib et al. 2017; Wu et al. 2019). For this reason, it has been shown that in most tissues, scRNA-seq is biased toward recovery of some cell types (e.g., immune cell types not anchored in tissue) over others (Kim et al. 2023) – in adult human neocortex, non-neuronal cells have been shown to be over-represented in single-cell suspensions (Darmanis et al. 2015). Furthermore, current scRNA-seq protocols require fresh tissue and cannot be performed with cells from archived frozen tissues or lightly fixed tissues (Habib et al. 2017).

Single-nucleus RNA-seq (snRNA-seq) was developed as an alternative to scRNA-seq for obtaining transcriptomic data from cells that are difficult to dissociate successfully (Kim et al. 2023). For snRNA-seq, only the nuclei of single cells, which are more resistant to freeze-thaw stress than whole cells, are isolated to generate gene expression data, rather than the entire cells, thus bypassing the issues inherent to whole-cell isolation. snRNA-seq primarily measures nuclear transcripts, although some transcripts attached to the rough endoplasmic reticulum might also be partially recovered (Bakken et al. 2018). It has been shown that neuronal nuclei in adult mouse primary visual cortex contain enough informative transcripts to identify highly related neuronal cell types with resolution similar to whole cells (Bakken et al. 2018). But comparisons of the type and amount of mRNA obtained from brain cell types using scRNA-seq and snRNA-seq methods have reported systematic differences in enrichment of specific types of mRNA transcripts (Skene et al. 2018; Santiago et al. 2023). This is an important consideration when determining the optimal sequencing method for a specific study.

In a snRNA-seq experiment, fresh or frozen, fragile, tightly interconnected, or multinucleated tissue is dissected, then chemically and mechanically lysed, and filtered to obtain a single-nucleus suspension. Enzymes for reverse transcription and barcodes are added, and after reverse transcription and amplification, a cDNA library is generated for sequencing (Kim et al. 2023; Iqbal et al. 2021). High-throughput sequencing is performed, and reads are aligned to the reference genome.

2.4.9 Spatial Transcriptomics

Due to cellular signaling between neighboring cells and soluble signals released in a cell's surroundings, the position of a cell relative to neighboring cells and non-cellular structures within a tissue influences the cell's type, state, phenotype, and function (Williams et al. 2022). Additionally, within individual cells, a gene's function directs the subcellular localization of its mRNA transcripts, which regulates where its protein product is synthesized and trafficked to (Holt and Bullock 2009; Williams et al. 2022). These spatial relationships between and within cells ultimately define tissue function. Despite the versatility and resolving power of current scRNA-seq methods, isolation of single cells necessitates dissociation of tissue samples, which destroys cellular interconnections, leading to loss of positional information about intercellular spatial context. Furthermore, isolation of total RNA from single cells for scRNA-seq results in loss of subcellular mRNA localization information (Ahmed et al. 2022).

To retain this spatial information and measure gene expression in individual cells within the context of the structural organization of the tissue section, spatially resolved methods for RNA sequencing that bypass tissue dissociation have been developed (Piwecka et al. 2023). The first techniques for combining spatial information with single-cell sequencing were introduced in 2012, and the term "spatial transcriptomics" was introduced in 2016 (Oevermans 2021). Spatial transcriptomics methods capture polyadenylated mRNA transcripts from frozen or formalin-fixed, paraffin-embedded tissue sections onto mRNA capture slides (i.e., arrays) containing non-specific, spatially barcoded single-stranded DNA probes. When a tissue section is attached to a slide, DNA probes bind to the poly-A tails of mRNA transcripts. Captured mRNA is reverse-transcribed to cDNA, which incorporates the spatial barcode of the probe, and sequenced using NGS technology (Illumina n.d.; Williams et al. 2022). Reconstruction of the relationship between transcripts captured by barcoded probes and their locations in the tissue is performed computationally during data analysis (Williams et al. 2022). By preserving the spatial context of cells, spatial transcriptomics allows for the broad characterization of transcriptional patterning and regulation within tissues and the local tissue features contributing to disease.

2.4.10 Third-Generation Sequencing and Current Innovations

Despite the immense utility of NGS, various technical challenges inherent to these technologies complicate their use in all cases (Shi et al. 2021). Specifically, the inability of short-read RNA-seq to sequence long continuous stretches of cDNA and sometimes to generate sufficient overlap between cDNA fragments makes it challenging to completely sequence, align, and assemble a complex transcriptome (Stark et al. 2019). Furthermore, PCR amplification of cDNA fragments prior to sequencing results in unequal amplification of all cDNA molecules (due to factors such as

GC bias, PCR stochasticity, and polymerase errors), which introduces artifacts and base composition bias during library construction, and can lead to distortions and inaccuracies in sequencing data (Baumann and Doerge 2012; Deshpande et al. 2023).

To address these challenges, "third-generation" sequencing (i.e., long-read sequencing) technologies, which employ fundamentally different sequencing approaches than NGS platforms, have been described beginning in 2008. These methods, which are still under active development, work to sequence longer single cDNA or RNA molecules with read lengths of 5,000–30,000 base pairs without the need for chemical modification or PCR amplification of the molecules (Soneson et al. 2019; Hong et al. 2020). Current long-read single-molecule sequencing platforms include Pacific Biosciences (PacBio) single-molecule real-time (SMRT) sequencing, Helicos single-molecule fluorescent sequencing, and Oxford Nanopore Technologies (ONT) nanopore sequencing (Hong et al. 2020). At present, however, long-read sequencing technologies tend to be lower throughput and much more error-prone than short-read sequencing technologies (Deshpande et al. 2023).

3 Computational Analysis of Transcriptomic Data

Since their introduction, high-throughput sequencing technologies have become a source of massive quantities of genomic data. Sophisticated computational tools, methods, and pipelines have been developed to process, manage, and analyze the millions of datapoints that result from high-throughput sequencing and microarray experiments and to overcome the significant technical challenges associated with these activities (Quackenbush 2001; Deshpande et al. 2023). As many of the tools and analyses developed for microarray have been adapted for use with NGS technology and data, this section will focus only on the major steps and associated challenges of computational analysis of RNA-seq data, from processing and data quality control (QC) to downstream quantitative analysis of gene expression.

3.1 Data Processing and Quality Control

Computational processing and analysis for RNA-seq data has many variations for which a wide variety of computational tools have been developed. A general computational workflow for bulk RNA-seq and scRNA-seq experiments consists of raw read generation, read alignment to reference genome or transcriptome, gene expression quantification, data normalization, confounder identification and removal, and downstream analysis.

First, base calls made by the sequencing technology are organized into raw, unaligned sequencing reads stored as FASTQ files. These reads are then aligned to the reference genome or reference transcriptome and sorted based on desired parameters for downstream applications (e.g., position, read ID). Duplicate reads are identified and marked, technical metrics for quality control of the sequencing data are compiled, and gene expression is quantified at the gene, transcript, or exon level to generate the raw counts expression matrix. Lowly expressed genes are defined and removed from the raw counts expression matrix, and the matrix is normalized using statistical models for approximating the distribution of the count-based expression data.

3.2 Biases, Batch Effects, and Importance of Quality Control of RNA-seq Data

RNA-seq allows for accurate quantitative characterization of relative gene expression in samples, including for protein-coding genes, non-coding genes, alternative splicing isoforms, and previously unannotated transcripts. But data from high-throughput sequencing technologies, including RNA-seq, is vulnerable to experimental and technical biases and batch effects that can impact downstream analyses and lead to incorrect interpretation of the data (Shi et al. 2021).

A bias arises when a systematic distorting effect results in an observed measurement that does not accurately reflect the quantity that was measured (Taub et al. 2010). RNA-seq measurements are often subject to experimental biases, including deficient experimental design, poor sample quality, inconsistent or inadequate library construction, and technical biases related to sequencing platform factors (Li et al. 2014; Hernandez et al. 2021; Shi et al. 2021). A few of the most common biases introduced by technical factors are described in Table 2 and include GC content bias, 3′-end and 5′-end capture bias, library complexity bias (i.e., differential PCR amplification bias), library size or sequencing depth bias, and gene length bias (Love 2016; Benjamini and Speed 2012; Love et al. 2016; Hansen et al. 2010; Taub et al. 2010).

A batch effect arises when a technical variable (e.g., processing date, sequencing date, technician) is correlated with variability in the data. Batch effects can be introduced at many steps in the experimental process, and avoiding batch effects requires careful experimental design. Upon detection of batch effects in RNA-seq data, computational correction following data generation can be accomplished through the use of covariates in linear models or more complex analyses such as surrogate variable analysis (Taub et al. 2010). As the accuracy of computational methods for batch correction depends on the relationship between technical variables and variables of interest, careful experimental design that minimizes the correlation between confounding variables and variables of interest is critical.

Identifying and accounting for biases and batch effects in RNA-seq data during data processing and quality control is critical for obtaining high-quality, interpretable data leading to accurate results of downstream analyses. But as commonly used

Table 2 Common biases observed in RNA-seq data

Bias	Description
GC content bias	Dependence between the proportion of guanine and cytosine bases (i.e., "GC content") in cDNA fragments and the read coverage (proportion of reads mapped to regions of the reference genome) in sequencing data GC-neutral cDNA fragments are able to be amplified more than GC-rich or AT-rich cDNA fragments, leading to underrepresentation of reads from GC-rich and GC-poor cDNA fragments in sequencing data (Benjamini and Speed 2012)
3′-end capture bias and 5′-end capture bias (i.e., positional bias, gene body coverage bias)	Higher mean read coverage at the 3′ end or 5′ end of a transcript as compared to the whole transcript Higher likelihood of capture of (1) 3′ ends of transcripts due to binding of poly-dT primer oligomers to poly-A tails during reverse transcription, or (2) 5′ ends of transcripts due to reverse transcription with random hexamers, which reduces the number of priming positions from which the reverse transcriptase enzyme can initiate cDNA synthesis (Finotello et al. 2014)
Library complexity bias (i.e., differential PCR amplification bias)	For a variety of technical reasons, PCR amplification during library preparation amplifies different sequences with unequal probabilities, resulting in uneven amplification of cDNA molecules (Kebschull and Zador 2015)
Library size or sequencing depth bias	Library size or sequencing depth is not always equivalent across samples
Gene length bias	Longer genes are more likely to be observed in sequencing data; the longer the gene, the higher the read count tends to be for a given expression level of the gene

computational approaches can lead to the removal of true biological signal (Hernandez et al. 2021), balancing under-correction and over-correction of experimental data is a formidable challenge to quality control of RNA-seq data.

3.3 Downstream Analysis of Transcriptomic Data

High-quality bulk RNA-seq and scRNA-seq data allows for a wide range of downstream statistical analyses to address a variety of research questions. scRNA-seq data can be used similarly to bulk RNA-seq data, and additionally in cell-type-specific analyses that can provide a higher resolution of cellular differences and the roles of individual cells in their microenvironments (Sampled 2023). This section presents the major types of analyses of transcriptomic data that are most commonly applied to studying the complexity of psychiatric disorders (in the context of this chapter, the term "psychiatric disorders" includes neuropsychiatric, neurodevelopmental, and neurodegenerative disorders).

Exploratory Gene Expression Analysis

For every gene expression dataset, exploratory analysis is performed following completion of RNA-seq data processing and quality control procedures and prior to initiation of other analyses. Exploratory analysis includes dimensionality reduction of the dataset using principal component analysis (PCA) or another method such as uniform manifold approximation and projection (UMAP) in order to compress the dataset containing thousands of dimensions (i.e., gene expression measurements) per sample into a much smaller dataset that retains most of the information from the large dataset. The reduced dataset can be visualized using plots and heatmaps to identify patterns in gene expression across samples and sample groups (Amezquita et al. 2021; Xiang et al. 2021).

Clustering Analysis

For scRNA-seq data, clustering analysis is used to (1) generate cell-type-specific clusters, (2) use known cell-type-specific marker genes to identify individual cell types representing generated clusters, and (3) confirm whether clusters represent true cell types or are the result of biological or technical variation (e.g., clusters based on batches, specific phases of the cell cycle, or high mitochondrial content) (Piper et al. 2019). Clustering analysis can be performed using a variety of methods including hierarchical clustering methods [i.e., grouping of genes with similar expression patterns], feature-selection methods, model-based methods, and dimensionality reduction techniques (Andrews and Hemberg 2018).

Cell-Type Deconvolution Analysis

Bulk RNA-seq data represents the average gene expression of individual genes across all cells of a sample. Cell-type deconvolution analysis is performed to estimate the proportions of different cell types in a bulk RNA-seq sample (Im and Kim 2023). During quality control of bulk RNA-seq data, the resulting cell-type fraction estimates are often included as covariates in statistical models of the data. In this way, differences in cell-type composition across samples can be corrected, as variations in cell-type proportion across samples affect bulk gene expression and may confound downstream analyses. A variety of deconvolution methods and computational tools have been developed, which can be reference-based (i.e., requiring a previously defined reference matrix representing expected values for cell-type specific marker genes for each cell type known to be present in the samples), reference-free, or partially reference-free (Keel et al. 2023). All three types of methods utilize statistical approaches to break down the contribution of each cell type to the bulk expression data.

Differential Gene Expression Analysis

The prevailing application of RNA-seq in studies of the human brain is the identification of differentially expressed genes (DEGs) between distinct sample groups (i.e., genes with a statistically significant difference in expression between two groups). Differentially expressed genes can provide baseline insight into the biological processes associated with a condition of interest and can be used in many downstream analyses. As a result, differential gene expression analysis is often used to understand the biological differences between cases and controls of psychiatric disorders in postmortem human brain tissue. Many algorithms have been developed for the identification of differentially expressed genes in RNA-seq data, including *limma* (Ritchie et al. 2015), *dream* (Hoffman and Roussos 2021), *DESeq* (Love et al. 2014), and *edgeR* (Robinson et al. 2010). These models are based on specific assumptions that the observed data may violate, and drawing meaningful conclusions from RNA-seq data requires an understanding of the parameters and constraints of the models used (Bullard et al. 2010; Hoffman et al. 2019; Fromer et al. 2016; Gandal et al. 2018b, 2022).

Enrichment Analysis
Obtaining a list of differentially expressed genes is the first step toward gaining biological insight from RNA-seq data. Pathway enrichment analysis provides information on the biological context of a provided set of genes (e.g., DEGs). Pathway enrichment analysis relies on curated annotation databases such as Gene Ontology (GO) (Gene Ontology Consortium 2015), Kyoto Encyclopedia of Genes and Genomes (KEGG) (Kanehisa et al. 2004), and DAVID (Huang et al. 2009). Starting from the gene list of interest, pathway enrichment analysis applies statistical methods (e.g., Fisher's exact test) to test for enrichment of the gene list in each annotated gene set representing a biological process or pathway. Another popular approach to enrichment analysis is gene set enrichment analysis (GSEA), which uses expression information from all expressed genes to determine sets of related genes in the dataset that exhibit statistically significant coordinated differences in expression between experimental groups (Wu et al. 2017). Pathways differ from gene sets in that pathways are models that represent established interactions and relationships between genes, while gene sets are unordered and unstructured collections of genes that can be defined based on many (sometimes seemingly arbitrary) criteria (Draghici 2018).

Gene Co-expression Network Analysis
A gene co-expression network is constructed in order to determine correlation patterns of genes across samples and identify groups of genes with similar expression patterns. Pairwise correlations of gene expression values are calculated for all possible gene pairs and represented as a network. Modules of co-expressed genes are defined based on the calculated gene–gene relationships. Upon module identification, hub genes (i.e., highly connected genes) within a module and across the network can be determined. Pathway enrichment analysis can be conducted on modules to functionally annotate modules to biological processes and pathways. Differential co-expression analysis can be performed to identify genes with differential co-expression across experimental conditions (Van Dam et al. 2017).

Alternative Splicing Analysis
Alternative splicing of mRNA transcripts allows for multiple protein isoforms to be produced from the same gene. Differentially spliced sites are determined from RNA-seq data using statistical models that estimate the expression of alternatively spliced exons and isoforms and provide statistical significance of estimates (Gondane and Itkonen 2023). Given that the human brain has been shown to exhibit more splicing than other tissues, alternative splicing analysis is used to investigate the association between splicing events and the development of psychiatric disorders (Wu et al. 2017).

Expression Quantitative Trait Loci (eQTL) Analysis
An eQTL is a genetic variant (i.e., SNP) whose allelic configuration (i.e., actual genotype) is associated with variation in gene expression of one or more genes. eQTL analysis uses genomic and transcriptomic data for the same individuals to identify genetic variants associated with the expression of one or more genes. In eQTL analysis, a gene's expression level is compared between groups with different copy numbers of the minor allele (i.e., 0, 1, or 2 copies) (Jung et al. 2020; Deshpande et al. 2023). eQTL analysis is complementary to differential gene expression analysis and allows for characterization of the functional effects of genomic variants and insight into the processes involved in regulation of gene expression.

4 Insights into Human Brain Biology Gained to Date

The primary tissue of interest for studies of psychiatric disorders is the living human brain. This section discusses current insights into human brain biology in the context of psychiatric disease that have been derived from studies utilizing transcriptomic and genomic data. Due to the current lack of such studies utilizing living human brain tissue, the findings discussed in this section originate primarily from studies utilizing postmortem human brain tissue, but also from some studies utilizing other postmortem human tissue as well as living tissue from experimental animals.

4.1 Human Brain Tissue Sources

Due to the inaccessibility of the living human brain for research purposes, postmortem brain tissue has long been used as a proxy for living brain tissue in transcriptomic studies of psychiatric disease. The expansion of postmortem brain banking programs combined with advances in high-throughput sequencing technology has led to an explosion of transcriptomic studies of psychiatric disorders in postmortem human brain, including schizophrenia, Alzheimer's disease, bipolar disorder, Parkinson's disease, and autism. But results across studies have been highly variable and difficult to reproduce, especially differential gene expression profiles of individual genes. For example, a recent metanalysis of the literature on gene expression in schizophrenia found little consensus across multiple studies for genes identified as differentially expressed, including inconsistent directions of effect reported in different studies (Merikangas et al. 2022). Discrepant results across studies have been attributed to discrepant methodological and analytic choices, cohort effects, and batch effects (Hernandez et al. 2021; Horvath and Mirnics 2015; Wu et al. 2017).

To address such issues of reproducibility and standardization, large-scale genomic data consortia (e.g., CommonMind Consortium [CMC], Psychiatric Genomics Consortium [PGC], PsychENCODE Consortium, BrainSeq Consortium, Accelerating Medicines Partnership [AMP]) have been formed to collect and uniformly process, analyze, and meta-analyze large numbers of brain samples from individuals with psychiatric disorders (Hernandez et al. 2021). At the time of writing, in addition to transcriptomic data, the vast majority of these consortia collect multiple data types to allow for integrated analyses of human brain function (Dong et al. 2021). For example, the PsychENCODE Consortium, which features the largest collection of postmortem brain data from control donors and donors with psychiatric disorders, currently offers bulk RNA-seq data for all brain tissues, and scRNA-seq, ChIP-seq, ATAC-seq, Ribo-seq, Hi-C, histone markers, proteomics, and DNA methylation data are available for some brain tissues (Dong et al. 2021).

To date, virtually all studies of psychiatric disease in the human brain have been conducted using postmortem human brain tissue.

4.2 Genetics of Psychiatric Disorders

It has been firmly established that psychiatric disorders are highly polygenic, with individual risk reflecting the combined effects of genetic variation at hundreds to thousands of risk alleles throughout the genome (i.e., many common alleles with small effect sizes contribute to all major psychiatric disorders) (Rees and Owen 2020; Stahl et al. 2019; Green et al. 2016; Grove et al. 2019; Wray et al. 2018; Lee et al. 2012; International Schizophrenia Consortium 2009). Large-scale genome-wide association studies (GWAS) have identified hundreds of individual genetic variants (i.e., SNPs and copy-number variants [CNVs]) significantly associated with risk for psychiatric disorders. Identified risk loci have been shown to be widely pleiotropic (i.e., associated with multiple traits or multiple psychiatric disorders) (Cross-Disorder Group of the Psychiatric Genomics Consortium 2013, 2019) and to reside predominantly in non-coding regions of the genome (Rees and Owen 2020; Ripke et al. 2014; Welter et al. 2014; Pardiñas et al. 2018). But psychiatric disorders have also been shown to be multifactorial (i.e., influenced by the interaction of genetic contributions and environmental exposure; e.g., early-life stress, trauma, substance use, social isolation, pregnancy and birth complications, advanced parental age) (Stilo and Murray 2019).

4.3 Gene Expression in Brain Compared with Gene Expression in Other Tissues

The human brain exhibits complex gene expression patterns as a result of the high number of genes expressed and high level of alternative splicing (Naro et al. 2021; Yeo et al. 2004; Merkin et al. 2012). Brain has been shown to exhibit a distinct gene expression pattern as compared with other tissues (Melé et al. 2015), and more accessible tissues do not recapitulate the complexities of brain gene expression (Hernandez et al. 2021; Melé et al. 2015). Given the relative ease of measuring gene expression in blood, the most accessible tissue, as compared to brain, characterizing the extent to which brain gene expression and transcriptomic organization (i.e., gene co-expression) is recapitulated in blood is a key research question. A review of transcriptomic studies of postmortem human, non-human primate, and rodent brain tissue utilizing various methodologies found that between 35% and 80% of known transcripts have been reported to be expressed in both brain and blood (Tylee et al. 2013; Cai et al. 2010). Brain has been reported to exhibit distinct alternative splicing as compared to all other tissues (Melé et al. 2015). A comparison of three postmortem human brain gene expression datasets to two living human blood gene expression datasets identified low correlations for mean gene expression levels of 8,799 genes between brain and blood (Cai et al. 2010). Estimates of the correlation of gene expression levels for individual genes in brain and blood were found to be highly variable across studies, with subsets of genes exhibiting stronger correlations than others (Tylee et al. 2013). Transcriptomic organization between brain and blood was found to be poorly preserved – only a few large brain co-expression modules contained subsets of genes (seemingly involved in basic cellular processes like metabolism) with preserved co-expression relationships in blood (Cai et al. 2010).

Despite the variety of results reported across studies, these findings indicate that expression of a portion of genes appears to be comparable in brain and blood tissue, and co-expression relationships of certain subsets of genes may be preserved across these two tissue types. This suggests the potential for identification of blood biomarkers for certain psychiatric disorders.

4.4 Regional Differences and Cell-Type-Specific Differences in Gene Expression in Postmortem Human Brain

Postmortem studies indicate that over 80% of human genes are expressed in adult human brain (Negi and Guda 2017; Hawrylycz et al. 2012). As brain function is impacted by specific gene expression regulation within anatomically distinct brain structures, knowledge of regional differential gene expression within the human brain can inform understanding of susceptibility to psychiatric disease. Large transcriptional differences between major brain structures have been reported, as well as similarities between district structures of larger regions, suggesting underlying functional differences across regions and similarities within regions (Chappell et al. 2018; Hawrylycz et al. 2012). Clustering patterns of gene expression in postmortem brain have been classified into three broad groups – genes expressed largely across the brain, genes expressed in functionally related structures, and genes expressed only in discrete locations. Most genes expressed across brain regions have been found to exhibit non-uniform expression across multiple regions (Chappell et al. 2018; Hawrylycz et al. 2012).

The large variety of morphologically and functionally heterogeneous cell types comprising the human brain exhibit distinct gene expression patterns (Zhang et al.

2014; McKenzie et al. 2018). Brain-wide variation of transcriptional patterns is reflective of the spatial distribution of the major cell types (e.g., neurons, oligodendrocytes, astrocytes, and microglia) (Hawrylycz et al. 2012). Among the genes with the highest expression in each brain cell type are the well-known marker genes for the cell type (McKenzie et al. 2018). Of all brain regions, the cerebellum has been shown to exhibit the most distinct gene expression pattern with the least internal heterogeneity of expression (Chappell et al. 2018; Hawrylycz et al. 2012).

4.5 Variations in Brain Gene Expression across the Lifespan

Transcriptomic studies comparing postmortem human brain samples from individuals of different ages have consistently identified broad age-associated transcriptomic variations (Ham and Lee 2020). Namely, multiple studies have reported that (1) genes upregulated in postmortem human brain during aging are enriched for genes involved in inflammatory and immune functions (Ham and Lee 2020; Lu et al. 2004; Erraji-Benchekroun et al. 2005; Berchtold et al. 2008; Cribbs et al. 2012; Primiani et al. 2014), and (2) genes downregulated in postmortem human brain during aging are enriched for genes involved in synaptic transmission and neuronal plasticity (Ham and Lee 2020; Lu et al. 2004; Erraji-Benchekroun et al. 2005; Berchtold et al. 2008, 2013; Naumova et al. 2012; Primiani et al. 2014).

Furthermore, a consistent pattern of gene expression has been identified in transcriptomic studies of postmortem human tissues including brain and in mouse and other vertebrate tissues – in older animals, longer transcripts are less abundant than shorter transcripts (Stoeger et al. 2022). In a variety of mouse and postmortem human tissues, the top 5% of genes with the shortest identified transcripts were found to be enriched for genes linked to shorter lifespans (e.g., genes involved in maintenance of telomere length and in immune function), while the top 5% of genes with the longest identified transcripts were found to be enriched for genes associated with longevity (Stoeger et al. 2022). These findings are consistent with other work in mice showing that accumulation of DNA damage during aging affects longer genes more than shorter genes, with transcription-stalling DNA lesions leading to reduced transcripts from longer genes (Vermeij et al. 2016).

Building upon these findings, recent work in mice provides robust evidence of the functional status of the basal transcriptional process underlying variations in gene expression during aging. Strong age-related transcriptional decline and skewing of transcriptional output are observed as a result of accumulating DNA damage. These age-related transcriptional variations are shown to be a general aging phenotype not specific to brain, which particularly affects lifespan-determining hallmark aging pathways (Stoeger et al. 2022).

Consistent with these results, a study of postmortem brain tissue found that Alzheimer's disease cases display lower expression of long genes as compared with age-matched controls and that these long genes were enriched for synaptic pathways thought to be less active in Alzheimer's disease (Soheili-Nezhad et al. 2021). These finding suggest that age-related transcriptional changes in the human brain may be involved in Alzheimer's disease pathogenesis and more broadly in age-related disease etiology.

5 Remaining Challenges

5.1 Concordance Between Genetic and Transcriptomic Findings

5.1.1 GWAS Loci and Differentially Expressed Genes

Due to the multifactorial and highly polygenic nature of psychiatric disorders, the underlying molecular biology of these conditions is among the most challenging to understand. Advances in high-throughput sequencing technology and computational approaches and the establishment of postmortem human brain consortia have resulted in many large-scale genetic and transcriptomic studies aiming to identify genes, genetic variants, and networks of genes associated with complex brain disorders. These studies have identified (1) hundreds to thousands of common genetic variants of small effect sizes and unknown function and rare de novo variants and CNVs with larger effect sizes that confer risk for psychiatric disorders (MacArthur et al. 2017), and (2) hundreds of differentially expressed genes between postmortem psychiatric disease cases and controls (Darnell et al. 2011; Fromer et al. 2016; Gandal et al. 2018a, b; Hoffman et al. 2019; Jaffe et al. 2018; Richards et al. 2012).

The majority of variants identified through genetic studies as conferring risk for psychiatric disorders are thought to act through altered gene expression (Bray and O'Donovan 2018; Huo et al. 2019; Richards et al. 2012). Therefore, case–control differential gene expression studies should be able to identify genes with higher disease associations and help to determine the most likely causal genes within risk-associated loci. Indirect support for this has been seen in case–control analyses of genomic and transcriptomic data, with associations between disease and genes with similar biological functions identified in both data types (Gandal et al. 2018b; Jaffe et al. 2018). At the time of writing, however, the hypothesis that genes identified as differentially expressed in postmortem brain tissue are enriched for genes that show genetic association with disease risk is not directly supported by the scientific literature (Clifton et al. 2023). Studies aiming to quantify the relationships between differential gene expression, GWAS summary statistics, associated genomic loci, and genes prioritized within implicated loci have found inconsistent evidence for robust relationships between differential gene expression and measures of genetic association (Fromer et al. 2016; Clifton et al. 2023). This fundamental discrepancy between transcriptomic and genetic findings in case–control studies of psychiatric disease (i.e., the lack of consensus for genes identified as being implicated in disease) is illustrated here using findings in schizophrenia.

Schizophrenia is a severe and complex brain disorder of multifactorial etiology that has been shown to be highly heritable (Owen et al. 2016a, b). Common and rare genetic variants, specifically rare copy-number variants (CNVs) and rare coding variants, as well as environmental factors have been associated with the etiology of schizophrenia (Ripke et al. 2014; Kirov 2015; Singh et al. 2017; Rees et al. 2020; Owen et al. 2016a, b; Rees and Kirov 2021). But no one-to-one Mendelian mapping exists between schizophrenia risk alleles and diagnosis, indicating a complex polygenic genetic architecture for the disorder arising from the cumulative effects of many genetic variants that each confer a small amount of risk (Fromer et al. 2016; International Schizophrenia Consortium 2009; Purcell et al. 2014). Due to linkage disequilibrium (LD), most associated risk loci also encompass more than one gene, preventing understanding of genes that are functionally impacted by associated risk alleles (Clifton et al. 2023). Furthermore, many identified risk loci are located in non-coding regions of the genome, potentially regulating cell-type-specific and context-specific gene expression (Ripke et al. 2014). This type of inheritance complicates identification of the causal variants that drive these associations, and understanding of the influence of these variants on the biological processes underpinning the disease (Fromer et al. 2016). In fact, despite the large number of genetic associations implicated in schizophrenia to date, most have not led to any useful clinical applications or advancements in the treatment of the disease (Owen et al. 2016a, b).

As noted previously, given the non-coding genomic locations of many common variants associated with schizophrenia, and the fact that few of these variants have been identified as being non-synonymous (i.e., variants that change the sequence of a protein) (Trubetskoy et al. 2022), it is thought that most common risk loci are likely exerting their effects by altering gene expression (Bray and O'Donovan 2018; Huo et al. 2019; Richards et al. 2012). If this is true, then genetically implicated schizophrenia risk genes should show evidence of differential gene expression, and this could inform prioritization of genes identified by GWAS (Clifton et al. 2023). Large-scale transcriptomic studies of postmortem human brain have identified many genes as differentially expressed between schizophrenia cases and controls. Despite results across these studies being inconsistent and highly variable (Merikangas et al. 2022), comparisons between genomic and transcriptomic findings have identified that genes implicated by both types of data converge on common biological pathways (e.g., genes involved with neuronal signaling) (Gandal et al. 2018b; Jaffe et al. 2018). Based on these findings, multiple studies have directly tested the relationships between differential gene expression and genetic association in postmortem human brain tissue. Results from such studies using bulk RNA-seq data have consistently reported no robust evidence for case–control differential gene expression in GWAS-implicated schizophrenia risk genes (Merikangas et al. 2022; Clifton et al. 2023). Furthermore, genes mapping to schizophrenia-associated loci have not been found to be more likely to be differentially expressed than other brain-expressed genes not mapping to these loci (Clifton et al. 2023). Of 160 consensus genes found to be differentially expressed between schizophrenia cases and controls in a recent metanalysis, six were reported in the largest schizophrenia case–control GWAS to date (Merikangas et al. 2022). In contrast, a recent schizophrenia case–control analysis of scRNA sequencing data for 25 brain cell types from postmortem human brain reported significant associations between common genetic variants and rare genetic variants and differentially expressed genes within specific excitatory neuronal populations (Ruzicka et al. 2022). In the same study, gene co-expression analysis identified a transcription factor module associated with differentially expressed neuronal genes, with seven out of 24 genes in the module overlapping with transcription factors encoded by schizophrenia GWAS genes (Ruzicka et al. 2022).

These conflicting finds illustrate that a comprehensive understanding of the functional relationship between genetic variation and psychiatric disease is currently lacking. This is a major impediment to progress in the treatment of psychiatric disorders (Hernandez et al. 2021). As most studies to date have been conducted using bulk RNA-seq approaches that average gene expression across postmortem tissue from adult brains and do not account for expression dynamics of individual cell types or developmental context of gene expression, future work expanding on scRNA-seq and snRNA-seq approaches and accounting for developmental context (Patel et al. 2022) may reveal stronger patterns of concordance between genomic and transcriptomic results.

5.1.2 GWAS Loci and eQTLs

An expression quantitative trait locus (eQTL) is a genetic variant (i.e., SNP) that has been identified as being significantly associated with variation in gene expression (Nica and Dermitzakis 2013). eQTL mapping studies utilize genomic and transcriptomic data from the same individuals to identify SNPs that influence gene expression. By linking SNPs to changes in gene expression, eQTL mapping studies can aid in functionally annotating the thousands of non-coding GWAS loci that have been identified as being associated with complex brain disorders (Jung et al. 2020). Similarly to the lack of convergence between findings from GWAS and differential gene expression studies, however, results from studies integrating GWAS findings with eQTL mapping demonstrate that most genetic variants identified through GWAS cannot be explained by existing eQTLs (GTEx Consortium 2017; GTEx Consortium et al. 2020; 145. Umans et al. 2021; Connally et al. 2022). In one recent study, only 43% of GWAS-identified loci were also identified as eQTLs in data from the Genome Tissue Expression (GTEx) Project (GTEx Consortium et al. 2020).

Various explanations have been proposed for the observed limited concordance between GWAS loci and eQTLs, including context specificity of eQTLs (e.g., developmental specificity, cell-type specificity, environmental specificity) (Mostafavi et al. 2023; Umans et al. 2021; Strober et al. 2019; Walker et al. 2019; Zhernakova et al. 2017; Yazar et al. 2022). Increased statistical power to detect effects and deeper sampling of specific cell types and contexts using scRNA-seq have been proposed to resolve these discrepancies. A recent study has proposed that a systematic difference exists between identified eQTLs and GWAS loci (Mostafavi et al. 2023). Namely, it is shown that GWAS loci are (1) at the gene level, located near selectively constrained genes (i.e., genes intolerant to damaging variants), while eQTLs are not, suggesting that natural selection has purged eQTLs with large effects on constrained genes, and (2) at the variant level, located further away from transcription start sites (TSSs) than eQTLs, which are enriched at promoter regions and cluster near TSSs (Mostafavi et al. 2023). The study suggests that this "colocalization gap" will be resolved using a combination of approaches including collection of larger samples with more cell types from a diversity of developmental stages and other contexts, integration with data from other molecular QTL assays and functional assays, and development of computational models.

5.2 *Multiomics, Neuroimaging, and Integrative Data Analyses*

Advances in high-throughput sequencing technologies have led to an explosion of data generation across multiple "omics" (i.e., biological layers and systems) – genomics (genetic variants, copy-number variations), transcriptomics (gene expression), epigenomics (DNA methylation, chromatin, histone modifications), proteomics (protein expression), metabolomics (metabolite expression), and spatial patterns of gene expression (Sathyanarayanan et al. 2023).

The availability of large-scale datasets including various types of omics and non-omics data (e.g., neuroimaging, epidemiological, and clinical data) has motivated efforts to integrate multiple data types into investigations, creating a new field of systems biology called "multiomics."

Given the complex and multifactorial nature of psychiatric disorders, multiomics approaches are particularly appropriate for elucidating the biological mechanisms underlying these conditions (Dong et al. 2021). At present, however, approaches to integrating multiple, large-scale, heterogeneous datasets of variable completeness and quality originating from various modalities are complicated by existing technical, analytical, and computational limitations (Flores et al. 2023). The possibilities and current challenges of integrative multiomics methods are illustrated in brief with the example of imaging transcriptomics. Chapter "Human Brain" provides a more thorough discussion of the field of imaging transcriptomics.

5.2.1 Imaging Transcriptomics

It has been argued that psychiatric disorders can be characterized as disorders associated with brain connectivity (i.e., the functional integration of brain processes) (Fornito et al. 2015; Stephan et al. 2009; Xia et al. 2018). Non-invasive neuroimaging has been used to characterize the structural and functional variations associated with the various stages of psychiatric disease (Arnatkeviciute et al. 2022). In this way, neuroimaging data can serve as an intermediate phenotype that links clinical manifestations of disease with their underlying molecular mechanisms (Mandal et al. 2022). In recent years, the field of imaging transcriptomics has emerged to bridge the gap between imaging-derived phenotypes (IDPs) (Mandal et al. 2022) and anatomically comprehensive spatial maps of brain gene expression (e.g., the Allen Human Brain Atlas) (Hawrylycz et al. 2012).

A common study design in imaging transcriptomics involves comparing spatial maps of case–control differences in a given IDP measure (e.g., structural connectivity) to spatial gene expression patterns (Romme et al. 2017; Shin et al. 2023). Such analyses can be hypothesis-driven and focusing on specific sets of previously identified genes, or data-driven and aiming to identify patterns across many genes. Studies across psychiatric disorders and imaging modalities have identified overlap between genes spatially correlated with disease-associated structural or functional brain changes and transcriptionally dysregulated genes in postmortem brain samples of clinical cases (Romme et al. 2017; Morgan et al. 2019). Some of these transcriptional associations have shown evidence for diagnostic specificity, but others have exhibited consistency across disorders, suggesting common genetics of disease risk (The Writing Committee for the Attention-Deficit/Hyperactivity Disorder, Autism Spectrum Disorder, Bipolar Disorder, Major Depressive Disorder, Obsessive-Compulsive Disorder, and Schizophrenia ENIGMA Working Group, et al. 2021).

Despite the rapid advances in the field of imaging transcriptomics, current methodological limitations to the integration of neuroimaging and spatial transcriptomic data and subsequent analyses present important challenges. Challenges include (1) selection of processing steps within data types, which can influence the resulting data (Arnatkeviciute et al. 2019), (2) selection of statistical methods to account for (a) intrinsic spatial autocorrelation of both transcriptomic and neuroimaging data (i.e., similar gene expression profiles or similar IDP values in nearby brain regions) (Arnatkeviciute et al. 2022; Markello and Misic 2021) and (b) disease specificity of any identified associations between IDPs and brain-specific transcriptomic variation (Arnatkeviciute et al. 2022), (3) interpretation of results of gene set enrichment analyses (which employ methods designed for case–control differential gene expression) (Fulcher et al. 2021), and (4) selection of more appropriate methodology for interpretation of results depending on the context. Furthermore, a fundamental limitation of current capabilities in imaging transcriptomics is the lack of availability of both types of data from within the same cohorts. Future work is required to overcome these limitations.

6 Conclusion

Over the past three decades, rapid technological and methodological advancements in transcriptomics have repeatedly transformed the field and revolutionized understanding of how genes are expressed. New developments are continuously overcoming existing challenges in measurement and analysis of gene expression and expanding the possibilities of what can be studied. This has created endless opportunities for understanding the biology of the human brain and led to key insights about brain function in healthy and diseased states.

The breakneck pace of technological evolution has also introduced complexities and challenges into the experimental application of transcriptomics approaches and computational methods. As new technologies are phased in and older technologies are phased out, one important consideration that served as the motivation for this chapter is the importance of retaining and building upon the knowledge gained from previous approaches, so that it can be used to guide and inform applications of new technology.

In order to validate the use of new technologies and evaluate their utility against existing approaches, a thorough understanding of (1) the nature and limitations of the biological information that can be obtained from each technology, and (2) the relationship between respective data types (e.g., bulk RNA-seq and snRNA-seq data for the same brain biopsy) is required. Furthermore, a thorough understanding of the similarities and differences in data generated by different technologies is critical for applying appropriate computational methods during data analysis and for ensuring that computational methods are updated as technology evolves.

Given that the functioning of the human brain occurs through complex interactions between different types of biomolecules whose activity is captured by different omics technologies, a comprehensive understanding of brain function is only possible through the integration of multiomics knowledge. The ability to integrate heterogeneous data types generated by different technologies into robust multiomics analyses that provide meaningful biological insights requires an extensive understanding of the characteristics, quality, depth, and limitations of each data type (e.g., unknown signals, missing data).

This chapter traces the historical development of approaches to gene expression quantification to provide an overall view of the current state of the field of transcriptomics, its applications to the study of the human brain, and its place in the broader emerging multiomics landscape. To enable a more complete understanding of the impact of gene expression on function, future work in transcriptomics should be informed by a thorough understanding of the evolution of transcriptomics technology and computational methods.

Glossary

Allele One of two or multiple versions of a DNA sequence (a single base or a segment of bases) at a given location; in the genome; alleles can differ at single positions or can have insertions or deletions of up to several thousand base pairs

Alternative splicing Cellular process occurring during or after transcription by which different combinations of exons are included or excluded from a final processed (i.e., mature) RNA transcript, allowing for the production of multiple different (alternative) RNA transcripts from the same DNA sequence, which can be translated into multiple protein isoforms

Autoradiography Photochemical technique that utilizes X- ray film to visualize the spatial distribution of molecules or fragments of molecules that have been radioactively labeled

Base (i.e., nitrogenous base; nucleobase; nucleotide base) Nitrogen-containing biological compound that comprises one of the four building blocks of a nucleotide; the four bases in DNA are guanine, cytosine, adenine, and thymine; in RNA, uracil replaces thymine

Base call The identity of the base at a specific position in the DNA or RNA sequence that is inferred from the raw physical signal produced by a sequencing platform

Colocalization State of two traits sharing a causal genetic variant

Copy-number variation (CNV) Type of genetic variance comprising variation in the number of repeats of specific DNA sequences across individuals' genomes

Deoxyribonucleic acid (DNA) Polymer molecule composed of two polynucleotide chains coiled around each other to form a double helix; the sequence of individual (deoxyribo)nucleotides in the polynucleotide chains encodes all of the biological information necessary for an organism to exist

Deoxynucleoside triphosphate (dNTP) One of the four nucleotide precursors comprising the building blocks of DNA, each of which uses a different DNA base (i.e., adenine, cytosine, guanine, thymine): dATP (deoxyadenosine triphosphate), dCTP (deoxycytidine triphosphate), dGTP (deoxyguanosine triphosphate), and dTTP (deoxythymidine triphosphate); dNTPs form the substrates for DNA polymerases

De novo variant Variant present in an individual's genome that was not inherited from a parent

Differential gene expression analysis Computational analysis in which the expression level of every gene is tested for association with a trait of interest using a regression model; a separate regression model is fit to each gene

Differentially expressed gene In differential expression analysis, a gene with a statistically significant association with the trait of interest

DNA clones Many copies of a DNA sequence produced by repeated cycles of replication (often by insertion of a particular DNA fragment into the purified DNA genome of a self-replicating genetic element (e.g., virus, plasmid)) that produces copies of the DNA as it replicates itself

DNA polymerase A type of enzyme that catalyzes the synthesis of DNA molecules from (deoxy)nucleoside triphosphates (the molecular precursors of DNA) during DNA replication

Electrophoresis Laboratory technique utilizing an electronic current to move DNA, RNA, or protein molecules through a gel or other matrix in order to separate the molecules based on their size and electrical charge

Expression quantitative trait locus (eQTL) Genetic variant that affects the expression of one or more genes

Exon Any nucleotide sequence within a gene that is included in a final mature mRNA transcript produced from the gene after introns have been removed by RNA splicing

Fragment (i.e., DNA/cDNA fragment) In high-throughput sequencing, comprises the piece of DNA/cDNA of interest and its two sequencing adapters (each of which is ligated to one end of the piece of DNA/cDNA)

Fusion gene Two genes that are joined such that they are transcribed and translated as a single unit; hybrid gene formed from two previously independent genes due to chromosomal rearrangement

Fractionation Encompasses various methods for separating subcellular components and organelles for the purpose of studying the structures, functions, and molecular compositions of the isolated components

Gene A sequence of nucleotides forming part of a chromosome in DNA that is transcribed to produce a functional RNA or set of closely related isoforms

Genetic architecture The genetic basis of a phenotypic trait (e.g., disease) encompassing the total number of functional variants that affect the trait and their frequencies and effect sizes

Genotype The combination of two alleles for a particular gene or at a particular locus that an individual possesses

Genomics The large-scale study of genomes, including structure, function, evolution, mapping, and editing

Genetics The study of individual genes, genetic variation, and heredity

Gene expression Process by which genetic information encoded in DNA gives rise to a phenotype; consists of transcription of a DNA sequence into mRNA and translation of mRNA into a sequence of amino acids during protein synthesis; in gene expression analysis, refers to the quantification of the RNA transcript levels of a particular gene or DNA sequence

Gene fusion Genomic event that generates a novel gene from two independent genes

Genome-wide association study (GWAS) Approach used in genetics research to identify genetic variants associated with a particular phenotype by testing hundreds of thousands of variants across many genomes to find variants with statistically significant associations with the specific trait or disease

GWAS summary statistics The aggregate p-values and association data for every variant analyzed in a genome-wide association study (GWAS)

Gene set enrichment analysis Method to identify whether a group of genes is over-represented or under-represented in a subset of genes as compared to the background of all genes

High throughput In sequencing, able to process large amounts of DNA or cDNA simultaneously in parallel

Hybridization Process by which two complementary single-stranded DNA and/or RNA molecules bond together to form a double-stranded molecule

Hybridization probe Nucleotide fragment of variable length that can be radioactively or fluorescently labeled and used to detect the presence of complementary DNA or RNA nucleotide sequences in a sample

Imaging-derived phenotype (IDP) In imaging transcriptomics, interpretable statistical map of regional brain properties (e.g., functional connectivity, cortical thickness, myelination) derived from raw brain imaging data

Intron Any nucleotide sequence within a gene that is not included in a final mature mRNA transcript produced from the gene

Library complexity Quantity of unique cDNA fragments before PCR amplification

Library size Either the total number of reads that were sequenced during a sequencing run or the total number of mapped reads

Linkage disequilibrium (LD) Degree to which inheritance of an allele is associated with inheritance of a nearby allele within a given population

Locus (i.e., genetic locus) The specific physical location of a gene or other DNA sequence on a region of a chromosome; alleles are genetic variants at a particular locus

Low throughput In sequencing, able to process small amounts of DNA or cDNA per unit of time

Massively parallel sequencing technology Any of several high-throughput approaches to DNA and RNA sequencing that determine the entire genome or transcriptome of an organism by processing millions of clonally amplified DNA or cDNA reads in parallel

Mendelian inheritance Refers to the inheritance of traits controlled by a single gene with two alleles, one of which may be completely dominant to the other; inheritance patterns of single-gene diseases are often referred to as Mendelian

Messenger RNA (mRNA) Single-stranded RNA molecule created from a template protein-coding DNA sequence during transcription (i.e., a transcript) that corresponds to the genetic sequence of the DNA template and is in turn used by a ribosome as the template for synthesizing a protein

Microarray Hybridization-based laboratory tool used to detect gene expression or genetic variation in many genes at the same time

Module (i.e., co-expression network module) Group of genes in a co-expression network that show a coordinated expression pattern across samples

Multiomics Field of systems biology that integrates multiple biological data types into analyses to enable more comprehensive understanding of the molecular basis of health and disease

Nucleoside Organic molecule composed of a nitrogenous base and a pentose sugar; addition of one or multiple phosphate groups to a nucleoside creates a nucleotide

Nucleotide Organic molecule composed of a nitrogenous base, a pentose sugar, and a phosphate that comprises the monomeric building block of DNA and RNA molecules, which are polymers consisting of long chains of nucleotides

Psychiatric disorders (i.e., psychiatric disease) In the context of this chapter, include neuropsychiatric, neurodevelopmental, and neurodegenerative disorders

Oligonucleotide Short, single-stranded or double-stranded DNA or RNA molecule often used as a probe for detecting a specific complementary sequence

Omics Any of a group of biological sciences that aim to quantify and describe the entire collection of biological molecules of a particular type (e.g., genome, transcriptome, proteome) in an organism or system, and how this determines the structure, function, and interactions of the organism or system (e.g., transcriptomics, genomics, proteomics, metabolomics)

Polynucleotide Polymer chain composed of several nucleotides covalently joined together; DNA and RNA are types of polynucleotide molecules

Read In DNA and RNA sequencing, a portion of a DNA/cDNA fragment that has been sequenced

Read count In DNA and RNA sequencing, the number of reads that map (i.e., align) to a given genomic feature such as a gene using computational methods

Read length In DNA and RNA sequencing, the number of bases of a DNA/cDNA fragment that are sequenced at a time

Restriction enzyme Enzyme that cleaves DNA at sequence-specific sites

Ribonucleic acid (RNA) Polymer molecule composed of a chain of (ribo)nucleotides that serves a multitude of diverse cellular functions required for survival

RNA-seq (i.e., RNA sequencing) Refers to the next-generation sequencing technologies and associated computational methods for capturing and quantifying RNA transcripts present in a sample

RNA splicing The process of removing introns and ligating together exons of a newly synthesized eukaryotic precursor RNA transcript in order to form the mature RNA transcript

RNA transcript (i.e., transcript) Single-stranded RNA molecule synthesized by transcription of a DNA sequence; through the process of alternative splicing, multiple different RNA transcripts can be produced from the same DNA sequence

Sequencing adapters In DNA and RNA sequencing, short oligonucleotides comprising known sequences that are ligated to the ends of the DNA/cDNA of interest to form the DNA/cDNA fragment; contain binding sites for sequencing primers for PCR amplification

Sequencing library In DNA and RNA sequencing, collection of similarly sized DNA/cDNA fragments with known adapter sequences ligated to their $5'$ and $3'$ ends

Sequencing depth (i.e., read depth; depth of coverage) In DNA and RNA sequencing, the average number of times that each base is sequenced during a sequencing run; i.e., average number of times a given position in the genome is sequenced

Sequencing coverage (i.e., coverage; read coverage) In DNA and RNA sequencing, the proportion or percentage of a genome or genomic region that has been sequenced at a certain depth; calculated based on the number of unique reads mapped/aligned to a region in a reference genome

Single-nucleotide polymorphism (SNP) A genomic variant at a single position in a DNA sequence comprising a change in one nucleotide among individuals; SNPs are the most common type of genetic variants among humans and can be used as markers for certain diseases

Single-pass sequence Sequence generated from a DNA clone by a one-directional sequencing reaction occurring only once

Throughput In DNA and RNA sequencing, the rate at which DNA/cDNA can be processed per unit time

Transcription The process by which a segment of DNA is copied into a new RNA molecule by an RNA polymerase enzyme with assistance from other transcription factor proteins

Transcriptome The set of all RNA molecules (i.e., transcripts: protein-coding (mRNA) and non-coding RNA, including rRNA, tRNA, lncRNA, pri-miRNA, and others) expressed in an entire organism or in a specific tissue type or cell type at a given time

Transcriptomics The study of the transcriptome; encompasses all aspects of RNA synthesis, structure, function, regulation, localization, trafficking, splicing, post-transcriptional modification, and degradation

Translation The process by which a protein (i.e., sequence of amino acids) is synthesized by a ribosome using the sequence of nucleotides in an mRNA molecule as a template

Transcript expression The expression of a specific RNA transcript of a DNA sequence (multiple different (i.e., alternative) RNA transcripts can be produced from the same DNA sequence through the process of alternative splicing)

Variant (i.e., genetic variant) A specific region of the genome that differs between individuals in a population; genetic variants include structural variants (i.e., SNPs, insertions, deletions, CNVs) and chromosomal rearrangements (i.e., translocations and inversions)

References

Adams G (2020) A beginner's guide to RT-PCR, qPCR and RT-qPCR. Biochem 42(3):48–53. https://doi.org/10.1042/BIO20200034

Adams MD, Kelley JM, Gocayne JD, Dubnick M, Polymeropoulos MH, Xiao H, Merril CR, Wu A, Olde B, Moreno RF (1991) Complementary DNA sequencing: expressed sequence tags and human genome project. Science 252(5013):1651–1656. https://doi.org/10.1126/science.2047873

Ahmed R, Zaman T, Chowdhury F, Mraiche F, Tariq M, Ahmad IS, Hasan A (2022) Single-cell RNA sequencing with spatial transcriptomics of cancer tissues. Int J Mol Sci 23(6):3042. https://doi.org/10.3390/ijms23063042

Alwine JC, Kemp DJ, Stark GR (1977) Method for detection of specific RNAs in agarose gels by transfer to diazobenzyloxymethyl-paper and hybridization with DNA probes. Proc Natl Acad Sci U S A 74(12):5350–5354. https://doi.org/10.1073/pnas.74.12.5350

Amezquita R, Lun A, Hicks S, Gottardo R (2021) Chapter 4: dimensionality reduction. In: Basics of single-cell analysis with bioconductor version 1. https://github.com/OSCA-source/OSCA.basic

Andrews TS, Hemberg M (2018) Identifying cell populations with scRNASeq. Mol Aspects Med 59:114–122. The emerging field of single-cell analysis. https://doi.org/10.1016/j.mam.2017.07.002

Arabameri N, Tseng BS (2023) Chapter 8 - studying gene expression in biofilms. In: Gurtler V, Patrauchan M (eds) Methods in microbiology, vol 53. Academic Press, pp 235–271. (Biofilms). https://doi.org/10.1016/bs.mim.2023.04.003

Arnatkeviciute A, Fulcher BD, Fornito A (2019) A practical guide to linking brain-wide gene expression and neuroimaging data. Neuroimage 189:353–367. https://doi.org/10.1016/j.neuroimage.2019.01.011

Arnatkeviciute A, Fulcher BD, Bellgrove MA, Fornito A (2022) Imaging transcriptomics of brain disorders. Biol Psychiatry Global Open Sci 2(4):319–331. https://doi.org/10.1016/j.bpsgos.2021.10.002

Atkinson MR, Deutscher MP, Kornberg A, Russell AF, Moffatt JG (1969) Enzymatic synthesis of deoxyribonucleic acid. XXXIV. Termination of chain growth by a 2′,3′-dideoxyribonucleotide. Biochemistry 8(12):4897–4904. https://doi.org/10.1021/bi00840a037

Bakken TE, Hodge RD, Miller JA, Yao Z, Nguyen TN, Aevermann B, Barkan E, Bertagnolli D, Casper T, Dee N et al (2018) Single-nucleus and single-cell transcriptomes compared in matched cortical cell types. PloS One 13(12):e0209648. https://doi.org/10.1371/journal.pone.0209648

Barba M, Czosnek H, Hadidi A (2014) Historical perspective, development and applications of next-generation sequencing in plant virology. Viruses 6(1):106–136. https://doi.org/10.3390/v6010106

Baumann D, Doerge RW (2012) Correcting for amplification bias in next-generation sequencing data. Conf Appl Stat Agric. https://doi.org/10.4148/2475-7772.1026

Behjati S, Tarpey PS (2013) What is next generation sequencing? Arch Dis Child Educ Pract Ed 98(6):236–238. https://doi.org/10.1136/archdischild-2013-304340

Belmont JW (2008) 101- molecular methods. In: Rich RR, Fleisher TA, Shearer WT, Schroeder HW, Frew AJ, Weyand CM (eds) Clinical immunology, 3rd edn. Mosby, Edinburgh, pp 1485–1502. https://doi.org/10.1016/B978-0-323-04404-2.10101-0

Benjamini Y, Speed TP (2012) Summarizing and correcting the GC content bias in high-throughput sequencing. Nucleic Acids Res 40(10):e72. https://doi.org/10.1093/nar/gks001

Berchtold NC, Cribbs DH, Coleman PD, Rogers J, Head E, Kim R, Beach T, Miller C, Troncoso J, Trojanowski JQ et al (2008) Gene expression changes in the course of normal brain aging are sexually dimorphic. Proc Natl Acad Sci 105(40):15605–15610. https://doi.org/10.1073/pnas.0806883105

Berchtold NC, Coleman PD, Cribbs DH, Rogers J, Gillen DL, Cotman CW (2013) Synaptic genes are extensively downregulated across multiple brain regions in normal human aging and Alzheimer's disease. Neurobiol Aging 34(6):1653–1661. https://doi.org/10.1016/j.neurobiolaging.2012.11.024

Bio-Rad (n.d.) What is real-time PCR (qPCR)? Accessed 5 Dec 2023. https://www.bio-rad.com/en-us/applications-technologies/what-real-time-pcr-qpcr?ID=LUSO4W8UU

BLAST. Sequence tags. 2017 Nov 20. Accessed 5 Dec 2023. https://web.cas.org/help/BLAST/topics/seq_tags.htm#:~:text=Short%20(usually%20about%20300%2D500,for%20a%20given%20cDNA%20library

Bray NJ, O'Donovan MC (2018) The genetics of neuropsychiatric disorders. Brain Neurosci Adv 2:239821281879927. https://doi.org/10.1177/2398212818799271

Brown A, Ampratwum PO, Ray SD (2024) Microarray analysis. In: Wexler P (ed) Encyclopedia of toxicology, 4th edn. Academic Press, Oxford, pp 385–392. https://doi.org/10.1016/B978-0-12-824315-2.00210-4

Brownlee GG, Sanger F, Barrell BG (1967) Nucleotide sequence of 5S-ribosomal RNA from Escherichia coli. Nature 215(5102):735–736. https://doi.org/10.1038/215735a0

Bullard JH, Purdom E, Hansen KD, Dudoit S (2010) Evaluation of statistical methods for normalization and differential expression in mRNA-Seq experiments. BMC Bioinformatics 11:94. https://doi.org/10.1186/1471-2105-11-94

Bumgarner R (2013) Overview of DNA microarrays: types, applications, and their future. Curr Protoc Mol Biol 101(1). https://doi.org/10.1002/0471142727.mb2201s101

Cai C, Langfelder P, Fuller TF, Oldham MC, Luo R, van den Berg LH, Ophoff RA, Horvath S (2010) Is human blood a good surrogate for brain tissue in transcriptional studies? BMC Genomics 11(1):589. https://doi.org/10.1186/1471-2164-11-589

Chang JC, Kan YW (1982) A sensitive new prenatal test for sickle-cell anemia. N Engl J Med 307(1):30–32. https://doi.org/10.1056/NEJM198207013070105

Chappell S, Patel T, Guetta-Baranes T, Sang F, Francis PT, Morgan K, Brookes KJ (2018) Observations of extensive gene expression differences in the cerebellum and potential relevance to Alzheimer's disease. BMC Res Notes 11:646. https://doi.org/10.1186/s13104-018-3732-8

Clark DP, Pazdernik NJ, McGehee MR (2019) Molecular biology. 3rd edn. Academic Press, London, an imprint of Elsevier

Clifton NE, Schulmann A, Schizophrenia Working Group of the Psychiatric Genomics Consortium, Holmans PA, O'Donovan MC, Vawter MP (2023) The relationship between case–control differential gene expression from brain tissue and genetic associations in schizophrenia. Am J Med Genet B Neuropsychiatr Genet 192(5–6):85–92. https://doi.org/10.1002/ajmg.b.32931

Connally NJ, Nazeen S, Lee D, Shi H, Stamatoyannopoulos J, Chun S, Cotsapas C, Cassa CA, Sunyaev SR. The missing link between genetic association and regulatory function. Flint J, Przeworski M, editors. Elife. 2022;11:e74970. https://doi.org/10.7554/eLife.74970

Cribbs DH, Berchtold NC, Perreau V, Coleman PD, Rogers J, Tenner AJ, Cotman CW (2012) Extensive innate immune gene activation accompanies brain aging, increasing vulnerability to cognitive decline and neurodegeneration: a microarray study. J Neuroinflammation 9(1):179. https://doi.org/10.1186/1742-2094-9-179

Crick F, Watson J (1954) The complementary structure of deoxyribonucleic acid. Proc Roy Soc Lond Series A Math Phys Sci 223(1152):80–96. https://doi.org/10.1098/rspa.1954.0101

Cross-Disorder Group of the Psychiatric Genomics Consortium (2013) Identification of risk loci with shared effects on five major psychiatric disorders: a genome-wide analysis. Lancet 381(9875):1371–1379. https://doi.org/10.1016/S0140-6736(12)62129-1

Cross-Disorder Group of the Psychiatric Genomics Consortium (2019) Genomic relationships, novel loci, and pleiotropic mechanisms across eight psychiatric disorders. Cell 179(7): 1469–1482.e11. https://doi.org/10.1016/j.cell.2019.11.020

Darmanis S, Sloan SA, Zhang Y, Enge M, Caneda C, Shuer LM, Hayden Gephart MG, Barres BA, Quake SR (2015) A survey of human brain transcriptome diversity at the single cell level. Proc Natl Acad Sci 112(23):7285–7290. https://doi.org/10.1073/pnas.1507125112

Darnell JC, Van Driesche SJ, Zhang C, Hung KYS, Mele A, Fraser CE, Stone EF, Chen C, Fak JJ, Chi SW et al (2011) FMRP stalls ribosomal translocation on mRNAs linked to synaptic function and autism. Cell 146(2):247–261. https://doi.org/10.1016/j.cell.2011.06.013

Deepak S, Kottapalli K, Rakwal R, Oros G, Rangappa K, Iwahashi H, Masuo Y, Agrawal G (2007) Real-time PCR: revolutionizing detection and expression analysis of genes. Curr Genomics 8(4):234–251

Deshpande D, Chhugani K, Chang Y, Karlsberg A, Loeffler C, Zhang J, Muszyńska A, Munteanu V, Yang H, Rotman J et al (2023) RNA-seq data science: From raw data to effective interpretation. Front Genet 14:997383. https://doi.org/10.3389/fgene.2023.997383

Dong X, Liu C, Dozmorov M (2021) Review of multi-omics data resources and integrative analysis for human brain disorders. Brief Funct Genomics 20(4):223–234. https://doi.org/10.1093/bfgp/elab024

Draghici S (2018) Pathway analysis vs gene set analysis: what is the difference and when should I use each? Advaita Bioinformatics. Accessed 6 Dec 2023. https://advaitabio.com/science/pathway-analysis-vs-gene-set-analysis/

Erraji-Benchekroun L, Underwood MD, Arango V, Galfalvy H, Pavlidis P, Smyrniotopoulos P, Mann JJ, Sibille E (2005) Molecular aging in human prefrontal cortex is selective and continuous throughout adult life. Biol Psychiatry 57(5):549–558. https://doi.org/10.1016/j.biopsych.2004.10.034

Fehr JE, Trotter GW, Oxford JT, Hart DA (2000) Comparison of northern blot hybridization and a reverse transcriptase-polymerase chain reaction technique for measurement of mRNA expression of metalloproteinases and matrix components in articular cartilage and synovial membrane from horses with osteoarthritis. Am J Vet Res 61(8):900–905. https://doi.org/10.2460/ajvr.2000.61.900

Finotello F, Lavezzo E, Bianco L, Barzon L, Mazzon P, Fontana P, Toppo S, Di Camillo B (2014) Reducing bias in RNA sequencing data: a novel approach to compute counts. BMC Bioinformatics 15(1):S7. https://doi.org/10.1186/1471-2105-15-S1-S7

Flores JE, Claborne DM, Weller ZD, Webb-Robertson B-JM, Waters KM, Bramer LM (2023) Missing data in multi-omics integration: recent advances through artificial intelligence. Front Artif Intell 6:1098308. https://doi.org/10.3389/frai.2023.1098308

Fornito A, Zalesky A, Breakspear M (2015) The connectomics of brain disorders. Nat Rev Neurosci 16(3):159–172. https://doi.org/10.1038/nrn3901

Fromer M, Roussos P, Sieberts SK, Johnson JS, Kavanagh DH, Perumal TM, Ruderfer DM, Oh EC, Topol A, Shah HR et al (2016) Gene expression elucidates functional impact of polygenic risk for schizophrenia. Nat Neurosci 19(11):1442–1453. https://doi.org/10.1038/nn.4399

Fulcher BD, Arnatkeviciute A, Fornito A (2021) Overcoming false-positive gene-category enrichment in the analysis of spatially resolved transcriptomic brain atlas data. Nat Commun 12:2669. https://doi.org/10.1038/s41467-021-22862-1

Gandal MJ, Haney JR, Parikshak NN, Leppa V, Ramaswami G, Hartl C, Schork AJ, Appadurai V, Buil A, Werge TM et al (2018a) Shared molecular neuropathology across major psychiatric disorders parallels polygenic overlap. Science 359(6376):693–697. https://doi.org/10.1126/science.aad6469

Gandal MJ, Zhang P, Hadjimichael E, Walker RL, Chen C, Liu S, Won H, van Bakel H, Varghese M, Wang Y, et al. Transcriptome-wide isoform-level dysregulation in ASD, schizophrenia, and bipolar disorder. Science . 2018b;362(6420):eaat8127. doi:https://doi.org/10.1126/science.aat8127

Gandal MJ, Haney JR, Wamsley B, Yap CX, Parhami S, Emani PS, Chang N, Chen GT, Hoftman GD, de Alba D et al (2022) Broad transcriptomic dysregulation occurs across the cerebral cortex in ASD. Nature 611(7936):532–539. https://doi.org/10.1038/s41586-022-05377-7

Gene Ontology Consortium (2015) Gene ontology consortium: going forward. Nucleic Acids Res 43(Database issue):D1049–D1056. https://doi.org/10.1093/nar/gku1179

Gergen JP, Stern RH, Wensink PC (1979) Filter replicas and permanent collections of recombinant DNA plasmids. Nucleic Acids Res 7(8):2115–2136

Gondane A, Itkonen HM (2023) Revealing the history and mystery of RNA-Seq. Curr Issues Mol Biol 45(3):1860–1874. https://doi.org/10.3390/cimb45030120

Govindarajan R, Duraiyan J, Kaliyappan K, Palanisamy M (2012) Microarray and its applications. J Pharm Bioallied Sci 4(Suppl 2):S310–S312. https://doi.org/10.4103/0975-7406.100283

Green EK, Rees E, Walters JTR, Smith K-G, Forty L, Grozeva D, Moran JL, Sklar P, Ripke S, Chambert KD et al (2016) Copy number variation in bipolar disorder. Mol Psychiatry 21(1): 89–93. https://doi.org/10.1038/mp.2014.174

Gross A, Schoendube J, Zimmermann S, Steeb M, Zengerle R, Koltay P (2015) Technologies for single-cell isolation. Int J Mol Sci 16(8):16897–16919. https://doi.org/10.3390/ijms160816897

Grove J, Ripke S, Als TD, Mattheisen M, Walters RK, Won H, Pallesen J, Agerbo E, Andreassen OA, Anney R et al (2019) Identification of common genetic risk variants for autism spectrum disorder. Nat Genet 51(3):431–444. https://doi.org/10.1038/s41588-019-0344-8

Grunstein M, Hogness DS (1975) Colony hybridization: a method for the isolation of cloned DNAs that contain a specific gene. Proc Natl Acad Sci U S A 72(10):3961–3965. https://doi.org/10.1073/pnas.72.10.3961

GTEx Consortium (2017) Genetic effects on gene expression across human tissues. Nature 550(7675):204–213. https://doi.org/10.1038/nature24277

GTEx Consortium, Aguet F, Anand S, Ardlie KG, Gabriel S, Getz GA, Graubert A, Hadley K, Handsaker RE, Huang KH et al (2020) The GTEx consortium atlas of genetic regulatory effects across human tissues. Science 369(6509):1318–1330. https://doi.org/10.1126/science.aaz1776

Gupta AK, Gupta UD (2014) Chapter 19- next generation sequencing and its applications. In: Verma AS, Singh A (eds) Animal biotechnology. Academic Press, pp 345–367. https://doi.org/10.1016/B978-0-12-416002-6.00019-5

Habib N, Avraham-Davidi I, Basu A, Burks T, Shekhar K, Hofree M, Choudhury SR, Aguet F, Gelfand E, Ardlie K et al (2017) Massively parallel single-nucleus RNA-seq with DroNc-seq. Nat Methods 14(10):955–958. https://doi.org/10.1038/nmeth.4407

Ham S, Lee S-JV (2020) Advances in transcriptome analysis of human brain aging. Exp Mol Med 52(11):1787–1797. https://doi.org/10.1038/s12276-020-00522-6

Hansen KD, Brenner SE, Dudoit S (2010) Biases in Illumina transcriptome sequencing caused by random hexamer priming. Nucleic Acids Res 38(12):e131. https://doi.org/10.1093/nar/gkq224

Hawrylycz MJ, Lein ES, Guillozet-Bongaarts AL, Shen EH, Ng L, Miller JA, van de Lagemaat LN, Smith KA, Ebbert A, Riley ZL et al (2012) An anatomically comprehensive atlas of the adult human brain transcriptome. Nature 489(7416):391–399. https://doi.org/10.1038/nature11405

Heather JM, Chain B (2016) The sequence of sequencers: the history of sequencing DNA. Genomics 107(1):1–8. https://doi.org/10.1016/j.ygeno.2015.11.003

Hernandez LM, Kim M, Hoftman GD, Haney JR, de la Torre-Ubieta L, Pasaniuc B, Gandal MJ (2021) Transcriptomic insight into the polygenic mechanisms underlying psychiatric disorders. Biol Psychiatry 89(1):54–64. https://doi.org/10.1016/j.biopsych.2020.06.005

Hitzemann R, Darakjian P, Walter N, Dan Iancu O, Searles R, McWeeney S (2014) Introduction to sequencing the brain transcriptome. In: International review of neurobiology, vol 116. Elsevier, pp 1–19. https://doi.org/10.1016/B978-0-12-801105-8.00001-1

Hoebeeck J, Speleman F, Vandesompele J (2007) Real-time quantitative PCR as an alternative to Southern blot or fluorescence in situ hybridization for detection of gene copy number changes. Methods Mol Biol 353:205–226. https://doi.org/10.1385/1-59745-229-7

Hoffman GE, Roussos P. Dream: powerful differential expression analysis for repeated measures designs. Gorodkin J, editor. Bioinformatics 2021;37(2):192–201. doi:https://doi.org/10.1093/bioinformatics/btaa687

Hoffman GE, Bendl J, Voloudakis G, Montgomery KS, Sloofman L, Wang Y-C, Shah HR, Hauberg ME, Johnson JS, Girdhar K et al (2019) CommonMind consortium provides transcriptomic and epigenomic data for schizophrenia and bipolar disorder. Sci Data 6(1):180. https://doi.org/10.1038/s41597-019-0183-6

Holley RW (1965) Structure of an alanine transfer ribonucleic acid. JAMA 194(8):868–871. https://doi.org/10.1001/jama.1965.03090210032009

Holt CE, Bullock SL (2009) Subcellular mRNA localization in animal cells and why it matters. Science 326(5957):1212–1216. https://doi.org/10.1126/science.1176488

Hong M, Tao S, Zhang L, Diao L-T, Huang X, Huang S, Xie S-J, Xiao Z-D, Zhang H (2020) RNA sequencing: new technologies and applications in cancer research. J Hematol Oncol 13(1):166. https://doi.org/10.1186/s13045-020-01005-x

Horvath S, Mirnics K (2015) Schizophrenia as a disorder of molecular pathways. Biol Psychiatry 77(1):22–28. https://doi.org/10.1016/j.biopsych.2014.01.001

Huang DW, Sherman BT, Lempicki RA (2009) Systematic and integrative analysis of large gene lists using DAVID bioinformatics resources. Nat Protoc 4(1):44–57. https://doi.org/10.1038/nprot.2008.211

Huo Y, Li S, Liu J, Li X, Luo X-J (2019) Functional genomics reveal gene regulatory mechanisms underlying schizophrenia risk. Nat Commun 10(1):670. https://doi.org/10.1038/s41467-019-08666-4

Illumina (n.d.) Compare technologies: RNA-seq vs microarrays. Accessed 5 Dec 2023. https://www.illumina.com/science/technology/next-generation-sequencing/microarray-rna-seq-comparison.html

Im Y, Kim Y (2023) A comprehensive overview of RNA deconvolution methods and their application. Mol Cells 46(2):99–105. https://doi.org/10.14348/molcells.2023.2178

International Schizophrenia Consortium (2009) Common polygenic variation contributes to risk of schizophrenia that overlaps with bipolar disorder. Nature 460(7256):748–752. https://doi.org/10.1038/nature08185

Iqbal F, Lupieri A, Aikawa M, Aikawa E (2021) Harnessing single-cell RNA sequencing to better understand how diseased cells behave the way they do in cardiovascular disease. Arterioscler Thromb Vasc Biol 41(2):585–600. https://doi.org/10.1161/ATVBAHA.120.314776

Jaffe AE, Straub RE, Shin JH, Tao R, Gao Y, Collado-Torres L, Kam-Thong T, Xi HS, Quan J, Chen Q et al (2018) Developmental and genetic regulation of the human cortex transcriptome illuminate schizophrenia pathogenesis. Nat Neurosci 21(8):1117–1125. https://doi.org/10.1038/s41593-018-0197-y

Jung S, Liu W, Baek J, Moon JW, Ye BD, Lee H-S, Park SH, Yang S-K, Han B, Liu J et al (2020) Expression quantitative trait loci (eQTL) mapping in Korean patients with Crohn's disease and identification of potential causal genes through integration with disease associations. Front Genet 11. https://doi.org/10.3389/fgene.2020.00486

Kanehisa M, Goto S, Kawashima S, Okuno Y, Hattori M (2004) The KEGG resource for deciphering the genome. Nucleic Acids Res 32(Database issue):D277–D280. https://doi.org/10.1093/nar/gkh063

Kang RB, Li Y, Rosselot C, Zhang T, Siddiq M, Rajbhandari P, Stewart AF, Scott DK, Garcia-Ocana A, Lu G (2023) Single-nucleus RNA sequencing of human pancreatic islets identifies novel gene sets and distinguishes β-cell subpopulations with dynamic transcriptome profiles. Genome Med 15:30. https://doi.org/10.1186/s13073-023-01179-2

Kebschull JM, Zador AM (2015) Sources of PCR-induced distortions in high-throughput sequencing data sets. Nucleic Acids Res 43(21):e143. https://doi.org/10.1093/nar/gkv717

Keel BN, Lindholm-Perry AK, Rohrer GA, Oliver WT (2023) Estimation of cell type proportions from bulk RNA-Seq of porcine whole blood samples using partial reference-free deconvolution. Anim Genet 30:200159. https://doi.org/10.1016/j.angen.2023.200159

Kim N, Kang H, Jo A, Yoo S-A, Lee H-O (2023) Perspectives on single-nucleus RNA sequencing in different cell types and tissues. J Pathol Transl Med 57(1):52–59. https://doi.org/10.4132/jptm.2022.12.19

Kirov G (2015) CNVs in neuropsychiatric disorders. Hum Mol Genet 24(R1):R45–R49. https://doi.org/10.1093/hmg/ddv253

Koltai H, Weingarten-Baror C (2008) Specificity of DNA microarray hybridization: characterization, effectors and approaches for data correction. Nucleic Acids Res 36(7):2395–2405. https://doi.org/10.1093/nar/gkn087

Lake BB, Codeluppi S, Yung YC, Gao D, Chun J, Kharchenko PV, Linnarsson S, Zhang K (2017) A comparative strategy for single-nucleus and single-cell transcriptomes confirms accuracy in predicted cell-type expression from nuclear RNA. Sci Rep 7(1):6031. https://doi.org/10.1038/s41598-017-04426-w

Lee SH, DeCandia TR, Ripke S, Yang J, Sullivan PF, Goddard ME, Keller MC, Visscher PM, Wray NR (2012) Estimating the proportion of variation in susceptibility to schizophrenia captured by common SNPs. Nat Genet 44(3):247–250. https://doi.org/10.1038/ng.1108

Li S, Łabaj PP, Zumbo P, Sykacek P, Shi W, Shi L, Phan J, Wu L, Wang M, Wang C et al (2014) Detecting and correcting systematic variation in large-scale RNA sequencing data. Nat Biotechnol 32(9):888–895. https://doi.org/10.1038/nbt.3000

Love M (2016) RNA-seq fragment sequence bias. Accessed 6 Dec 2023. https://mikelove.wordpress.com/2016/09/26/rna-seq-fragment-sequence-bias/

Love MI, Huber W, Anders S (2014) Moderated estimation of fold change and dispersion for RNA-seq data with DESeq2. Genome Biol 15(12):550. https://doi.org/10.1186/s13059-014-0550-8

Love MI, Hogenesch JB, Irizarry RA (2016) Modeling of RNA-seq fragment sequence bias reduces systematic errors in transcript abundance estimation. Nat Biotechnol 34(12):1287–1291. https://doi.org/10.1038/nbt.3682

Lowe R, Shirley N, Bleackley M, Dolan S, Shafee T (2017) Transcriptomics technologies. PLoS Comput Biol 13(5):e1005457. https://doi.org/10.1371/journal.pcbi.1005457

Lu T, Pan Y, Kao S-Y, Li C, Kohane I, Chan J, Yankner BA (2004) Gene regulation and DNA damage in the ageing human brain. Nature 429(6994):883–891. https://doi.org/10.1038/nature02661

MacArthur J, Bowler E, Cerezo M, Gil L, Hall P, Hastings E, Junkins H, McMahon A, Milano A, Morales J et al (2017) The new NHGRI-EBI catalog of published genome-wide association studies (GWAS Catalog). Nucleic Acids Res 45(D1):D896–D901. https://doi.org/10.1093/nar/gkw1133

Mandal AS, Gandal M, Seidlitz J, Alexander-Bloch A (2022) A critical appraisal of imaging transcriptomics. Biol Psychiatry Global Open Sci 2(4):311–313. https://doi.org/10.1016/j.bpsgos.2022.08.001

Markello RD, Misic B (2021) Comparing spatial null models for brain maps. Neuroimage 236: 118052. https://doi.org/10.1016/j.neuroimage.2021.118052

Maxam AM, Gilbert W (1977) A new method for sequencing DNA. Proc Natl Acad Sci 74(2): 560–564. https://doi.org/10.1073/pnas.74.2.560

McKenzie AT, Wang M, Hauberg ME, Fullard JF, Kozlenkov A, Keenan A, Hurd YL, Dracheva S, Casaccia P, Roussos P et al (2018) Brain cell type specific gene expression and co-expression network architectures. Sci Rep 8(1):8868. https://doi.org/10.1038/s41598-018-27293-5

Melé M, Ferreira PG, Reverter F, DeLuca DS, Monlong J, Sammeth M, Young TR, Goldmann JM, Pervouchine DD, Sullivan TJ et al (2015) Human genomics. The human transcriptome across tissues and individuals. Science 348(6235):660–665. https://doi.org/10.1126/science.aaa0355

Merikangas AK, Shelly M, Knighton A, Kotler N, Tanenbaum N, Almasy L (2022) What genes are differentially expressed in individuals with schizophrenia? A systematic review. Mol Psychiatry 27(3):1373–1383. https://doi.org/10.1038/s41380-021-01420-7

Merker JD, Arber DA (2010) Chapter 21- molecular diagnostics of hematopoietic disorders. In: Grody WW, Nakamura RM, Strom CM, Kiechle FL (eds) Molecular diagnostics. Academic Press, San Diego, pp 243–260. https://doi.org/10.1016/B978-0-12-369428-7.00021-5

Merkin J, Russell C, Chen P, Burge CB (2012) Evolutionary dynamics of gene and isoform regulation in mammalian tissues. Science 338(6114):1593–1599. https://doi.org/10.1126/science.1228186

Morgan SE, Seidlitz J, Whitaker KJ, Romero-Garcia R, Clifton NE, Scarpazza C, Van Amelsvoort T, Marcelis M, Van Os J, Donohoe G et al (2019) Cortical patterning of abnormal morphometric similarity in psychosis is associated with brain expression of schizophrenia-related genes. Proc Natl Acad Sci 116(19):9604–9609. https://doi.org/10.1073/pnas.1820754116

Mostafavi H, Spence JP, Naqvi S, Pritchard JK (2023) Systematic differences in discovery of genetic effects on gene expression and complex traits. Nat Genet 55(11):1866–1875. https://doi.org/10.1038/s41588-023-01529-1

Mullis K, Faloona F, Scharf S, Saiki R, Horn G, Erlich H (1986) Specific enzymatic amplification of DNA in vitro: the polymerase chain reaction. Cold Spring Harb Symp Quant Biol 51(Pt 1): 263–273. https://doi.org/10.1101/sqb.1986.051.01.032

Nadelmann ER, Gorham JM, Reichart D, Delaughter DM, Wakimoto H, Lindberg EK, Litviňukova M, Maatz H, Curran JJ, Gutierrez DI et al (2021) Isolation of nuclei from mammalian cells and tissues for single-nucleus molecular profiling. Curr Protoc 1(5):e132. https://doi.org/10.1002/cpz1.132

Naro C, Cesari E, Sette C (2021) Splicing regulation in brain and testis: common themes for highly specialized organs. Cell Cycle 20(5–6):480–489. https://doi.org/10.1080/15384101.2021.1889187

National Human Genome Research Institute. Human genome project fact sheet. Genomegov. 24 Aug 2022a. https://www.genome.gov/about-genomics/educational-resources/fact-sheets/human-genome-project

National Human Genome Research Institute. Polymerase chain reaction (PCR) fact sheet. Genome. 17 Aug 2022b. https://www.genome.gov/about-genomics/fact-sheets/Polymerase-Chain-Reaction-Fact-Sheet

Naumova OY, Palejev D, Vlasova NV, Lee M, Rychkov SY, Babich ON, Vaccarino FM, Grigorenko EL (2012) Age-related changes of gene expression in the neocortex: preliminary data on RNA-seq of the transcriptome in three functionally distinct cortical areas. Dev Psychopathol 24(4):1427–1442. https://doi.org/10.1017/S0954579412000818

Negi SK, Guda C (2017) Global gene expression profiling of healthy human brain and its application in studying neurological disorders. Sci Rep 7(1):897. https://doi.org/10.1038/s41598-017-00952-9

Nica AC, Dermitzakis ET (2013) Expression quantitative trait loci: present and future. Philos Trans R Soc B Biol Sci 368(1620):20120362. https://doi.org/10.1098/rstb.2012.0362

Nickle T, Barrette-Ng I (2016) 8.7: DNA analysis- blotting and hybridization. In: Open access genetics textbook. LibreTexts Project. (LibreTexts Biology). https://bio.libretexts.org/Bookshelves/Genetics/Online_Open_Genetics_(Nickle_and_Barrette-Ng)/08%3A_Techniques_of_Molecular_Genetics/8.07%3A__DNA_Analysis-_Blotting_and_Hybridization

Odelberg W, Nobel Foundation (eds) (1981) Les Prix Nobel: 1980 : Nobel prizes, presentations, biographies and lectures. Almquist & Wiksell International, Stockholm

Oevermans H (2021) What is spatial transcriptomics? Single cell discoveries. Accessed 6 Dec 2023. https://www.scdiscoveries.com/blog/knowledge/what-is-spatial-transcriptomics/

Okoniewski MJ, Miller CJ (2006) Hybridization interactions between probesets in short oligo microarrays lead to spurious correlations. BMC Bioinformatics 7(1):276. https://doi.org/10.1186/1471-2105-7-276

Owen MJ, Sawa A, Mortensen PB (2016a) Schizophrenia. Lancet 388(10039):86–97. https://doi.org/10.1016/S0140-6736(15)01121-6

Owen MJ, Sawa A, Mortensen PB (2016b. Accessed 6 Dec 2023; https://linkinghub.elsevier.com/retrieve/pii/S0140673615011216) Schizophrenia. Lancet 388(10039):86–97. https://doi.org/10.1016/S0140-6736(15)01121-6

Pardiñas AF, Holmans P, Pocklington AJ, Escott-Price V, Ripke S, Carrera N, Legge SE, Bishop S, Cameron D, Hamshere ML et al (2018) Common schizophrenia alleles are enriched in mutation-intolerant genes and in regions under strong background selection. Nat Genet 50(3):381–389. https://doi.org/10.1038/s41588-018-0059-2

Patel Y, Shin J, Abé C, Agartz I, Alloza C, Alnæs D, Ambrogi S, Antonucci LA, Arango C, Arolt V et al (2022. Accessed 9 Jan 2024; https://linkinghub.elsevier.com/retrieve/pii/S0006322322010678) Virtual ontogeny of cortical growth preceding mental illness. Biol Psychiatry 92(4):299–313

Piper M, Pantano L, Mistry M, Khetani R. Single-cell RNA-seq: clustering analysis. Introduction to single-cell RNA-seq. 2019 Nov 21. Accessed 6 Dec 2023. https://hbctraining.github.io/scRNA-seq/lessons/07_SC_clustering_cells_SCT.html

Piwecka M, Rajewsky N, Rybak-Wolf A (2023) Single-cell and spatial transcriptomics: deciphering brain complexity in health and disease. Nat Rev Neurol 19(6):346–362. https://doi.org/10.1038/s41582-023-00809-y

Primiani CT, Ryan VH, Rao JS, Cam MC, Ahn K, Modi HR, Rapoport SI. Coordinated gene expression of neuroinflammatory and cell signaling markers in dorsolateral prefrontal cortex during human brain development and aging. Sirén A-L, editor. PloS One. 2014;9(10):e110972. doi:https://doi.org/10.1371/journal.pone.0110972

Purcell SM, Moran JL, Fromer M, Ruderfer D, Solovieff N, Roussos P, O'Dushlaine C, Chambert K, Bergen SE, Kähler A et al (2014) A polygenic burden of rare disruptive mutations in schizophrenia. Nature 506(7487):185–190. https://doi.org/10.1038/nature12975

Quackenbush J (2001) Computational analysis of microarray data. Nat Rev Genet 2(6):418–427. https://doi.org/10.1038/35076576

Rees E, Kirov G (2021) Copy number variation and neuropsychiatric illness. Curr Opin Genet Dev 68:57–63. Molecular and genetic basis of disease. https://doi.org/10.1016/j.gde.2021.02.014

Rees E, Owen MJ (2020) Translating insights from neuropsychiatric genetics and genomics for precision psychiatry. Genome Med 12(1):43. https://doi.org/10.1186/s13073-020-00734-5

Rees E, Han J, Morgan J, Carrera N, Escott-Price V, Pocklington AJ, Duffield M, Hall L, Legge SE, Pardiñas AF et al (2020) De novo mutations identified by exome sequencing implicate rare missense variants in SLC6A1 in schizophrenia. Nat Neurosci 23(2):179–184. https://doi.org/10.1038/s41593-019-0565-2

Richards C. How next-generation sequencing came to be: a brief history Drug Target Rev 2015 Jan 31. https://www.drugtargetreview.com/article/39292/how-next-generation-sequencing-came-to-be-a-brief-history/

Richards AL, Jones L, Moskvina V, Kirov G, Gejman PV, Levinson DF, Sanders AR, Purcell S, Visscher PM, Craddock N et al (2012) Schizophrenia susceptibility alleles are enriched for alleles that affect gene expression in adult human brain. Mol Psychiatry 17(2):193–201. https://doi.org/10.1038/mp.2011.11

Ripke S, Neale BM, Corvin A, Walters JTR, Farh K-H, Holmans PA, Lee P, Bulik-Sullivan B, Collier DA, Huang H et al (2014) Biological insights from 108 schizophrenia-associated genetic loci. Nature 511(7510):421–427. https://doi.org/10.1038/nature13595

Ritchie ME, Silver J, Oshlack A, Holmes M, Diyagama D, Holloway A, Smyth GK (2007) A comparison of background correction methods for two-colour microarrays. Bioinformatics 23(20):2700–2707. https://doi.org/10.1093/bioinformatics/btm412

Ritchie ME, Phipson B, Wu D, Hu Y, Law CW, Shi W, Smyth GK (2015) Limma powers differential expression analyses for RNA-sequencing and microarray studies. Nucleic Acids Res 43(7):e47–e47. https://doi.org/10.1093/nar/gkv007

Robinson MD, McCarthy DJ, Smyth GK (2010) edgeR: a bioconductor package for differential expression analysis of digital gene expression data. Bioinformatics 26(1):139–140. https://doi.org/10.1093/bioinformatics/btp616

Romme IAC, de Reus MA, Ophoff RA, Kahn RS, van den Heuvel MP (2017) Connectome disconnectivity and cortical gene expression in patients with schizophrenia. Biol Psychiatry 81(6):495–502. Schizophrenia: Circuitry and Mechanisms. https://doi.org/10.1016/j.biopsych.2016.07.012

Ruzicka WB, Mohammadi S, Fullard JF, Davila-Velderrain J, Subburaju S, Tso DR, Hourihan M, Jiang S, Lee H-C, Bendl J et al (2022) Single-cell multi-cohort dissection of the schizophrenia transcriptome. https://doi.org/10.1101/2022.08.31.22279406

Saccone C, Pesole G (2005) Handbook of comparative genomics: principles and methodology. Wiley

Sampled. Bulk RNA sequencing versus single-cell RNAseq: 2 powerful... 2023 Mar 15. Accessed 6 Dec 2023. https://sampled.com/bulk-rna-sequencing-vs-single-cell-rna-sequencing/

Sanger F, Coulson AR (1975) A rapid method for determining sequences in DNA by primed synthesis with DNA polymerase. J Mol Biol 94(3):441–448. https://doi.org/10.1016/0022-2836(75)90213-2

Sanger F, Thompson EOP, Kitai R (1955) The amide groups of insulin. Biochem J 59(3):509–518. https://doi.org/10.1042/bj0590509

Sanger F, Donelson JE, Coulson AR, Kössel H, Fischer D (1973) Use of DNA polymerase I primed by a synthetic oligonucleotide to determine a nucleotide sequence in phage fl DNA. Proc Natl Acad Sci U S A 70(4):1209–1213. https://doi.org/10.1073/pnas.70.4.1209

Sanger F, Air GM, Barrell BG, Brown NL, Coulson AR, Fiddes CA, Hutchison CA, Slocombe PM, Smith M (1977a) Nucleotide sequence of bacteriophage phi X174 DNA. Nature 265(5596): 687–695. https://doi.org/10.1038/265687a0

Sanger F, Nicklen S, Coulson AR (1977b) DNA sequencing with chain-terminating inhibitors. Proc Natl Acad Sci 74(12):5463–5467. https://doi.org/10.1073/pnas.74.12.5463

Santiago CP, Gimmen MY, Lu Y, McNally MM, Duncan LH, Creamer TJ, Orzolek LD, Blackshaw S, Singh MS (2023) Comparative analysis of single-cell and single-nucleus RNA-sequencing in a rabbit model of retinal detachment-related proliferative vitreoretinopathy. Ophthalmol Sci 3(4):100335. https://doi.org/10.1016/j.xops.2023.100335

Sathyanarayanan A, Mueller TT, Ali Moni M, Schueler K, Baune BT, Lio P, Mehta D, Baune BT, Dierssen M, Ebert B et al (2023) Multi-omics data integration methods and their applications in psychiatric disorders. Eur Neuropsychopharmacol 69:26–46. https://doi.org/10.1016/j.euroneuro.2023.01.001

Schadt EE, Turner S, Kasarskis A (2010) A window into third-generation sequencing. Hum Mol Genet 19(R2):R227–R240. https://doi.org/10.1093/hmg/ddq416

Schena M, Shalon D, Davis RW, Brown PO (1995) Quantitative monitoring of gene expression patterns with a complementary DNA microarray. Science 270(5235):467–470. https://doi.org/10.1126/science.270.5235.467

Shendure J (2008) The beginning of the end for microarrays? Nat Methods 5(7):585–587. https://doi.org/10.1038/nmeth0708-585

Shendure J, Ji H (2008) Next-generation DNA sequencing. Nat Biotechnol 26(10):1135–1145. https://doi.org/10.1038/nbt1486

Shi H, Zhou Y, Jia E, Pan M, Bai Y, Ge Q (2021) Bias in RNA-seq library preparation: current challenges and solutions. Biomed Res Int:2021. https://doi.org/10.1155/2021/6647597

Shin J, Patel Y, Parker N, Paus T, Pausova Z (2023) Prediabetic HbA1c and cortical atrophy: underlying neurobiology. Diabetes Care 46(12):2267–2272. https://doi.org/10.2337/dc23-1105

Singh T, Walters JTR, Johnstone M, Curtis D, Suvisaari J, Torniainen M, Rees E, Iyegbe C, Blackwood D, McIntosh AM et al (2017) The contribution of rare variants to risk of schizophrenia in individuals with and without intellectual disability. Nat Genet 49(8):1167–1173. https://doi.org/10.1038/ng.3903

Skene NG, Bryois J, Bakken TE, Breen G, Crowley JJ, Gaspar HA, Giusti-Rodriguez P, Hodge RD, Miller JA, Muñoz-Manchado AB et al (2018) Genetic identification of brain cell types underlying schizophrenia. Nat Genet 50(6):825–833. https://doi.org/10.1038/s41588-018-0129-5

Soheili-Nezhad S, Van Der Linden RJ, Olde Rikkert M, Sprooten E, Poelmans G (2021) Long genes are more frequently affected by somatic mutations and show reduced expression in Alzheimer's disease: implications for disease etiology. Alzheimers Dement 17(3):489–499. https://doi.org/10.1002/alz.12211

Soneson C, Yao Y, Bratus-Neuenschwander A, Patrignani A, Robinson MD, Hussain S (2019) A comprehensive examination of nanopore native RNA sequencing for characterization of complex transcriptomes. Nat Commun 10(1):3359. https://doi.org/10.1038/s41467-019-11272-z

Southern EM (1975) Detection of specific sequences among DNA fragments separated by gel electrophoresis. J Mol Biol 98(3):503–517. https://doi.org/10.1016/s0022-2836(75)80083-0

Stahl EA, Breen G, Forstner AJ, McQuillin A, Ripke S, Trubetskoy V, Mattheisen M, Wang Y, Coleman JRI, Gaspar HA et al (2019) Genome-wide association study identifies 30 loci associated with bipolar disorder. Nat Genet 51(5):793–803. https://doi.org/10.1038/s41588-019-0397-8

Stark R, Grzelak M, Hadfield J (2019) RNA sequencing: the teenage years. Nat Rev Genet 20(11):631–656. https://doi.org/10.1038/s41576-019-0150-2

Stephan KE, Friston KJ, Frith CD (2009) Dysconnection in schizophrenia: from abnormal synaptic plasticity to failures of self-monitoring. Schizophr Bull 35(3):509–527. https://doi.org/10.1093/schbul/sbn176

Stilo SA, Murray RM (2019) Non-genetic factors in schizophrenia. Curr Psychiatry Rep 21(10):100. https://doi.org/10.1007/s11920-019-1091-3

Stoeger T, Grant RA, McQuattie-Pimentel AC, Anekalla KR, Liu SS, Tejedor-Navarro H, Singer BD, Abdala-Valencia H, Schwake M, Tetreault M-P et al (2022) Aging is associated with a systemic length-associated transcriptome imbalance. Nat Aging 2(12):1191–1206. https://doi.org/10.1038/s43587-022-00317-6

Strober BJ, Elorbany R, Rhodes K, Krishnan N, Tayeb K, Battle A, Gilad Y (2019) Dynamic genetic regulation of gene expression during cellular differentiation. Science 364(6447): 1287–1290. https://doi.org/10.1126/science.aaw0040

Sturla L-M, Fernandez-Teijeiro A, Pomeroy SL (2003) Application of microarrays to neurological disease. Arch Neurol 60(5):676. https://doi.org/10.1001/archneur.60.5.676

Taub MA, Corrada Bravo H, Irizarry RA (2010) Overcoming bias and systematic errors in next generation sequencing data. Genome Med 2(12):87. https://doi.org/10.1186/gm208

The Writing Committee for the Attention-Deficit/Hyperactivity Disorder, Autism Spectrum Disorder, Bipolar Disorder, Major Depressive Disorder, Obsessive-Compulsive Disorder, and Schizophrenia ENIGMA Working Group, Patel Y, Parker N, Shin J, Howard D, French L, Thomopoulos SI, Pozzi E, Abe Y, Abé C et al (2021) Virtual histology of cortical thickness and shared neurobiology in 6 psychiatric disorders. JAMA Psychiatry 78(1):47. https://doi.org/10.1001/jamapsychiatry.2020.2694

ThermoFisher Scientific (n.d.) The basics: RT-PCR. Accessed 5 Dec 2023. https://www.thermofisher.com/us/en/home/references/ambion-tech-support/rtpcr-analysis/general-articles/rt%2D%2Dpcr-the-basics.html

Thompson SD, Prahalad S, Colbert RA (2016) Chapter 5- integrative genomics. In: Petty RE, Laxer RM, Lindsley CB, Wedderburn LR (eds) Textbook of pediatric rheumatology, 7th edn. W.B. Saunders, Philadelphia, pp 43–53.e3. https://doi.org/10.1016/B978-0-323-24145-8.00005-3

Trubetskoy V, Pardiñas AF, Qi T, Panagiotaropoulou G, Awasthi S, Bigdeli TB, Bryois J, Chen C-Y, Dennison CA, Hall LS et al (2022) Mapping genomic loci implicates genes and synaptic biology in schizophrenia. Nature 604(7906):502–508. https://doi.org/10.1038/s41586-022-04434-5

Tylee DS, Kawaguchi DM, Glatt SJ (2013) On the outside, looking in: a review and evaluation of the comparability of blood and brain "-omes". Am J Med Genet B Neuropsychiatr Genet 162(7): 595–603. https://doi.org/10.1002/ajmg.b.32150

Umans BD, Battle A, Gilad Y (2021) Where are the disease-associated eQTLs? Trends Genet 37(2): 109–124. https://doi.org/10.1016/j.tig.2020.08.009

Vaga S (2022) Understanding single cell sequencing, how it works and its applications. Genomics Res Technol Networks. 20 Jan 2022

Valihrach L, Androvic P, Kubista M (2018) Platforms for single-cell collection and analysis. Int J Mol Sci 19(3):807. https://doi.org/10.3390/ijms19030807

Van Dam S, Vosa U, Van Der Graaf A, Franke L, De Magalhaes JP (2017) Gene co-expression analysis for functional classification and gene–disease predictions. Brief Bioinform:bbw139. https://doi.org/10.1093/bib/bbw139

Venter JC, Adams MD, Myers EW, Li PW, Mural RJ, Sutton GG, Smith HO, Yandell M, Evans CA, Holt RA et al (2001) The sequence of the human genome. Science 291(5507):1304–1351. https://doi.org/10.1126/science.1058040

Vermeij WP, Dollé MET, Reiling E, Jaarsma D, Payan-Gomez C, Bombardieri CR, Wu H, Roks AJM, Botter SM, van der Eerden BC et al (2016) Restricted diet delays accelerated ageing and genomic stress in DNA-repair-deficient mice. Nature 537(7620):427–431. https://doi.org/10.1038/nature19329

Walker RL, Ramaswami G, Hartl C, Mancuso N, Gandal MJ, de la Torre-Ubieta L, Pasaniuc B, Stein JL, Geschwind DH (2019) Genetic control of expression and splicing in developing human brain informs disease mechanisms. Cell 179(3):750–771.e22. https://doi.org/10.1016/j.cell.2019.09.021

Wang Z, Gerstein M, Snyder M (2009) RNA-Seq: a revolutionary tool for transcriptomics. Nat Rev Genet 10(1):57–63. https://doi.org/10.1038/nrg2484

Watson JD, Crick FHC (1953) Molecular structure of nucleic acids: a structure for deoxyribose nucleic acid. Nature 171(4356):737–738. https://www.nature.com/articles/171737a0. https://doi.org/10.1038/171737a0

Welter D, MacArthur J, Morales J, Burdett T, Hall P, Junkins H, Klemm A, Flicek P, Manolio T, Hindorff L et al (2014) The NHGRI GWAS Catalog, a curated resource of SNP-trait associations. Nucleic Acids Res 42(Database issue):D1001–D1006. https://doi.org/10.1093/nar/gkt1229

Williams CG, Lee HJ, Asatsuma T, Vento-Tormo R, Haque A (2022) An introduction to spatial transcriptomics for biomedical research. Genome Med 14(1):68. https://doi.org/10.1186/s13073-022-01075-1

Wray NR, Ripke S, Mattheisen M, Trzaskowski M, Byrne EM, Abdellaoui A, Adams MJ, Agerbo E, Air TM, Andlauer TMF et al (2018) Genome-wide association analyses identify 44 risk variants and refine the genetic architecture of major depression. Nat Genet 50(5): 668–681. https://doi.org/10.1038/s41588-018-0090-3

Wu R, Kaiser AD (1968) Structure and base sequence in the cohesive ends of bacteriophage lambda DNA. J Mol Biol 35(3):523–537. https://doi.org/10.1016/s0022-2836(68)80012-9

Wu C, Bendriem RM, Garamszegi SP, Song L, Lee C (2017) RNA sequencing in post-mortem human brains of neuropsychiatric disorders. Psychiatry Clin Neurosci 71(10):663–672. https://doi.org/10.1111/pcn.12550

Wu H, Kirita Y, Donnelly EL, Humphreys BD (2019) Advantages of single-nucleus over single-cell RNA sequencing of adult kidney: rare cell types and novel cell states revealed in fibrosis. J Am Soc Nephrol 30(1):23–32. https://doi.org/10.1681/ASN.2018090912

Xia CH, Ma Z, Ciric R, Gu S, Betzel RF, Kaczkurkin AN, Calkins ME, Cook PA, García de la Garza A, Vandekar SN et al (2018) Linked dimensions of psychopathology and connectivity in functional brain networks. Nat Commun 9(1):3003. https://doi.org/10.1038/s41467-018-05317-y

Xiang R, Wang W, Yang L, Wang S, Xu C, Chen X (2021) A comparison for dimensionality reduction methods of single-cell RNA-seq data. Front Genet 12

Yang T, Zhang M, Zhang N (2022) Modified northern blot protocol for easy detection of mRNAs in total RNA using radiolabeled probes. BMC Genomics 23(1):66. https://doi.org/10.1186/s12864-021-08275-w

Yazar S, Alquicira-Hernandez J, Wing K, Senabouth A, Gordon MG, Andersen S, Lu Q, Rowson A, Taylor TRP, Clarke L et al (2022) Single-cell eQTL mapping identifies cell type–specific genetic control of autoimmune disease. Science 376(6589):eabf3041. https://doi.org/10.1126/science.abf3041

Yeo G, Holste D, Kreiman G, Burge CB (2004) Variation in alternative splicing across human tissues. Genome Biol 5(10):R74. https://doi.org/10.1186/gb-2004-5-10-r74

Yu X, Abbas-Aghababazadeh F, Chen YA, Fridley BL (2021) Statistical and bioinformatics analysis of data from bulk and single-cell RNA sequencing experiments. In: Markowitz J (ed) Translational bioinformatics for therapeutic development, vol 2194. Springer US, New York, pp 143–175. https://doi.org/10.1007/978-1-0716-0849-4_9

Zhang Y, Chen K, Sloan SA, Bennett ML, Scholze AR, O'Keeffe S, Phatnani HP, Guarnieri P, Caneda C, Ruderisch N et al (2014) An RNA-sequencing transcriptome and splicing database of glia, neurons, and vascular cells of the cerebral cortex. J Neurosci 34(36):11929–11947. https://doi.org/10.1523/JNEUROSCI.1860-14.2014

Zhang X, Li T, Liu F, Chen Y, Yao J, Li Z, Huang Y, Wang J (2019) Comparative analysis of droplet-based ultra-high-throughput single-cell RNA-Seq systems. Mol Cell 73(1):130–142.e5. https://doi.org/10.1016/j.molcel.2018.10.020

Zhernakova DV, Deelen P, Vermaat M, van Iterson M, van Galen M, Arindrarto W, van't Hof P, Mei H, van Dijk F, Westra H-J et al (2017) Identification of context-dependent expression quantitative trait loci in whole blood. Nat Genet 49(1):139–145. https://doi.org/10.1038/ng.3737

Zhou X, Li Y eds (2015) Chapter 2- techniques for oral microbiology. Atlas of oral microbiology. Academic Press, Oxford, pp 15–40. https://doi.org/10.1016/B978-0-12-802234-4.00002-1

Large-Scale Population-Based Studies of Blood Metabolome and Brain Health

Zdenka Pausova and Eeva Sliz

Contents

Abstract Metabolomics technologies enable the quantification of multiple metabolomic measures simultaneously, which provides novel insights into molecular aspects of human health and disease. In large-scale, population-based studies, blood is often the preferred biospecimen. Circulating metabolome may relate to brain health either by affecting or reflecting brain metabolism. Peripheral metabolites may act at or cross the blood–brain barrier and, subsequently, influence brain metabolism, or they may reflect brain metabolism if similar pathways are engaged.

Z. Pausova
Centre hospitalier universitaire Sainte-Justine and Department of Pediatrics, University of Montreal, Montreal, QC, Canada

E. Sliz (✉)
Research Unit of Population Health, Faculty of Medicine, University of Oulu, Oulu, Finland
e-mail: eeva.sliz@oulu.fi

Curr Topics Behav Neurosci (2024) 68: 177–220
https://doi.org/10.1007/7854_2024_463
Published Online: 21 March 2024

Peripheral metabolites may also include those penetrating the circulation from the brain, indicating, for example, brain damage. Most brain health-related metabolomics studies have been conducted in the context of neurodegenerative disorders and cognition, but some studies have also focused on neuroimaging markers of these disorders. Moreover, several metabolomics studies of neurodevelopmental disorders have been performed. Here, we provide a brief background on the types of blood metabolites commonly assessed, and we review the literature describing the relationships between human blood metabolome ($n > 50$ metabolites) and brain health reported in large-scale studies ($n > 500$ individuals).

Keywords Brain health · Lipidomics · Metabolomics

1 Introduction

In the field of molecular epidemiology, the traditional biochemical assays and analytical techniques for measuring blood metabolites have given way to an increasingly common and more holistic approach. The development of metabolomics technologies (e.g., nuclear magnetic resonance [NMR] spectroscopy, mass spectrometry [MS]) has facilitated analyses of multiple blood metabolites in large-scale population-based cohorts, which aids in the discovery of molecular mechanisms of human health and disease. While different biospecimens can be used in metabolomics applications, blood is commonly preferred in large-scale studies due to the ease of the sample collection, but also the fact that blood contains metabolites originating from multiple tissues and, thus, provides an insight into the whole-body metabolism.

Blood metabolome may be related to brain health by either reflecting or affecting brain metabolism: if similar pathways are engaged, the blood metabolomic profile may reflect brain metabolism, while some peripheral metabolites may act at or cross the blood–brain barrier (BBB) and affect brain metabolism. In some cases, brain metabolites can also cross the BBB in the opposite direction and be detected in the circulation. This review offers a foundational overview of commonly studied circulating metabolites, and it surveys the published literature on blood metabolomics of brain health conducted in large-scale (n > 500 participants) population-based studies. Furthermore, smaller-scale investigations are included in the discussion to supplement the findings from the large-scale studies. In our definition of brain health-related phenotypes, we include common neurological disorders, such as stroke and Alzheimer's disease and related dementias (ADRD), as well as neuroimaging markers of these disorders. We have provided a description of the studies we review here, including their respective sample sizes and the metabolomic platforms used, in Table 1.

In blood, the most abundant molecules are typically buffering agents (e.g., bicarbonate) and electrolytes (e.g., sodium and potassium), whereas the less

Table 1 Summary of metabolomic studies of brain health

First author	Reference	Metabolomics platform	Biospecimen	Fasting status	Phenotype(s)	Sample size[a]
Amin	Amin et al. (2023)	Nightingale Health Ltd (NMR)	Plasma	Not fasted	Lifetime major depressive disorder	58,257
					Major depressive disorder	62,508
					Recurrent major depressive disorder	66,878
Anand	Anand et al. (2021)	Harvard-MIT Broad Institute Metabolite Profiling Laboratory (LC-MS)	Plasma	Not fasted	Child ADHD vs. cord metabolites	626[b]
Baloni	Baloni et al. (2022)	Biocrates, p180 (MS)	Serum	Fasted and non-fasted	AD, brain atrophy, cognition (ADAS-Cog13)	754
Bernath	Bernath et al. (2020)	NIH West Coast Metabolomics Center (MS)	Serum	Fasted	ADRD, MCI, AD biomarkers (CSF Aβ, CSF p-tau, hippocampal volume, enorhinal thickness, FDG global cortex, CSF t-tau, white matter hyperintensity)	689
Chouraki	Chouraki et al. (2017)	AB SCIEX 4000 QTRAP (LC-MS)	Plasma	Fasted	Incident ADRD	2,067
Cui	Cui et al. (2020)	Bruker Biospin (NMR)	Serum	Fasted	Incident dementia	1,440
de Leeuw	de Leeuw et al. (2021)	Nightingale Health Ltd (NMR)	Plasma	Not fasted	Brain volume	3,962
					Hippocampus volume	3,962
					White matter hyperintensities	3,962
Harshfield	Harshfield and Markus (2023)	Nightingale Health Ltd (NMR)	Plasma	Not fasted	Fractional anisotropy	9,779

(continued)

Table 1 (continued)

First author	Reference	Metabolomics platform	Biospecimen	Fasting status	Phenotype(s)	Sample size[a]
					Incident all-cause dementia	118,021
					Incident ADRD	118,021
					Incident intracerebral hemorrhage	118,021
					Incident ischemic stroke	118,021
					Incident stroke	118,021
					Incident vascular dementia	118,021
					Mean diffusivity	9,779
					White matter hyperintensities	9,998
Harshfield	Harshfield et al. (2022)	Bruker Biospin (NMR) Waters Corp. (UPLC-MS)	Serum	Not mentioned	Brain atrophy as the inverse of total brain volume	624
					Executive function	624
					Global cognition	624
					Lacune count	624
					Lacune presence	624
					Mean diffusivity normalized histogram peak height	624
					Microbleeds	624
					Processing speed	624
					Peak width of skeletonized mean diffusivity	624
					Simple SVD score	624
					White matter hyperintensities	624
Holmes	Holmes et al. (2018)	Nightingale Health Ltd (NMR)	Plasma	Not fasted	Intracerebral hemorrhage	1,138
					Ischemic stroke	1,146

Huo	Huo et al. (2020)	Biocrates, p180 (MS)	Serum	Not mentioned	Incident ADRD	530
Hyunh	Huynh et al. (2020)	Agilent (MS)	Serum and plasma	Fasted and non-fasted	ADRD	1,912
Machado-Fragua	Machado-Fragua et al. (2022)	Nightingale Health Ltd (NMR)	Serum	Fasted	Dementia	5,374
MahmoudianDehkordi	Mahmoudian Dehkordi et al. (2019)	Biocrates, Bile Acids kit (MS)	Serum	Fasted	ADRD, early MCI, late MCI	1,446
Mapstone	Mapstone et al. (2014)	MS: (UPLC-ESI-QTOF-MS)	Plasma	Fasted	ADRD	106
Marx	Marx et al. (2022)	Nightingale Health Ltd (NMR)	Serum	Not fasted	Child cognition vs. maternal metabolites	662[b]
Neuffer	Neuffer et al. (2022)	Agilent (UHPLC-MS/MS)	Serum	Fasted	Cognition	838
Nho	Nho et al. (2019)	Biocrates, Bile Acids kit (MS)	Serum	Fasted	CSF biomarkers (Aβ_{1-42}, t-tau, p-tau)	1,562
Palacios	Palacios et al. (2020)	Metabolon (MS)	Plasma	Fasted	Cognitive function	736
Prince	Prince et al. (2023)	Metabolon (MS)	Plasma	Not fasted	Autism spectrum disorder (subgroups)	2,001
Sliz	Sliz et al. (2022)	Nightingale Health Ltd, National Phenome Center, Bruker Biospin (NMR); Metabolon, Biocrates Life Sciences AG, Broad Institute (MS)	Serum and plasma	Mostly fasted	White matter hyperintensity	9,290
Sliz	Sliz et al. (2021)	Nightingale Health Ltd (NMR)	Serum	Fasted	*DHCR24* gene variant	931
Sliz	Sliz et al. (2020)	Metabolon (NMR)	Serum	Fasted	Thickness of the cerebral cortex	441

(continued)

Table 1 (continued)

First author	Reference	Metabolomics platform	Biospecimen	Fasting status	Phenotype(s)	Sample size[a]
Sun	Sun et al. (2023)	Nightingale Health Ltd (NMR)	Plasma	Not fasted	White matter hyperintensity	8,190
Sun	Sun et al. (2019)	Metabolon (MS)	Serum	Fasted	Ischemic stroke	3,904
Syme	Syme et al. (2019)	LC-ESI-MS (in house)	Serum	Fasted	White matter microstructure (T1W-SI), processing speed	872
Tian	Tian et al. (2023)	Biocrates MxP Quant 500 (MS)	Plasma	Fasted	MRI-derived brain atrophy index (SPARE-BA)	475
Toledo	Toledo et al. (2017)	Biocrates p180 (MS)	Serum	Fasted	MCI, ADRD, $A\beta_{1-42}$, T-tau/$A\beta_{1-42}$, cognition (ADAS-Cog13), brain atrophy related to AD (SPARE-AD)	730
Tynkkynen	Tynkkynen et al. (2018)	Nightingale Health Ltd (NMR) + a subset using LC-MS	Serum	Not fasted	Incident ADRD	22,623
van der Lee	van der Lee et al. (2018)	Nightingale Health Ltd (NMR)	Plasma	Fasted	ADRD	25,868
					General cognitive ability	5,182
Vojinovic	Vojinovic et al. (2020)	Nightingale Health Ltd (NMR) + a subset using LipoScience, Raleigh, NC (NMR)	Serum and plasma	Fasted	Incident stroke	38,797
Wang	Wang et al. (2020)	MS: ultra-performance liquid chromatography triple quadrupole mass spectrometry (Waters XEVO TQ-S, Milford, USA) gas chromatography time-of-flight mass spectrometry (Leco Corporation, St Joseph, USA)	Serum	Not fasted	Cognitive impairment	566

Wang	Wang et al. (2022)	Nightingale Health Ltd (NMR)	Serum	Mostly fasted	All stroke	521,612[c]
					Brain microbleeds	25,862[c]
					Cardioembolic stroke	463,456[c]
					Fractional anisotropy	17,663[c]
					Intracerebral hemorrhage	3,036[c]
					Ischemic stroke	514,791[c]
					Lacunar stroke	262,136[c]
					Large-artery atherosclerotic stroke	461,138[c]
					Mean diffusivity	17,663[c]
					White matter hyperintensity	18,381[c]
Zhang	Zhang et al. (2022)	Nightingale Health Ltd (NMR)	Plasma	Not fasted	ADRD	110,655
					Incident dementia	110,655

NMR nuclear magnetic resonance, *MS* mass spectrometry

ADHD attention deficit hyperactivity disorder, *AD* Alzheimer's disease, *ADAS-Cog13* Alzheimer's Disease Assessment Scale-Cognitive Subscale with 13 items, *ADRD* Alzheimer's disease and related dementias, *MCI* mild cognitive impairment, *CSF* cerebrospinal fluid, *p-tau* phosphorylated tau, *t-tau* total tau, *FDG* fluorodeoxyglucose, *ADRD* Alzheimer's disease and related dementias, *DHCR24* 24-dehydrocholesterol reductase

[a] If a single value is provided, it is applicable to all study phenotypes. Otherwise, sample sizes for individual phenotypes are specified separately. If the study presents results for multiple tissue types, the sample size is reported based on the number of individuals who contributed blood samples

[b] The number of mother–child pairs

[c] This is a Mendelian randomization study; sample sizes are given for GWAS of the respective phenotypes

abundant molecules are hormones or other signaling molecules (Psychogios et al. 2011). Non-lipid metabolites that circulate in high concentrations include amino acids (e.g., glutamine, isoleucine, and phenylalanine), glucose metabolism-related molecules (e.g., glucose, citrate, and lactate), and different waste or catabolic by-products (e.g., urea and creatinine) (Psychogios et al. 2011). In terms of the number of molecular species, lipids are the most plentiful class of metabolites, indicating their importance and also the complexity of involvement in multiple physiological processes: lipids serve as structural cell elements, signaling molecules, and a source of energy (Quehenberger and Dennis 2011).

Due to the diversity in chemical properties of the circulating metabolites, no single analytical method can capture the entire blood metabolome. The two most common analytical methods are mass spectrometry (MS) combined with different chromatography techniques, and nuclear magnetic resonance (NMR) spectroscopy. Briefly, MS measures the compounds present in a biospecimen by their mass-to-charge ratio, whereas NMR-based platforms quantify the molecules based on spin properties of their isotope nuclei (e.g., ^{1}H and ^{13}C), which absorb radiation and resonate at a specific frequency when placed in a magnetic field. MS is highly sensitive, and its applications have been especially fruitful in the field of lipidomics. In turn, NMR spectroscopy is reproducible, quantitative, and sample non-destructive due to minimal sample preparation. The technical details of these methods are described elsewhere (Soininen et al. 2015; Sas et al. 2015; Lind et al. 2016; Dunn et al. 2010).

2 Lipids

The brain is the second most fatty organ in the body after adipose tissue – lipids comprise approximately 50–60% of the brain's dry weight (Hamilton et al. 2007). The brain has the capability to synthesize many but not all lipids De Novo. For example, the predominant *fatty acids* (FAs) in the brain, docosahexaenoic acid (DHA) and arachidonic acid (ARA) that constitute ~20% of the dry weight of the human brain (Fraser et al. 2010), are derived from essential FAs α-linolenic acid (ALA) and linoleic acid (LA), respectively, that must be imported into the brain from the circulation.

Fatty acids are the simplest type of lipids, consisting of a carboxylate group attached to a hydrocarbon chain that varies in length and degree of saturation. Regarding FA nomenclature, the form *C:D* is often used, where C indicates the number of carbons in the hydrocarbon chain, and D indicates the number of double bonds. Unsaturated FAs have one or more double bonds (monounsaturated FA, MUFA; polyunsaturated FA, PUFA), whereas saturated FAs (SFA) have no double bonds in their hydrocarbon chain. FAs are fundamental constituents of other lipids and important building blocks of *cholesteryl esters* (CEs), *phospholipids* (PLs), and *triacylglycerols* (TAGs): each molecule of CEs, PLs, and TAGs contains one to three FAs (Fig. 1a). FA composition regulates the biochemical properties of the

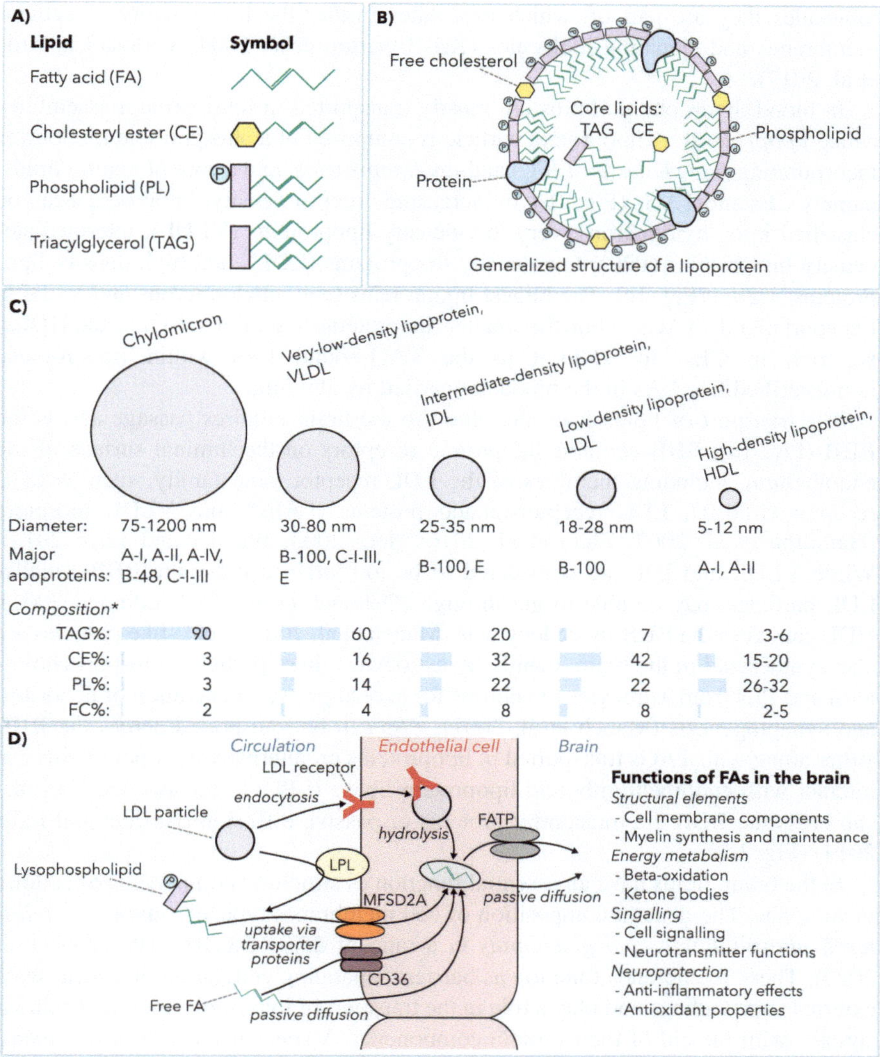

Fig. 1 Major lipid classes, lipoproteins, and their interactions with the endothelial cells of the blood–brain barrier upon brain fatty acid uptake. (**a**) Fatty acids (FAs) are important building blocks of more complex lipids, including cholesteryl esters (CEs), phospholipids (PLs), and triacylglycerols (TAGs). (**b**) A lipoprotein particle consists of the spherical shell of PLs, free cholesterol (FC), and proteins encapsulating the core lipids, namely TAGs and CEs. (**c**) The common lipoprotein classification in an increasing order of density and their characteristics (adapted from Bhagavan and Ha 2011). * Percent of mass. (**d**) Transport of FAs from the circulation to the brain across the blood–brain barrier (BBB). FAs entering the brain can originate from several circulating pools of FAs, including those of lipoproteins, lysophospholipids, and free FAs. After being internalized by endothelial cells, FAs bind with FA-binding proteins and are transported across to the brain parenchyma. Figure 1d is modified from (Chen and Bazinet 2015 and Bazinet and Layé 2014). LDL, low-density lipoprotein; LPL, lipoprotein lipase; MFSD2A, major facilitator superfamily domain containing 2A (official full name MFSD2 lysolipid transporter A, lysophospholipid); CD36, CD36 molecule (also known as fatty acid translocase); FATP, fatty acid transporter protein

molecules they are part of, which modulates further the functionality of cellular membranes and signaling molecules (Kris-Etherton et al. 2004; Tortosa-Caparrós et al. 2017).

In blood, hydrophobic lipids are mostly transported in lipid–protein assemblies called *lipoproteins*. A lipoprotein particle is composed of a phospholipid monolayer incorporating *free cholesterol* (FC) and *apolipoproteins*, and a core of neutral lipids, namely CEs and TAGs (Fig. 1b). In increasing order of density, lipoproteins can be classified into chylomicrons, very low-density lipoproteins (VLDL), intermediate-density lipoproteins (IDL), low-density lipoproteins (LDL), and high-density lipoproteins (HDL) (Fig. 1c). The largest lipoproteins (e.g., chylomicrons and VLDLs) transport mainly TAGs, while the smaller lipoproteins (e.g., IDLs, LDLs, and HDLs) are rich in CEs. In addition to the TAG-bound FAs within lipoproteins, non-esterified/free FAs in the blood are carried by albumin.

The transport of lipids from the blood to the brain requires passage across the BBB (Fig. 1d). BBB contains lipoprotein receptors on the luminal surface of the endothelium, including members of the LDL receptor gene family, such as LDL receptor (LDLR), LDL receptor-related proteins (LRPs), and VLDL receptors (Hamilton et al. 2007; Zhao et al. 2016; Herz 2003; Bazinet and Layé 2014). While VLDL and IDL are considered to be too large to cross the BBB, smaller LDL particles may be able to get through (Dehouck et al. 1997; Edmond 2001). HDL can cross the BBB by endocytosis (Mauch et al. 2001). HDL-like particles are also synthesized in the brain, mainly by astrocytes; these particles transport cholesterol and PLs from astrocytes to neurons for axonal growth, generation of dendrites, and synaptogenesis (Mauch et al. 2001). FAs can be transported across the BBB either along with TAGs transported in lipoproteins or, alternatively, lipoproteins can interact with endothelium-bound lipoprotein lipase (LPL) to release free FAs that can cross the BBB via transporter proteins or passive diffusion (Bazinet and Layé 2014) (Fig. 1d).

In the brain, lipids have an essential function as structural components of cellular membranes. The general composition of cell membranes typically involves *cholesterol, phospholipids, and glycolipids* in a ratio of 25%:65%:10% (Poitelon et al. 2020), These membranes function as barriers separating cellular environment from external surroundings and play a role in the transport of ions, molecules, and cellular signals, with the aid of their protein components. A specialized type of cell membrane in the brain is the myelin sheath – myelin sheaths are cell membranes of oligodendrocytes that are wrapped around neuronal axons. Myelin is of a particularly high lipid content: approximately 70–85% of the dry mass of myelin consists of lipids (Poitelon et al. 2020). The thickening and maintenance of myelin sheaths require a high level of lipid synthesis and uptake of extracellular lipids. Myelin sheaths have a distinct composition with relatively higher proportions of cholesterol and glycolipids, with the ratio of cholesterol, phospholipids, and glycolipids being 40%:40%:20% (Poitelon et al. 2020).

In addition to serving as structural components of membranes, lipids in the brain function as signaling molecules in the regulation of numerous ion pumps, channels, and transporters (Hardie and Muallem 2009; Ridone et al. 2018). They also function as a source of energy; for this purpose, the brain uses FAs and their metabolic products, ketone bodies (Panov et al. 2014; Mergenthaler et al. 2013). Further, FAs in the brain regulate lipid synthesis and degradation (Nakamura et al. 2004), modulate neuroinflammation (Desale and Chinnathambi 2020; Lancaster et al. 2018; Bright et al. 2008), and exhibit antioxidant properties (Shi et al. 2016; Díaz et al. 2021).

2.1 *Polyunsaturated Fatty Acids*

The normal structure and function of the brain are highly dependent on high amounts of long-chain PUFAs, principally DHA (FA22:6n-3), and ARA (FA20:4n-6). The demand for PUFAs is high in the developing brain, particularly in the first year of life when a large proportion of De Novo myelination occurs (Deoni et al. 2018). In the adult brain, replacing daily consumed PUFAs is sufficient (Bazinet and Layé 2014). Estimates suggest that each day, 2–8% of esterified DHA and 3–5% of esterified ARA in the adult brain undergo replacement. The adult brain's daily uptake of DHA and ARA from the circulating pool of free FAs is approximately 4.6 mg and 17.8 mg, respectively (Rapoport et al. 2007).

Higher circulating DHA levels have been associated with better cognitive functioning and lower risk of Alzheimer's disease and related dementias (ADRD) in a meta-analysis of large-scale population-based studies (van der Lee et al. 2018), which supported the results of several previous smaller-scale studies (Mielke et al. 2011; Schaefer et al. 2006; Tan et al. 2012; Li et al. 2016). Likewise, a recent extensive study based on the UK Biobank reported a negative association between incident ADRD and DHA, but also with total PUFA, total omega-6 FAs, LA, and total omega-3 FAs (Zhang et al. 2022). It is worth noting that the study by Tynkkynen et al. did not replicate the association between DHA and ADRD (Tynkkynen et al. 2018). The authors speculate, however, that their conservative strategy to minimize reporting false-positive associations could contribute to missing some true associations (Tynkkynen et al. 2018). Notably, Marx et al. reported that the maternal omega-3 FAs, specifically the ratios of DHA to total FAs and omega-3 to total FAs measured at approximately twenty-eighth week of gestation, are associated with child's cognition at 2 years of age (Marx et al. 2022). This adds to the growing body of evidence affirming the pivotal role of DHA in promoting normal cognition even from a very early age.

Wang et al. found evidence of altered FA profiles in both blood and brain tissue samples of patients with ADRD (Wang et al. 2020). Their findings suggest that lower circulating levels of LA, petroselinic acid (FA18:1cis-6), myristic acid (FA14:0), palmitic acid (FA16:0), and palmitoleic acid (FA16:1n-7) are associated with a higher risk of mild cognitive impairment (MCI) and/or ADRD. Consistently, the higher circulating levels of the same FA species were associated with better global cognitive functioning (Wang et al. 2020). Interestingly, brain tissue samples from ADRD cases contained higher levels of these FAs compared with controls, suggesting a greater brain tissue accumulation of these FAs in cases than in controls (Wang et al. 2020). The authors suggest that the underlying mechanisms may involve a shift in energy supply from glucose to alternative sources, including FAs, which may be due to the excess transfer of free FAs from blood to the brain induced by either damage of the BBB or excess FA transporters present at the BBB (Wang et al. 2020).

2.2 Cholesterol and Lipoproteins

Cholesterol is a sterol lipid that can be acquired from both endogenous and exogenous sources. In the brain, cholesterol is mainly synthesized by astrocytes and, to a lesser degree, by oligodendrocytes and neurons. In the adult human body, 23% of total body cholesterol resides in the brain (Chang et al. 2017). Most brain cholesterol is synthesized in the brain. Both astrocytes and neurons have a high demand for cholesterol, and both cell types can synthesize it. Oligodendrocytes can also synthesize cholesterol, with cholesterol availability being a limiting factor for the growth of myelin membranes during brain development (Chang et al. 2017). Cholesterol transport between different cell types occurs via lipoprotein-like particles containing apolipoprotein E (APOE) and phospholipids (Chang et al. 2017).

Associations between levels of circulating cholesterol and health outcomes are typically reported as cholesterol within different lipoprotein fractions. The direction of the associations differs depending on the lipoprotein carrier: for instance, dyslipidemia associated with metabolic syndrome and cardiovascular disease comprises elevation in apolipoprotein B (APOB)-containing lipoproteins and related lipids (namely, VLDL, IDL, and LDL, and cholesterol and TAGs within these particles), and reduction in HDL and related lipids (i.e., mostly cholesterol) (Grundy et al. 2006). The so-called "bad cholesterol," predisposing to atherosclerosis, refers particularly to the cholesterol within the LDL fraction (LDL-C), while the "good cholesterol" refers to the cholesterol within the HDL fraction (HDL-C). Nonetheless, studies on detailed lipoprotein subfraction measures have indicated heterogeneity regarding HDL fractions: the very large and large HDL (XL-HDL, L-HDL) particles (and lipids within) show typically negative associations, whereas the medium and small HDL (M-HDL, S-HDL) particles (and lipids within) show typically positive associations with cardiovascular disease (Raheem et al. 2022; Würtz et al. 2014; Holmes et al. 2018). Compared with cardiovascular outcomes, the associations between brain-related endpoints and lipoprotein measures have been scarcer and partially less consistent, as described next.

Holmes et al. determined associations between circulating metabolites and stroke subtypes, including ischemic stroke and intracerebral hemorrhage, in the China Kadoorie Biobank (Holmes et al. 2018). In association with ischemic stroke, lipoprotein measures showed a typical association pattern commonly seen in association with a higher risk of cardiovascular disease. This pattern included high levels of APOB-containing lipoproteins and related lipids (i.e., cholesterol and TAGs), and larger VLDL particle size and smaller LDL particle size (Holmes et al. 2018). Compared with VLDL measures, the associations between ischemic stroke and LDL measures were overall weaker in terms of both statistical significance and effect sizes (Holmes et al. 2018). Intracerebral hemorrhage was not associated with any of the lipoprotein measures (Holmes et al. 2018). But in a large-scale metabolomic study involving participants of European origin (Vojinovic et al. 2020), most of the associations observed with VLDL and LDL measures in the Chinese sample were not replicated. The authors discuss that the discrepancies could be due to environmental and ethnic differences or differing covariate adjustments (Vojinovic et al. 2020). Regarding lipoprotein measures, in the European sample, the risk of stroke was associated with cholesterol in M-HDL particles (a negative association) and TAGs in L- and M-LDL particles (a positive association). Similar to the Chinese sample, no significant associations were observed between hemorrhagic stroke and lipoprotein measures in the European sample (Vojinovic et al. 2020).

In their metabolomic study of cognitive ability and dementia, van der Lee et al. reported associations of lipids in HDL subfractions with better cognitive ability and, consistently, with lower risk of all-cause dementia (van der Lee et al. 2018). The associations were the most consistent for the total cholesterol, CE, and PL concentrations in the M-HDL subfraction. Another study by the same authors did not, however, replicate some of the key findings – including the M-HDL-related results – in association with incident ADRD (Tynkkynen et al. 2018). In the latter study, Tynkkynen et al. reported associations mostly on VLDL subfraction measures, suggesting lower cholesterol concentration in the S-VLDL subfraction and a higher relative proportion of cholesterol and lower relative proportion of TAGs in the very large VLDL subfraction (XL-VLDL-C% and XL-VLDL-TAG%, respectively) in association with incident ADRD (Tynkkynen et al. 2018). In addition, a higher relative proportion of CE in L-HDL (L-HDL-CE%) was associated with incident ADRD (Tynkkynen et al. 2018). A recent extensive study by Zhang et al., conducted in the UK Biobank, reported negative associations between multiple lipoprotein lipid measures and incident ADRD, including those of total lipids and TAGs within L-VLDL and M-VLDL, as well as CE, FC, and PL within S-HDL (Zhang et al. 2022). Additionally, the study reported negative associations for total lipids and CE within L-LDL subfraction (Zhang et al. 2022). In their work, van der Lee et al. speculate that the lipids – namely PLs and FCs – within HDL particles may serve as a buffer helping to repair damaged neuronal membranes and/or modify lipid raft quantity and composition, thus helping to protect the brain from neurodegeneration (van der Lee et al. 2018). Consistent with the metabolomic associations, VLDL receptor polymorphism has been reported to constitute a genetic susceptibility factor for ADRD, especially among patients with vascular risk factors (Helbecque et al. 2001). It was hypothesized that the VLDL receptor may contribute to the metabolism of APOE and APOE–amyloid β (Aβ) complexes. Alternatively, in interaction with APOE isoforms, the VLDL receptor may play a role in the neurodegenerative processes involving reelin, which is a glycoprotein involved in neuronal migration, regulation of synaptic plasticity, development of dendritic spines, and modulation of the release of neurotransmitters, and which also binds to the VLDL receptor (Helbecque et al. 2001). Further investigations are needed, however.

Recent studies have explored the associations between blood metabolic measures and white matter hyperintensities (WMH), a neuroimaging marker of cerebral small vessel disease and a major risk factor of ADRD (Debette et al. 2019). Typically, WMH do not exhibit strong associations with lipoprotein measures in pooled samples (Sliz et al. 2022; Harshfield and Markus 2023; de Leeuw et al. 2021). But in our study of a general population of middle-aged and older individuals, we discovered a significant association between larger LDL particle size and a greater volume of WMH (Sliz et al. 2022). This association was particularly strong in males. We proposed that this association may arise from higher relative TAG and lower relative CE content within LDL particles (Sliz et al. 2022), potentially due to altered interactions between lipoproteins and their lipid composition-modulating enzymes – such as the endothelium-bound LPL (Fig. 1d). This association has not reached significance in other studies, however, with some studies reporting their nonsignificant effect sizes in the matching (de Leeuw et al. 2021) and some in the opposing direction (Harshfield and Markus 2023; Sun et al. 2023). The discrepancies could be attributed to differences in fasting status; our sample was predominantly fasted, while other studies included partially fasted (de Leeuw et al. 2021) or non-fasted samples (Harshfield and Markus 2023; Sun et al. 2023). Similar to our study (Sliz et al. 2022), the findings by Sun et al. suggested that the metabolomic associations, the lipid-related ones in particular, are more significant in males than in females, and they further reported particularly strong associations in individuals aged under 50 years (Sun et al. 2023). While the association between WMH and lower total cholesterol reached only nominal significance in our study (Sliz et al. 2022), another study reported this association as significant (Harshfield et al. 2022). One could argue about the confounding effect of statin treatment on these associations. Based on our sensitivity analyses, however, statin use does not seem to drive the negative associations, as we found both minimal attenuation after adjusting for statin use, and consistent negative effect estimates between WMH and relative CE content within LDL subfractions persisted in all the analytical subgroups, including individuals without statin use (Sliz et al. 2022).

Findings of our recent study suggest that a shared biological mechanism could influence peripheral lipid metabolism, brain structure and cognitive functioning in adolescents: we identified a locus harboring 24-dehydrocholesterol reductase (*DHCR24*; also known as selective Alzheimer's disease indicator-1, *SELADIN1*) that was associated with the shared variance of visceral fat, a novel circulating marker of systemic inflammation (a glycerophospholipid), and white matter microstructural properties on brain MRI (Sliz et al. 2021). The lead variant was also associated with particle concentrations of the five largest (XXL-S) VLDL subfractions as well as total lipids within these subfractions (Sliz et al. 2021). By comparing the metabolomic profiles of known genetic modulators of lipoprotein metabolism, we found that the metabolomic profile of the *DHCR24* variant was similar to that of a transmembrane 6 superfamily member 2 (*TM6SF2*) variant impairing hepatic secretion of VLDL due to shortage in phosphatidylcholines

(Luukkonen et al. 2017). This was unexpected considering the well-described function of DHCR24 in converting desmosterol to cholesterol. Our findings further suggested that the role of DHCR24 in TAG-rich lipoprotein metabolism may be specific to childhood and adolescence, a period of human lifespan when the final maturational changes of the brain (e.g., myelination) occur, and the demand for cholesterol by the brain may still be high (Sliz et al. 2021).

Observational studies have limitations when it comes to establishing causal relationships rather than mere correlations. Wang et al. employed Mendelian randomization to study causal inferences concerning stroke and its subtypes, as well as MRI features of cerebral small vessel disease, including WMH, and various NMR-based lipoprotein subfraction measures (Wang et al. 2022). In contrast to conventional epidemiological associations, which are often reported for VLDL and HDL subfraction measures when it comes to brain-related outcomes, this study demonstrated a significant causal association between S-LDL particle concentration and overall stroke risk, ischemic stroke, and large-artery atherosclerotic stroke (Wang et al. 2022). Additionally, a significant causal effect was found between APOB concentration and overall stroke risk and the ischemic subtype. The study did not identify significant causal associations between lipoprotein lipids and MRI phenotypes, namely WMH, fractional anisotropy, mean diffusivity, and brain microbleeds (Wang et al. 2022). Also, Amin et al. explored bidirectional causal associations between mostly lipoprotein-lipid-related metabolic measures and major depressive disorder (MDD) (Amin et al. 2023). On top of the epidemiological associations between various VLDL and HDL subfraction measures and lipids within, they found MDD to be causal for changes in 25 VLDL, 5 IDL, and 8 HDL-related measures, suggesting that alterations in these measures occur in the course of the disease but are not on the causal pathway leading to MDD (Amin et al. 2023). It is important to emphasize that, while exploring the causal relationships between circulating metabolic measures and brain health outcomes is essential to differentiate between biomarkers and causal molecules, the selection of instruments for such investigations warrants careful consideration. This is particularly crucial due to the widespread pleiotropy observed in lipid loci, making it challenging to select instruments specific to a particular lipid or lipoprotein measure. Additionally, it is worth highlighting the need for subgroup-specific study settings. As mentioned previously, for instance, the associations between circulating metabolic measures and WMH seem to exhibit sex specificity (Sliz et al. 2022; Sun et al. 2023). Therefore, customizing the study approach to account for such variations is essential for a comprehensive understanding of these relationships.

2.3 Glycerophospholipids and Sphingolipids

Glycerophospholipids (GPLs) consist of a glycerol backbone to which one or two FAs and a head group are attached; the head group consists of a phosphate and an alcohol, such as choline and ethanolamine (Harayama and Riezman 2018). GPLs with a choline-containing head are glycerophosphocholines (GPCs), and GPLs with an ethanolamine-containing head are glycerophosphoethanolamines (GPEs). GPLs that have one of the two FAs attached with an ether or vinyl-ether bond (instead of an ester bond) are termed *ether GPLs*, and the ether GPLs with the vinyl-ether bond are termed *plasmalogens* (Harayama and Riezman 2018; Brites et al. 2004). *Sphingolipids* consist of a sphingoid backbone to which one FA and a head group are attached. The head groups include phosphocholine and oligosaccharides. Sphingolipids with a phosphocholine head are *sphingomyelins*, and sphingolipids with an oligosaccharide head are *gangliosides* (Harayama and Riezman 2018). Sphingolipids show a higher content in some specialized cell membranes, such as myelin (Poitelon et al. 2020; Schmitt et al. 2015).

GLPs and sphingolipids are key structural elements of cell membranes (Harayama and Riezman 2018). Variations in the molecular structure of these lipid species and in their proportions within cell membranes modulate cell membrane stability, permeability, and fluidity (Van Meer et al. 2008); these variations also alter tissue magnetic resonance (MR) properties, and thus, for example, signal intensity of T1-weighted MR images of the brain (Kucharcyk et al. 1994). Further, cell membranes are continuously remodeled, releasing second messenger signaling molecules, such as *lyso-GPCs* (also known as LPCs) and *platelet-activating factors* (PAFs), which have diverse cell functions, including immunomodulation and the regulation of neuron survival (Xu et al. 2013; Marathe et al. 2014; Sevastou et al. 2013). Enzymes of cell membrane remodeling have been implicated in the pathobiology of ADRD (Oliveira and Di Paolo 2010; Sanchez-Mejia et al. 2008).

Several large-scale metabolomic studies have identified associations of GPLs and sphingolipids with ADRD or related phenotypes. One notable study performed targeted blood lipidomic profiling of 32 lipid classes, including neutral lipids, GPLs, and sphingolipids, comprising 569 species, in 2 longitudinal cohorts of older adults enriched in patients with ADRD (Huynh et al. 2020). The results showed consistent negative associations of ether GPCs and GPEs, including plasmalogens and specific subclasses of gangliosides, with both prevalent and incident ADRD in both cohorts. This was especially remarkable for ether lipids esterified with omega-3 FAs, including DHA (Huynh et al. 2020). A study conducted by Huo et al. in 2 community-based longitudinal cohorts of older adults showed that higher baseline blood levels of several ether and ester GPCs were associated with cognitive decline (Huo et al. 2020). Intriguingly, brain tissue levels of some of these metabolites were also associated with AD pathobiology. Toledo et al. (Toledo et al. 2017) performed targeted blood profiling of the same metabolites as Huo et al. (Huo et al. 2020) in a longitudinal cohort of older adults enriched in MCI and ADRD cases. The results showed that higher baseline levels of several ether GPCs and sphingomyelins were associated with cognitive decline, brain atrophy, AD pathology, and/or conversion from MCI to ADRD (Toledo et al. 2017). In their study, Baloni et al. presented multi-omic evidence implicating dysregulation of sphingomyelin and ceramide metabolism in AD, including the association between the SM(d34:1) to SM(d43:1) ratio and all clinical AD diagnosis, brain atrophy in AD-related regions, and cognitive function (Baloni et al. 2022). In a pioneering study in the field, comprising a small community-based cohort of older adults, a set of 10 lipids (ester and ether GPCs and LPC among others) predicted phenoconversion to either amnestic MCI or ADRD within a 2- to 3-year timeframe with over 90% accuracy (Mapstone et al. 2014). The authors suggested that this biomarker panel may reflect cell membrane integrity, and may be sensitive to early neurodegeneration of preclinical AD (Mapstone et al. 2014). Additionally, lower levels of several LPCs and hydroxysphingomyelins have been associated with higher WMH load in our study of a general population of middle-aged and older adults (Sliz et al. 2022). We suggested that these associations may index BBB and neuron injury and disruption of myelin, which are key pathobiological features of WMH (Sliz et al. 2022).

In our study of a community-based sample of adolescents, we investigated a panel of ester and ether GPCs and their second messengers, including LPCs and PAFs. We identified a strong association between acyl-PAF (PC16:0/2:0) and various factors, such as body adiposity, C-reactive protein (a marker of systemic inflammation), as well as with cognitive functioning and variations in white matter microstructural properties on brain MRI, as assessed with normalized T1-weighted signal intensity and magnetization transfer ratio (Syme et al. 2016; Syme et al. 2019). Analyzing a shared variance among these variables, we identified a locus near *DHCR24* (24-dehydrocholesterol reductase) encoding a cholesterol synthesizing enzyme that was originally identified as differentially expressed in AD. With blood metabolomic profiling, we found the association of this locus may arise due to altered GPL rather than cholesterol metabolism (Sliz et al. 2021).

In a recent study, Prince et al. studied the metabolomic profiles of children aged to 4–17 years with autism spectrum disorder (ASD) (Prince et al. 2023). The authors employed unsupervised hierarchical clustering using 40 ASD phenotypes to identify distinct subgroups of children sharing similar phenotype patterns. They then explored the associations between these subgroups and blood metabolites, followed by enrichment analyses of the association results, which they presented as their primary findings. Subgroup 1, characterized by milder behavioral challenges, exhibited lower levels of metabolites related to lipid metabolism pathways, including those within the phosphatidylinositol and phosphatidylethanolamine pathways. In contrast, subgroup 2, characterized by the highest degree of impairment across all ASD phenotype domains, displayed opposing metabolic patterns for these metabolites. Subgroup 3, featuring the highest IQ scores while still reporting behavioral traits and co-occurring conditions, demonstrated increased levels in six lipid pathways associated with sphingolipid and FA metabolism, contrasting subgroup 1. The authors highlight specific metabolites, such as 1-linoleoyl-glycerophosphoethanolamine (18:2) from the lysophospholipid pathway and sphinganine from the sphingolipid synthesis pathway, as potential key drivers distinguishing subgroups and promising biomarkers for future research. The authors conclude that elevated levels of membrane lipids and increased oxidative stress appear to be pertinent factors in children with higher ASD impairment and alterations in lipid metabolism may contribute to a greater extent of maladaptive behavioral issues and the presence of co-occurring conditions in these children (Prince et al. 2023).

2.4 Triacylglycerols

In a TAG molecule, three FAs are esterified with a glycerol backbone (Fig. 1a). The varying combinations of different kinds of FAs contribute to multiple TAG species, and more than 6,000 combinations are encompassed by the term TAG (Bernath et al. 2020). TAGs originate from dietary sources (most of the dietary fat is in the form of TAG) and endogenous TAG synthesis, which takes place in the liver and in adipose tissue: hepatic TAGs can be secreted to circulation in VLDLs, while adipose tissue TAG is the major energy store of the body. To note, Sect. 2.2 describes the associations between brain health and TAGs within lipoprotein subfractions, while this section elaborates on findings specifically related to distinct TAG species.

Reports on the associations between circulating total-TAG and brain health have provided inconsistent results: both elevated and lowered serum total-TAG have been described in association with ADRD risk and in association with the thickness of the cerebral cortex (Krishnadas et al. 2013; Schwarz et al. 2018). More recently, more detailed TAG measures have been studied. In their study, Bernath et al. quantified 84 TAG species and explored TAG scores, derived from principal component analysis, in association with MCI and ADRD risk (Bernath et al. 2020). The authors found lower scores of the principal component loaded by PUFA-containing TAGs (PUTGs) in individuals with both MCI and ADRD compared with cognitively normal older adults. Lower levels of these PUTGs were also associated with the lower thickness of the entorhinal cortex and hippocampus, regions affected in the early stages of ADRD (Bernath et al. 2020). Given TAGs are synthesized in the liver or absorbed from the gut, the authors emphasize that their findings represent links in the gut–liver–brain axis and, considering the inflammation-modulating properties of PUFAs, it remains possible that decreases in PUTGs reflect reductions in PUFAs and loss of neuroprotective metabolites (Bernath et al. 2020).

In line with these findings, we demonstrated that the positive associations between PUTGs and cortical thickness extend to generally healthy middle-aged adults (Sliz et al. 2020): our results suggested that specific TAG species carrying 18-carbon unsaturated FAs may play a key role. In particular, five out of the seven significant TAG species carried FA18:3, which includes the essential ALA known to be crucial for normal neural development (Clark et al. 1992) and a precursor for DHA, a key structural FA in the brain (Bentsen 2017). Furthermore, the associations of these TAGs and cortical thickness appeared stronger in cortical regions with higher expression of genes regulating long-chain FA metabolism and cellular membranes, and cortical thickness in these regions was also most closely associated with better cognitive functioning (Sliz et al. 2020). The results of the two studies are consistent and suggest that specific subcategories rather than total TAGs may play a crucial role in brain health associations. Importantly, based on these studies, it emerges that the circulating TAG associations with brain health may appear early in the course of disease (Bernath et al. 2020) or even before its clinical onset (Sliz et al. 2020), underscoring the potential of these lipid markers in the development of future therapeutics.

3 Lipid Derivatives

3.1 Bile Acids

Cholesterol is cleared through the production of *bile acids* (BAs) in the liver. Subsequently, BAs are conjugated with taurine or glycine, secreted into the bile and stored in the gallbladder (Qi et al. 2015). Upon food intake, BAs are secreted from the gallbladder into the intestine where they facilitate the absorption of dietary lipids, steroids, and fat-soluble vitamins. In the intestine, gut microbiota plays a key role in converting primary BAs into secondary BAs through deconjugation and dehydroxylation processes, thereby increasing the diversity of BAs (Mertens et al. 2017). The majority of the intestinal BAs (~95%) are reabsorbed and transported back to the liver (Monteiro-Cardoso et al. 2021).

The widespread expression of BA receptors throughout tissues, including the brain, is an indicator of BAs being versatile signaling molecules (Mertens et al. 2017). The presence of multiple BA transporters at the BBB, including members of the solute carrier (SLC) and ATP-binding cassette (ABC) transporter families (Mertens et al. 2017), suggests that BAs could cross the BBB, which is supported by animal studies (Quinn et al. 2014; Lalić-Popović et al. 2013). In addition, excess circulating BAs have the potential to damage the lipid membranes of the BBB, given their detergent and lytic action, and, at lower concentrations, BAs may influence the BBB by modulating tight junctions and subsequently permeability of the BBB (Mertens et al. 2017). Further, parts of the BA biosynthetic pathway have been identified in the brain, likely playing a role in the cholesterol clearance (Mertens et al. 2017). In mice, taurine-conjugated ursodeoxycholic acid (a secondary BA) has demonstrated anti-inflammatory properties in acute brain inflammation (Yanguas-Casás et al. 2017), suggesting that BAs may have broader effects than those directly related to metabolism.

The numerous biologically relevant properties of BAs have raised research interest in their association with brain health. Metabolomic evidence consistently indicates alterations in the blood BA profile in association with cognitive impairment. When compared with cognitively normal adults, patients with ADRD have been reported to have lower blood levels of cholic acid (CA; a primary BA), and higher levels of deoxycholic acid (DCA; a bacterially produced secondary BA), along with its glycine (GDCA) and taurine (TDCA) conjugated forms (Mahmoudian Dehkordi et al. 2019). Higher levels of GDCA and two other secondary conjugated BAs, glycolithocholic acid (GLCA) and taurolithocholic acid (TLCA), were also associated with worse cognitive functioning as estimated using Alzheimer Disease

Assessment Scale 13-item cognitive subscale (ADAS-Cog13) (Mahmoudian Dehkordi et al. 2019). Moreover, higher ratios of secondary to primary BAs, exemplified by DCA:CA, was observed in association with AD, indicating potential alterations in bacterial 7α-hydroxylase activity and accumulation of cytotoxic secondary BAs (Mahmoudian Dehkordi et al. 2019). Confirmatory evidence from another study, comprising also longitudinal data, indicated that patients with MCI or ADRD had a higher level of GLCA and higher DCA:CA ratio in both blood and brain tissue samples compared with those without cognitive impairment (Wang et al. 2020). Circulating BA profile has also been investigated in association with biomarkers of AD pathophysiology. Consistent with the previous findings, the higher ratios of secondary to primary BAs, namely GDCA:CA, TDCA:CA, and GLCA: chenodeoxycholic acid (CDCA), were associated with lower CSF $A\beta_{1-42}$ levels, lower cortical glucose metabolism, and lower hippocampal volume (Nho et al. 2019). Levels of glycochenodeoxycholic acid (GCDCA), TLCA, and GLCA showed a positive association also with levels of CSF total tau (t-tau) and phosphorylated tau (p-tau) (Nho et al. 2019).

3.2 *Fatty Acid Derivatives*

In the brain, FAs, such as DHA and ARA, can be released from cell membranes through the action of phospholipases. The released FAs can serve multiple purposes, including oxidation for energy production (Dienel 2012a; White et al. 2020) and/or conversion into *lipid mediators* with the help of enzymes cyclooxygenase and lipooxygenase. These lipid mediators encompass prostaglandins, leukotrienes, resolvins, and neuroprotectins, among others, and they play a role in modulating neuroinflammation (Desale and Chinnathambi 2020). The phospholipases and lipid mediators have been implicated in the pathobiology of ADRD by research in human brain tissues and animal models (Oliveira and Di Paolo 2010; Sanchez-Mejia et al. 2008; Sundaram et al. 2012; Ebright et al. 2022). Further, a gene encoding one of the phospholipases, *PLCG2* (phospholipase C gamma 2), has been identified as a genome-wide significant locus of ADRD (Bellenguez et al. 2022), and rare variants in its coding region have been shown to impact microglial innate immune function (Sims et al. 2017).

Systemic metabolites of FA catabolism, including *ketone bodies*, can serve as alternate cerebral energy sources, especially during starvation or in the developing brain (Jensen et al. 2020). Prolonged fasting (>10 h) leads to depletion of liver glycogen and induces lipolysis of TAGs in adipocytes, thus reducing body adiposity and releasing free FA into the bloodstream. Free FAs are taken up by hepatocytes and metabolized via β-oxidation to acetyl CoA, which is used to generate ketone bodies, i.e., acetoacetate, β-hydroxybutyrate, and acetone. These ketone bodies are then transported through the bloodstream into the brain via monocarboxylic acid transporters in the membranes of the vascular endothelial cells. Within neurons, ketone bodies are metabolized to acetyl CoA, which enters the Krebs cycle in mitochondria, ultimately producing adenosine triphosphate (ATP), i.e., a molecule that carries energy within cells. Additionally, astrocytes are capable of ketogenesis, providing an important local source of β-hydroxybutyrate for neurons. β-Hydroxybutyrate can also upregulate the expression of neurotrophic factors, such as brain-derived neurotrophic factor (BDNF), promoting mitochondrial biogenesis, neural outgrowth/cell survival and cellular resistance to oxidative stress (Brocchi et al. 2022; Mattson et al. 2018). A recent plasma metabolomic study conducted in the UK Biobank identified ketone bodies as a part of the metabolomic signature of 38 metabolites associated with incident ADRD: there was a significant positive association between incident ADRD and the levels of acetoacetate, β-hydroxybutyrate, and acetone (Zhang et al. 2022).

In another study, Sun et al. identified prospective associations between ischemic stroke and two long-chain dicarboxylic acids, specifically tetradecanedioate and hexadecanedioate; the association was particularly strong between hexadecanedioate and cardioembolic stroke subtype (Sun et al. 2019). Both metabolites are derived from FAs by ϖ-oxidation, which takes a more prominent role when β-oxidation is compromised (Wanders et al. 2011). Previous evidence suggests that hexadecanedioate plays a causal role in the regulation of blood pressure (Menni et al. 2015); the association between hexadecanedioate and ischemic stroke remained significant, however, in the model adjusted for additional covariates including hypertension, suggesting that the role of hexadecanedioate in ischemic stroke may be independent of its impact on blood pressure (Sun et al. 2019). Experimental evidence suggests that hexadecanedioate has multiple physiological functions, including hypolipidemic, anti-obesity, and anti-diabetogenic properties (Bar-tana et al. 1989). It remains to be investigated if and how these functions modulate the risk of ischemic stroke, particularly its cardioembolic subtype (Sun et al. 2019). As summarized by the authors, genetic evidence hints at a yet undefined etiologic pathway that may involve solute carrier organic anion transporters (Sun et al. 2019).

The catabolism of FAs produces *acylcarnitines*, which are compounds where a FA is esterified to a carnitine molecule. Acylcarnitines are derived prominently from the β-oxidation of FAs (Thompson et al. 2012) but also from the catabolism of branched-chain amino acids (BCAAs) (Millington and Stevens 2011). Acylcarnitines possess the ability to cross cellular membranes and are detectable in blood, thus making the circulating acylcarnitine profile a reflection of the intramitochondrial acyl-coenzyme A status (Millington and Stevens 2011). The analysis of circulating acylcarnitine profiles has proven useful in identifying inborn errors of metabolism arising from issues in FA oxidation and organic acid metabolism (Jones et al. 2010). Concentrations of circulating acylcarnitines have been shown to predict cognitive decline and the risk of incident ADRD (Huo et al. 2020). In the study by Huo et al., three acylcarnitine species—decanoylcarnitine, pimelylcarnitine, and tetradecadienyl-carnitine – had a combined effect associated with a 60% lower risk of ADRD onset during the average 4.5-year follow-up (Huo et al. 2020). This relationship remained significant after accounting for age, sex, and education. Consistently, higher levels of these three acylcarnitines were associated with a slower cognitive decline over time. Additionally, acylcarnitine C14:2 displayed inverse associations with ADRD in both blood and brain tissue (Huo et al. 2020). In line with these findings, smaller studies have reported lower circulating concentrations of multiple acylcarnitines in patients with ADRD (Mapstone et al. 2014; Ciavardelli et al. 2016; Cristofano et al. 2016) or with schizophrenia (Cao et al. 2019) compared with healthy controls. Tian et al.'s findings further suggested that lower levels of short-chain acylcarnitines were associated with accelerated brain aging, as estimated using the MRI-derived Spatial Patterns of Atrophy for Recognition of Brain Aging (SPARE-BA) score, but this association was observed in females only and not in males (Tian et al. 2023). While acylcarnitines have been recognized for their multifaceted roles in neuroprotection, including involvement in lipid synthesis, cholinergic neurotransmission, and the modulation of membrane composition and gene expression, among others (reviewed in Jones et al. 2010), further research is needed to elucidate their exact physiological functions.

3.3 Miscellaneous

β-Cryptoxanthin is an organic compound under a superclass of lipids and lipid-like molecules, and it belongs to the class of carotenoids (Wishart et al. 2018). As a major source of vitamin A, second only to β-carotene, it has been recognized for its potential health benefits, including lower risk of inflammatory disorders (Pattison et al. 2005) or protection against lung cancer (Lian et al. 2006). Palacios et al. reported a positive association between circulating β-cryptoxanthin and better cognitive functioning (Palacios et al. 2020). This association remained significant in multivariable models adjusted for covariates, and in analyses restricted to participants without diabetes. Also, in Least Absolute Shrinkage and Selection Operator (LASSO) models, β-cryptoxanthin was retained as predictive of good cognition (Palacios et al. 2020). Considering that previous studies have indicated a deficiency of vitamin A to be associated with worse cognition in patients with ADRD (Zeng et al. 2017) and β-cryptoxanthin to be a predictor of healthy diet in postmenopausal women (McCullough et al. 2019), the authors highlight the important role for healthy diet promoting cognitive well-being (Palacios et al. 2020).

In their study, Neuffer et al. reported consistent associations between cognitive decline and *propionic acid* in two French samples of older adults (Neuffer et al. 2022). Propionic acid, a short-chain carboxylic acid, can originate from gut fermentation, oral microbiome, and dietary sources, particularly processed and packaged foods where it serves as a preservative (Neuffer et al. 2022). While 90% of the propionic acid absorbed from the gastrointestinal tract is metabolized by the liver (Peters et al. 1992), it likely can transverse the BBB via monocarboxylate transporters that are highly expressed in endothelial cells (Silva et al. 2020). Previous evidence on the role of propionic acid in cognitive aging suggests involvement of multiple mechanisms, including modulation of BBB integrity, microglial maturation and function, and inflammation-modulating properties (Neuffer et al. 2022). The authors observed a robust correlation between propionic acid and blood glucose (Pearson's $r = 0.79$), and the identification of a significant mediation effect led to the introduction of a novel concept that blood glucose may mediate the adverse association of propionic acid with cognitive decline (Neuffer et al. 2022). Dysfunctional glucose metabolism is indeed a recognized key risk factor for cognitive decline and dementia, as elaborated in more detail in Sect. 4.2.

4 Non-lipid Metabolites

Non-lipid metabolites originating from the circulation play a crucial role in normal brain physiology. *Amino acids*, such as *glutamine*, are critical for synthesis of proteins that function as important components of cell membranes, forming ion channels and various transporters, and as precursors of many neurotransmitters (Fernstrom 2005). Circulating amino acids can be quantified using both NMR and MS-based methods, and the most common commercially available platforms quantify several amino acids, including branched-chain amino acids (BCAAs) *isoleucine*, *leucine*, and *valine*, as well as aromatic amino acids (AAAs) *phenylalanine*, *tyrosine*, and *histidine*, among others.

The brain primarily relies on *glucose* as its predominant source of energy. Despite the adult brain accounting only for ~2% of the body weight, it consumes 20–25% of glucose-derived energy (Mergenthaler et al. 2013; Chen and Zhong 2013). Most of the energy is consumed by synaptic activity and maintenance of resting potential of neurons (Mergenthaler et al. 2013). Additionally, glucose metabolism provides precursors for neurotransmitters (Dienel 2012b), further underscoring its vital role in sustaining neural function.

4.1 *Amino Acids*

The findings regarding stroke and circulating amino acids are somewhat conflicting. A recent study conducted in European cohorts reported a negative association between circulating histidine levels and both ischemic stroke and stroke overall (Vojinovic et al. 2020). But a prior study conducted in the Chinese Kadoorie Biobank did not find a significant association in this regard (Holmes et al. 2018). Likewise, a positive association was reported between phenylalanine levels and risk of ischemic stroke in the European sample (Vojinovic et al. 2020) but this association was not significant in the Chinese sample (Holmes et al. 2018). The authors of the former study suggest that these discrepancies could be attributed to environmental and/or ethnic disparities, or differences in the analytical approaches when adjusting for potential confounders (Vojinovic et al. 2020).

In turn, investigations into the associations between dementia and amino acid variations have yielded relatively consistent results. Lower blood levels of BCAAs, including valine (Zhang et al. 2022; Tynkkynen et al. 2018; Cui et al. 2020), leucine (Zhang et al. 2022; Tynkkynen et al. 2018; Cui et al. 2020), and isoleucine (Zhang et al. 2022; Tynkkynen et al. 2018), have been associated with a higher risk of ADRD. Supportive to these findings, lower blood valine level has been associated with accelerated cognitive decline and greater ADRD-related cerebral atrophy (Toledo et al. 2017). Furthermore, BCAA derivatives and degradation products, like N-acetylisoleucine, N-acetylleucine, 3-methylglutaconate, and beta-hydroxyisovalerate, have been negatively associated with cognitive functioning (Palacios et al. 2020). Several potential pathways may underlie the associations between dementia and BCAAs and their derivatives. It has been proposed that lipid changes occurring early in the disease process could, over time, lead to altered membrane structures, including those in mitochondria; these changes may further disrupt lipid oxidation, membrane potentials, and membrane transport, ultimately affecting energy-substrate utilization and prompting the use of BCAAs as an energy source (Toledo et al. 2017). Alternatively, as dietary intake is the key source of circulating BCAAs, reduced BCAA levels may indicate nutritional deficiencies (Tynkkynen et al. 2018). Changes in gut microbiome may also influence levels of BCAAs, given the correlation with the gut microbiome's BCAA biosynthetic potential (Pedersen et al. 2016). Also, elevated blood levels of anthranilic acid – that is, an aminobenzoic acid and a metabolite of the tryptophan-kynurenine pathway catabolizing tryptophan – have been associated with a higher risk of ADRD (Chouraki et al. 2017). Interestingly, the metabolites of the tryptophan-kynurenine pathway have both direct and indirect effects on *glutamate* (reviewed in Schwarcz 2016), the major excitatory neurotransmitter in the brain, the blood levels of which were suggestively associated with the risk of incident dementia (Chouraki et al. 2017). Additionally, the study reported a suggestive negative association between the risk of incident dementia and *taurine*, an amino acid that can cross the BBB and may protect against glutamate toxicity, possibly by having beneficial vascular effects such as those involved in blood pressure regulation (Chouraki et al. 2017). Higher blood *glutamine* levels have also shown consistent associations with lower cognitive function and higher risk of all-cause dementia as well as ADRD (van der Lee et al. 2018), providing further evidence that alterations in the glutamine–glutamate signaling may play a role in dementia.

Higher levels of BCAAs, particularly leucine and isoleucine, in cord plasma have been associated with a higher risk of childhood diagnosis for attention-deficit hyperactivity disorder (ADHD) (Anand et al. 2021). This association remained significant when adjusted for maternal plasma BCAA levels. There was no evidence of an association between maternal plasma BCAA levels and ADHD risk (Anand et al. 2021). The authors speculate that the competition for the same transporters at the BBB between BCAAs and AAAs might lead to BCAA-induced deficiencies in tyrosine and other AAAs, potentially disrupting the balanced synthesis of the neurotransmitters dopamine, norepinephrine, and serotonin and, thereby contributing to a higher risk of ADHD (Anand et al. 2021).

In our metabolomic study of WHM, we detected a robust positive association between WMH and hydroxyphenylpyruvate (Sliz et al. 2022). In the male sub-sample of our study, circulating hydroxyphenylpyruvate explained up to 14.3% of the variance in WMH volume (Sliz et al. 2022), surpassing the variance explained by vascular risk factors, such as hypertension, which accounted for up to 2% (Wardlaw et al. 2014; Cox et al. 2019). Hydroxyphenylpyruvate is a potentially toxic metabolite on the catabolic pathway of phenylalanine and tyrosine. In the light of our findings, it appeared that low activity of hydroxyphenylpyruvate dioxygenase (HPD) would be the most likely mechanism underlying the association between WMH and circulating hydroxyphenylpyruvate (Sliz et al. 2022). As HPD requires oxygen in the reaction converting hydroxyphenylpyruvate to homogentisic acid and hypoxia has been shown to promote the accumulation of hydroxyphenylpyruvate (Jones and Mason 1978), we hypothesized that higher circulating hydroxyphenylpyruvate level may be an indicator of ischemic hypoxia promoting WMH (Sliz et al. 2022).

4.2 Glucose and Related Metabolites

Dysregulated brain glucose metabolism has been implicated in neurodegenerative diseases and in ADRD in particular (Chen and Zhong 2013; An et al. 2018). This dysregulation involves insulin resistance in brain tissue and hyperglycemia, both of which co-occur with type 2 diabetes, a major risk factor for ADRD (Kellar and Craft 2020; Arnold et al. 2018). Insulin resistance is defined by lower cell responsiveness to insulin, a hormone primarily produced in pancreas that promotes cellular uptake of glucose. As reviewed in (Kellar and Craft 2020) and (Arnold et al. 2018), most insulin in the brain originates from the circulating (pancreatic) insulin. In the brain, insulin receptors are present on all cell types, including neurons, with high expression levels in the hypothalamus, hippocampus, and cerebral cortex. Experimental work suggests that insulin enhances neurogenesis and inhibits neuronal apoptosis (Arnold et al. 2018). Some aspects of brain glucose metabolism, such as glucose transport from the peripheral circulation into the brain and its uptake by glial cells and neurons, are largely insulin independent (Benarroch 2014). In ADRD, brain tissue contains more glucose, and cell uptake and utilization of glucose (glycolysis) is impaired (An et al. 2018). Hyperglycemia promotes glycation of cellular proteins and lipids, forming advanced glycation end products that have been shown to increase oxidative stress and inflammation, and promote neuronal apoptosis (Rom et al. 2020).

Two large-scale longitudinal studies have shown that multiple metabolites, including glucose and glycolysis-related compounds, are associated with incident dementia (Zhang et al. 2022; Machado-Fragua et al. 2022). Additionally, these studies have also shown that adding some of these metabolites to the conventional prediction model improves the prediction of incident dementia (Zhang et al. 2022; Machado-Fragua et al. 2022). Palacios et al. reported associations between cognitive functioning and levels of several circulating carbohydrates, including the positive association with 1,5-anhydroglucitol, and the negative associations with glucose, mannose, mannitol/sorbitol, and ribitol (Palacios et al. 2020). 1,5-anhydroglucitol is a marker of short-term glycemic control (Dungan 2008) and it is strongly negatively correlated with glucose and other carbohydrates (Palacios et al. 2020). The study's overall findings implicate that metabolic dysregulation, including insulin resistance, diabetes, and the metabolic syndrome, is linked to cognition especially in older women making up the majority of the study (Palacios et al. 2020).

Two recent community-based studies have shown an association between a blood index of peripheral insulin resistance and lower thickness of the cerebral cortex (Lu et al. 2021; Shin et al. 2020). Additionally, a large-scale study has shown that a negative association between a clinical marker of chronic glycemia, i.e., glycated hemoglobin (HbA1c), and cortical thickness is non-continuous, emerging within the low prediabetic range (Shin et al. 2023). Cell type and gene enrichment analyses combining brain imaging data with In Silico transcriptomic data suggested that the negative Hb1Ac-cortical thickness association involves weak excitotoxicity due to glutamatergic dysregulation of excitatory neurons, and diminished counterregulatory functions of astrocytes and microglia. A genomic locus linking higher HbA1c to lower cortical thickness regulated cortical expression of genes involved in immune function, synaptic function, axonal transport, and ATP. As neurons and their synapse-related processes require high levels of ATP generated by mitochondria that are being continuously replenished via axonal transport, the identified locus may enhance the (pre)diabetes-related weak excitotoxicity by heightening the depletion of ATP. This study demonstrating that prediabetic levels of HbA1c are associated with cortical thinning highlights the need for effective glycemic control early in the course of diabetes to protect brain health (Shin et al. 2023).

In our recent work, we found a positive association between WMH and glucuronate, a derivative of glucose; this association was restricted to the female sub-sample of our study and remained significant after adjustment for type 2 diabetes (Sliz et al. 2022). Glucuronate is a key substance in a process called glucuronidation, a process vital for the protection of the brain from systemic toxic substances (Ouzzine et al. 2014). Glucuronate is also required for the synthesis of glycosaminoglycans, which are constituents of the glycocalyx layer tightening the endothelial barrier and limiting vascular permeability (Eelen et al. 2018). We hypothesized that alterations in endothelial cells' glucose metabolism might compromise these protective functions of the BBB and, thus, underlie the association between WMH and circulating glucuronate (Sliz et al. 2022). It is noteworthy that the findings regarding WMH association with glucose are conflicting, with some studies reporting a significant positive association (de Leeuw et al. 2021) while in other reports, including our recent study, the association did not reach statistical significance (Sliz et al. 2022; Harshfield et al. 2022).

In their work, Pan et al. demonstrated that patients with AD have lower levels of blood thiamine diphosphate, a key coenzyme for several enzymes involved in glucose catabolism, in comparison to controls (Pan et al. 2016). They further proposed that the levels of blood thiamine metabolites may serve as a promising biomarker for Alzheimer's disease with sensitivity and specificity over 80% and capacity to differentiate AD from vascular and frontotemporal dementias (Pan et al. 2016). Their findings further suggested that thiamine metabolites serve as a biomarker for AD independent of potential confounders, such as age and blood hemoglobin levels. The authors discuss that altered thiamine metabolism perturbs brain glucose metabolism and predisposes to cognitive impairment and neurodegeneration by inducing multiple pathogenic factors, including enhanced brain amyloid deposit (Pan et al. 2016).

4.3 Markers of Systemic Inflammation

Systemic, low-grade inflammation is a prominent feature of obesity and the metabolic syndrome. In obesity – especially in visceral obesity – adipose tissue produces excess amounts of circulating proinflammatory molecules that have direct effects on endothelium and vascular permeability. Consequently, systemic inflammation also induces multiple changes in the physiology of BBB that potentially compromise brain health, including neuroinflammation (Galea 2021; Woo et al. 2022), which involves proinflammatory transformations of microglia (Galea 2021) and astrocytes accompanied by lower – neuron protective – homeostatic functions, such as their positive contributions to maintenance myelination (Hoogland et al. 2015; Bocarsly et al. 2015).

Glycoprotein acetyls (GlycA) is an inflammatory biomarker that reflects the number of N-acetyl groups in circulating glycoproteins. It has a demonstrated predictive value for cardiovascular disease, type 2 diabetes mellitus, and all-cause mortality (Ahola-Olli et al. 2019; Würtz et al. 2015; Deelen et al. 2019). While circulating GlycA level has been associated with lower cognitive functioning, this association may be, at least partially, explained by the association of GlycA with various diseases that themselves impair cognitive abilities (van der Lee et al. 2018). Interestingly, GlycA levels do not appear to be associated with ADRD (van der Lee et al. 2018; Tynkkynen et al. 2018) but another NMR-based inflammatory marker, O-acetyl-glycoprotein level, has been reported to be associated with incident dementia and incident ADRD (Cui et al. 2020). Also, a higher GlycA level has been consistently associated with a higher risk of stroke (Holmes et al. 2018; Vojinovic et al. 2020) and higher WMH volume (Sliz et al. 2022; Harshfield and Markus 2023; de Leeuw et al. 2021; Sun et al. 2023).

Evidence from studies involving younger populations suggests that systemic inflammation could have detrimental effects on brain health already in youth. The findings of our lipidomics study provided evidence that circulating GPCs, most prominently PC16:0/2:0, may mediate the effect between visceral fat volume and white matter microstructure as well as between visceral fat volume and processing speed in a community-based sample of adolescents aged 12–18 years (Syme et al. 2019). In the same population of adolescents, the above-described *DHCR24* variant associated with white matter microstructure also showed a nominally significant association with lower serum GlycA levels (Sliz et al. 2021) providing additional evidence that systemic inflammation may play a role in influencing brain health from a young age. Notably, it appears that maternal inflammation, as measured by circulating GlycA levels around twenty-eighth gestational week, is associated negatively with child cognition at the age of 2 years (Marx et al. 2022).

The metabolomic evidence consistently indicates that circulating markers of inflammation are associated with adverse outcomes for cognitive abilities and brain tissue properties. This supports the perception that systemic inflammation likely plays a role in modulating brain health. More studies are warranted, however, to assess the detailed molecular pathways.

5 APOE Effect

Apolipoprotein E (APOE) is an apolipoprotein involved in lipid transfer between cells and tissues. APOE is mainly produced in the liver, but it is also synthesized in the brain. In the periphery, APOE serves as a ligand for LDL receptor family of cell surface receptors, promoting the clearance of TAG-rich lipoproteins from the circulation (Marais 2019). There are three common genetic variants of *APOE*, named *ε2*, *ε3*, and *ε4*, originating from two missense variants (rs429358 and rs7412), which encode three protein isoforms – APOE2, APOE3, and APOE4 (Bennet et al. 2007). Compared with individuals with two copies of *ε3* allele, those with at least one *ε2* exhibit lower LDL-cholesterol levels and lower risk of atherosclerotic coronary disease. Conversely, individuals with at least one *ε4* allele, compared with *ε3* homozygotes, tend to have slightly higher LDL-cholesterol level and a higher risk of atherosclerotic coronary disease (Bennet et al. 2007). Recent comprehensive blood metabolomic profiling has confirmed the principal blood lipid findings (Karjalainen et al. 2019). In the brain, APOE is predominantly expressed by astrocytes, with lower expression levels in other glial cells, neurons, and the cells of BBB (Marais 2019). It is a key molecule for transporting lipids between cells. Furthermore, In Vitro evidence suggests it plays a role in maintaining BBB integrity by controlling the activity of protein kinase C η (PKCη) and coordinating the phosphorylation of occludin, a protein crucial for the establishment and maintenance of cellular tight junctions (Marais 2019; Nishitsuji et al. 2011). *APOE* is the strongest genetic locus of late-onset ADRD; variant ε4 increases the risk of ARDR, and ε2 decreases it (Andrews et al. 2020). While the exact underlying biology is not fully understood, recent research suggests that ε4 variant promotes dysregulated lipid metabolism and proinflammatory signaling in astrocytes and microglia (Sienski et al. 2021; Julia et al. 2022), impairing their homeostatic functions in relation to neuronal network activity (Victor et al. 2022). Additionally, the ε4 variant promotes abnormal lipid accumulation in oligodendrocytes, leading to reduced myelination and axonal diameter (Blanchard et al. 2022).

Given the importance of the *APOE* variants in both systemic and brain lipid metabolism and in brain health, several studies have explored whether *APOE* alleles modify the associations between circulating metabolites and brain health-related phenotypes, such as neuroimaging markers of brain aging, cognition, and/or ADRD. Most studies have not identified, however, a significant modifying effect of *APOE*, as evidenced by the following findings. Adjusting for *APOE* alleles did not modify the associations between PUFA-containing TAGs and the thickness of the cerebral cortex in a population-based sample of middle-aged adults (Sliz et al. 2020). Similarly, in a sample of older adults including cases with MCI and ADRD, *APOE* adjustment did not alter the associations between TAGs and hippocampal volume, thickness of the entorhinal cortex, or total volume of WMH (Bernath et al. 2020). In the latter study, however, some associations were more pronounced in ε4 carriers (Bernath et al. 2020). Furthermore, *APOE* adjustment did not change the associations between 15 blood metabolites and general cognitive ability, but it did attenuate the associations between total cholesterol, CE, and PL within M-HDL and ADRD (van der Lee et al. 2018). Conversely, no major modifying effect of *APOE* on the associations between circulating metabolites and ADRD was observed (Tynkkynen et al. 2018; Machado-Fragua et al. 2022). In their study, Arnold et al. conducted a twofold stratification, considering both sex and *APOE* genotype (Arnold et al. 2020). Interestingly, their findings suggested that certain associations between circulating metabolites, such as PC ae C44:6, PC ae C44:5, and PC ae C42:4, and AD biomarkers, including CSF p-tau, CSF Aβ1–42, and FDG-PET, may be driven by, or be specific to, the group of *APOE* ε4-carrying females.

Diverse practices exist regarding the coding of *APOE*, with some studies categorizing individuals based on their ε4 carrier status (i.e., individuals carrying one or two copies of ε4 versus non-carriers) (Bernath et al. 2020; Machado-Fragua et al. 2022; Arnold et al. 2020), while others adjust for the *APOE* genotypes (van der Lee et al. 2018; Sliz et al. 2020). In metabolomic associations, the ε2 allele exhibits contrasting and larger effect sizes on circulating lipid traits compared with the ε4 allele (Karjalainen et al. 2019; Sebastiani et al. 2023), possibly arising from the fact that the APOE2 isoform has an approximately 50-fold lower binding affinity with LDL receptor compared with APOE3 and APOE4 (Bu 2009). Thus, future studies may benefit from incorporating ε2 effects/adjustments alongside ε4 effects/adjustments. Also, limitations in genotype data can reduce sample size, potentially reducing statistical power to detect associations and further complicating the interpretation of the *APOE* allele-adjusted results (Palacios et al. 2020). Considering the factors mentioned above, assessing the impact of *APOE* alleles in metabolomic associations with brain health poses a complex challenge due to their complicated and opposing effects on both brain tissue and peripheral metabolism.

6 Conclusions and Future Directions

Metabolomic studies indicate broad variations in the circulating metabolome in association with various brain health-related phenotypes (summarized in Table 2) and, often, these associations are independent of traditional vascular risk factors. The associations of circulating TAGs, bile acids, and BCAAs with brain health appear to be most consistent across different studies and phenotypes. On the contrary, the associations between circulating lipoprotein measures and brain health emerge to be less consistent than, for example, their associations with cardiovascular endpoints. Nonetheless, studies have reported insightful findings regarding WMH and dementia and relative proportions of lipids within lipoprotein subfractions (Tynkkynen et al. 2018; Sliz et al. 2022) suggesting that lipoprotein lipid composition rather than absolute concentrations may provide valuable information when it comes to investigating brain health. Moreover, the ratios of bile acids have shown consistent associations with cognitive impairment (Wang et al. 2020; Mahmoudian Dehkordi et al. 2019), the ratios of sphingomyelins with ADRD, brain atrophy, and cognitive functioning (Baloni et al. 2022), and the maternal ratio of omega-3 to total FAs with child cognition (Marx et al. 2022). The changes in the metabolite ratios may be indicative of changes in the activities of related enzymes (Sliz et al. 2022; Mahmoudian Dehkordi et al. 2019) and, thus, investigations of the relative measures may reveal different aspects of underlying molecular pathways compared with the absolute concentration values.

Typically, distinct studies reviewed in this chapter have investigated varying sets of metabolic measures that are often quantified using different metabolomic platforms. While this assures novelty with regard to the phenotypic associations, the downside is that many of the findings are based on one population and, therefore, several discoveries lack replication. In metabolomic studies, replication would be of high value, as these studies involve tens or typically hundreds of metabolic measures, thus increasing the probability of false-positive findings. Compared with stringent p-value thresholds to correct for multiple testing, consistent findings from multiple populations would be a more desirable solution to evaluate the probability of type I error. Multiple testing approaches may be harsh on individual metabolic measures, and may cause biologically relevant associations to be missed. Metabolic pathways, such as lipoprotein metabolism, are interconnected systems, and, hence, it may be more meaningful to evaluate the global patterns instead of interpreting the associations on the level of individual metabolic measures (Sliz et al. 2022; de Leeuw et al. 2021).

Elderly populations are overrepresented in the metabolomic studies of brain health and disease. Accumulating evidence suggests, however, that blood metabolomic deviations associated with brain structural features can be observed in adult populations considered to be generally healthy before the onset of disease (Sliz et al. 2022; Sliz et al. 2020). Importantly, children and adolescents are highly underrepresented in these studies. Studies involving young populations have the potential to improve understanding of the metabolic pathways involved in brain

Table 2 Summary of the key metabolomic associations of brain health-related phenotypes

	Lipids				Lipid derivatives		Non-lipid metabolites		
Brain phenotype	Fatty acids	Cholesterol and lipoproteins	Glycerophospholipids and sphingolipids	Triacylglycerols	Bile acids	Fatty acid derivatives	Amino acids	Glucose and related metabolites	Markers of systemic inflammation
AD and related dementias	• DHA ↓ • PUFA ↓ • Omega-6 FA ↓ • Linoleic acid ↓ • Omega-3 FA ↓ • Petroselinic acid ↓ • Myristic acid ↓ • Palmitic acid ↓ • Palmitoleic acid ↓	• Overall conflicting evidence • Lipids in some VLDL subfractions ↓	• Ether GPCs and GPEs; especially ether lipids esterified with omega-3 FAs ↓ • Dysregulation of SM and ceramide metabolism	• PUFA-containing TAGs ↓	• Cholic acid (CA) ↓ • Deoxycholic acid (DCA) ↑ • GDCA ↑ • TDCA ↑ • DCA:CA ↑	• Decanoylcarnitine ↓ • Pimelylcarnitine ↓ • Tetradecadienyl-carnitine ↓ • Ketone bodies ↑	• Valine ↓ • Leucine ↓ • Isoleucine ↓ • Glutamine ↑	• Thiamine diphosphate ↓	• O-acetyl-glycoprotein ↑
Cognitive function	• DHA ↑ • Linoleic acid ↑ • Petroselinic acid ↑ • Myristic acid ↑ • Palmitic acid ↑ • Palmitoleic acid ↑	• Lipids in M-HDL ↑	• Several GPCs ↑ • PC16:0/2:0 ↑ in adolescents		• GDCA ↓ • GLCA ↓ • TLCA ↓	• Decanoylcarnitine ↑[a] • Pimelylcarnitine ↑[a] • Tetradecadienyl-carnitine ↑[a]	• Valine ↓[b] • BCAA derivatives and degradation products ↓ • Glutamine ↓	• 1,5 anhydroglucitol ↑ • Glucose ↓ • Mannose ↓ • Mannitol/sorbitol ↓ • Ribitol ↓	• Glycoprotein acetylation ↓
Thickness of the cerebral cortex				• TAGs carrying 18-carbon FAs ↑				• HOMA-IR ↓ • HbA_{1c} ↓ in prediabetic individuals	

(continued)

Table 2 (continued)

	Lipids				Lipid derivatives		Non-lipid metabolites		
Brain phenotype	Fatty acids	Cholesterol and lipoproteins	Glycerophospholipids and sphingolipids	Triacylglycerols	Bile acids	Fatty acid derivatives	Amino acids	Glucose and related metabolites	Markers of systemic inflammation
White matter hyperintensities		• No strong associations in pooled samples • Consistent suggestive associations between WMH and total-C ↓, especially in males	• LPC ↓ in males • SM-OH ↓ in males				• Hydroxyphenylpyruvate ↑ in males	• Glucuronate ↑ in females	• Glycoprotein acetylation ↑
Stroke		• Overall conflicting evidence • Consistently no associations with intracerebral hemorrhage				• Tetradecanedioate ↑[c] • Hexadecanedioate ↑[c]	• Conflicting evidence		• Glycoprotein acetylation ↑

↑ Positive association
↓ Negative association
[a] Slower cognitive decline
[b] Accelerated cognitive decline
[c] Ischemic stroke, cardioembolic subtype

maturation (Sliz et al. 2021) and, also, in early pathogenic processes modulating the risk of neurological disorders (Anand et al. 2021). Thus, more research is highly warranted in this field. Also, in future studies, considerations should be given to assessing the metabolomic associations in sex-specific subgroups (see Chapter "Sex and Gender in Population Neuroscience" for general discussion of sex and gender in population neuroscience). Sex hormones have direct effects on the endothelium (Stanhewicz et al. 2018), on the brain cells, including oligodendrocytes (Cerghet et al. 2009) and neurons (Pesaresi et al. 2015), as well as on the peripheral metabolism (Crook and Seed 1990; Wawrzkiewicz-Jałowiecka et al. 2021). Intriguingly, multiple sex-specific metabolomic associations regarding brain health have already been reported (Sliz et al. 2022; Sun et al. 2023; Tian et al. 2023; Palacios et al. 2020; Arnold et al. 2020). Additionally, when planning the studies and interpreting the results, it is essential to carefully consider the fasting status (Lehmann 2021).

Finally, the associations between brain health and circulating metabolic measures are typically investigated in cross-sectional settings. To be able to infer causality, longitudinal and/or Mendelian randomization-based evidence would be warranted. Summary statistics from genome-wide association studies are increasingly available, which facilitates future investigations in two-sample Mendelian randomization settings. It is crucial to emphasize that careful considerations are required when choosing instruments for Mendelian randomization, due to the pleiotropy observed in most lipid-associated loci. Moreover, one must pay close attention to the composition of the study sample, taking into consideration factors such as possible subgroup specificities. Understanding the causal directions would be helpful in distinguishing disease biomarkers from the causal molecules that could potentially serve as targets for future therapeutics. More work should also be done to identify truly predictive biomarkers, as most currently identified biomarkers have a limited predictive value.

In summary, the evolving landscape of research on circulating metabolic measures and brain health has revealed promising but also complex insights. While some associations are consistent, other findings require critical consideration. Further exploration in diverse populations, including younger age groups, and the application of longitudinal and Mendelian randomization approaches hold the key to enhancing our understanding of the metabolic background of neurological disorders. Collectively, these endeavors move us closer to a hopeful future where we are making progress toward improved brain health.

Conflict of Interest The authors declare that the research was conducted in the absence of any commercial or financial relationships that could be construed as a potential conflict of interest.

Funding The present work was funded by the National Institutes of Health (R01AG056726) and the Research Council of Finland (338229).

References

Ahola-Olli AV, Mustelin L, Kalimeri M, Kettunen J, Jokelainen J, Auvinen J et al (2019) Circulating metabolites and the risk of type 2 diabetes: a prospective study of 11,896 young adults from four Finnish cohorts. Diabetologia 62(12):2298–2309. [Internet]. 2022 Aug 8 [cited 2023 Jan 9]. Available from: https://link.springer.com/article/10.1007/s00125-019-05001-w

Amin N, Liu J, Bonnechere B, Mahmoudiandehkordi S, Arnold M, Batra R et al (2023) Interplay of metabolome and gut microbiome in individuals with major depressive disorder vs control individuals. JAMA Psychiatry 80(6):597–609. [Internet]. 2023 Jun 7 [cited 2023 Sep 27]. Available from: https://pubmed.ncbi.nlm.nih.gov/37074710/

An Y, Varma VR, Varma S, Casanova R, Dammer E, Pletnikova O et al (2018) Evidence for brain glucose dysregulation in Alzheimer's disease. Alzheimers Dement 14(3):318–329. [Internet]. 2018 Mar 1 [cited 2023 Jan 3]. Available from: https://pubmed.ncbi.nlm.nih.gov/29055815/

Anand NS, Ji Y, Wang G, Hong X, van der Rijn M, Riley A et al (2021) Maternal and cord plasma branched-chain amino acids and child risk of attention-deficit hyperactivity disorder: a prospective birth cohort study. J Child Psychol Psychiatry 62(7):868–875. [Internet]. 2021 Jul 1 [cited 2023 Nov 2]. Available from: https://pubmed.ncbi.nlm.nih.gov/32960988/

Andrews SJ, Fulton-Howard B, Goate A (2020) Interpretation of risk loci from genome-wide association studies of Alzheimer's disease. Lancet Neurol 19(4):326–335. [Internet]. 2020 Apr 1 [cited 2023 Jan 9]. Available from: https://pubmed.ncbi.nlm.nih.gov/31986256/

Arnold SE, Arvanitakis Z, Macauley-Rambach SL, Koenig AM, Wang HY, Ahima RS et al (2018) Brain insulin resistance in type 2 diabetes and Alzheimer disease: concepts and conundrums. Nat Rev Neurol 14(3):168–181. [Internet]. 2018 Jan 29 [cited 2023 Jan 3]. Available from: https://www.nature.com/articles/nrneurol.2017.185

Arnold M, Nho K, Kueider-Paisley A, Massaro T, Huynh K, Brauner B et al (2020) Sex and APOE ε4 genotype modify the Alzheimer's disease serum metabolome. Nat Commun 11(1) [Internet]. 2020 Dec 1 [cited 2023 Oct 4]. Available from: https://pubmed.ncbi.nlm.nih.gov/32123170/

Baloni P, Arnold M, Buitrago L, Nho K, Moreno H, Huynh K et al (2022) Multi-Omic analyses characterize the ceramide/sphingomyelin pathway as a therapeutic target in Alzheimer's disease. Commun Biol 5(1):1074. [Internet]. 2022 Oct 8 [cited 2023 Sep 27]. Available from: https://pubmed.ncbi.nlm.nih.gov/36209301/

Bar-tana J, Ben-shoshan S, Blum J, Migron Y, Hertz R, Bar-tana-rose-khan G et al (1989) Synthesis and hypolipidemic and antidiabetogenic activities of β,β,β',β'-tetrasubstituted, long-chain dioic acids. J Med Chem 32(9):2072–2084

Bazinet RP, Layé S (2014) Polyunsaturated fatty acids and their metabolites in brain function and disease. Nat Rev Neurosci 15(12):771–785. [Internet]. 2014 Dec 11 [cited 2023 Sep 29]. Available from: https://pubmed.ncbi.nlm.nih.gov/25387473/

Bellenguez C, Küçükali F, Jansen IE, Kleineidam L, Moreno-Grau S, Amin N et al (2022) New insights into the genetic etiology of Alzheimer's disease and related dementias. Nat Genet 54(4): 412–436. [Internet]. 2022 Apr 1 [cited 2023 Sep 27]. Available from: https://pubmed.ncbi.nlm.nih.gov/35379992/

Benarroch EE (2014) Brain glucose transporters. Neurology 82(15):1374–1379. [Internet]. 2014 Apr 15 [cited 2023 Jan 3]. Available from: https://n.neurology.org/content/82/15/1374

Bennet AM, Di Angelantonio E, Ye Z, Wensley F, Dahlin A, Ahlbom A et al (2007) Association of apolipoprotein E genotypes with lipid levels and coronary risk. JAMA 298(11):1300–1311. [Internet]. 2007 Sep 19 [cited 2023 Jan 9]. Available from: https://pubmed.ncbi.nlm.nih.gov/17878422/

Bentsen H (2017) Dietary polyunsaturated fatty acids, brain function and mental health. Microb Ecol Health Dis 28(sup1):1281916. [Internet]. 2017 Feb 24 [cited 2023 Oct 30]. Available from: /pmc/articles/PMC5445635/

Bernath MM, Bhattacharyya S, Nho K, Barupal DK, Fiehn O, Baillie R et al (2020) Serum triglycerides in Alzheimer disease: relation to neuroimaging and CSF biomarkers. Neurology

94(20):E2088–E2098. [Internet]. 2020 May 19 [cited 2023 Oct 4]. Available from: https://pubmed.ncbi.nlm.nih.gov/32358220/

Bhagavan NV, Ha CE (2011) Essentials of medical biochemistry. Academic Press, Amsterdam

Blanchard JW, Akay LA, Davila-Velderrain J, von Maydell D, Mathys H, Davidson SM et al (2022) APOE4 impairs myelination via cholesterol dysregulation in oligodendrocytes. Nature 611(7937):769–779. [Internet]. 2022 Nov 16 [cited 2023 Jan 9]. Available from: https://www.nature.com/articles/s41586-022-05439-w

Bocarsly ME, Fasolino M, Kane GA, Lamarca EA, Kirschen GW, Karatsoreos IN et al (2015) Obesity diminishes synaptic markers, alters microglial morphology, and impairs cognitive function. Proc Natl Acad Sci USA 112(51):15731–15736

Bright JJ, Kanakasabai S, Chearwae W, Chakraborty S (2008) PPAR regulation of inflammatory signaling in CNS diseases. PPAR Res

Brites P, Waterham HR, Wanders RJA (2004) Functions and biosynthesis of plasmalogens in health and disease. Biochim Biophys Acta Mol Cell Biol Lipids 1636(2–3):219–231

Brocchi A, Rebelos E, Dardano A, Mantuano M, Daniele G (2022) Effects of intermittent fasting on brain metabolism. Nutrients 14(6) [Internet]. 2022 Mar 1 [cited 2023 Sep 27]. Available from: https://pubmed.ncbi.nlm.nih.gov/35334932/

Bu G (2009) Apolipoprotein E and its receptors in Alzheimer's disease: pathways, pathogenesis and therapy. Nat Rev Neurosci 10(5):333–344. [Internet]. 2009 Apr 2 [cited 2023 Oct 4] Available from: https://www.nature.com/articles/nrn2620

Cao B, Wang D, Pan Z, Brietzke E, McIntyre RS, Musial N et al (2019) Characterizing acyl-carnitine biosignatures for schizophrenia: a longitudinal pre- and post-treatment study. Transl Psychiatry 9(1)

Cerghet M, Skoff RP, Swamydas M, Bessert D (2009) Sexual dimorphism in the white matter of rodents. J Neurol Sci 286(1–2):76–80

Chang TY, Yamauchi Y, Hasan MT, Chang C (2017) Cellular cholesterol homeostasis and Alzheimer's disease. J Lipid Res 58(12):2239–2254. [Internet]. 2017 [cited 2023 Sep 27]. Available from: https://pubmed.ncbi.nlm.nih.gov/28298292/

Chen CT, Bazinet RP (2015. 2015 Jan 1) β-Oxidation and rapid metabolism, but not uptake regulate brain eicosapentaenoic acid levels. Prostaglandins Leukot Essent Fatty Acids 92:33–40

Chen Z, Zhong C (2013) Decoding Alzheimer's disease from perturbed cerebral glucose metabolism: implications for diagnostic and therapeutic strategies. Prog Neurobiol 108:21–43

Chouraki V, Preis SR, Yang Q, Beiser A, Li S, Larson MG et al (2017) Association of amine biomarkers with incident dementia and Alzheimer's disease in the Framingham study. Alzheimers Dement 13(12):1327–1336. [Internet]. 2017 Dec 1 [cited 2023 Nov 6]. Available from: https://pubmed.ncbi.nlm.nih.gov/28602601/

Ciavardelli D, Piras F, Consalvo A, Rossi C, Zucchelli M, Di Ilio C et al (2016) Medium-chain plasma acylcarnitines, ketone levels, cognition, and gray matter volumes in healthy elderly, mildly cognitively impaired, or Alzheimer's disease subjects. Neurobiol Aging 43:1–12

Clark KJ, Makrides M, Neumann MA, Gibson RA (1992) Determination of the optimal ratio of linoleic acid to alpha-Iinolenic acid in infant formulas. J Pediatr 120(4 Pt 2):S151–S158

Cox SR, Lyall DM, Ritchie SJ, Bastin ME, Harris MA, Buchanan CR et al (2019) Associations between vascular risk factors and brain MRI indices in UK biobank. Eur Heart J 40(28): 2290–2299

Cristofano A, Sapere N, La Marca G, Angiolillo A, Vitale M, Corbi G et al (2016) Serum levels of acyl-carnitines along the continuum from normal to Alzheimer's dementia. PLoS One 11(5): 1–16

Crook D, Seed M (1990) Endocrine control of plasma lipoprotein metabolism: effects of gonadal steroids. Bailliere Clin Endocrinol Metab 4(4):851–875. [Internet]. 1990 [cited 2023 Oct 4] Available from: https://pubmed.ncbi.nlm.nih.gov/2082909/

Cui M, Jiang Y, Zhao Q, Zhu Z, Liang X, Zhang K et al (2020) Metabolomics and incident dementia in older Chinese adults: the Shanghai aging study. Alzheimers Dement 16(5):

779–788. [Internet]. 2020 May 1 [cited 2023 Nov 6]. Available from: https://pubmed.ncbi.nlm.nih.gov/32270572/

de Leeuw FA, Karamujić-Čomić H, Tijms BM, Peeters CFW, Kester MI, Scheltens P et al (2021) Circulating metabolites are associated with brain atrophy and white matter hyperintensities. Alzheimers Dement 17(2):205–214. [Internet]. 2021 Feb 1 [cited 2023 Sep 27]. Available from: https://onlinelibrary.wiley.com/doi/full/10.1002/alz.12180

Debette S, Schilling S, Duperron MG, Larsson SC, Markus HS (2019) Clinical significance of magnetic resonance imaging markers of vascular brain injury: a systematic review and meta-analysis. JAMA Neurol 76(1):81–94. [Internet]. 2019 Jan 1 [cited 2023 Jan 2]. Available from: https://pubmed.ncbi.nlm.nih.gov/30422209/

Deelen J, Kettunen J, Fischer K, van der Spek A, Trompet S, Kastenmüller G et al (2019) A metabolic profile of all-cause mortality risk identified in an observational study of 44,168 individuals. Nat Commun 10(1):1–8. [Internet]. 2019 Aug 20 [cited 2023 Sep 27]. Available from: https://www.nature.com/articles/s41467-019-11311-9

Dehouck B, Fenart L, Dehouck MP, Pierce A, Torpier G, Cecchelli R (1997) A new function for the LDL receptor: transcytosis of LDL across the blood-brain barrier. J Cell Biol 138(4):877–889. [Internet]. [cited 2023 Sep 29]. Available from: http://www.jcb.org

Deoni S, Dean D, Joelson S, O'Regan J, Schneider N (2018. Sep 1) Early nutrition influences developmental myelination and cognition in infants and young children. NeuroImage 178:649–659

Desale SE, Chinnathambi S (2020) Role of dietary fatty acids in microglial polarization in Alzheimer's disease. J Neuroinflammation 17(1) [Internet]. 2020 Mar 24 [cited 2023 Sep 28]. Available from: https://pubmed.ncbi.nlm.nih.gov/32209097/

Díaz M, Mesa-Herrera F, Marín R (2021) DHA and its elaborated modulation of antioxidant Defenses of the brain: implications in aging and AD neurodegeneration. Antioxidants 10(6):907. [Internet]. 2021 Jun 3 [cited 2023 Oct 5]. Available from: https://www.mdpi.com/2076-3921/10/6/907/htm

Dienel GA (2012a) Brain lactate metabolism: the discoveries and the controversies. J Cereb Blood Flow Metab 32(7):1107–1138. [Internet]. 2012 Jul [cited 2023 Sep 27]. Available from: https://pubmed.ncbi.nlm.nih.gov/22186669/

Dienel GA (2012b) Fueling and imaging brain activation. ASN Neuro 4(5):267–321

Dungan KM (2008) 1,5-anhydroglucitol (GlycoMark™) as a marker of short-term glycemic control and glycemic excursions. Expert Rev Mol Diagn 8(1):9–19

Dunn WB, Broadhurst DI, Atherton HJ, Goodacre R, Griffin JL (2010) Systems level studies of mammalian metabolomes: the roles of mass spectrometry and nuclear magnetic resonance spectroscopy. Chem Soc Rev 40(1):387–426. [Internet]. 2010 Dec 15 [cited 2023 Oct 5]. Available from: https://pubs.rsc.org/en/content/articlehtml/2011/cs/b906712b

Ebright B, Assante I, Poblete RA, Wang S, Duro MV, Bennett DA et al (2022) Eicosanoid lipidome activation in post-mortem brain tissues of individuals with APOE4 and Alzheimer's dementia. Alzheimers Res Ther 14(1):1–13. [Internet]. 2022 Dec 1 [cited 2023 Sep 27]. Available from: https://alzres.biomedcentral.com/articles/10.1186/s13195-022-01084-7

Edmond J (2001) Essential polyunsaturated fatty acids and the barrier to the brain: the components of a model for transport. J Mol Neurosci 16(2–3):181–193. [Internet]. [cited 2023 Sep 29]. Available from: https://link.springer.com/article/10.1385/JMN:16:2-3:181

Eelen G, de Zeeuw P, Treps L, Harjes U, Wong BW, Carmeliet P (2018) Endothelial cell metabolism. Physiol Rev 98(1):3–58

Fernstrom JD (2005) Branched-chain amino acids and brain function. J Nutr 135(6):1539S–1546S

Fraser T, Tayler H, Love S (2010) Fatty acid composition of frontal, temporal and parietal neocortex in the normal human brain and in Alzheimer's disease. Neurochem Res 35(3):503–513

Galea I (2021) The blood–brain barrier in systemic infection and inflammation. Cell Mol Immunol 18(11):2489–2501

Grundy SM, Cleeman JI, Daniels SR, Donato KA, Eckel RH, Franklin BA et al (2006) Diagnosis and management of the metabolic syndrome: An American Heart Association/National Heart, Lung, and Blood Institute scientific statement. Circulation 21(1):1–6

Hamilton JA, Hillard CJ, Spector AA, Watkins PA (2007) Brain uptake and utilization of fatty acids, lipids and lipoproteins: application to neurological disorders. J Mol Neurosci 33(1):2–11

Harayama T, Riezman H (2018) Understanding the diversity of membrane lipid composition. Nat Rev Mol Cell Biol 19(5):281–296. [Internet]. 2018 Feb 7 [cited 2022 Dec 29]. Available from: https://www.nature.com/articles/nrm.2017.138

Hardie RC, Muallem S (2009) Lipids in Ca2+ signalling-An introduction. Cell Calcium 45(6): 517–520

Harshfield EL, Markus HS (2023) Association of Baseline Metabolomic Profiles with Incident Stroke and Dementia and with imaging markers of cerebral small vessel disease. Neurology 101(5):E489–E501. [Internet] 2023 Aug 1 [cited 2023 Sep 27]. Available from: https://pubmed.ncbi.nlm.nih.gov/37290969/

Harshfield EL, Sands CJ, Tuladhar AM, De Leeuw FE, Lewis MR, Markus HS (2022) Metabolomic profiling in small vessel disease identifies multiple associations with disease severity. Brain 145(7):2461–2471. [Internet]. 2022 Jul 29 [cited 2023 Nov 3]. Available from: https://doi.org/10.1093/brain/awac041

Helbecque N, Berr C, Cottel D, Fromentin-David I, Sazdovitch V, Ricolfi F et al (2001) VLDL receptor polymorphism, cognitive impairment, and dementia. Neurology 56(9):1183–1188

Herz J (2003) LRP: a bright beacon at the blood-brain barrier. J Clin Invest 112(10):1483–1485

Holmes MV, Millwood IY, Kartsonaki C, Hill MR, Bennett DA, Boxall R et al (2018. Feb 13) Lipids, lipoproteins, and metabolites and risk of myocardial infarction and stroke. J Am Coll Cardiol 71(6):620–632

Hoogland ICM, Houbolt C, van Westerloo DJ, van Gool WA, van de Beek D (2015) Systemic inflammation and microglial activation: systematic review of animal experiments. J Neuroinflammation 12(1):1–13

Huo Z, Yu L, Yang J, Zhu Y, Bennett DA, Zhao J (2020. Feb 1) Brain and blood metabolome for Alzheimer's dementia: findings from a targeted metabolomics analysis. Neurobiol Aging 86: 123–133

Huynh K, Lim WLF, Giles C, Jayawardana KS, Salim A, Mellett NA et al (2020) Concordant peripheral lipidome signatures in two large clinical studies of Alzheimer's disease. Nat Commun 11(1):1–11. [Internet]. 2020 Nov 10 [cited 2023 Jan 2]. Available from: https://www.nature.com/articles/s41467-020-19473-7

Jensen NJ, Wodschow HZ, Nilsson M, Rungby J (2020) Effects of ketone bodies on brain metabolism and function in neurodegenerative diseases. Int J Mol Sci 21(22):1–17

Jones DP, Mason HS (1978) Metabolic hypoxia: accumulation of tyrosine metabolites in hepatocytes at low pO2. Biochem Biophys Res Commun 80(3):477–483

Jones LL, McDonald DA, Borum PR (2010) Acylcarnitines: role in brain. Prog Lipid Res 49(1): 61–75

Julia TCW, Qian L, Pipalia NH, Chao MJ, Liang SA, Shi Y et al (2022) Cholesterol and matrisome pathways dysregulated in astrocytes and microglia. Cell 185(13):2213–2233.e25. [Internet]. 2022 Jun 23 [cited 2023 Jan 9]. Available from: https://pubmed.ncbi.nlm.nih.gov/35750033/

Karjalainen JP, Mononen N, Hutri-Kähönen N, Lehtimäki M, Juonala M, Ala-Korpela M et al (2019) The effect of apolipoprotein E polymorphism on serum metabolome-a population-based 10-year follow-up study. 9(1) [cited 2023 Oct 4]; Available from: https://doi.org/10.1038/s41598-018-36450-9

Kellar D, Craft S (2020. Sep 1) Brain insulin resistance in Alzheimer's disease and related disorders: mechanisms and therapeutic approaches. Lancet Neurol 19(9):758–766

Kris-Etherton PM, Hecker KD, Binkoski AE (2004) Polyunsaturated fatty acids and cardiovascular health. Nutr Rev 62(11):414–426

Krishnadas R, McLean J, Batty DG, Burns H, Deans KA, Ford I et al (2013) Cardio-metabolic risk factors and cortical thickness in a neurologically healthy male population: results from the psychological, social and biological determinants of ill health (pSoBid) study. Neuroimage Clin 2(1):646–657

Kucharcyk W, Macdonald M, Stanisz GHRM (1994) Relaxivity at MR imaging: of cerebrosides and pH. Radiology:521–529

Lalić-Popović M, Vasović V, Milijašević B, Goločorbin-Kon S, Al-Salami H, Mikov M (2013) Deoxycholic acid as a modifier of the permeation of gliclazide through the blood brain barrier of a rat. J Diabetes Res 598603:1–9

Lancaster GI, Langley KG, Berglund NA, Kammoun HL, Reibe S, Estevez E et al (2018) Evidence that TLR4 is not a receptor for saturated fatty acids but mediates lipid-induced inflammation by reprogramming macrophage metabolism. Cell Metab 27(5):1096–1110.e5. [Internet]. 2018 May 1 [cited 2023 Sep 28]. Available from: http://www.cell.com/article/S155041311830192X/fulltext

Lehmann R (2021) From bedside to bench-practical considerations to avoid pre-analytical pitfalls and assess sample quality for high-resolution metabolomics and lipidomics analyses of body fluids. Anal Bioanal Chem. [cited 2023 Oct 4] Available from: https://doi.org/10.1007/s00216-021-03450-0

Li D, Misialek JR, Boerwinkle E, Gottesman RF, Sharrett AR, Mosley TH et al (2016) Plasma phospholipids and prevalence of mild cognitive impairment and/or dementia in the ARIC neurocognitive study (ARIC-NCS). Alzheimers Dement (Amst) 3:73–82. [Internet]. 2016 [cited 2023 Sep 29]. Available from: https://pubmed.ncbi.nlm.nih.gov/27408938/

Lian F, Hu KQ, Russell RM, Wang XD (2006) B-cryptoxanthin suppresses the growth of immortalized human bronchial epithelial cells and non-small-cell lung cancer cells and up-regulates retinoic acid receptor B expression. Int J Cancer 119(9):2084–2089

Lind MV, Savolainen OI, Ross AB (2016) The use of mass spectrometry for analysing metabolite biomarkers in epidemiology: methodological and statistical considerations for application to large numbers of biological samples. Eur J Epidemiol 31(8):717–733. [Internet]. 2016 Aug 1 [cited 2023 Oct 5]. Available from: https://pubmed.ncbi.nlm.nih.gov/27230258/

Lu R, Aziz NA, Diers K, Stöcker T, Reuter M, MMB B (2021) Insulin resistance accounts for metabolic syndrome-related alterations in brain structure. Hum Brain Mapp 42(8):2434–2444. [Internet]. 2021 Jun 1 [cited 2023 Jan 3]. Available from: https://onlinelibrary.wiley.com/doi/full/10.1002/hbm.25377

Luukkonen PK, Zhou Y, Nidhina Haridas PA, Dwivedi OP, Hyötyläinen T, Ali A et al (2017) Impaired hepatic lipid synthesis from polyunsaturated fatty acids in TM6SF2 E167K variant carriers with NAFLD. J Hepatol 67(1):128–136

Machado-Fragua MD, Landré B, Chen M, Fayosse A, Dugravot A, Kivimaki M et al (2022) Circulating serum metabolites as predictors of dementia: a machine learning approach in a 21-year follow-up of the Whitehall II cohort study. BMC Med 20(1):1–12. [Internet]. 2022 Dec 1 [cited 2023 Jan 3]. Available from: https://bmcmedicine.biomedcentral.com/articles/10.1186/s12916-022-02519-6

Mahmoudian Dehkordi S, Arnold M, Nho K, Ahmad S, Jia W, Xie G et al (2019) Altered bile acid profile associates with cognitive impairment in Alzheimer's disease-An emerging role for gut microbiome. Alzheimers Dement 15(1):76–92. [Internet]. 2019 Jan 1 [cited 2023 Nov 3]. Available from: https://pubmed.ncbi.nlm.nih.gov/30337151/

Mapstone M, Cheema AK, Fiandaca MS, Zhong X, Mhyre TR, Macarthur LH et al (2014) Plasma phospholipids identify antecedent memory impairment in older adults. Nat Med 20(4):415–418. [Internet]. 2014 Mar 9 [cited 2023 Sep 27]. Available from: https://www.nature.com/articles/nm.3466

Marais AD (2019) Apolipoprotein E in lipoprotein metabolism, health and cardiovascular disease. Pathology 51(2):165–176

Marathe GK, Pandit C, Lakshmikanth CL, Chaithra VH, Jacob SP, D'Souza CJM (2014) To hydrolyze or not to hydrolyze: the dilemma of platelet-activating factor acetylhydrolase. J Lipid Res 55(9):1847–1854. [Internet]. 2014 Sep 1 [cited 2023 Jan 2]. Available from: http://www.jlr.org/article/S0022227520356601/fulltext

Marx W, Thomson S, O'Hely M, Symeonides C, Collier F, Tang MLK et al (2022) Maternal inflammatory and omega-3 fatty acid pathways mediate the association between socioeconomic disadvantage and childhood cognition. Brain Behav Immun 100:211–218. [Internet]. 2022 Feb 1 [cited 2023 Sep 26]. Available from: https://pubmed.ncbi.nlm.nih.gov/34896180/

Mattson MP, Moehl K, Ghena N, Schmaedick M, Cheng A (2018) Intermittent metabolic switching, neuroplasticity and brain health. Nat Rev Neurosci 19(2):81–94. [Internet]. 2018 Feb 1 [cited 2023 Sep 27]. Available from: https://pubmed.ncbi.nlm.nih.gov/29321682/

Mauch DH, Nägier K, Schumacher S, Göritz C, Müller EC, Otto A et al (2001) CNS synaptogenesis promoted by glia-derived cholesterol. Science (1979) 294(5545):1354–1357

McCullough ML, Maliniak ML, Stevens VL, Carter BD, Hodge RA, Wang Y (2019) Metabolomic markers of healthy dietary patterns in US postmenopausal women. Am J Clin Nutr 109(5): 1439–1451

Menni C, Graham D, Kastenmüller G, Alharbi NHJ, Alsanosi SM, Mcbride M et al (2015) Metabolomic identification of a novel pathway of blood pressure regulation involving hexadecanedioate. Hypertension 66(2):422–429

Mergenthaler P, Lindauer U, Dienel GA, Meisel A (2013) Sugar for the brain: the role of glucose in physiological and pathological brain function. Trends Neurosci 36(10):587. [Internet]. 2013 Oct [cited 2023 Oct 5]. Available from: /pmc/articles/PMC3900881/

Mertens KL, Kalsbeek A, Soeters MR, Eggink HM (2017) Bile acid signaling pathways from the enterohepatic circulation to the central nervous system. Front Neurosci 11:617

Mielke MM, Haughey NJ, Bandaru VVR, Weinberg DD, Darby E, Zaidi N et al (2011) Plasma sphingomyelins are associated with cognitive progression in Alzheimer's disease. J Alzheimers Dis 27(2):259–269. [Internet]. 2011 [cited 2023 Sep 29]. Available from: https://pubmed.ncbi.nlm.nih.gov/21841258/

Millington DS, Stevens RD (2011) Acylcarnitines: analysis in plasma and whole blood using tandem mass spectrometry. Methods Mol Biol 708:55–72

Monteiro-Cardoso VF, Corlianò M, Singaraja RR (2021) Bile acids: a Communication Channel in the gut-brain Axis. NeuroMolecular Med 23(1):99–117

Nakamura MT, Cheon Y, Li Y, Nara TY (2004) Mechanisms of regulation of gene expression by fatty acids. Lipids 39(11):1077–1083. [Internet]. 2004 Nov 1 [cited 2023 Sep 28]. Available from: https://onlinelibrary.wiley.com/doi/full/10.1007/s11745-004-1333-0

Neuffer J, González-Domínguez R, Lefèvre-Arbogast S, Low DY, Driollet B, Helmer C et al (2022) Exploration of the gut-brain Axis through metabolomics identifies serum propionic acid associated with higher cognitive decline in older persons. Nutrients 14(21) [Internet]. 2022 Nov 1 [cited 2023 Sep 27]. Available from: https://pubmed.ncbi.nlm.nih.gov/36364950/

Nho K, Kueider-Paisley A, MahmoudianDehkordi S, Arnold M, Risacher SL, Louie G et al (2019) Altered bile acid profile in mild cognitive impairment and Alzheimer's disease: relationship to neuroimaging and CSF biomarkers. Alzheimers Dement 15(2):232–244. [Internet]. 2019 Feb 1 [cited 2023 Nov 3]. Available from: https://onlinelibrary.wiley.com/doi/full/10.1016/j.jalz.2018.08.012

Nishitsuji K, Hosono T, Nakamura T, Bu G, Michikawa M (2011. May 20) Apolipoprotein E regulates the integrity of tight junctions in an isoform-dependent manner in an in vitro blood-brain barrier model. J Biol Chem 286(20):17536–17542

Oliveira TG, Di Paolo G (2010) Phospholipase D in brain function and Alzheimer's disease. Biochim Biophys Acta 1801(8):799–805. [Internet]. 2010 Aug [cited 2023 Sep 27]. Available from: https://pubmed.ncbi.nlm.nih.gov/20399893/

Ouzzine M, Gulberti S, Ramalanjaona N, Magdalou J, Fournel-Gigleux S (2014) The UDP-glucuronosyltransferases of the blood-brain barrier: their role in drug metabolism and detoxication. Front Cell Neurosci 8(October):1–12

Palacios N, Lee JS, Scott T, Kelly RS, Bhupathiraju SN, Bigornia SJ et al (2020) Circulating plasma metabolites and cognitive function in a Puerto Rican cohort. J Alzheimers Dis 76(4):1267–1280. [Internet]. 2020 [cited 2023 Nov 6]. Available from: https://pubmed.ncbi.nlm.nih.gov/32716356/

Pan X, Fei G, Lu J, Jin L, Pan S, Chen Z et al (2016) Measurement of blood thiamine metabolites for Alzheimer's disease diagnosis. EBioMedicine 3:155–162

Panov A, Orynbayeva Z, Vavilin V, Lyakhovich V (2014) Fatty acids in energy metabolism of the central nervous system. Biomed Res Int 2014:472459

Pattison DJ, Symmons DPM, Lunt M, Welch A, Bingham SA, Day NE et al (2005) Dietary β-cryptoxanthin and inflammatory polyarthritis: results from a population-based prospective study. Am J Clin Nutr 82(2):451–455

Pedersen HK, Gudmundsdottir V, Nielsen HB, Hyotylainen T, Nielsen T, Jensen BAH et al (2016) Human gut microbes impact host serum metabolome and insulin sensitivity. Nature 535(7612): 376–381

Pesaresi M, Soon-Shiong R, French L, Kaplan DR, Miller FD, Paus T (2015) Axon diameter and axonal transport: in vivo and in vitro effects of androgens. NeuroImage 115:191–201

Peters SG, Pomare EW, Fisher CA (1992) Portal and peripheral blood short chain fatty acid concentrations after caecal lactulose instillation at surgery. Gut 33(9):1249–1252. [Internet]. [cited 2023 Sep 27]. Available from: https://pubmed.ncbi.nlm.nih.gov/1427380/

Poitelon Y, Kopec AM, Belin S (2020) Myelin fat facts: An overview of lipids and fatty acid metabolism. Cell 9(4) [Internet]. 2020 Mar 27 [cited 2023 Sep 27]. Available from: https://pubmed.ncbi.nlm.nih.gov/32230947/

Prince N, Chu SH, Chen Y, Mendez KM, Hanson E, Green-Snyder LA et al (2023) Phenotypically driven subgroups of ASD display distinct metabolomic profiles. Brain Behav Immun 111:21–29. [Internet]. 2023 Jul 1 [cited 2023 Oct 2]. Available from: https://pubmed.ncbi.nlm.nih.gov/37004757/

Psychogios N, Hau DD, Peng J, Guo AC, Mandal R, Bouatra S et al (2011) The human serum metabolome. PLoS One 6(2):e16957

Qi Y, Jiang C, Cheng J, Krausz KW, Li T, Ferrell JM et al (2015) Bile acid signaling in lipid metabolism: metabolomic and lipidomic analysis of lipid and bile acid markers linked to anti-obesity and anti-diabetes in mice. Biochim Biophys Acta Mol Cell Biol Lipids 1851(1):19–29

Quehenberger O, Dennis EA (2011) The human plasma lipidome. N Engl J Med 365(19): 1812–1823

Quinn M, McMillin M, Galindo C, Frampton G, Pae HY, DeMorrow S (2014) Bile acids permeabilize the blood brain barrier after bile duct ligation in rats via Rac1-dependent mechanisms. Dig Liver Dis 46(6):527–534

Raheem J, Sliz E, Shin J, Holmes MV, Gaudet D, Paus T et al (2022) Visceral adiposity is associated with metabolic profiles predictive of type 2 diabetes and myocardial infarction. Commun Med 2(81):1–7

Rapoport SI, Rao JS, Igarashi M (2007) Brain metabolism of nutritionally essential polyunsaturated fatty acids depends on both the diet and the liver. Prostaglandins Leukot Essent Fatty Acids 77(5–6):251–261

Ridone P, Grage SL, Patkunarajah A, Battle AR, Ulrich AS, Martinac B (2018) "Force-from-lipids" gating of mechanosensitive channels modulated by PUFAs. J Mech Behav Biomed Mater 79 (December 2017):158–167

Rom S, Heldt NA, Gajghate S, Seliga A, Reichenbach NL, Persidsky Y (2020) Hyperglycemia and advanced glycation end products disrupt BBB and promote occludin and claudin-5 protein secretion on extracellular microvesicles. Sci Rep 10(1):1–14. [Internet]. 2020 Apr 29 [cited 2023 Jan 3]. Available from: https://www.nature.com/articles/s41598-020-64349-x

Sanchez-Mejia RO, Newman JW, Toh S, Yu GQ, Zhou Y, Halabisky B et al (2008) Phospholipase A2 reduction ameliorates cognitive deficits in a mouse model of Alzheimer's disease. Nat Neurosci 11(11):1311–1318. [Internet], 2008 Oct 19 [cited 2023 Sep 27]. Available from: https://www.nature.com/articles/nn.2213

Sas KM, Karnovsky A, Michailidis G, Pennathur S (2015) Metabolomics and diabetes: analytical and computational approaches. Diabetes 64(3):718–732. [Internet]. 2015 Mar 1 [cited 2023 Oct 5] Available from: https://pubmed.ncbi.nlm.nih.gov/25713200/

Schaefer EJ, Bongard V, Beiser AS, Lamon-Fava S, Robins SJ, Au R et al (2006) Plasma phosphatidylcholine docosahexaenoic acid content and risk of dementia and Alzheimer disease: the Framingham heart study. Arch Neurol 63(11):1545–1550. [Internet]. Nov [cited 2023 Sep 29]. Available from: https://pubmed.ncbi.nlm.nih.gov/17101822/

Schmitt S, Cantuti Castelvetri L, Simons M (2015. Aug 1) Metabolism and functions of lipids in myelin. Biochim Biophys Acta (BBA) - Mol Cell Biol Lipids 1851(8):999–1005

Schwarcz R (2016) Kynurenines and glutamate: multiple links and therapeutic implications, Advances in pharmacology, vol 76, 1st edn. Elsevier Inc., Burlington, pp 13–37

Schwarz NF, Nordstrom LK, Pagen LHG, Palombo DJ, Salat DH, Milberg WP et al (2018) Differential associations of metabolic risk factors on cortical thickness in metabolic syndrome. Neuroimage Clin 17(April 2017):98–108

Sebastiani P, Song Z, Ellis D, Tian Q, Schwaiger-Haber M, Stancliffe E et al (2023) A metabolomic signature of the APOE2 allele. Geroscience 45(1):415–426. [Internet]. 2023 Feb 1 [cited 2023 Oct 4] Available from: https://link.springer.com/article/10.1007/s11357-022-00646-9

Sevastou I, Kaffe E, Mouratis MA, Aidinis V (2013. Jan 1) Lysoglycerophospholipids in chronic inflammatory disorders: the PLA2/LPC and ATX/LPA axes. Biochim Biophys Acta (BBA) - Mol Cell Biol Lipids 1831(1):42–60

Shi Z, Ren H, Luo C, Yao X, Li P, He C et al (2016. Nov 1) Enriched endogenous Omega-3 polyunsaturated fatty acids protect cortical neurons from experimental ischemic injury. Mol Neurobiol 53(9):6482–6488

Shin J, Pelletier S, Richer L, Pike GB, Gaudet D, Paus T et al (2020) Adiposity-related insulin resistance and thickness of the cerebral cortex in middle-aged adults. J Neuroendocrinol 32(12): e12921. [Internet]. 2020 Dec 1 [cited 2023 Jan 3]. Available from: https://onlinelibrary.wiley.com/doi/full/10.1111/jne.12921

Shin J, Patel Y, Parker N, Paus T, Pausova Z (2023) Prediabetic HbA1c and cortical atrophy: underlying neurobiology. Diabetes Care. [Internet]. 2023 Oct 12 [cited 2023 Nov 2]; Available from: https://doi.org/10.2337/dc23-1105

Sienski G, Narayan P, Bonner JM, Kory N, Boland S, Arczewska AA et al (2021) APOE4 disrupts intracellular lipid homeostasis in human iPSC-derived glia. Sci Transl Med 13(583) [Internet]. 2021 Mar 3 [cited 2023 Jan 9]. Available from: https://pubmed.ncbi.nlm.nih.gov/33658354/

Silva YP, Bernardi A, Frozza RL (2020) The role of short-chain fatty acids from gut microbiota in gut-brain communication. Front Endocrinol (Lausanne) 11:25. Internet]. 2020 Jan 31 [cited 2023 Sep 27]. Available from: /pmc/articles/PMC7005631/

Sims R, Van Der Lee SJ, Naj AC, Bellenguez C, Badarinarayan N, Jakobsdottir J et al (2017) Rare coding variants in PLCG2, ABI3, and TREM2 implicate microglial-mediated innate immunity in Alzheimer's disease. Nat Genet 49(9):1373–1384. [Internet]. 2017 Jul 17 [cited 2023 Sep 27]. Available from: https://www.nature.com/articles/ng.3916

Sliz E, Shin J, Syme C, Black S, Seshadri S, Paus T et al (2020) Thickness of the cerebral cortex shows positive association with blood levels of triacylglycerols carrying 18-carbon fatty acids. Commun Biol 3(1) [Internet]. 2020 Dec 1 [cited 2023 Jan 14]. Available from: https://pubmed.ncbi.nlm.nih.gov/32820227/

Sliz E, Shin J, Syme C, Patel Y, Parker N, Richer L et al (2021) A variant near DHCR24 associates with microstructural properties of white matter and peripheral lipid metabolism in adolescents. Mol Psychiatry 26(8):3795–3805. [Internet]. 2021 Aug 1 [cited 2023 Nov 6]. Available from: https://pubmed.ncbi.nlm.nih.gov/31900429/

Sliz E, Shin J, Ahmad S, Williams DM, Frenzel S, Gauß F et al (2022) Circulating metabolome and White matter hyperintensities in women and men. Circulation 145(14):1040–1052. [Internet]. 2019 Jan 1 [cited 2023 Jan 2]. Available from: https://www.ahajournals.org/doi/abs/10.1161/CIRCULATIONAHA.121.056892

Soininen P, Kangas AJ, Würtz P, Suna T, Ala-Korpela M (2015) Quantitative serum nuclear magnetic resonance metabolomics in cardiovascular epidemiology and genetics. Circ Cardiovasc Genet 8(1):192–206. [Internet]. 2015 Feb 4 [cited 2023 Oct 5]. Available from: https://pubmed.ncbi.nlm.nih.gov/25691689/

Stanhewicz AE, Wenner MM, Stachenfeld NS (2018) Sex differences in endothelial function important to vascular health and overall cardiovascular disease risk across the lifespan. Am J Physiol Heart Circ Physiol 315(6):H1569–H1588

Sun D, Tiedt S, Yu B, Jian X, Gottesman RF, Mosley TH et al (2019) A prospective study of serum metabolites and risk of ischemic stroke. Neurology 92(16):E1890–E1898. [Internet]. 2019 Apr 16 [cited 2023 Nov 6]. Available from: https://pubmed.ncbi.nlm.nih.gov/30867269/

Sun Y, Guo Y, Li HQ, Tan L, Feng JF, Cheng W et al (2023) Associations of circulating metabolites with cerebral white matter hyperintensities. J Neurochem 166(2):414–423. [Internet]. 2023 Jul 1 [cited 2023 Sep 27]. Available from: https://pubmed.ncbi.nlm.nih.gov/37222503/

Sundaram JR, Chan ES, Poore CP, Pareek TK, Cheong WF, Shui G et al (2012) Cdk5/p25-induced cytosolic PLA2-mediated lysophosphatidylcholine production regulates neuroinflammation and triggers neurodegeneration. J Neurosci 32(3):1020–1034. [Internet]. 2012 Jan 18 [cited 2023 Sep 27]. Available from: https://pubmed.ncbi.nlm.nih.gov/22262900/

Syme C, Czajkowski S, Shin J, Abrahamowicz M, Leonard G, Perron M et al (2016) Glycerophosphocholine metabolites and cardiovascular disease risk factors in adolescents: a cohort study. Circulation 134(21):1629–1636. [Internet]. 2016 Nov 22 [cited 2023 Sep 27]. Available from: https://www.ahajournals.org/doi/abs/10.1161/circulationaha.116.022993

Syme C, Pelletier S, Shin J, Abrahamowicz M, Leonard G, Perron M et al (2019) Visceral fat-related systemic inflammation and the adolescent brain: a mediating role of circulating glycerophosphocholines. Int J Obes 43(6):1223–1230. [Internet]. 2019 Jun 1 [cited 2023 Sep 27]. Available from: https://pubmed.ncbi.nlm.nih.gov/30206338/

Tan ZS, Harris WS, Beiser AS, Au R, Himali JJ, Debette S et al (2012) Red blood cell ω-3 fatty acid levels and markers of accelerated brain aging. Neurology 78(9):658–664. [Internet]. 2012 Feb 28 [cited 2023 Sep 29]. Available from: https://pubmed.ncbi.nlm.nih.gov/22371413/

Thompson DK, Sloane R, Bain JR, Stevens RD, Newgard CB, Pieper CF et al (2012) Daily variation of serum Acylcarnitines and amino acids. Metabolomics 8(4):556–565

Tian Q, Mitchell BA, Erus G, Davatzikos C, Moaddel R, Resnick SM et al (2023. Sep 1) Sex differences in plasma lipid profiles of accelerated brain aging. Neurobiol Aging 129:178–184

Toledo JB, Arnold M, Kastenmüller G, Chang R, Baillie RA, Han X et al (2017) Metabolic network failures in Alzheimer's disease: a biochemical road map. Alzheimers Dement 13(9):965–984. [Internet]. 2017 Sep 1 [cited 2023 Jan 2]. Available from: https://onlinelibrary.wiley.com/doi/full/10.1016/j.jalz.2017.01.020

Tortosa-Caparrós E, Navas-Carrillo D, Marín F, Orenes-Piñero E (2017) Anti-inflammatory effects of omega 3 and omega 6 polyunsaturated fatty acids in cardiovascular disease and metabolic syndrome. Crit Rev Food Sci Nutr 57(16):3421–3429

Tynkkynen J, Chouraki V, van der Lee SJ, Hernesniemi J, Yang Q, Li S et al (2018) Association of branched-chain amino acids and other circulating metabolites with risk of incident dementia and Alzheimer's disease: a prospective study in eight cohorts. Alzheimers Dementx 14(6):723–733. [Internet]. 2018 Jun 1 [cited 2023 Oct 4]. Available from: https://pubmed.ncbi.nlm.nih.gov/29519576/

van der Lee SJ, Teunissen CE, Pool R, Shipley MJ, Teumer A, Chouraki V et al (2018) Circulating metabolites and general cognitive ability and dementia: evidence from 11 cohort studies. Alzheimers Dement 14(6):707–722. [Internet]. 2018 Jun 1 [cited 2023 Oct 4]. Available from: https://pubmed.ncbi.nlm.nih.gov/29316447/

Van Meer G, Voelker DR, Feigenson GW (2008) Membrane lipids: where they are and how they behave. Nat Rev Mol Cell Biol 9(2):112–124

Victor MB, Leary N, Luna X, Meharena HS, Scannail AN, Bozzelli PL et al (2022. Aug 4) Lipid accumulation induced by APOE4 impairs microglial surveillance of neuronal-network activity. Cell Stem Cell 29(8):1197–1212.e8

Vojinovic D, Kalaoja M, Trompet S, Fischer K, Shipley MJ, Li S et al (2020) Association of circulating metabolites in plasma or serum and risk of stroke: meta-analysis from seven

prospective cohorts. Neurology 96(8):E1110–E1123. [Internet]. 2020 Feb 23 [cited 2023 Nov 6]. Available from: https://pubmed.ncbi.nlm.nih.gov/33268560/
Wanders RJA, Komen J, Kemp S (2011) Fatty acid omega-oxidation as a rescue pathway for fatty acid oxidation disorders in humans. FEBS J 278(2):182–194
Wang J, Wei R, Xie G, Arnold M, Kueider-Paisley A, Louie G et al (2020) Peripheral serum metabolomic profiles inform central cognitive impairment. Sci Rep 10(1) [Internet]. 2020 Dec 1 [cited 2023 Nov 6]. Available from: https://pubmed.ncbi.nlm.nih.gov/32820198/
Wang RZ, Huang SY, Li HQ, Yang YX, Chen SD, Yu JT (2022) Genetic determinants of circulating metabolites and the risk of stroke and its subtypes. Eur J Neurol 29(12): 3711–3719. [Internet]. 2022 Dec 1 [cited 2023 Sep 27]. Available from: https://pubmed.ncbi.nlm.nih.gov/36086915/
Wardlaw JM, Allerhand M, Doubal FN, Hernandez MV, Morris Z, Gow AJ et al (2014) Vascular risk factors, large-artery atheroma, and brain white matter hyperintensities. Neurology 82(15): 1331–1338
Wawrzkiewicz-Jałowiecka A, Lalik A, Soveral G (2021) Recent update on the molecular mechanisms of gonadal steroids action in adipose tissue. Int J Mol Sci 22(10) [Internet]. 2021 May 2 [cited 2023 Oct 4]. Available from: https://pubmed.ncbi.nlm.nih.gov/34069293/
White CJ, Lee J, Choi J, Chu T, Scafidi S, Wolfgang MJ (2020) Determining the bioenergetic capacity for fatty acid oxidation in the mammalian nervous system. Mol Cell Biol 40-(10) [Internet]. 2020 Apr 28 [cited 2023 Sep 27]. Available from: https://www.tandfonline.com/doi/abs/10.1128/MCB.00037-20
Wishart DS, Feunang YD, Marcu A, Guo AC, Liang K, Vázquez-Fresno R et al (2018) HMDB 4.0: the human metabolome database for 2018. Nucleic Acids Res 46(D1):D608–D617
Woo A, Botta A, SSW S, Paus T, Pausova Z (2022) Obesity-related neuroinflammation: magnetic resonance and microscopy imaging of the brain. Int J Mol Sci 23(15):8790. [Internet]. 2022 Aug 8 [cited 2023 Jan 9]. Available from: https://www.mdpi.com/1422-0067/23/15/8790/htm
Würtz P, Wang Q, Kangas AJ, Richmond RC, Skarp J, Tiainen M et al (2014) Metabolic signatures of adiposity in young adults: mendelian randomization analysis and effects of weight change. PLoS Med 11(12):e1001765
Würtz P, Havulinna AS, Soininen P, Tynkkynen T, Prieto-Merino D, Tillin T et al (2015) Metabolite profiling and cardiovascular event risk: a prospective study of 3 population-based cohorts. Circulation 131(9):774–785. [Internet]. 2015 [cited 2023 Sep 27]. Available from: https://www.ahajournals.org/doi/abs/10.1161/CIRCULATIONAHA.114.013116
Xu H, Valenzuela N, Fai S, Figeys D, Bennett SAL (2013) Targeted lipidomics—advances in profiling lysophosphocholine and platelet-activating factor second messengers. FEBS J 280(22):5652–5667. [Internet]. 2013 Nov 1 [cited 2022 Dec 30]. Available from: https://onlinelibrary.wiley.com/doi/full/10.1111/febs.12423
Yanguas-Casás N, Barreda-Manso MA, Pérez-Rial S, Nieto-Sampedro M, Romero-Ramírez L (2017) TGFβ contributes to the anti-inflammatory effects of Tauroursodeoxycholic acid on an animal model of acute neuroinflammation. Mol Neurobiol 54(9):6737–6749
Zeng J, Chen L, Wang Z, Chen Q, Fan Z, Jiang H et al (2017) Marginal vitamin a deficiency facilitates Alzheimer's pathogenesis. Acta Neuropathol 133(6):967–982
Zhang X, Hu W, Wang Y, Wang W, Liao H, Zhang X et al (2022) Plasma metabolomic profiles of dementia: a prospective study of 110,655 participants in the UK Biobank. BMC Med 20(1): 1–10. [Internet]. 2022 Dec 1 [cited 2023 Jan 3]. Available from: https://bmcmedicine.biomedcentral.com/articles/10.1186/s12916-022-02449-3
Zhao Y, Li D, Zhao J, Song J, Zhao Y (2016) The role of the low-density lipoprotein receptor-related protein 1 (LRP-1) in regulating blood-brain barrier integrity. Rev Neurosci 27(6): 623–634

Part IV
Environment

Integrating the Physical Environment Within a Population Neuroscience Perspective

Lindsey Smith

Contents

Abstract Population neuroscience recognises the role of the environment in shaping brain, behaviour, and mental health. An overview of current evidence from neuroscientific and epidemiological studies highlights the protective effects of nature on cognitive function and stress reduction, the detrimental effects of urban living on mental health, and emerging concerns relating to extreme weather events and eco-anxiety. Despite the growing body of evidence in this area, knowledge gaps remain due to inconsistent measures of exposure and a reliance on small samples. In this chapter, attention is given to the physical environment and population-level studies as a necessary starting point for exploring the long-term impacts of environmental exposures on mental health, and for informing future research that may capture immediate emotional and neural responses to the environment. Key data sources, including remote sensing imagery, administrative, sensor, and social media data, are outlined. Appropriate measures of exposure are advocated for, recognising

L. Smith (✉)
Department of Geography and Planning, University of Toronto, Toronto, ON, Canada
e-mail: lindsey.smith@utoronto.ca

Curr Topics Behav Neurosci (2024) 68: 223–238
https://doi.org/10.1007/7854_2024_477
Published Online: 2 May 2024

the value of area-level measures for estimating exposure over large study samples and spatial and temporal scales. Although integrating data from multiple sources requires consideration for data quality and completeness, deep learning and the increasing availability of high-resolution data present opportunities to build a more complete picture of physical environments. Advances in leveraging detailed locational data are discussed as a subsequent approach for building upon initial observations from population studies and improving understanding of the mechanisms underlying behaviour and human–environment interactions.

Keywords Exposure assessment · GIS · Neuroscience · Physical environment · Population studies

1 Introduction

Population science focuses on the study of human populations including behaviours, health outcomes, and the factors that influence them. A population perspective within neuroscience recognises the complexity of multiple and interacting factors on the development and functioning of the brain (Paus 2010). These factors can be broadly categorised into the *genome* and the *environment*, with the latter concerned with macrolevel structures, such as physical and social environments.

Broad environmental-level factors have been recognised in socioecological models as contributing determinants of behaviour and health inequality (Preiser 1983; Whitehead and Dahlgren 1991; Barton and Grant 2006). Though such models have been critiqued and revised to address limitations in scope and applicability in different contexts (Dahlgren and Whitehead 2021; Dyar et al. 2022), environmental factors are consistently acknowledged as having important high-level influences on populations. Frameworks that give prominence to the environment identify key intersecting categories including sociocultural, economic, political, and physical environments.

In this chapter, attention is given specifically to the physical environment and its influence on the brain and mental health. Spatial and temporal dimensions of the physical environment are used to distinguish between different types of relationships and respective methodologies for studying them. For example, area-level measures such as neighbourhood boundaries provide an appropriate method for measuring exposures and outcomes at the population level. These provide value when exploring the impacts of long-term exposures on mental health outcomes and form the focus of the chapter. The use of high-frequency geo-located data is proposed as a follow-up to population studies to enable a more precise identification of environments individuals engage with, and to explore more immediate responses that short-term environmental interactions may elicit. Subsequent chapters of this book focus on urban and social environments.

2 Mapping Connections Between the Physical Environment, Human Brain, and Behaviour: A Population Perspective

The physical environment encompasses physical surroundings such as air, water, vegetation, climate conditions, and features of urban form including transport infrastructure, housing, and urban parks. Both natural and built elements may vary in quality through, for example, the presence of substances or absence of services. Elements of the physical environment are typically broad in scale, both in terms of spatial area and the number of people they impact. These large-scale influences and impacts hold relevance in population science where key goals are to understand the extent to which observations may be applicable to population groups, and to build evidence to inform interventions that can have impact at scale.

Figure 1 illustrates potential ways relationships between the physical environment, human brains, and behaviours exist. Differences in spatial and temporal dimensions of the physical environment, human brain, and behaviour contribute to different types of relationships and warrant relevant methodologies for studying them. For example, long-term exposures are important for mental health and may be targeted through area- or population-level interventions. Examples are provided to demonstrate connections and do not represent a comprehensive list of environmental features, outcomes, or moderating factors.

Humans interact with their physical environment using multiple sensory organs, from which brains generate conscious and unconscious emotional and behavioural responses (Edelstein and Macagno 2012). Exposure to environments may elicit short-term responses such as joy or fear, while longer term exposures may shape regular behaviours, mental health, and life experience. As examples, interactions with natural environments can trigger positive emotions, urban design shapes decisions around how places are navigated, and neighbourhood facilities may provide health services, education, and employment opportunities. Relationships between the brain and the physical environment do not therefore conform to simplistic linear models of cause and effect; rather interactions with the physical environment provide a background to events through the life course, and environmental influences are often experienced via behavioural mechanisms.

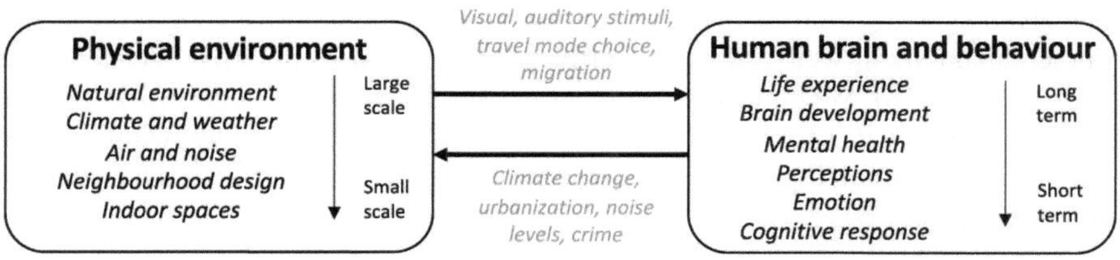

Fig. 1 Conceptual map of connections between the physical environment, brain, and behaviour

Additionally, humans have the capacity to shape their environments through large-scale processes such as urbanisation and agriculture, and through behaviours or services that may change more localised spaces such as contributions to noise levels, waste collection, or urban design. Recognising the interconnection between built and social environments, facilitated by examples such as service provision and the presence of cultural amenities, is essential for developing a comprehensive understanding of the physical environment's role as a determinant of health and behaviour outcomes. The ability to adapt and modify elements of the physical environment has led to it being consistently picked up in public health research and policy as a target for increasing healthy, habitual behaviours at the population level (Marteau et al. 2012; Giles-Corti et al. 2016; Wilkie et al. 2018). For example, social and health scientists have long been interested in the relationship between built environments and behaviours manifested in choices of diet, travel, and physical activity (Charreire et al. 2014; van de Coevering et al. 2015; Pontin et al. 2022). Studies are typically conducted in large, representative samples with a view to generalising to broader, theoretically relevant populations (for definitions of sample, source, and target populations, see the chapter on "Population neuroscience: understanding concepts of generalizability and transportability and their application to improving the public's health" in this book). Population strategies for behaviour change are considered to require low effort on behalf of individuals and address systemic risk factors.

3 Physical Environments and Neuroscience Research: An Overview of Current Evidence

Neuroscientific and epidemiological studies have explored relationships with physical environments of variable scale including natural environments, urban spaces, and indoor environments.

A growing body of evidence recognises the psychological benefits of exposure to natural environments, and green and blue spaces (Jimenez et al. 2021; Bolouki 2023). Observational studies measure exposure through frequency of use of green spaces, proximity to or time spent near natural features including private gardens and public parks, and levels of greenness in residential neighbourhoods. Positive associations have been found for self-reported mental health outcomes including lower perceived stress, lower risk of depressive symptoms, and improved cognitive development and overall well-being (McCormick 2017; Sarkar et al. 2018; Vanaken and Danckaerts 2018). While many observational studies are cross-sectional in design, findings have been shown for population groups of different ages with benefits perceived differentially based on socioeconomic, neighbourhood, and identity characteristics.

Emerging longitudinal evidence further suggests that cumulative contact with nature may have a protective impact on mental health through the life course

(Jimenez et al. 2021). While typically carried out with smaller, less representative samples, experimental studies measure short-term contact with nature through in situ engagement with the environment, such as walking through a designated space, or exposure to videos and images of nature in a laboratory setting. Using varying techniques to measure outcomes, including blood pressure, self-report, and neuroimaging such as electroencephalography (EEG) and functional magnetic resonance imaging (fMRI), such laboratory-based studies find positive associations with reduced stress, feelings of relaxation, happiness, and improved attention (Song et al. 2016; Kondo et al. 2018; Bolouki 2023). Even short-term, simulated, or distal exposures, such as views of nature through a window, may have beneficial influences.

Two key theories provide insight into the mechanisms through which natural environments affect the human brain: attention restoration theory (ART) and stress reduction theory (SRT). The former hypothesises that spending time in nature reduces mental fatigue experienced from modern lifestyles and helps restore directed attention (Kaplan 1995). The latter posits that humans' innate connection to nature elicits positive emotions when interacting with natural environments due to the activation of the parasympathetic nervous system to reduce stress and autonomic arousal (Ulrich et al. 1991). Additionally, natural spaces provide inclusive spaces for social interaction, physical activity, and play and present opportunities for different aspects of brain development including discovery, creativity, and risk taking (Kahn and Kellert 2002). Taken together, these theories and study findings suggest implementing policies aimed at enhancing access to green and natural spaces could yield sustainable population health benefits.

While exposure to natural environments has largely been shown to have positive associations with cognitive functioning and mental health, climate change poses challenges for these relationships. Climate change, defined here as human-induced changes to the atmosphere and unsustainable depletion of natural systems, has wide-reaching implications for populations. For example, experiencing extreme weather events, such as high ambient temperatures, has been associated with emotional distress and an increased risk of suicide (Fernandez et al. 2015; Thompson et al. 2018). An increased awareness of the threat of the climate crisis may also create eco-anxiety (Coffey et al. 2021).

Contributing factors to climate change, such as urbanisation, create challenges for public health with the global proportion of urban dwellers projected to reach 70% by 2050 (United Nations Department of Economic and Social Affairs 2018). Access to resources such as transport infrastructure, food retail, and healthcare facilities in urban settings has been studied at scale (Higgs 2004; Badland and Schofield 2005; Lytle and Sokol 2017). Although methods to assess access and exposure in observational studies are heterogeneous and have contributed to mixed findings, associations have been shown for outcomes such as physical activity and diet that, altogether, have important implications for mental health and brain development (Dauncey 2009; Stillman and Erickson 2018). Additionally, a higher risk of exposure to noise and air pollution with urban living has been associated with higher stress, often through sleep disturbance, and cognitive decline (Clifford et al. 2016).

Experimental studies utilising neuroimaging techniques for both laboratory and real-life settings suggest exposure to urban built environments triggers fear and stress-related responses (Lederbogen et al. 2011; Costa e Silva and Steffen 2019; Ancora et al. 2022). This may limit capacity for directed attention and recovery, as suggested by the ART concept. At the smaller scale of features that characterise the built environment, factors such as indoor air quality, lighting, and noise have been related to cognitive performance and heart rate variability as demonstrated through daily performance tests (Edelstein and Macagno 2012; Allen et al. 2016).

Evidence for influences of the physical environment is building. Nevertheless, experimental studies, while important for gaining an important understanding for neural mechanisms, rely on small samples and short-term exposures that, in their current form, limit potential for informing population strategies. A large body of observational research highlights an association between physical environments and mental health. Consistency and patterns have the potential to inform the development of environments and mitigation strategies that improve psychological functioning and mental health at scale. The present broad range of methodological approaches, however, make it difficult to draw strong conclusions for policy. In the following section, available tools, data, and methods for measuring exposure are outlined, and priorities for advancing measures to further population-level studies of the physical environment and neuroscience are identified.

4 Measuring the Physical Environment

Geographic Information Science (GIS) provides a powerful means for mapping and analysing data that pertain to a spatial location. In the context of neuroscience, GIS allows for measures of the environment with which an individual is exposed to, or interacts with, to be matched to measures of behavioural and mental-health outcomes (see chapter on "Population Neuroscience: Principles and Advances" in this book).

4.1 Data Sources

Secondary data allow for features of the physical environment to be characterised and quantified. As examples, this may include the density of greenspace, concentration of air pollution, or distance to mental health clinics within a specified area of interest. In Fig. 2, examples of physical environments and corresponding data sources are provided. The list is not designed to be complete and different examples of features and data sources may be more appropriate in alternative contexts.

Data acquired with remote sensing provide information about the earth's surface and are valuable for obtaining large-scale measures of land cover and climate conditions. Data are typically recorded aboard satellites and have the advantage of repeat global coverage over long temporal scales, enabling changes to be detected

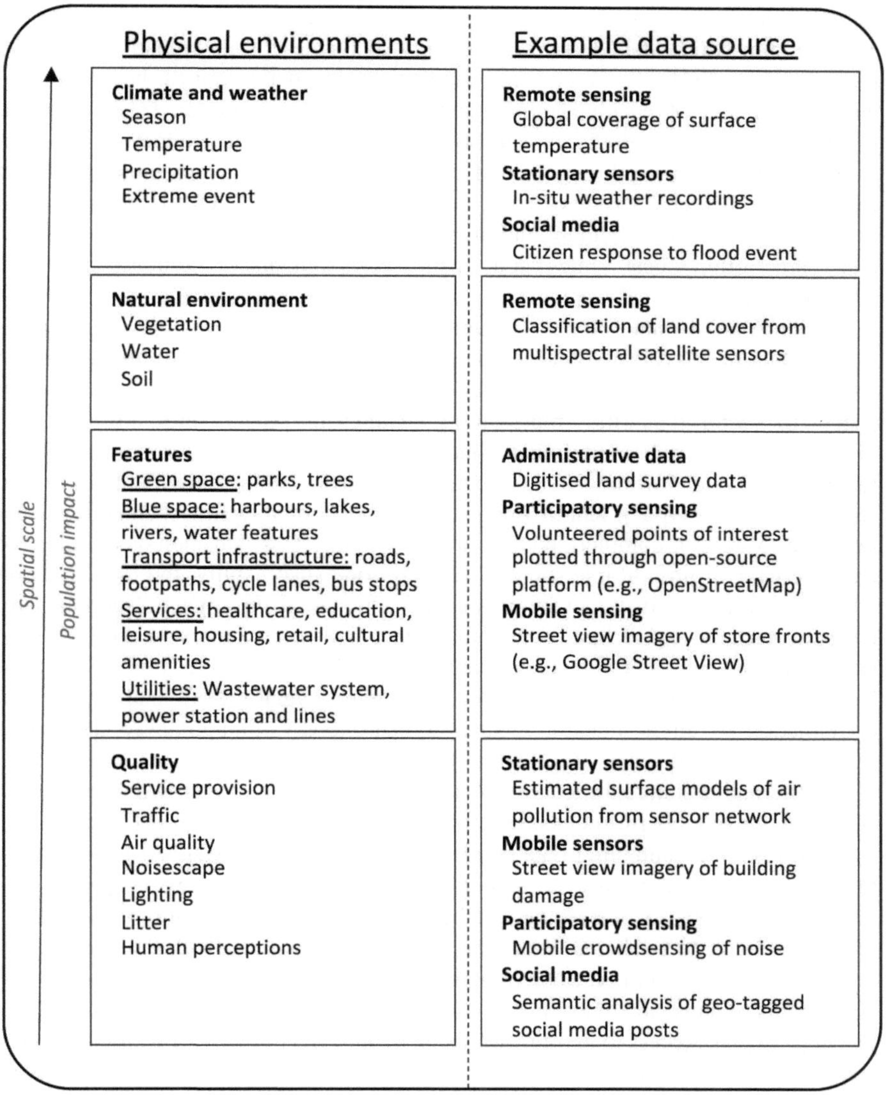

Fig. 2 Examples of physical environments and corresponding data sources (adapted from Paus and Kum 2024)

(Wulder et al. 2019). Satellite data have consequently been used to monitor land-surface temperature, meteorological conditions, and greenness. This holds relevance for studies interested in exploring the effects of changes in land cover, such as urbanisation and deforestation, as well as events such as floods or wildfires, on the brain and behaviour.

A characteristic example of satellites used to monitor land surface is the Landsat program for which data have been openly available for download since 2008 (Hemati et al. 2021). Processing raw remote sensing data requires, however, specialist knowledge of machine learning and classification techniques. To some extent, this is addressed through the availability of pre-classified data products, and access to data and algorithms through cloud computing platforms such as Google Earth Engine (Gorelick et al. 2017; Gong et al. 2019; Li et al. 2020). Given the interaction of environmental factors and limitations with classification, however, remote sensing data are often best integrated with complementary sources such as administrative data.

Administrative data sources contain high-quality data at regional or national scales. Data have typically been prepared by organisations or institutions, and are available for immediate download. As examples, local governments may provide access to vector files (files representing features as points, lines, or polygons) through an open-data portal that characterise local infrastructure or municipal facilities (City of Toronto 2018; The City of Barrie 2019). Individual files often represent a single type of feature only, such as educational facilities, and effort must be made to build a complete picture of urban landscapes that includes a range of characteristics from multiple sources. Even if done successfully, the time frame at which recordings for different features were made may be misaligned. National level datasets, such as DMTI Spatial in Canada and Ordnance Survey in the UK, help provide consistency across mapped features (DMTI Spatial 2019; The University of Edinburgh 2023). This is beneficial for analysing national cohorts and exploring a combination of environmental features synchronously, though these data often require institutional access from within their source country.

An alternative source for features of natural and urban form is crowd-sourced data such as that available from OpenStreetMap (OSM) (OpenStreetMap 2023). OSM is a global resource containing data contributed by volunteers. Mapped features may include high-resolution data, for example, rivers, locations of trees by species, bus routes, transit stops, cycle parking, and benches. Although community input enables local knowledge to be shared, data users must check for issues with standardisation, completeness, and validity. Social media data may also provide context relating to perceptions of physical environments and extreme events (Kovacs-Györi et al. 2018; Fan et al. 2020). Access to geo-tagged social media data through APIs often requires approval, and there have been concerns around geo-privacy (disclosure of individuals' sensitive locations) and representativeness, particularly with changing governance associated with platforms such as X, formally known as Twitter.

Lastly, sensor data, such as environmental sensors and static cameras, are suitable for recording traffic data, including volumes and modes, monitoring weather conditions, and detecting air quality and noise. Data may be available in real time or available for download as historic records for monitoring sites. These data are valuable for cohort studies exploring the impacts of weather effects at the daily and local level, or for building surface models of air pollution, such as the model generated from 36 sites across Europe as part of the European Study of Cohorts for Air Pollution Effects project (ESCAPE) (Beelen et al. 2013). Interpolated surface model values may then be attributed to home addresses of study participants. To limit uncertainty with estimated values in between monitoring sites, data from mobile sensors can improve spatial coverage of sensor networks. Photographic datasets, such as those from Google Street View, have also proven useful for identifying and validating the presence of features through deep learning techniques. This includes the identification of features that may affect the quality of environments and behavioural responses such as broken windows, billboard advertising, potholes, and shadow from buildings (Lu 2018; Huang et al. 2021). It should be noted, however, that data often represent a snapshot of environments and do not capture the dynamics of spaces over time.

While a summary of key data types that are openly available to researchers has been provided here, this list is not exhaustive, and it is important to acknowledge that the quality and availability of environmental data is constantly evolving. In general, when using data to estimate environmental exposure, it is recommended that measurements of a range of environmental features and appropriate time frames for the outcome of interest are used. For example, longer term exposures to urban built environments and noise are important for mental health outcomes. Where possible, the use longitudinal data that represents exposure over time is encouraged. Consideration should also be given to quality of the data including data sensitivity (e.g. ensuring that features present in the real world are included in the data), data precision (e.g. ensuring that features represented in the data are present in the real world), and potential biases (e.g. differences in completeness of data across different spatial or cultural contexts).

4.2 Conceptualising Environmental Exposure

The multidisciplinary field of *exposure assessment* focuses on understanding and quantifying the ways populations come into contact with physical environments, and how magnitude of exposure and outcomes varies for population groups (Patel et al. 2017). Despite the necessity of exposure assessment in population health research and the emerging field of environmental neuroscience (Berman et al. 2019), there has been little agreement in the conceptualisation of environmental exposure neither within nor between disciplines.

Quantitative metrics for exposure include static and proximity measures. Static measures have commonly been used in large-scale studies of exposure and typically focus on the residential neighbourhood. Administrative boundaries or buffers around home locations are applied to estimate, for example, the density of nearby features or concentration of pollution. Static measures of the environment provide suitable estimates for exploring exposures over extended periods, larger spatial scales, and among larger populations. Such measures provide an important first step in identifying potential exposure-outcome relationships at the population level and for estimating long-term exposures important for mental health outcomes.

Static measures are subject to misclassification, however, as they do not characterise environments that individuals are exposed to beyond their home address as a result of their daily activities. Additionally, static measures assume constant exposure for all individuals and household members within the same administrative or neighbourhood boundary, without acknowledging variation in individuals' spatial and temporal habits. This gives rise to the *Uncertain Geographic Context Problem (UGCP)* whereby measures of exposure do not match the spatial context of behaviours; for example, if behaviours occur beyond the home, relevant environments are missed (Kwan 2012). Furthermore, use of administrative data that have been aggregated to different boundaries, such as a census tracts and postal code areas, limits analyses through the *Modifiable Areal Unit Problem (MAUP)* (Wong 2004). Repeating analyses using a different spatial unit, for example, can yield different results. Similarly, relying on aggregate data to make inferences can lead to *ecological fallacy* whereby observations of relationships from group data do not hold true for individuals within the group (Piantadosi et al. 1988). Considering the differences in environmental experiences based on social and cultural context, as well as variability in spatial behaviour within and between individuals, caution is therefore warranted when interpreting findings from studies using static measures.

Static measures of exposure are concerned with the amount of opportunities within a specified area, such as the proportion of greenspace within a 1 km network buffer of a residential address. Conversely, proximity measures rely on time or distance to the closest opportunity, such as distance from home to the nearest park. A study that compared both types of measure found those that quantify features within a specified area to be a more reliable predictor of health (Ekkel and de Vries 2017). In addition, rudimentary proximity measures typically focus on a single environmental feature, which limits the ability to explore a range of environments in combination. Given exposures to features and characteristics of the physical environment do not occur in isolation, measures and studies must start to account for the collective exposure of diverse elements of the physical environment. This progression and other considerations for future directions are discussed in the following section.

5 Opportunities and Future Directions

Accessing spatial data and quantifying environmental features within a chosen metric form the initial stages of exposure assessment. While this provides opportunities for population studies, limitations in current approaches exist. Further research is needed to develop more accurate and conceptually relevant measures of environmental exposure that best predict neural, behavioural, and mental health outcomes while allowing for comparison across studies. As such, key directions for future research are highlighted below.

5.1 *Identify Relevant and Inter-Dependent Factors of Influence*

The physical environment comprises multiple features that interact with each other, with social environments, and with multiple individual-level factors, such as genetic variations and comorbidities. As behavioural outcomes emerge as a product of multiple factors, single short-term interventions are unlikely to promote positive change. Yet to date, elements of the physical environment have often been explored in isolation. Research is therefore needed to understand better which features are most relevant to brain and behaviour, particularly in combination, and how changes in the environment feedback to neural and behavioural responses.

Systems thinking and systems mapping provide ways to help examine and visualise a multitude of factors and the relationships between them, and have been used to build understanding around public health challenges and solutions (Peters 2014; Allender et al. 2015; Rutter et al. 2017). When integrating data from multiple sources to represent a range of features, consideration must be given to differing time frames used for collecting data, and disparate scales and coverage. Here, deep learning may provide an opportunity to bridge gaps between different data types and sources (Zhang et al. 2019).

5.2 *Build Consensus on Exposure Metrics*

A range of metrics have been used to assess exposure and features of the physical environment may be quantified in a number of ways, including a binary presence/absence score, absolute number, density, mean value for an area, or a proportion. While different characteristics of the environment warrant relevant measures (e.g. average concentration of air pollutants over 1 month vs presence of a park), comparative work is needed to understand which measures and delineations of exposure provide the most valid representation for specific features and exposure pathways relevant for mental health. Alongside appropriate metrics, more consideration is needed for which elements of exposure need to be measured. In the case of greenspace, for example, there is little understanding around how an urban park may confer different influences when compared with a forest, and whether biodiversity, time of day/year, or different species hold importance.

5.3 *Capture Variations in Experiences of Exposure*

Technological advances have enabled the collection of various passively generated datasets. Aggregated geo-located data, such as traffic volumes, store check-in data, and mobile device usage, provide opportunities to investigate people's mobility and activity dynamics at scale (Xu et al. 2015). While a static measure of exposure centred on an individual's home location provides an important anchor point, considering common patterns of mobility and use of space around home locations may address issues associated with UGCP and MAUP.

Building on this approach, cohort studies that collect Global Positioning System (GPS) data from individuals over a specified study period enable a dynamic measure of exposure to be estimated. The activity space concept is used to delineate the areas an individual interacts with and measures may include daily path areas (buffer of all trajectories of movement by the individual), key locations (buffers of anchor points such as home, work, food outlets, or social meeting places where time is spent), convex hulls (smallest possible bounding box of an individual's recorded GPS points), or exact locations of behaviour (Smith et al. 2019). Weight may also be given to the frequency, regularity, and duration of visits to locations (Liu et al. 2020). The degree of exposure experienced by individuals may vary due to proximity, time spent exposed, level of shelter or protection, or distraction (e.g. exposure when walking through a neighbourhood vs driving). Though presently uncommon, the increasing availability of high-volume location data allows for the inclusion of temporal patterning in measuring cumulative exposures and distance decay effects, beyond spatial delineations (Wei et al. 2023).

Activity spaces provide an advance on static metrics and an appropriate measure of exposure when mobility data such as individual-level GPS traces are available. These measures provide little information, however, on how individuals engage with environments. Better knowledge around how people experience environments, particularly in relation to different people's social and cultural backgrounds (Mezuk et al. 2013), and how sensory factors such as smells and sounds contribute to environmental interactions, is needed. Given immediate emotional responses to spaces, the incorporation of ecological momentary assessment (EMA) to collect real-time assessments based on prompts via a smartphone device holds promise for capturing behaviours, moods, and experiences as individuals navigate their environments (Shiffman et al. 2008). For example, a recent study explored the synchronous association between greenspace and stress (Mennis et al. 2018). The time and location of self-reported psychological stress was matched with a measure of greenness in an individual's activity space to estimate a relationship.

Exposure measures should always align with the outcome being studied. Immediate emotional responses to environments are well suited to high-frequency, granular location data recorded from GPS tracking, while static measures may provide a suitable estimation when exploring the impacts of long-term exposures on mental health. These static measures may also be developed to incorporate duration and frequency of exposure to better reflect exposure dosage. Associations observed in large-scale population studies may be used to inform the collection of granular location data from smaller samples, as well as experimental studies which use neuroimaging to measure brain structure and function. In the case of the latter, the use of virtual reality (VR) is emerging as a method for simulating environments (Schutte et al. 2017), offering promise for improving exposure measures in controlled settings and complementing findings from population studies. Integrating findings from multiple data sources and study types presents an opportunity to deepen understanding of relationships between different environments, the human brain, behaviour, and mental health.

6 Conclusions

Observational studies taking advantage of static measures of the physical environment for large samples, time frames, and geographic scales mark a significant starting point for exploring the impact of long-term exposures on mental health and advancing the field of population neuroscience. Furthermore, in conjunction with the growing availability and continual refinement of datasets encompassing diverse physical environments, opportunities are presented to deepen understanding of the ways physical surroundings impact the human brain and behaviour. This chapter concludes with prospective directions for enhancing exposure assessments, particularly focusing on capturing the variations in individuals' experiences of environmental interactions through GPS and EMA data. Pursuing these directions necessitates an interdisciplinary approach but can ultimately play a role in shaping the design of healthier and more liveable environments.

References

Allen JG et al (2016) Associations of cognitive function scores with carbon dioxide, ventilation, and volatile organic compound exposures in office workers: a controlled exposure study of green and conventional office environments. Environ Health Perspect 124(6):805–812. https://doi.org/10.1289/ehp.1510037

Allender S et al (2015) A community based systems diagram of obesity causes. PLoS One 10(7): e0129683. https://doi.org/10.1371/journal.pone.0129683

Ancora LA et al (2022) Cities and neuroscience research: a systematic literature review. Front Psych 13. https://www.frontiersin.org, Accessed 30 Aug 2023. https://doi.org/10.3389/fpsyt.2022.983352

Badland H, Schofield G (2005) Transport, urban design, and physical activity: an evidence-based update. Transp Res Part D: Transp Environ 10(3):177–196. https://doi.org/10.1016/j.trd.2004.12.001

Barton H, Grant M (2006) A health map for the local human habitat. J R Soc Promot Health 126(6): 252–253. https://doi.org/10.1177/1466424006070466

Beelen R et al (2013) Development of NO2 and NOx land use regression models for estimating air pollution exposure in 36 study areas in Europe – the ESCAPE project. Atmos Environ 72:10–23. https://doi.org/10.1016/j.atmosenv.2013.02.037

Berman MG et al (2019) The promise of environmental neuroscience. Nat Hum Behav 3(5): 414–417. https://doi.org/10.1038/s41562-019-0577-7

Bolouki A (2023) Neurobiological effects of urban built and natural environment on mental health: systematic review. Rev Environ Health 38(1):169–179. https://doi.org/10.1515/reveh-2021-0137

Charreire H et al (2014) Using remote sensing to define environmental characteristics related to physical activity and dietary behaviours: a systematic review (the SPOTLIGHT project). Health Place 25:1–9. https://doi.org/10.1016/j.healthplace.2013.09.017

City of Toronto (2018) Open data. City of Toronto.

Clifford A et al (2016) Exposure to air pollution and cognitive functioning across the life course – a systematic literature review. Environ Res 147:383–398. https://doi.org/10.1016/j.envres.2016.01.018

Coffey Y et al (2021) Understanding eco-anxiety: a systematic scoping review of current literature and identified knowledge gaps. J Clim Change Health 3:100047. https://doi.org/10.1016/j.joclim.2021.100047

Costa e Silva JA, Steffen RE (2019) Urban environment and psychiatric disorders: a review of the neuroscience and biology. Metabolism 100:153940. https://doi.org/10.1016/j.metabol.2019.07.004

Dahlgren G, Whitehead M (2021) The Dahlgren-Whitehead model of health determinants: 30 years on and still chasing rainbows. Public Health 199:20–24. https://doi.org/10.1016/j.puhe.2021.08.009

Dauncey MJ (2009) New insights into nutrition and cognitive neuroscience: symposium on "Early nutrition and later disease: current concepts, research and implications". Proc Nutr Soc 68(4): 408–415. https://doi.org/10.1017/S0029665109990188

DMTI Spatial (2019) Location data products – DMTI spatial.

Dyar OJ et al (2022) Rainbows over the world's public health: determinants of health models in the past, present, and future. Scand J Public Health 50(7):1047–1058. https://doi.org/10.1177/14034948221113147

Edelstein EA, Macagno E (2012) Form follows function: bridging neuroscience and architecture. In: Rassia ST, Pardalos PM (eds) Sustainable environmental design in architecture: impacts on health. Springer (Springer Optimization and Its Applications), New York, pp 27–41. https://doi.org/10.1007/978-1-4419-0745-5_3

Ekkel ED, de Vries S (2017) Nearby green space and human health: evaluating accessibility metrics. Landsc Urban Plan 157:214–220. https://doi.org/10.1016/j.landurbplan.2016.06.008

Fan C, Jiang Y, Mostafavi A (2020) Social sensing in disaster city digital twin: integrated textual–visual–geo framework for situational awareness during built environment disruptions. J Manag Eng 36(3):04020002. https://doi.org/10.1061/(ASCE)ME.1943-5479.0000745

Fernandez A et al (2015) Flooding and mental health: a systematic mapping review. PLoS One 10(4):e0119929. https://doi.org/10.1371/journal.pone.0119929

Giles-Corti B et al (2016) City planning and population health: a global challenge. Lancet 388(10062):2912–2924. https://doi.org/10.1016/S0140-6736(16)30066-6

Gong P et al (2019) Stable classification with limited sample: transferring a 30-m resolution sample set collected in 2015 to mapping 10-m resolution global land cover in 2017. Sci Bull 64(6): 370–373. https://doi.org/10.1016/j.scib.2019.03.002

Gorelick N et al (2017) Google earth engine: planetary-scale geospatial analysis for everyone. Remote Sens Environ 202:18–27. https://doi.org/10.1016/j.rse.2017.06.031

Hemati M et al (2021) A systematic review of Landsat data for change detection applications: 50 years of monitoring the earth. Remote Sens (Basel) 13(15):2869. https://doi.org/10.3390/rs13152869

Higgs G (2004) 'A literature review of the use of GIS-based measures of access to health care services. Health Serv Outcomes Res Methodol 5(2):119–139. https://doi.org/10.1007/s10742-005-4304-7

Huang Y, Parker D, Minaker L (2021) Identifying latent demand for transit-oriented development neighbourhoods: evidence from a mid-sized urban area in Canada. J Transp Geogr 90:102940

Jimenez MP et al (2021) Associations between nature exposure and health: a review of the evidence. Int J Environ Res Public Health 18(9):4790. https://doi.org/10.3390/ijerph18094790

Kahn PH Jr, Kellert SR (2002) Children and nature: psychological, sociocultural, and evolutionary investigations. MIT Press

Kaplan S (1995) The restorative benefits of nature: toward an integrative framework. J Environ Psychol 15(3):169–182. https://doi.org/10.1016/0272-4944(95)90001-2

Kondo MC, Jacoby SF, South EC (2018) Does spending time outdoors reduce stress? A review of real-time stress response to outdoor environments. Health Place 51:136–150. https://doi.org/10.1016/j.healthplace.2018.03.001

Kovacs-Györi A et al (2018) Beyond spatial proximity – classifying parks and their visitors in London based on spatiotemporal and sentiment analysis of Twitter data. ISPRS Int J Geo Inf 7(9):378. https://doi.org/10.3390/ijgi7090378

Kwan M-P (2012) How GIS can help address the uncertain geographic context problem in social science research. Ann GIS 18(4):245–255. https://doi.org/10.1080/19475683.2012.727867

Lederbogen F et al (2011) City living and urban upbringing affect neural social stress processing in humans. Nature 474(7352):498–501. https://doi.org/10.1038/nature10190

Li W et al (2020) Integrating Google earth imagery with Landsat data to improve 30-m resolution land cover mapping. Remote Sens Environ 237:111563. https://doi.org/10.1016/j.rse.2019.111563

Liu B et al (2020) Association between time-weighted activity space-based exposures to fast food outlets and fast food consumption among young adults in urban Canada. Int J Behav Nutr Phys Act 17(1):62. https://doi.org/10.1186/s12966-020-00967-y

Lu Y (2018) The association of urban greenness and walking behavior: using Google street view and deep learning techniques to estimate residents' exposure to urban greenness. Int J Environ Res Public Health 15(8):1576. https://doi.org/10.3390/ijerph15081576

Lytle LA, Sokol RL (2017) Measures of the food environment: A systematic review of the field, 2007–2015. Health Place 44:18–34. https://doi.org/10.1016/j.healthplace.2016.12.007

Marteau TM, Hollands GJ, Fletcher PC (2012) Changing human behavior to prevent disease: the importance of targeting automatic processes. Science 337(6101):1492–1495. https://doi.org/10.1126/science.1226918

McCormick R (2017) Does access to green space impact the mental well-being of children: a systematic review. J Pediatr Nurs 37:3–7. https://doi.org/10.1016/j.pedn.2017.08.027

Mennis J, Mason M, Ambrus A (2018) Urban greenspace is associated with reduced psychological stress among adolescents: a geographic ecological momentary assessment (GEMA) analysis of activity space. Landsc Urban Plan 174:1–9. https://doi.org/10.1016/j.landurbplan.2018.02.008

Mezuk B et al (2013) "White Box" epidemiology and the social neuroscience of health Behaviors: the environmental affordances model. Soc Mental Health 3(2):79–95. https://doi.org/10.1177/2156869313480892

OpenStreetMap (2023) OpenStreetMap.

Patel CJ et al (2017) Opportunities and challenges for environmental exposure assessment in population-based studies. Cancer Epidemiol Biomarkers Prev 26(9):1370–1380. https://doi.org/10.1158/1055-9965.EPI-17-0459

Paus T (2010) Population neuroscience: why and how. Hum Brain Mapp 31(6):891–903. https://doi.org/10.1002/hbm.21069

Paus T, Kum H-C (2024) Digital ethology: human behavior in geospatial context. The MIT Press (Strüngmann Forum Reports), Cambridge

Peters DH (2014) The application of systems thinking in health: why use systems thinking? Health Res Policy Syst 12(1):51. https://doi.org/10.1186/1478-4505-12-51

Piantadosi S, Byar DP, Green SB (1988) The ecological fallacy. Am J Epidemiol 127(5):893–902

Pontin FL et al (2022) Objectively measuring the association between the built environment and physical activity: a systematic review and reporting framework. Int J Behav Nutr Phys Act 19(1):119. https://doi.org/10.1186/s12966-022-01352-7

Preiser WFE (1983) The habitability framework: a conceptual approach towards linking human behaviour and physical environment. Des Stud 4(2):84–91. https://doi.org/10.1016/0142-694X(83)90038-8

Rutter H et al (2017) The need for a complex systems model of evidence for public health. Lancet 390(10112):2602–2604. https://doi.org/10.1016/S0140-6736(17)31267-9

Sarkar C, Webster C, Gallacher J (2018) Residential greenness and prevalence of major depressive disorders: a cross-sectional, observational, associational study of 94 879 adult UK biobank participants. Lancet Planetary Health 2(4):e162–e173. https://doi.org/10.1016/S2542-5196(18)30051-2

Schutte NS et al (2017) The impact of virtual environments on restorativeness and affect. Ecopsychology 9(1):1–7. https://doi.org/10.1089/eco.2016.0042

Shiffman S, Stone AA, Hufford MR (2008) Ecological momentary assessment. Annu Rev Clin Psychol 4(1):1–32. https://doi.org/10.1146/annurev.clinpsy.3.022806.091415

Smith L, Foley L, Panter J (2019) Activity spaces in studies of the environment and physical activity: a review and synthesis of implications for causality. Health Place 58:102113. https://doi.org/10.1016/j.healthplace.2019.04.003

Song C, Ikei H, Miyazaki Y (2016) Physiological effects of nature therapy: a review of the research in Japan. Int J Environ Res Public Health 13(8):781. https://doi.org/10.3390/ijerph13080781

Stillman CM, Erickson KI (2018) Physical activity as a model for health neuroscience. Ann N Y Acad Sci 1428(1):103–111. https://doi.org/10.1111/nyas.13669

The City of Barrie (2019) City of Barrie open data.

The University of Edinburgh (2023) Digimap. Available at: https://digimap.edina.ac.uk/ Accessed 31 Aug 2023

Thompson R et al (2018) Associations between high ambient temperatures and heat waves with mental health outcomes: a systematic review. Public Health 161:171–191. https://doi.org/10.1016/j.puhe.2018.06.008

Ulrich RS et al (1991) Stress recovery during exposure to natural and urban environments. J Environ Psychol 11(3):201–230. https://doi.org/10.1016/S0272-4944(05)80184-7

United Nations Department of Economic and Social Affairs (2018) 68% of the world population projected to live in urban areas by 2050, says UN. Available at: https://www.un.org/development/desa/en/news/population/2018-revision-of-world-urbanization-prospects.html Accessed 30 Aug 2023

van de Coevering P, Maat K, van Wee B (2015) Multi-period research designs for identifying causal effects of built environment characteristics on travel behaviour. Transp Rev 35(4):512–532. https://doi.org/10.1080/01441647.2015.1025455

Vanaken G-J, Danckaerts M (2018) Impact of green space exposure on children's and adolescents' mental health: a systematic review. Int J Environ Res Public Health 15(12):2668. https://doi.org/10.3390/ijerph15122668

Wei L et al (2023) Measuring environmental exposures in people's activity space: the need to account for travel modes and exposure decay. J Expo Sci Environ Epidemiol:1–9. https://doi.org/10.1038/s41370-023-00527-z

Whitehead M, Dahlgren G (1991) Policies and strategies to promote social equity in health. Institute for Future Studies, Stockholm

Wilkie S et al (2018) Restructuring the built environment to change adult health behaviors: a scoping review integrated with behavior change frameworks. Cities Health 2(2):198–211. https://doi.org/10.1080/23748834.2019.1574954

Wong DWS (2004) The modifiable areal unit problem (MAUP). In: Janelle DG, Warf B, Hansen K (eds) WorldMinds: geographical perspectives on 100 problems: commemorating the 100th anniversary of the Association of American Geographers 1904–2004. Springer, Dordrecht, pp 571–575. https://doi.org/10.1007/978-1-4020-2352-1_93

Wulder MA et al (2019) Current status of Landsat program, science, and applications. Remote Sens Environ 225:127–147. https://doi.org/10.1016/j.rse.2019.02.015

Xu Y et al (2015) Understanding aggregate human mobility patterns using passive mobile phone location data: a home-based approach. Transportation 42(4):625–646. https://doi.org/10.1007/s11116-015-9597-y

Zhang C et al (2019) Heterogeneous graph neural network'. In: Proceedings of the ACM SIGKDD international conference on knowledge discovery and data mining. https://doi.org/10.1145/3292500.3330961

Leveraging Generative AI Models in Urban Science

J. Balsa-Barreiro, M. Cebrián, M. Menéndez, and K. Axhausen

Contents

J. Balsa-Barreiro (✉)
MIT Media Lab, Massachusetts Institute of Technology, Cambridge, MA, USA

CITIES, New York University Abu Dhabi, Abu Dhabi, UAE
e-mail: jobalbar@media.mit.edu

M. Cebrián
Center for Automation and Robotics, Spanish National Research Council, Madrid, Spain

M. Menéndez
CITIES, New York University Abu Dhabi, Abu Dhabi, UAE

K. Axhausen
Institute for Transport Planning and Systems, Zurich, Switzerland

Curr Topics Behav Neurosci (2024) 68: 239–276
https://doi.org/10.1007/7854_2024_482
Published Online: 23 July 2024

Abstract Since the late 2000s, cities have emerged as the primary human habitat across the globe, and this trend is anticipated to continue strengthening in the coming decades. As we increasingly inhabit human-designed urban spaces, it becomes crucial to understanding better how these environments influence human behavior and how individuals perceive the city. In this chapter, we begin by examining the interplay between urban form and social behavior, highlighting key indicators of urban morphology, and presenting state-of-the-art methodologies for data collection. Subsequently, we harness the computational capability of foundation models, the latest Artificial Intelligence (AI) generation, to simulate interactions between individuals and urban built environments in a diverse group of 21 cities across the globe. Through this exploration, we scrutinize the models' capacity to encapsulate the intricate complexities of how individuals behave and perceive cities. These examples demonstrate the potential of advanced AI systems to assist urban scientists in understanding cities, emphasizing the necessity for a meticulous evaluation of their capabilities and limitations for the optimal application of Generative AI in urban research and policymaking.

Keywords Artificial intelligence · Built environments · Compact cities · Foundation model · Generative AI · Human behavior · Large language model · Machine perception · Urban dynamics · Urban science

1 Introduction

According to the United Nations (2023), urban areas were home to less than one-third of the global population (29.6%) by 1950. Currently, this rate has nearly doubled (57.5% in 2023) and is expected to continue growing, surpassing two-thirds of the global population (68.4%) by 2050. Behind this global trend, significant regional disparities are evident, with a noticeable acceleration in the urbanization process observed in the world's poorest regions (Balsa-Barreiro et al. 2019). *Low-income countries* (LICs)[1] are experiencing a swift transition, transitioning from

[1]The methodology employed by the United Nations Committee for Development Policy (2020) classifies countries based on their Gross National Income per capita as defined by the World Bank. This classification results in four categories: *low-income* (LIC), *lower-middle income* (LMIC), *upper-middle income* (UMIC), and *high-income* (HIC) countries. According to the World Bank (2023), low-income countries (LIC) had a Gross National Income per capita below \$1,085, while high-income countries (HIC) had a figure exceeding \$13,205.

historically low urbanization rates (9.3% in 1950) to levels expected to surpass 50% by 2050, representing a growth of +16.6 percentage points from 2023 (34.6%). Conversely, *high-income countries* (HICs) are further along in the urbanization process, having begun with relatively high urbanization rates in the past (58.5% in 1950) and currently exhibiting even higher rates (82.4% in 2023). Over the next two decades, these countries are projected to experience a modest growth of +6 percentage points, reaching 88.4% by 2050 (United Nations 2023).

Two different perspectives help us to understand large-scale urbanization across the globe. On the one hand, the most traditional one emphasizes the tendency for people and economic activities to concentrate and develop *economies of scale* (Gill and Goh 2010; Wheaton and Shishido 1981). Cities act as epicenters where job supply and demand come together by minimizing the physical distance between both types of agents. This spatial proximity strengthens the labor market, creating new opportunities, favoring aspects related to innovation, and building economies that are more diverse and resilient (Pentland 2014). Dense cities boast vast and versatile labor markets, providing businesses with access to a substantial workforce pool, while employees can expand their job prospects with a wider and more diverse job market. Furthermore, proximity between agents reduces commuting times and encourages the use of collective transportation systems. As a result, cities facilitate resource optimization and decrease travel expenses. This explains why the world's largest economies are predominantly *urban economies* (Frick and Rodríguez-Pose 2018), and why major cities increasingly tend to concentrate a share of global wealth far exceeding their demographic weight, as demonstrated by Dobbs and Remes (2013).

While the previous perspective assumes that countries accumulate wealth as they urbanize (Hyde 2023), this is not always the case. We find a clear example in many African countries (Collier 2006, 2007) where *pathological urbanization* occurs, referring to substantial population shifts without the accompanying income growth (Spence et al. 2009). A second perspective explains this trend in many cities undergoing rapid urbanization today. Just a decade ago, as of 2015, the *Global South* accounted for the majority of megacities (27 out of 33), most of the world's urban population (75%), and the highest rates of recent urbanization (94% of the increase in urban population worldwide between 2010 and 2015) (Smit 2021). In *Global South* countries, rapid urbanization is driven more by expulsion and lack of opportunities in rural areas than by new job opportunities in cities. This leads to massive processes of *pseudo-* or *false-urbanization*, with informal settlements dominating urban landscapes (Hashimov et al. 2013). In Africa, approximately 60% of the urban dwellers live in unplanned informal settlements with severe shortages of critical infrastructure and basic services (UNDP 2017; Bandauko et al. 2021). This urbanization process is associated with many negative externalities related to environmental degradation (Erdoğan et al. 2022), heightened vulnerability to climate change and extreme hazards (He et al. 2021), adverse health effects due to inadequate sanitation and housing (Kuddus et al. 2020), and increased risk of social exclusion for large segments of the population (Williams et al. 2019).

These two perspectives highlight how the urbanization process evolves over time and space, demonstrating that while high population-densities in many thriving cities offer significant competitive advantages, they also give rise to challenges known as *urban diseconomies* (Balsa-Barreiro et al. 2021). Historically, early nineteenth-century industrial cities faced issues such as fires, social unrest, and the rapid spread of diseases, resulting in shorter lifespans for residents (Berghauser and Haupt 2010). Today, densely populated areas continue to face challenges, primarily stemming from conflicting land uses that result in pollution, traffic congestion, and various social tensions associated with crime and insecurity (Kotkin 2020). These complexities spark discussions on the global urbanization process, which is expected to increase wealth and reduce social inequalities (Wan et al. 2022; Bloom et al. 2008), even though this might not always be the case (Shutters et al. 2022).

Cities embody constructed and artificial environments crafted by humans, hence *built environments*, exerting a significant influence on human behavior (Balsa-Barreiro and Menendez 2024). Within these urban landscapes, various elements such as infrastructure, public spaces, housing arrangements, and transportation systems play a pivotal role in shaping the perceptions and experiences of urban residents (Park 1915; Xia et al. 2020; Mouratidis 2021). Given that cities have become the predominant habitat for humans and considering the substantial challenges anticipated in the forthcoming decades, concerted efforts are essential to gain a deeper understanding of our interaction with these spaces. This understanding is crucial for unraveling how cities shape the behavior and perception of urban dwellers, and vice versa, to optimize the positive attributes of urbanization and minimize adverse outcomes. This will contribute to creating healthier and more attractive built environments, enhancing the well-being and life experience of their inhabitants.

On the other hand, the major breakthrough in *Generative AI* has positioned it as an emerging analytical tool. In recent years, this technology has been under evaluation for a wide range of applications across diverse industries, spanning healthcare, entertainment, marketing, education, and more (Ooi et al. 2023; Bommasani et al. 2021). Their practical value in real-world scenarios, coupled with continuous improvements in accuracy and refinement, equips researchers with potent tools to push the boundaries of knowledge, enabling them to tackle intricate problems and address complex challenges (Bender et al. 2021). For this purpose, we delve into the potential of *Generative AI* within urban science, identifying notable opportunities for its application.

This chapter adopts a structured approach. Initially, we conduct a comprehensive literature review examining the interplay between urban form and social behavior, particularly focusing on compact cities. Subsequently, we delve into the scientific principles underlying urban morphology, highlighting key indicators and presenting state-of-the-art methodologies for data collection. After that, we explore the concept and potential application of *Generative AI* in simulating and perceiving cities. We then present findings from our case study, which encompasses 21 cities across various global regions, discussing both the advantages and limitations of our analysis, as well as the implications of our observations. Finally, we provide concluding remarks for effectively applying *Generative AI* tools in planning sustainable, efficient, and resilient cities in the next decades.

2 Urban Form and Social Behavior in Compact Cities

The complexity of cities is apparent in their diverse urban morphology, reflecting a blend of historical forms, volumes, and shapes (Marshall 2004). These morphologies have evolved over time and continue to adapt, albeit in unique ways depending on the geographical context. In high-income countries, cities typically evolve gradually through regulated expansion dynamics, supported by well-established infrastructure and ongoing urban development assessments. Conversely, urbanization in low-income countries often unfolds rapidly and chaotically, resulting in challenges such as overcrowding, the proliferation of informal settlements, and inadequate infrastructure.

Throughout history, the prevailing technology and modes of transportation have been fundamental factors in shaping the city form (Balsa-Barreiro and Menéndez 2021, 2022). Medieval or pre-industrial cities, emerging as trade hubs amidst expansive agricultural lands, featured compact, maze-like layouts with narrow streets optimized for pedestrian traffic or small horse-drawn carriages. Characterized by dense urbanization, these cities had limited green spaces embedded into the urban fabric. Since the end of the eighteenth century, the Industrial Revolution sparked urban growth and dramatic transformations in European cities. Industrial facilities and neighborhoods attracted significant rural-to-urban migration, leading to uncontrolled urban sprawl and the emergence of marginalized communities facing poor living conditions. The principles of *hygienism* influenced significantly urban planning in the nineteenth century. These principles advocated for the implementation of sewer systems to enhance sanitation, the creation of green areas such as parks to improve public health, and zoning strategies to separate industrial zones from residential areas, thereby safeguarding inhabitants from pollution (Peterson 1979). During that century, urban reforms addressed challenges posed by industrialization through organized residential planning, the provision of public parks, and the introduction of essential services such as clean water and sewage systems. The *industrial city* featured wider streets tailored to accommodate public transportation, such as tram systems. In the twentieth century, Western cities underwent rapid urbanization, expanding suburbs and established constructing modern infrastructure. While these cities optimized layouts (*grids*) for private automobiles on broad streets and robust communication routes, public transportation remained vital, particularly in Europe and East-Asian countries (Ortigosa and Menendez 2014).

In recent decades, the principles of sustainable development and ecological design have prompted a reevaluation of city form. Jabareen (2006), who advocates for *compact city* models, encapsulates this debate. Compact cities feature high built densities and mixed-use environments, supported by efficient public transportation systems and infrastructure that promote walking and cycling. Among their advantages, four fundamental aspects stand out. Firstly, compact cities minimize commuting distances between workplaces and homes while also fostering the development of efficient public transportation modes. Secondly, compact cities promote land reuse in less developed areas, thereby curbing urban sprawl and easing strain on suburban areas. This approach to urban compactness enhances accessibility to goods and services due to their proximity. Thirdly, shorter distances and mixed land uses foster diversity and promote social cohesion among residents through increased interactions. Lastly, compact cities contribute to minimizing infrastructural costs, making them more viable by increasing demand and favoring the use of collective transportation systems. In accordance with the *EU Green Paper on the Urban Environment* (CEC 1990), the advantages of compact cities have attracted significant attention from both experts and policymakers. This has resulted in a heightened focus on compact urban models both in Europe and in the majority of Western countries since the 1990s (Burton et al. 1996).

Balsa-Barreiro and Menendez (2024) introduced a framework for examining the reciprocal relationship between cities and social behavior, drawing from an extensive literature review. This interplay can be evaluated from diverse perspectives. For instance, the mobility patterns of urban residents are distinct, tending to drive less and use public transportation more frequently. Knowles (2023) found that residents of Tokyo own vehicles at a much lower rate compared with other parts of Japan, largely due to the inconvenience of driving caused by discouraging policies and the availability of efficient public transportation systems. The impact can vary, however, depending on the geographical context. In Sweden, Stojanovski (2019) observed that despite efforts to promote sustainable mobility, the modal share of car travel has remained relatively constant over the past two decades influenced by factors like fragmented urban regions and specific road layouts. Yet, urban environments do not only affect the transportation modes we use, but also shape our spatiotemporal mobility patterns. Car-centric urban sprawl frequently results in extended commuting distances and worsened traffic congestion (Travisi et al. 2010), whereas compact cities typically favor shorter commuting distances, while also encouraging alternative modes of transportation such as walking or cycling (Mouratidis et al. 2019).

The benefits of compact cities extend beyond mobility, although these can be closely interconnected. Dense cities encourage the use of public and alternative transportation modes, leading to environmental and health benefits for urban dwellers. Chamberlain et al. (2023) found notable health enhancements, particularly in relation to cardiovascular diseases, within specific neighborhoods in London, UK shortly after the implementation of measures discouraging private vehicle usage and the establishment of low-emission zones. Burton (2000) noted that compact cities foster social equity among residents, primarily attributed to the availability of better public transportation systems and improved accessibility to facilities. Moreover, the prioritization of accessible public transportation models and pedestrian-friendly designs in compact cities facilitates social interactions among residents, thereby nurturing livable and community-oriented urban environments (Appleyard 1981). These findings align those published in seminal studies advocating for the creation of more human-centric cities through urban design (Jacobs 1961; Gehl 1971; Whyte

1980). All these reasons contribute to residents of compact cities often showing higher levels of well-being and personal satisfaction (Montgomery 2013).

As Western societies age, there is increasing concern about the influence of urban life on both physical and mental health. In compact cities, the direct impacts of pollution or the *heat-island* effect are more pronounced, potentially leading to severe health effects among vulnerable groups (Marquez and Smith 1999; McCarty and Kaza 2015; Obradovich and Fowler 2017). Thompson et al. (2023) conducted a systematic review of 144 studies, finding that ambient outdoor temperature was positively associated with suicidal behavior and other negative outcomes for community mental health. Therefore, urban design of compact cities must incorporate green spaces, considering not only the extension and structure but also their spatial fragmentation and widespread distribution, to ensure that their benefits are evenly distributed (Fisher et al. 2022; McDonald et al. 2020). From an environmental standpoint, prevalence of green spaces helps mitigate heatwaves, thereby reducing the risk of heat-related illnesses for vulnerable populations (Iungman et al. 2023). These spaces not only encourage physical activity but also provide mental-health benefits such as psychological relaxation and stress reduction (Braubach et al. 2017). A meta-analysis of 21 articles showed that a 10% increase in green space exposure was linked to about a 4% decrease in depression risk, along with reduced anxiety (Liu et al. 2023).

3 Urban Morphology

Understanding contemporary cities necessitates evaluating various qualitative aspects, often assessed across different scales and levels of granularity. The categorization of city forms encompasses concepts related to morphology, geometry, and typology of built environments. This includes physical characteristics like shape, size, density, and spatial relationships among elements, as well as aspects related to land use, including green spaces. The analysis of these aspects has been approached differently by various disciplines, with different names ranging from a traditionally qualitative and visual perspective (*Urban Morphology*) to a more quantitative approach (*Urban Morphometrics*), which has gained prominence in recent years due to the increased availability of data and advanced computational capabilities (Kropf 2009; Dibble et al. 2019). The development of quantitative methods has led to the implementation of different indicators considering various spatial units, scales, and levels of data aggregation. In this context, notable contributions include the studies conducted by Hermosilla et al. (2014), Boeing (2019), Biljecki and Chow (2022), and Bourdic et al. (2012). To standardize criteria amidst the wide variety of methodologies employed, Fleischmann et al. (2020) conducted a comprehensive review of indicators, encompassing 72 quantitative studies and identifying an initial list of 465 measurable indicators of urban form, which was subsequently refined to 361 due to terminological inconsistencies and vague methodological descriptions. The authors classified urban indicators across six categories (dimension, shape,

Table 1 Summary of some relevant morphological indicators collected by Fleischmann et al. (2020), featuring categories, definitions, and relevant indicators

Category	Definition	Indicators
Dimension	Geometric properties of individual objects	Length
		Height
		Number of floors
		Mesh size
		Area
		Length
Shape	Geometric dimensions' mathematical properties	Height-to-width ratio
		Compactness index
		Form factor
		Fractal dimension
		Rectangularity index
		Complexity index
Distribution	Spatial distribution of objects in space	Built front ratio
		Distance
		Continuity
		Concentration index
Intensity	Density of elements by unit area	Covered area ratio
		Floor area ratio
		Number of plots
		Weighted number of intersections
Connectivity	Spatial interconnection of street networks	Closeness centrality
		Clustering coefficient
		Node/edge connectivity
		Node connectivity
Diversity	Diversity and complexity of the elements	Power law distribution of areas
		Plot area heterogeneity
		Plot area diversity
		Intersection type proportion

spatial distribution, intensity, connectivity, and diversity) and three conceptual scales (small, medium, and large). Table 1 displays some of the most relevant indicators included in Fleischmann et al. (2020).

4 State-of-the-Art Methodologies

Characterizing built environments and understanding the interplay between physical and human factors in cities necessitates the consideration of various technologies and methodologies offering diverse types of data. In Table 2, we summarize a

Table 2 List of potential sources and technologies for data collection related to urban morphology and social behavior. In scope, *Phys.* refers to the physical space, while *Virt.* pertains to the virtual space

Data source	Information	Data description	Scope
Air quality stations	Data records	Environmental data	Phys.
Meteorological stations	Data records	Environmental data	Phys.
Banking	Credit card transactions	Purchase behavior	Phys.
Social networks	Social interactions	Social patterns	Phys./Virt.
Mobile phones	Apps (profile, type)	Social/Purchase behavior	Phys./Virt.
	Call detail records	Social/Communication/Mobility patterns	Phys./Virt.
Crowdsourcing	Volunteer data	Social/Communication/Mobility patterns	Phys./Virt.
	Points-of-Interest	Social/ Communication/Mobility patterns	Phys.
Observational data	Various	Social/Communication/Mobility patterns	Phys./Virt.
Simulation analysis	Experim. Analysis	Social/Communication/Mobility patterns	Phys./Virt.
Surveys	Experim. Surveys	Social/Communication/Mobility patterns	Phys.
	Subjective perception	Urban experience/Urban perception	Phys./Virt.
Health reports	Data records	Social/Health patterns	Phys.
Personal wearables	Various	Social/Health patterns	Phys./Virt.
Mobility flows	GPS traces	Social/Mobility patterns	Phys.
	Loop detectors	Social/Mobility patterns	Phys.
	Public transport reports	Social/Mobility patterns	Phys.
Census data	Household data	Socioeconomic data	Phys.
Cadastral plans	Thematic data	Urban form	Phys.
Historical maps	Thematic data	Urban form	Phys.
Laser scanner	Point cloud	Urban form	Phys.
Photogrammetry	Imagery/Point cloud	Urban form	Phys.
Official reports	Thematic data	Urban form/Socioeconomic data	Phys.
Aerial imagery	Imagery	Urban form/Urban environment	Phys.
Remote sensing	Imagery	Urban form/Urban environment	Phys.

comprehensive list of potential data sources for conducting an extensive examination of cities, utilizing state-of-the-art methodologies, with recent studies serving as the baseline (Balsa-Barreiro and Menendez 2024; Weigle et al. 2024). In recent years, our capacity to gather vast amounts of data has significantly expanded. We now utilize fine-grained data collected at high temporal frequencies, resulting in more abundant and precise data than ever before (Mayer-Schonberger and Cukier 2013).

Additionally, standardized data exchange protocols and enhanced information accessibility are ensuring broader access to data for users. These advancements are unlocking the potential for more ambitious studies that integrate both physical and human factors within urban environments.

The adoption of state-of-the-art methodologies comes with significant challenges, however. Firstly, the widespread use of data collection sources employing different technologies leads to diverse data formats and structures, i.e., different ways to characterize and measure similar features. This diversity often results in different levels of granularity or spatial-time resolution across databases, leading to lack of compatibility between different geographic areas. This demands extensive efforts during subsequent steps for data processing. These discrepancies can impede the use of some data sources in particular applications. Secondly, despite the technological advancements for massive data collection, our capacity for processing and analysis of huge datasets often lags behind (Balsa-Barreiro et al. 2018). Many tasks related to collecting and/or processing information require specialized knowledge and expertise, posing challenges in integrating databases from various sources and assembling multidisciplinary teams capable of conducting comprehensive studies of cities. Thirdly, costs associated with the utilization of certain technologies for data collection may be prohibitive, limiting studies to specific regions or demographic groups. This could ultimately lead to biased or incomplete results. Moreover, significant regional disparities exist both in the public availability of data and in their reliability. Finally, concerns about data availability can arise due to ethical or legal considerations, which limit public access to certain datasets and restrict opportunities for widespread sharing and utilization of this data.

5 Potential of Generative AI in Urban Science

Generative AI refers to artificial intelligence capable of producing text, imagery, or other data based on given *prompts*, i.e., a natural language text describing a desired output (Feuerriegel et al. 2024). These AI systems are often powered by *Foundation Models*, which are large-scale machine learning models utilizing *transformers*. These models acquire their understanding of language structures, semantics, and contextual nuances through extensive self-supervised and semi-supervised training on massive datasets collected from multiple sources including, predominantly, Internet sources. Models' abilities are improved through better algorithms and training methods, making their responses more accurate and contextually fitting over time (Dhamani and Engler 2024).

Since the early 2020s, advancements in transformer-based deep neural networks and massive training data have fueled the rapid development of all kinds of *Generative AI models*, including *Large Language Model* (LLM) chatbots – *ChatGPT*, *Copilot*, and *Gemini* (Myers et al. 2024; Abdullah et al. 2022), as well as text-to-image generation systems – *Stable Diffusion*, *Midjourney*, and *DALL-E* (Zhang et al. 2023), among others. These developments have been driven by the expansion of massive datasets, the scaling up of model sizes, and innovations in model architectures, resulting in unprecedented capabilities offered by foundation models.

Several studies have started to explore the potential of *Generative AI* across different disciplines in real-world scenarios. For instance, Zhang and Kamel Boulos (2023) provided an overview of various *Generative AI* applications in medicine and healthcare. Moor et al. (2023) introduced a groundbreaking approach to *Generalist Medical AI* (GMAI), developed through self-supervision on diverse medical datasets. This enables the interpretation of various medical data-types with advanced reasoning capabilities. Tiu et al. (2022) implemented a foundation model (*CheXzero*) capable of detecting numerous diseases in chest X-rays without training labels. In environmental sciences, Rane, Choudhary, and Rane et al. (2024) assessed how LLMs accelerate the implementation of mitigation strategies addressing climate change. Cooper (2023) conducted an exploratory study on the potential impact of *Generative AI* on teaching pedagogy. In civil engineering, Shoemaker et al. (2023) examined the utility of *ChatGPT-4* for common geotechnical engineering tasks, demonstrating reasonable responses in many cases.

Next, we explore the potential of Generative AI in urban science, identifying significant opportunities for its application. Thus far, only initial studies have explored this potential, such as the work by Yan et al. (2023), which introduces *UrbanCLIP*, the first framework enhanced by LLMs integrating textual knowledge into urban imagery profiling. Results from predicting three urban indicators in four major Chinese metropolises demonstrated how *UrbanCLIP* performance was 6.1% better in R^2 compared with state-of-the-art methods.

6 Case Study

We assess various factors defining the perception and experience of urban residents by using LLMs. In this case study, we input a textual prompt to the language model, simulating the experience of a particular individual in different urban environments. For that, we analyze several factors related to how urban environments are experienced by a randomly chosen individual who typifies the sociodemographic profile of an *average person*. This profile is derived from the demographic characteristics of a city, encompassing age, sex, occupation, socioeconomic status, and other pertinent factors. We establish a list of 52 indicators concerning urban elements and factors that we consider relevant. These indicators are grounded in established criteria derived from a comprehensive literature review, as well as our own criteria. These indicators cover both the physical and human aspects of the city, many of which are mutually interconnected. Through these indicators, we aim to obtain a comprehensive and multifaceted understanding of cities, based on the perceptions of their residents.

The aforementioned indicators in Tables 1 and 2 are reinterpreted and grouped into a list of seven categories presented sequentially as follows: *Urban morphology* (uMOR), *Urban environment* (uENV), *Spatial dynamics* (sDYN), *Urban health* (uHEA), *Urban perception* (uPER), *Urban social* (uSOC), and *Urban economy* (uECO). We briefly summarize each of these groups below, but we encourage readers to consult the comprehensive list of all indicators available in the Appendix (Sect. A1).

1. The group 1 – *Urban morphology* (uMOR) evaluates the urban form and city structure. In accordance with most of the studies previously introduced, we assign positive value to *compact cities* given their promotion of proximity living, which fosters social interactions among residents and encourages local businesses, thus enhancing economic diversity. In summary, this group consists of 9 indicators assessing aspects related to building density and other factors associated with the extent of public space conducive to the implementation of human-centric and pedestrian-friendly cities.
2. The group 2 – *Urban environment* (uENV) encompasses 13 indicators, incorporating factors related to pollution (acoustic and atmospheric), as well as spatial planning considerations such as the density and distribution of green spaces throughout the city. Additionally, we consider physical determinants like sunlight and weather conditions, which reflect the city's suitability for outdoor activities. Many of these factors interconnect with indicators from other groups. For instance, air quality not only affects health and well-being but also influences nighttime illumination, which in turn can affect the city's safety. All indicators that influence well-being and contribute to a pleasant sensation receive positive assessments.
3. The group 3 – *Spatial dynamics* (sDYN) comprises a set of 10 indicators that offer a comprehensive insight into a city's mobility ecosystem, unveiling the efficiency, accessibility, and adaptability of transportation networks and road infrastructures. These indicators encompass various aspects of urban mobility, ranging from the duration of daily commutes to the quality and significance of public transportation. They also appraise the city's provisions for alternative modes of transportation, such as cycling and micromobility modes. Moreover, these indicators assess the connectivity and accessibility within cities via public transit, walking, cycling, or driving. We positively evaluate all indicators that promote efficient, rapid, and sustainable mobility.
4. The group 4 – *Urban health* (uHEA) comprises 5 indicators that evaluate health and well-being of urban dwellers. These indicators delve into both physical and mental aspects of health, as well as quantitative factors such as the expected lifespan of city dwellers. Additionally, we include subjective indicators related to mental resilience and life satisfaction of residents. In this group, we positively assess all factors contributing to collective health and well-being.
5. The group 5 – *Urban perception* (uPER) assesses how residents and visitors perceive the city, both aesthetically and functionally. It encompasses 5 indicators ranging from the visual appeal of the city and its architectural coherence to the ease of living and the city's ability to attract tourists. These indicators enable us to determine if a city provides a harmonious and attractive environment with

potential to improve the quality of life and draw visitors, aspects that are evaluated positively.

6. The group 6 – *Urban social* (uSOC) assesses the urban vibrancy, diversity, and social cohesion through a set of 8 indicators. Essentially, this group reflects the dynamic, harmonious, and safe atmosphere of a city as perceived by its residents and visitors. In addition, it gauges whether the city offers a vibrant environment conducive to fostering strong community connections. For that, we include indicators related to the presence of multicultural amenities, alongside others that estimate the degree of social integration of residents, as well as daytime and nighttime activity. These indicators also offer insights into the community's social cohesion and the perceived level of safety within the city. We positively assess aspects related to vibrant, diverse, and safe community life.
7. The group 7 – *Urban economy* (uECO) comprises 8 indicators to assess the prosperity and economic diversity of a city. Among other indicators, we evaluate the overall wealth of a city, but also factors such as housing affordability, economic diversification, and income equality. In a way, this group paints a comprehensive picture of the economic landscape of a city, reflecting its robustness, fairness, and labor opportunities. All indicators related to aspects of prosperity and economic stability receive positive assessments.

Each indicator is evaluated according to the established criteria within its respective group. We use a scoring scale ranging from 0 to 10 points for all indicators, where higher values indicate a positive assessment. The score for each group is calculated by summing the scores of all indicators considered within that group, while the overall score for each city is determined by adding up the scores across all indicator groups. Through this evaluation methodology, our objective is to understand comprehensively the urban experience of urban dwellers by quantifying all aspects related to their quality of life, safety, prosperity, and physical and mental well-being. As a result, cities with higher scores are generally associated with better livability, prosperity, and overall quality of life, while lower scores suggest the opposite.

We assess the aforementioned indicators across a diverse group of cities, considering two key factors: (a) city size and (b) the geographical region where each city is located. We compile a list of 21 cities distributed across seven predefined regions: Africa (AFR), China (CHN), Europe (EUR), India (IND), Latin America (LAT), the Middle East (MEA), and the United States (USA). Within each region, we choose three cities with significantly different population sizes, categorized into three size groups: *large-populated cities* (LAC) with over five million inhabitants, *mid-sized cities* (MSC) with populations ranging from 500,000 to one million, and *least-populated cities* (LEC) with fewer than 150,000 inhabitants.

Once all the indicators and criteria are defined, we run a two-step input prompt. In the first step, we ask the language model to generate a random list of cities meeting specific population and geographical location criteria within each region. In the second step, we query the model about each of the predefined indicators within each group. The prompts implemented in both steps are provided in the Appendix (Sect. A2). Our analysis employs *GPT-4*, the latest version of *ChatGPT* developed by OpenAI (2023). The prompts and results were formulated and obtained on August 20th, 2023.

6.1 Results

The 21 selected cities by GPT-4 that meet the criteria established are shown in Fig. 1.

Figure 2 summarizes the scores obtained for the selected cities based on different aggregation levels. Each indicator is represented by an individual box, grouped by color to denote the 7 previously introduced groups. Although all groups have the same number of boxes, their sizes vary proportionally based on the score. A comprehensive and detailed review of the scores obtained, both grouped (by population size and/or geographical location) and broken down by individual cities, is presented in the Appendix (Sect. A3).

6.2 Analysis of Results by Cities' Population Size

As we mentioned earlier, our analysis categorizes cities based on population size into three groups: *large-populated cities* (LAC) exceeding five million inhabitants, *mid-sized cities* (MSC) ranging from 500,000 to one million residents, and *least-populated cities* (LEC) with fewer than 150,000 inhabitants. Each group includes 7 cities, with representation across all regions. Upon evaluating city scores by population size, we observe that *large-populated cities* achieve the highest score, totaling 2889 points (sum across the seven cities). The groups of *mid-sized* and *least-populated cities* weigh at a similar level, around 2670 points, which is 7.5% lower. If we consider these categories by region, the *large-populated cities* located in America and Europe are clearly the best ranked, with very similar scores (480 and 477 points per city, respectively). Within this same category, Chinese cities rank next (448 points), while cities located in the other regions present much lower scores, below 400 points per city. For the category of *mid-sized cities*, the trends are similar, although there is a significant decrease in scores across all regions except for Europe, where scores remain stable (455 points per city). In contrast, the rest of the regions exhibit scores below 400 points per city, indicating a notable decline, particularly in the USA. Similar trends are observed for the category of *least-populated cities*, with Europe again obtaining the highest scores (446 points). In the case of the USA, *least-populated cities* received a higher score compared to the previous category (425 points), while the third best-rated are the Indian cities (401 points). African cities consistently rank at the bottom with scores below 340 points across all categories, while Latin American cities are the second lowest-rated for the mid-sized and least-populated categories. The cities in the Middle East receive slightly better scores, especially when they are less populated.

We also observe significant differences across city sizes according to different indicators. The group of *large-populated cities* shows much higher scores in aspects related to social life, perception, economy, and especially urban morphology. When comparing the scores by indicators' groups with the worst scored group (*least-populated cities* in all cases), *large-populated cities* scored +163 points in urban morphology (486 vs 323), +83 points in urban social (442 vs 359), and + 65 points in

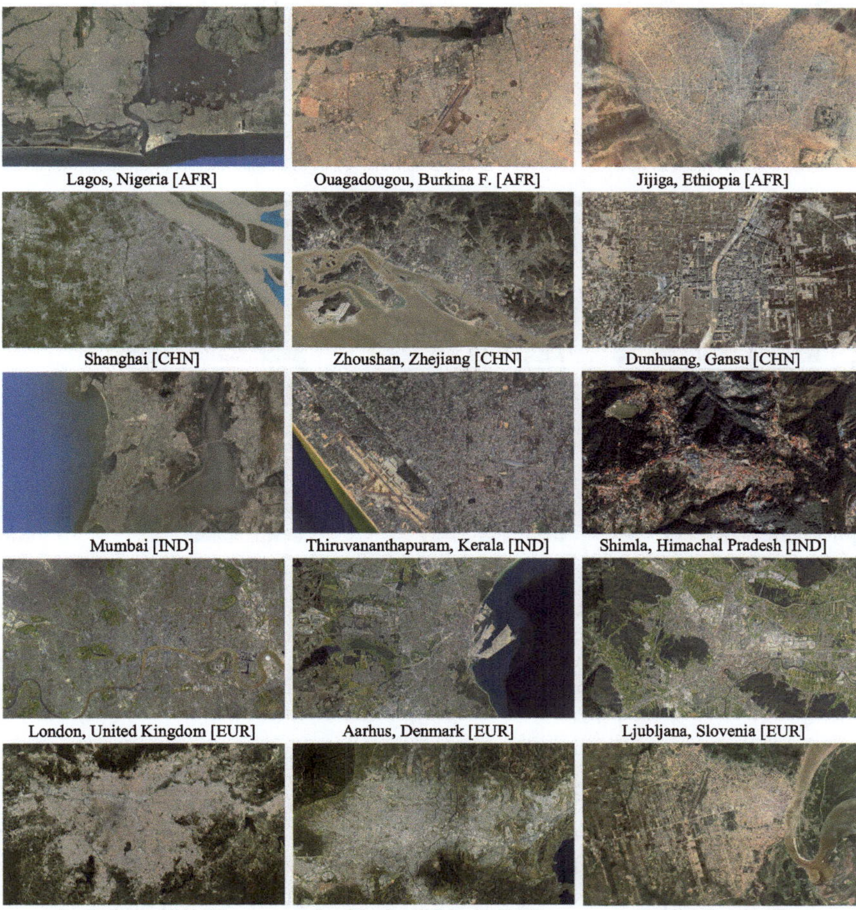

Fig. 1 Aerial imagery of the cities selected randomly by the language model according to geographical location and population criteria within each region. This panel of cities is arranged based on each city's population category (X-axis) and region (Y-axis). In short, 21 cities are represented in a 7 × 3 matrix

Fig. 1 (continued)

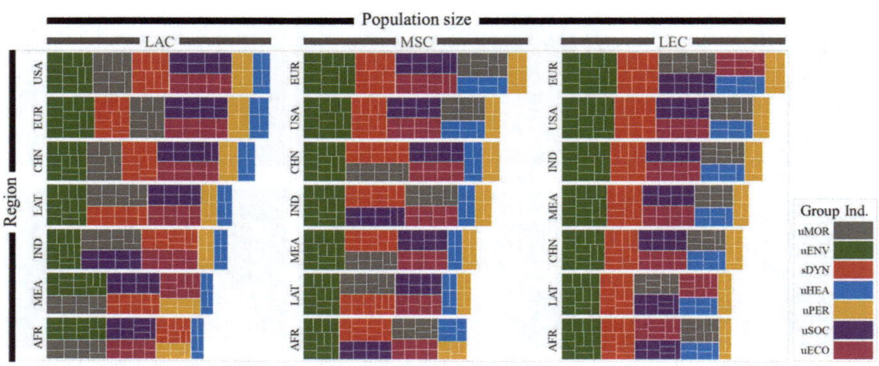

Fig. 2 Score results for the different indicator groups. Each indicator corresponds to an individual box, while boxes of the same color belong to the same indicator group. The size of each box represents the score, whether evaluated individually (indicator) or aggregated (indicator groups and/or regions). This results panel is organized based on each city's population category (*X-axis*) and region (*Y-axis*). In summary, a 7 × 3 matrix represents 21 cities

urban economy (416 vs 351). On the other hand, *least-populated cities* were rated higher than *large-populated cities* in urban health (258 vs 238), urban environment (643 vs 594), and spatial dynamics (510 vs 453). On the other hand, *mid-sized cities* hold intermediate positions in almost all indicator groups, except for the urban perception's group where they emerge as the lowest scored. Figure 3 (left) represents the results aggregated by population size, while detailed results can be found in the Appendix (Sect. A3).

These results reveal several insights. *Large-populated cities* like New York and London excel in urban economy indicators, showcasing robust economic

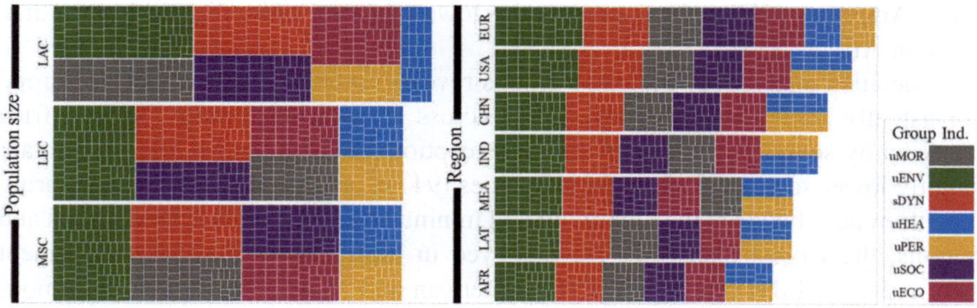

Fig. 3 Score results for the different indicator groups. These results aggregate cities by (**left**) population size and (**right**) region. Each indicator corresponds to an individual box, while boxes of the same color belong to the same indicator group. The size of each box represents the score, whether evaluated individually (indicator) or aggregated (indicator groups)

foundations despite challenges in urban health and environmental issues due to dense populations and industrial legacies. Economic diversity is evident globally, yet affordable housing remains a persistent challenge. While these cities enjoy positive perceptions driven by global prominence and cultural richness, issues like overcrowding and crime persist. Urban health indicators remain crucial, with pollution and limited green spaces affecting populous hubs. *Mid-sized cities*, like Aarhus in Europe, demonstrate economic health, while regions like Ouagadougou in Africa face challenges in affluence and job opportunities. Urban morphology varies, with European cities blending historical and modern elements, while Asian or African cities lean toward rapid urbanization or traditional layouts. Socially, community engagement varies, with developed regions boasting more structured systems. Urban perception in developed regions scores high due to superior infrastructure, while environmental concerns are universal, with wealthier regions having more resources to combat pollution. Lastly, *least-populated cities* like Ljubljana exhibit economic strength, translating to a high quality of life. But those located in regions like Latin America and Africa face significant limitations in urban health and spatial dynamics.

6.3 *Analysis of Results by Region*

As mentioned earlier, the cities under study are distributed across 7 regions, with 3 cities selected from each region representing various population sizes within the categories defined beforehand. Grouping cities by region (sum across the three cities in each region) unveils noteworthy insights, irrespective of population size. Aggregate scores highlight Western cities take the lead with 1378 points in Europe and 1295 points in the USA. In contrast, African cities trail behind with 1009 points, marking the lowest scores. The rest of the regions occupy intermediate positions. Chinese cities score relatively high with 1208 points, and Indian cities follow closely with 1173 points. In contrast, cities located in the Middle East (1083 points) and

Latin America (1079 points) present the lowest scores, only slightly higher than those in Africa.

A detailed analysis by indicator groups reveals several notable findings. Europe consistently achieves the highest scores across all indicator groups, while Africa consistently scores the lowest, with one exception: cities in the Middle East obtain slightly lower scores (93) than African cities (94) in urban health. When comparing the differences between the maximum and minimum values for indicator groups and regions, the largest disparities are observed in the groups of urban environment (99) and spatial dynamics (67). Latin American and Middle Eastern cities demonstrate similar scores across most groups, with the most significant differences found in the group of urban morphology, where Latin American cities score notably better (+20 points). Chinese cities rank the second-highest after those in Western countries, often surpassing Indian cities in many indicators. On the other hand, Latin American cities tend to score low in urban economy, urban environment, urban health, urban perception, and urban social, frequently falling below Middle Eastern cities and sometimes achieving similar scores to African cities. Figure 3 (right) visually represents the results aggregated by region, while detailed results can be found in the Appendix Sect. 3 (Figs. 5 and 6).

The analysis reveals that cities in the Western world, including North America and Europe, consistently exhibit strong economic indicators across all sizes. This economic strength typically translates into better-developed health infrastructure, urban planning, and more sustainable environmental management practices. There may be, however, a downside in terms of perceived community cohesiveness in these cities. Conversely, cities in emerging regions such as Latin America, Africa, and parts of Asia may show varying scores in urban economic indicators, influenced by their stages of industrialization and economic development within the context of globalization (Balsa-Barreiro et al. 2019). Despite this, they often boast rich cultural ties, fostering stronger community bonds and city perceptions. Additionally, they demonstrate notable resilience in overcoming challenges related to rapid urbanization without corresponding infrastructural development. Regarding spatial dynamics, Western cities tend to have higher levels of accessibility and connectivity due to their efficient transportation systems. Finally, while environmental concerns are global, cities in affluent regions are often viewed as more sustainable. This perception stems from their access to greater resources and their capacity to implement urban planning strategies aimed at addressing pollution and conserving green spaces.

6.4 AI-Generated Images for Cities

In order to assess the consistency and quality of the obtained results, we generated (artificially) images of the studied areas. We tried to reconstruct the built environments of each city based on the image that the AI has of these spaces. Initially, we attempted to create photorealistic images using the *ChatGPT* image-generating interface, prompting *DALL-E 3*. But the results consistently produced cartoonish images that lacked the realism necessary for urban science. These limitations arise from technical constraints, including insufficient training data for intricate details and the computational demands of generating high-resolution images. Additionally, ethical considerations could also play a significant role: many AI image generators impose restrictions to protect individual privacy and counteract the potential amplification of biases in order to mitigate the spread of deepfakes and misinformation.

Thus, we turned to an AI-Powered Image Generator (*Getimg*[2]), which has the capability to create new images based on an input text-prompt describing the elements to be included in the generated image. This text-to-image model is built upon a latent diffusion model known as *Stable Diffusion*, a type of deep generative artificial neural network trained to eliminate Gaussian noise from training images. We employ *DPM-Solver++* as the sampler for generating the final image (Rombach et al. 2022).

Generated images must encapsulate the most significant features based on the perception of an average individual in each city. This individual is shaped by the prevailing demographic attributes of the city, including age, sex, occupation, socio-economic status, and other relevant factors. The representation of the urban environment is simulated at a specific time and location, corresponding to 3 p.m. on a weekday at the street where the individual's workplace is located. The resulting images are depicted in Fig. 4, while the input prompt used for generating these images is provided in the Appendix (Sect. A2).

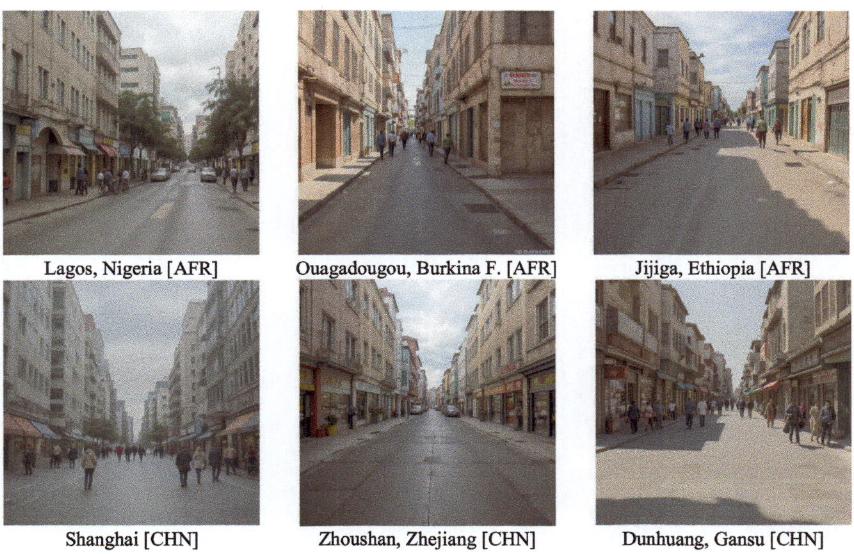

Fig. 4 Imagery generated for all the cities by using an AI-powered image generator. This panel of results is arranged based on each city's population category (*X-axis*) and region (*Y-axis*). In short, 21 cities are depicted in a 7 × 3 matrix

[2] Available online at https://getimg.ai (Accessed on Jan. 20, 2024).

Mumbai [IND] | Thiruvananthapuram, Kerala [IND] | Shimla, Himachal Pradesh [IND]

London, United Kingdom [EUR] | Aarhus, Denmark [EUR] | Ljubljana, Slovenia [EUR]

Sao Paulo, Brazil [LAT] | San Salvador, El Salvador [LAT] | Pucallpa, Peru [LAT]

Cairo, Egypt [MEA] | Erbil, Iraq [MEA] | Nizwa, Oman [MEA]

New York City, NY [USA] | Wichita, KS [USA] | Flagstaff, AZ [USA]

Fig. 4 (continued)

The results obtained offer striking insights from various angles, shedding light on how AI perceives cities and helping us understand better the scores assigned to each city. We observe highly realistic images that effectively portray each city, providing a glimpse into the unique architectural styles prevalent in different regions. Initially, it is noticeable that built environments dominate the images, although urbanization levels vary significantly. Surprisingly, most cities depicted lack of traffic, yet they still exhibit signs of life. Differences emerge, such as Middle Eastern cities featuring more trees, while African and Indian cities appear livelier compared with Western cities, which seem relatively empty. Interestingly, Indian cities exhibit similar characteristics across different population sizes, whereas in other regions, there is a clear link between building density and population size, with larger cities showing denser urban landscapes. Finally, it is important to reiterate that the depiction of each city relies on what we have considered as an average person, which varies in each city according to its particular characteristics. This could explain why, in some cases, the AI generates built environments that are quieter and less urbanized than initially expected.

7 Discussion

Understanding the dynamic interplay between built environments and human behavior involves examining a multitude of factors, many of which depend on the shape of cities. Exploring how cities are perceived is intricate, requiring consideration of various aspects across numerous indicators. Recent strides in *Generative AI* offer promising avenues to navigate this complexity. Hence, it is crucial to assess the capabilities of AI tools for urban science by evaluating their current effectiveness.

This case study illustrates a groundbreaking use of *Generative AI* in urban science. By integrating textual knowledge, AI models access swiftly diverse information sources, facilitating a comprehensive urban planning approach. They analyze extensive quantitative and qualitative data from various sources, including journal papers, reports, social media posts, and policy documents, to offer deep insights into urban environments. Their ability for synthetizing information is adaptable and scalable to user needs. Moreover, their cost-effectiveness and easy access provide a significant advantage over traditional methods, empowering researchers to address complex urban challenges effectively. Hence, in the era of global urbanization, harnessing *Generative AI* has the potential to enrich our understanding of cities and adopt policies to enhance the well-being of urban residents.

Here, we used *Generative AI* and language modeling to simulate experiences and perceptions in urban environments, incorporating physical and social indicators across a diverse range of cities. These cities vary in population size and geographical location. The *Generative AI* demonstrates its ability to evaluate and score indicators across all cities, even those least populated where data may be limited. Their consistent results across diverse urban settings highlight their ability to decode complex urban dynamics, offering valuable insights for researchers and policymakers. Importantly, these models can capture qualitative aspects like public sentiment, going beyond quantitative data analysis. While our study has limitations, we illustrate how AI tools enhance our understanding about how humans perceive and interact with built environments, potentially unveiling behavioral patterns and social interactions across multiple scales.

The use of *Generative AI* in urban science still faces several limitations. One major concern is the presence of inherent biases in the AI model's responses, which arise due to various factors such as limited data availability and disparities in training datasets across regions. Our findings may not capture fully the dynamics of lower-income regions or consider adequately cultural and historical factors, potentially leading to biased responses. Comparisons between cities across different regions highlight that Western cities tend to receive higher scores in most of the indicator groups, likely because of the availability of more extensive datasets coming from these regions. Therefore, it is crucial to enhance the model's training datasets with more representative samples from diverse regions to improve its reliability. Furthermore, spatial differences in the training datasets may reflect social disparities, as the internet user profile may not fully represent marginalized groups. This can lead to biases in the AI model's responses, potentially skewing the perceptions of urban dwellers. Surprisingly, some regions with expectedly lower scores in certain indicator groups received moderately high scores, suggesting potential biases. There may also be a Western-centric bias in how the model interprets cities, influenced by the majority of training datasets coming from Western tourists visiting cities in other regions. Noteworthy, discrepancies between the model's population estimations and official census data were observed for some cities, particularly in the group of least-populated cities. This is the case of *Pucallpa* in Latin America or *Ljubljana* in Europe, appearing to be more populated than what was estimated by the AI model. This could be attributed to ambiguities in defining city boundaries and highlight the need for improved data accuracy and consistency in future AI applications in urban science.

Regarding the results, there are four main points to consider. Firstly, the simulations conducted by *Generative AI* models positively score a city model characterized by compactness, concentration, diversity, and social vibrancy, aligning with perspectives proposed by authors such as Jacobs (1961), Jabareen (2006), Mouratidis (2019), and Montgomery (2013). This urban perspective may not, however, represent fully all types of urban layouts, potentially undervaluing alternative models such as urban sprawl cities, which might perform better in certain indicators. Secondly, we must emphasize that our results offer a snapshot of using the current (August 2023) version of *GPT4* via the *ChatGPT* interface. Therefore, as the volume of training datasets increases, AI models are expected to generate different and more nuanced responses over time. In addition, the rapid advancement of highly adaptable and reusable models is expected to introduce new capabilities and opportunities in this direction. Thirdly, this study marks the inaugural exploration of its kind. Our findings must be regarded within a highly circumscribed scope, where only one city per size category and region is considered. Therefore, to draw more robust conclusions for policymaking, it may be necessary to increase the sample size. Fourthly, while *Generative AI* models hold promise for real-world urban interventions, it is important to acknowledge that we are in a preliminary phase, and the responses may carry certain constraints or potential biases. Hence, we advocate for a supervised and judicious scrutiny of the responses to ensure a responsible and ethical utilization of AI models in urban planning and governance.

8 Conclusions

Understanding how urban dwellers perceive and experience cities is complex and influenced by various factors, often shaped by the city's form. These factors can be broken down into individual indicators covering both the physical and human aspects of the city, many of which are interconnected. By examining these indicators, we aim to gain a comprehensive understanding of cities based on the perceptions of their residents.

Recent advancements in *Generative AI* offer promising opportunities to explore this complexity, highlighting the importance of evaluating its effectiveness in urban science. *Generative AI* models offer significant opportunities to integrate information from diverse data sources, reshaping our understanding of cities. As these models undergo further training, they become increasingly reliable in generating and predicting complex patterns of human behavior. However, integrating these models encounters challenges such as data validation, inherent biases, and ethical considerations. To address these challenges, it is crucial to ensure that training datasets are not only large but also diverse, accurately representing social and spatial diversity. Moving forward, the urban science community must carefully consider these limitations to ensure the reliability of results. This will enable the adoption of appropriate urban policies that benefit all citizens.

Acknowledgments The results of this work are part of the *MIT International Science and Technology Initiatives* (MISTI) Project titled "*COST-OMIZE: Understanding Customer Behavior under Urban Segregation for Cost-reducing and Sustainable Logistics*" funded by the *MIT-Spain INDITEX Sustainability Seed Fund* in 2020. J. Balsa-Barreiro acknowledges support from the *MIT-Spain INDITEX Sustainability Seed Fund* and *MISTI*. With M. Menendez, they acknowledge support from *NYUAD CITIES*, funded by *Tamkeen* under *NYUAD Research Institute Award CG001*. M. Cebrian acknowledges the *Spanish Ministry of Science, Innovation, and Universities* and the Call for Grants from Carlos III University of Madrid for the requalification of the Spanish university system for 2021–23, dated July 1, 2021 (R.D. 289/2021, Apr.20), which regulates the direct granting of subsidies to public universities for the requalification of the Spanish university system. Furthermore, he is thankful for the support of project number *PID2022-137243OB-I00 funded by the MCIN/AEI/10.13039/501100011033* and "*FEDER Una manera de hacer Europa.*"

Appendix

Throughout this appendix, we present relevant information related to the manuscript. This information includes:

- A1. Indicator definition and evaluation criteria. This section provides detailed descriptions of each indicator group and the individual indicators within each group. In total, we define 52 indicators distributed across 7 groups. Additionally, we present a brief description of the criteria defined for evaluating each indicator.
- A2. Conversational prompts. In this section, we transcribe the conversational prompts used in this study.
- A3. Detailed results. This section provides a more detailed and comprehensive review of the results presented in the book chapter, including charts and tables.

A1. Indicator Definition and Evaluation Criteria

We present the list of indicators used in our case study, grouped into 7 categories: Urban Morphology (uMOR), Urban Environment (uENV), Spatial Dynamics (sDYN), Urban Health (uHEA), Urban Perception (uPER), Urban Social (uSOC), and Urban Economy (uECO). Each group includes a general description of the indicators followed by a table detailing each one, providing a brief description and specific evaluation criteria.

Group 1. Urban Morphology (uMOR)

This indicator group evaluates key urban metrics that define the city's physical structure. It includes population density, building height profiles, and street widths. For instance, skyscraper prevalence reflects vertical growth, while the "Street-building Height Ratio" gauges street openness. "Spatial Order" evaluates urban planning coherence. Together, these metrics provide insights into the city's development trends, planning strategies, and aesthetic appeal.

Code	Indicator	Description	High Score	Low Score
uMOR-01	Building compactness	Built-up to total land ratio	High ratio	Low ratio
uMOR-02	Population density	Residents per unit area	High density	Low density
uMOR-03	Mean building height	Verticality of urban form	Tall	Short
uMOR-04	Mean built volume	Building footprints & heights	High volume	Low volume
uMOR-05	High-rise density	Tall buildings prevalence	Many	Few
uMOR-06	Street width	Streetscape dimension	Wide	Narrow
uMOR-07	Sidewalk width	Pedestrian space width	Wide	Narrow
uMOR-08	Street-building height ratio	Street space enclosure	High enclosure	Low enclosure
uMOR-09	Spatial order	Urban fabric structure	Structured	Haphazard

Group 2. Urban Environment (uENV)

This indicator group assesses urban elements related to the environment, which directly influence the quality of life in cities. We evaluate factors such as the presence and distribution of green spaces, recreational areas, and the adequacy of city lighting. Additionally, broader aspects like weather conditions and air quality are considered. Essentially, these indicators reflect how well the city balances urban development with the preservation of natural resources, along with efforts to enhance lighting and air quality, thereby contributing to the well-being of its residents.

Code	Indicator	Description	High Score	Low Score
uENV-01	Green space	Urban green availability	Abundant greenery	Scarce greenery
uENV-02	Green distribution	Green spread	Evenly distributed	Clustered green areas
uENV-03	Leisure areas	Recreational spaces	Many areas available	Few areas
uENV-04	Daylight	Natural light during the day	Bright	Dim
uENV-05	Night light	Nighttime illumination	Well-lit	Poorly lit
uENV-06	Weather	Climatic conditions	Pleasant	Harsh/extreme
uENV-07	Air quality	Cleanliness of air	Fresh air	Polluted

Group 3. Spatial Dynamics (sDYN)

This indicator group evaluates various aspects of urban transportation, such as daily commute duration, public transportation quality and accessibility, and provisions for alternative modes like cycling and micromobility modes. It also evaluates the city's connectivity and ease of movement by different transport modes. In summary, these indicators provide a thorough understanding of a city's transportation system, revealing its efficiency, accessibility, and adaptability.

Code	Indicator	Description	High Score	Low Score
sDYN-01	Commute time	Duration of travel	Short trips	Long commutes
sDYN-02	Accessibility	Transport infrastructure ease	Easy access	Hard to access
sDYN-03	Connectivity	Transport network cohesion	Well-connected	Fragmented routes
sDYN-04	Transit access	Public transport availability	Frequent services	Infrequent/limited
sDYN-05	Congestion	Traffic flow	Smooth traffic	Heavy congestion

(continued)

Code	Indicator	Description	High Score	Low Score
sDYN-06	Public transport importance	Value of public transit	Highly valued	Not prioritized
sDYN-07	Public transport quality	Quality of public services	Highly satisfactory	Needs improvement
sDYN-08	Bike infrastructure	Cycling provisions	Excellent paths	Insufficient paths
sDYN-09	Micromobility	Usage of micro-modes	Widely adopted	Rarely used
sDYN-10	Walkability	Pedestrian-friendly areas	High walkability	Not pedestrian-friendly

Group 4. Urban Health (uHEA)

This indicator group focuses on the overall health and well-being of urban dwellers, encompassing indicators such as life expectancy, mental resilience, and overall life satisfaction. It evaluates the city's ability to support residents' health and well-being, considering physical longevity as well as mental and emotional health aspects.

Code	Indicator	Description	High Score	Low Score
uHEA-01	Life expectancy	Projected lifespan	Long lifespan	Shorter lifespan
uHEA-02	Mental resilience	Overcoming mental distress	Low resilience	High resilience
uHEA-03	Mental Well-being	Absence of depression	Fewer cases	More cases
uHEA-04	Well-being	Health and life satisfaction	High satisfaction	Low satisfaction
uHEA-05	Livability	Overall quality of life	Highly livable	Less livable

Group 5. Urban Perception (uPER)

This indicator group evaluates the visual and functional appeal of the city, considering factors such as its aesthetic attractiveness, architectural consistency, and ability to attract tourists. It provides insight into how appealing the city is to both residents and visitors, with positive assessments indicating a harmonious and visually appealing environment.

Code	Indicator	Description	High Score	Low Score
uPER-01	Perceived beauty	Aesthetic appeal of cityscape	Highly appealing	Less appealing

(continued)

Code	Indicator	Description	High Score	Low Score
uPER-02	Architectural harmony	Consistency of built styles	Cohesive designs	Mixed designs
uPER-03	City living comfort	Ease of living in the city	Highly comfortable	Less comfortable
uPER-04	Tourism appeal	City's attraction for visitors	Major tourist attraction	Less frequented
uPER-05	Aesthetics	Visual quality of urban area	Visually stunning	Less striking

Group 6. Urban Social (uSOC)

This indicator group evaluates the vibrancy, diversity, and social cohesion of a city, including its cultural richness, multicultural composition, and the integration of its residents. It assesses both the daytime and nighttime energy of the city, along with indicators related to social bonds and safety perceptions. Essentially, it provides insight into how lively, harmonious, and safe a city feels to its inhabitants and visitors, reflecting its ability to foster community connections and ensure a vibrant social environment.

Code	Indicator	Description	High Score	Low Score
uSOC-01	Cultural amenities	Presence of cultural venues/ events	Rich cultural scene	Limited culture
uSOC-02	Multiculturality	Ethnic & linguistic diversity	Highly diverse	Less diverse
uSOC-03	Residential integration	Mixing of social/ethnic groups	Well-integrated	Less integrated
uSOC-04	Nightlife vibrancy	Evening activity & pedestrian traffic	Bustling nightlife	Quieter nights
uSOC-05	Daylife vibrancy	Daytime activity & street usage	Active streets	Calmer daytime
uSOC-06	Social capital	Strength of community ties	Strong community bonds	Weaker ties
uSOC-07	Community safety	Incidence of reported crimes	Low crime rates	Higher crime rates
uSOC-08	Safety	Perceived security from harm	Feeling safe	Feeling insecure

Group 7. Urban Economy (uECO)

This indicator group evaluates the economic strength, inclusivity, and diversity of cities by assessing various aspects such as overall economic health, productivity, housing affordability, economic diversification, and income equality. It considers wealth generation, distribution, and job opportunities to provide a comprehensive overview of the economic landscape, reflecting its robustness, fairness, and opportunities for residents.

Code	Indicator	Description	High Score	Low Score
uECO-01	Wealth per capita	GDP per resident	High economic productivity	Lower productivity
uECO-02	Affluence rate	People above the economic comfort line	High rate above comfort line	High rate below comfort line
uECO-03	Unemployment rate	Job-seeking residents	Low joblessness	Elevated joblessness
uECO-04	Affordable housing	Homes priced for median income	Widespread affordability	Pricier housing
uECO-05	Economic diversity	Variety in industry sectors	Diverse economy	Single-sector reliance
uECO-06	Income equity	Earnings distribution	Even income spread	Larger income gaps
uECO-07	Community stability	Preservation of socio-economic structures	Strong community continuity	Rapid changes
uECO-08	Job opportunities	Job availability	Abundant opportunities	Limited job sectors

A2. Conversational Prompts

We transcribe here the conversational prompts used in this study. These are the following three:

Prompt 1. Generate a List of Cities

1. **Define Initial Conditions:**
   - Geographical Locations: Africa (AFR), China (CHN), Europe (EUR), India (IND), Latin America (LAT), the Middle East (MEA), and the United States (USA).
   - Population Size Categories:
     - Large-populated cities (LAC): Over five million inhabitants.
     - Mid-sized cities (MSC): Population ranging from 500,000 to one million.
     - Least-populated cities (LEC): Fewer than 150,000 inhabitants.
2. **Generate List:**
   - For each geographical region:

  - Randomly select one city from the large-populated cities (LAC) category.
  - Randomly select one city from the mid-sized cities (MSC) category.
  - Randomly select one city from the least-populated cities (LEC) category.

3. **Repeat for Each Region:**
   - Perform the above steps for all seven regions.
4. **Compile Final List:**
   - Once three cities are selected for each region, compile the final list totaling 21 cities.

Prompt 2. Score City Experience

1. **City Selection:**
   - Choose a specific city, referred to as {City}.
2. **Demographic Profile:**
   - Identify the predominant demographic attributes for the chosen city, including age, gender, occupation, socioeconomic status, and other relevant factors. This profile will serve as the archetype of our "demographically correct person."
3. **Simulate Daily Mobility:**
   - Simulate a typical day for the individual, outlining their movements from their residence to the workplace. This step should consider factors such as transportation mode, commute time, and any notable landmarks or challenges along the route.
4. **Urban Experience Simulation:**
   - Simulate the urban experience and perception of the individual based on a set of indicators provided in the list previously defined [Annex, A1].
5. **Scoring and Evaluation:**
   - Assign weights to each indicator to contribute to a score ranging from 0 to 10, where 0 represents the most negative value and 10 represents the most positive.
   - Evaluate the urban experience based on the weighted scores, using evaluation criteria provided in the [Annex, A1].

Prompt 3. Generate an Image Related to the Previous City Experience

1. **City Selection:**
   - Generate a photorealistic image of a street in {City}.
2. **Demographic Profile:**
   - Determine the most prevalent demographic attributes for the chosen city, including age, gender, occupation, socioeconomic status, and other relevant factors. This demographic profile will represent our "demographically correct person."
3. **Place:**
   - Depict the street where the demographically correct person works and commutes on any given workday.
4. **Time:**
   - Capture the street image at 3:00 PM on February 20th, 2024.
5. **Image Configuration:**
   - Display the street image realistically, capturing it from street level and ensuring it is centered.
6. **Urban Settings:**
   - Consider the most typical and particular settings and conditions that identify the chosen street and/or city. This may include architectural styles, street furniture, signage, and other urban elements characteristic of the area.
7. **Environmental Conditions:**
   - Take into account the most probable weather conditions at that time, including factors such as temperature, precipitation, and sky conditions.

A3. Detailed Results

A detailed breakdown of the results is provided considering different levels of aggregation, taking into account criteria such as population size and region. The results are presented in both charts and tables.

Results are summarized in the next two graphs. The first one displays the scores for indicator groups disaggregated by cities, while the second one displays them in an aggregated manner. More information is available in the respective figure captions (Figs. 5 and 6).

Results are summarized in the next four tables. These display the scores for each region in an aggregated form (Table 3) and broken down by city size (Tables 4, 5, and 6).

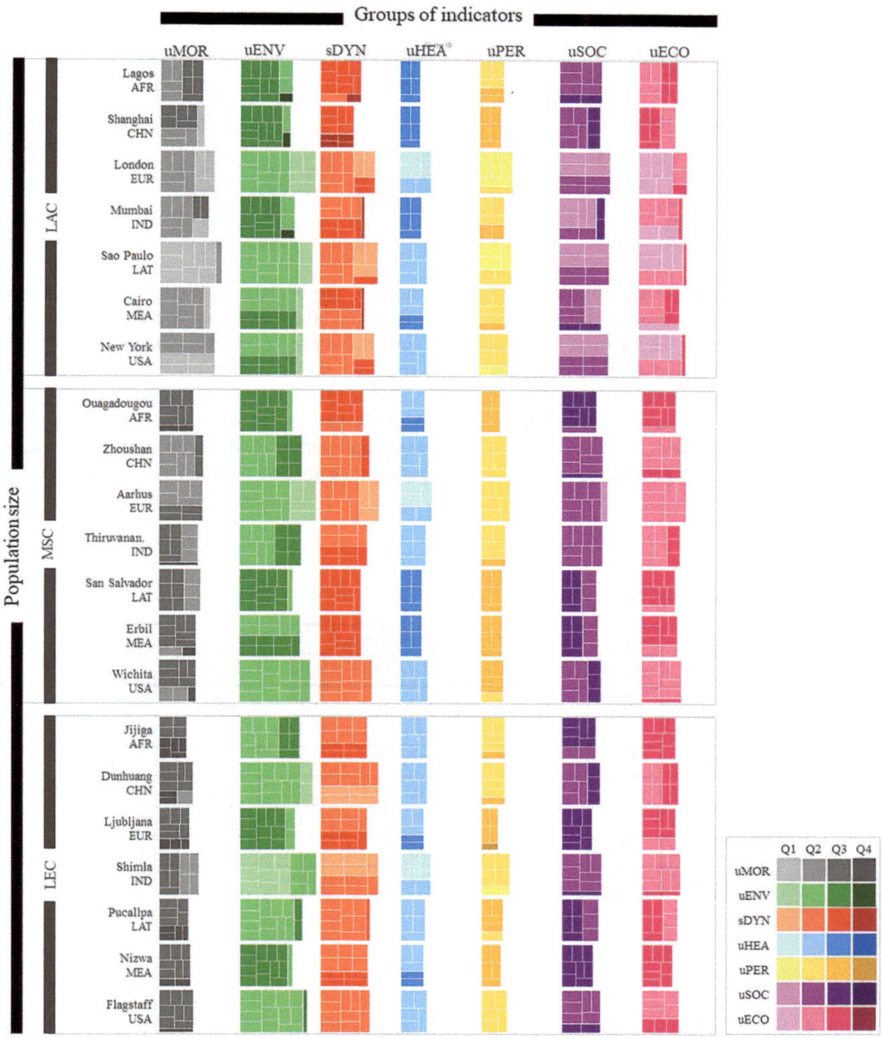

Fig. 5 Score results for the different indicator groups by cities. Each indicator corresponds to an individual box, while boxes of the same color belong to the same indicator group. The size of each box represents the score, whether evaluated individually (indicator) or aggregated (indicator groups and/or regions). Color legend categorizes scores into four quartiles, each represented by a distinct shade. Quartile Q1 denotes higher score levels (7.5–10 points) shown in lighter shades, while quartile Q4 indicates lower score levels (0–2.5 points) depicted in darker shades. Consequently, smaller boxes and darker shades represent lower scores. This results panel is organized based on indicator group (*X-axis*) and population category (*Y-axis*)

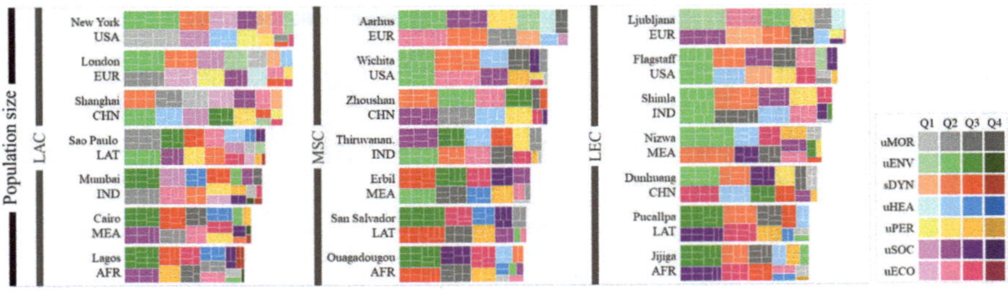

Fig. 6 Score results for the different indicator groups – aggregately presented – by individual cities. Each indicator corresponds to an individual box, while boxes of the same color belong to the same indicator group. The size of each box represents the score, whether evaluated individually (indicator) or aggregated (indicator groups and/or regions). Color legend categorizes scores into four quartiles, each represented by a distinct shade. Quartile Q1 denotes higher score levels (7.5–10 points) shown in lighter shades, while quartile Q4 indicates lower score levels (0–2.5 points) depicted in darker shades. Consequently, smaller boxes and darker shades represent lower scores. This results panel is organized based on indicator group (*X-axis*) and population category (*Y-axis*)

Table 3 Score for cities – aggregated by regions – and indicator groups

		Group of indicators							Total
		uMOR	uENV	sDYN	uHEA	uPER	uSOC	uECON	
All cities	EUR	191	320	239	129	126	191	182	1,378
	USA	183	303	234	114	109	179	173	1,295
	CHN	175	259	210	114	109	174	167	1,208
	IND	169	257	197	103	105	176	166	1,173
	MEA	151	246	184	93	93	162	154	1,083
	LAT	171	237	185	96	92	152	146	1,079
	AFR	150	221	172	94	79	148	145	1,009

Table 4 Score for Large-populated cities (LAC) – aggregated by regions – and indicator groups

		Group of indicators							Total
		uMOR	uENV	sDYN	uHEA	uPER	uSOC	uECON	
LAC	USA	84	100	79	38	44	69	66	480
	EUR	74	104	75	43	45	70	66	477
	CHN	75	87	75	38	40	68	65	448
	LAT	69	87	61	33	35	58	56	399
	IND	66	75	61	30	34	63	60	389
	MEA	58	72	56	28	33	58	53	358
	AFR	60	69	46	28	29	56	50	338
									2,889

Table 5 Score for Mid-sized cities (MSC) – aggregated by regions – and indicator groups

		Group of indicators							Total
		uMOR	uENV	sDYN	uHEA	uPER	uSOC	uECON	
MSC	EUR	60	106	81	43	40	64	61	455
	USA	51	98	72	38	31	55	55	400
	CHN	61	86	68	38	35	57	54	399
	IND	54	85	65	35	34	57	53	383
	MEA	51	84	57	30	29	51	50	352
	LAT	57	74	56	30	29	49	46	341
	AFR	47	73	59	33	26	48	47	333
									2,663

Table 6 Score for Least-populated cities (LEC) – aggregated by regions – and indicator groups

		Group of indicators							Total
		uMOR	uENV	sDYN	uHEA	uPER	uSOC	uECON	
LEC	EUR	57	110	83	43	41	57	55	446
	USA	48	105	83	38	34	55	52	415
	IND	49	97	71	38	37	56	53	401
	MEA	42	90	71	35	31	53	51	373
	CHN	39	86	67	38	34	49	48	361
	LAT	45	76	68	33	28	45	44	339
	AFR	43	79	67	33	24	44	48	338
									2,673

References

Abdullah M, Madain A, Jararweh Y (2022) ChatGPT: Fundamentals, applications and social impacts. In The Ninth International Conference on Social Networks Analysis, Management and Security (SNAMS), Milan, Italy, 2022, pp. 1–8. https://doi.org/10.1109/SNAMS58071.2022.10062688

Appleyard D (1981) Livable Streets. Univ. California Press, Oakland

Balsa-Barreiro J, Menéndez M (2021) Cómo son y cómo se mueven las redes urbanas. Los Angeles Times. https://www.latimes.com/espanol/opinion/articulo/2021-10-27/opinion-como-son-y-como-se-mueven-las-redes-urbanas (In Spanish). Accessed 20 Jan 2024

Balsa-Barreiro J, Menéndez M (2022) Fisionomía y flujos de tráfico ¿Cómo entender la movilidad en las ciudades? Foreign Affairs Latinoamérica. https://revistafal.com/fisionomia-urbana-y-flujos-de-trafico/ (In Spanish). Accessed 20 Jan 2024

Balsa-Barreiro J, Menendez M (2024) How cities influence social behavior. In: Paus T, Kum H-C (eds) Human behavior in geospatial context. The MIT Press, Boston, MA, pp 139–157

Balsa-Barreiro J, Morales AJ, Castelló E (2018) Datos, Inteligencia Artificial y Complejidad. Una Visión de la Sociedad del Futuro. Instituto de Ingeniería de España, Madrid. (In Spanish)

Balsa-Barreiro J, Li Y, Morales AJ, Pentland A (2019) Globalization and the shifting centers of gravity of world's human dynamics: implications for sustainability. J Clean Prod 239:117923. https://doi.org/10.1016/j.jclepro.2019.117923

Balsa-Barreiro J, Morales AJ, Lois-González RC (2021) Mapping population dynamics at local scales using spatial networks. Complexity 2021:8632086. https://doi.org/10.1155/2021/8632086

Bandauko E, Annan-Aggrey E, Arku G (2021) Planning and managing urbanisation in the twenty-first century: content analysis of selected African countries' national urban policies. Urban Res Pract 14(1):94–104

Bender EM, Gebru T, McMillan-Major A, et al. (2021) On the dangers of stochastic parrots: can language models be too big? In Conference on fairness, accountability, and transparency (FAccT '21), March 3–10, 2021, Virtual Event, Canada. ACM, New York, NY, USA

Berghauser M, Haupt P (2010) Space, density and urban form. NAi Publishers, Rotterdam

Biljecki F, Chow YS (2022) Global building morphology indicators. Comput Environ Urban Syst 95:101809. https://doi.org/10.1016/j.compenvurbsys.2022.101809

Bloom DE, Canning D, Fink G (2008) Urbanization and the wealth of nations. Science 319:772–775

Boeing G (2019) Urban spatial order: street network orientation, configuration, and entropy. Appl Netw Sci 4:67. https://doi.org/10.1007/s41109-019-0189-1

Bommasani, et al. (2021) On the opportunities and risks of foundation models. https://arxiv.org/pdf/2108.07258.pdf. Accessed 20 Jan 2024

Bourdic L, Salat S, Nowacki C (2012) Assessing cities: A new system of cross-scale spatial indicators. Build Res Inf 40:592–605

Braubach M, Egorov A, Mudu P et al (2017) Effects of urban green space on environmental health, equity and resilience. In: Kabisch N et al (eds) Theory and practice of urban sustainability transitions. Springer, Cham, pp 187–205

Burton E (2000) The Compact City: just or just compact? A preliminary analysis. Urban Stud 37(11):1969–2006

Burton E, Jenks M, Williams K (1996) The compact city: a sustainable urban form? Routledge, London

Chamberlain RC, Fecht D, Davies B, Laverty AA (2023) Health effects of low emission and congestion charging zones: A systematic review. Lancet Public Health 8(7):e559–e574

Collier P (2006) Africa: geography and growth. Oxford University Press, Oxford

Collier P (2007) The bottom billion: why the poorest countries are failing and what can be done about it. Oxford University Press, Oxford

Commission of the European Communities, CEC (1990) Green paper on the urban environment: communication from the commission to the council and the parliament. Office for Official Publications of the European Communities, Luxembourg

Cooper G (2023) Examining science education in ChatGPT: an exploratory study of generative artificial intelligence. J Sci Educ Technol 32:444–452

Dhamani N, Engler M (2024) Introduction to generative AI. Ed. Manning Publications, Shelter Island, NY

Dibble J, Prelorendjos A, Romice O et al (2019) On the origin of spaces: morphometric foundations of urban form evolution. Environ Plan B Urban Anal City Sci 46(4):707–730

Dobbs R, Remes J (2013) Trends: the shifting urban economic landscape: what does it mean for cities? In: World Bank's sixth urban research and knowledge symposium. World Bank, Washington, D.C.

Erdoğan S, Onifade ST, Altuntaş M et al (2022) Synthesizing urbanization and carbon emissions in Africa: how viable is environmental sustainability amid the quest for economic growth in a globalized world? Environ Sci Pollut Res 29:24348–24361

Feuerriegel S, Hartmann J, Janiesch C et al (2024) Generative AI. Bus Inf Syst Eng 66:111–126

Fisher J, Rankin E, Irvine K, Goddard M (2022) Can biodiverse streetscapes mitigate the effects of noise and air pollution on human wellbeing? Environ Res 212(A):113154. https://doi.org/10.1016/j.envres.2022.113154

Fleischmann M, Romice O, Porta S (2020) Measuring urban form: overcoming terminological inconsistencies for a quantitative and comprehensive morphologic analysis of cities. Environ Plan B Urban Anal City Sci 48:2133–2150
Frick SA, Rodríguez-Pose A (2018) Big or small cities? On City size and economic growth. Growth Chang 49:4–32
Gehl J (1971) Life between buildings: using public space. Ed. Island Press, Washington, DC
Gill IS, Goh C-C (2010) Scale economies and cities. World Bank Res Obs 25:235–262
Hashimov A, Pathiddinov I, Makhmutova M et al (2013) Urbanization in Central Asia: challenges, issues and prospects. Center for Economic Research, Tashkent
He C, Huang Q, Bai X et al (2021) A global analysis of the relationship between urbanization and fatalities in earthquake-prone areas. Int J Disaster Risk Sci 12:805–820
Hermosilla T, Palomar-Vázquez J, Balaguer-Beser A, Balsa-Barreiro J, Ruiz LA (2014) Using street based metrics to characterize urban typologies. Comput Environ Urban Syst 44:68–79
Hyde -with minor processing by Our World in Data (2023) Population share in urban areas [dataset]. PBL Netherlands Environmental Assessment Agency, History Database of the Global Environment 3.3 (original data). https://ourworldindata.org/grapher/urbanization-vs-gdp. Accessed 20 Jan 2024
Iungman T, Cirach M, Marando F et al (2023) Cooling cities through urban green infrastructure: a health impact assessment of European cities. Lancet 401(10376):577–589
Jabareen YR (2006) Sustainable urban forms: their typologies, models, and concepts. J Plan Educ Res 26(1). https://doi.org/10.1177/0739456X05285119
Jacobs J (1961) The death and life of great American cities. Random House, New York
Knowles D (2023) How Tokyo became an anti-car paradise. https://heatmap.news/economy/tokyo-anti-car-pedestrian-paradise. Accessed 20 Jan 2024
Kotkin J (2020) The coming age of dispersion. Quillette. https://quillette.com/2020/03/25/the-coming-age-of-dispersion/. Accessed 20 Jan 2024
Kropf K (2009) Aspects of urban form. Urban Morphol 13:105–120
Kuddus MA, Tynan E, McBryde E (2020) Urbanization: A problem for the rich and the poor? Public Health Rev 41., Article Number 1. https://doi.org/10.1186/s40985-019-0116-0
Liu Z, Chen X, Cui H et al (2023) Green space exposure on depression and anxiety outcomes: a meta-analysis. Environ Res 231(Pt 3):116303. https://doi.org/10.1016/j.envres.2023.116303
Marquez L, Smith N (1999) A framework for linking urban form and air quality. Environ Model Softw 14:541–548
Marshall S (2004) Street and patterns: the structure of urban geometry. Routledge, London
Mayer-Schonberger V, Cukier K (2013) Big data: a revolution that will transform how we live, work and think. Ed. Houghton Mifflin Harcourt, New York, NY
McCarty J, Kaza N (2015) Urban form and air quality in the United States. Landsc Urban Plan 139: 168–179
McDonald RI, Kroeger T, Zhang P, Hamel P (2020) The value of US urban tree cover for reducing heat-related health impacts and electricity consumption. Ecosystems 23:137–150
Montgomery C (2013) Happy City: transforming our lives through urban design. Farrar, Straus and Giroux, New York
Moor M, Banerjee O, Abad Z et al (2023) Foundation models for generalist medical artificial intelligence. Nature 616:259–265
Mouratidis K (2019) Compact City, urban sprawl, and subjective well-being. Cities 92:261–272
Mouratidis K (2021) Urban planning and quality of life: A review of pathways linking the built environment to subjective well-being. Cities 115:103229. https://doi.org/10.1016/j.cities.2021.103229
Mouratidis K, Ettema D, Næss P (2019) Urban form, travel behavior, and travel satisfaction. Transp Res A Policy Pract 129:306–320
Myers D, Mohawesh R, Chellaboina V et al (2024) Foundation and large language models: fundamentals, challenges, opportunities, and social impacts. Clust Comput 27:1–26. https://doi.org/10.1007/s10586-023-04203-7

Obradovich N, Fowler JH (2017) Climate change may alter human physical activity patterns. Nat Hum Behav 1:0097. https://doi.org/10.1038/s41562-017-0097

Ooi K-B, Tan GW, Al-Emran M et al (2023) The potential of generative artificial intelligence across disciplines: perspectives and future directions. J Comput Inf Syst. https://doi.org/10.1080/08874417.2023.2261010

OpenAI, et al. (2023) GPT-4 Technical Report. https://arxiv.org/abs/2303.08774

Ortigosa J, Menendez M (2014) Traffic performance on quasi-grid urban structures. Cities 36:18–27

Park RE (1915) The City: suggestions for the investigation of human behavior in the city environment. Am J Sociol 20(5):577–612

Pentland A (2014) Social physics: how social networks can make us smarter. Penguin Books, London

Peterson JA (1979) The impact of sanitary reform upon American urban planning, 1840–1890. J Soc Hist 13(1):83–103

Rane N, Choudhary S, Rane J (2024) Contribution of ChatGPT and similar generative Artificial Intelligence for enhanced climate change mitigation strategies. https://doi.org/10.2139/ssrn.4681720

Rombach R, Blattmann A Lorenz D, Esser P, Ommer B (2022) High-resolution image synthesis with latent diffusion models. In Computer Vision and Pattern Recognition CVPR 2022. https://arxiv.org/abs/2112.10752

Shoemaker TA, Beaino C, Centella DM, Zhao W, Tanissa C, Lawrence J, Hashash Y (2023) Generative AI: the new geotechnical assistant? J Geotech Geoenviron 149(10). https://doi.org/10.1061/JGGEFK.GTENG-11859

Shutters ST, Applegate JM, Wentz E et al (2022) Urbanization favors high wage earners. npj Urban Sustain 2:6. https://doi.org/10.1038/s42949-022-00049-x

Smit W (2021) Urbanization in the global south. Oxford Research Encyclopedia of Global Public Health, Oxford

Spence M, Annez PC, Buckley RM (2009) Urbanization and growth: commission on growth and development. World Bank Publications, Washington, DC. (number 2582)

Stojanovski T (2019) Urban form and mobility analysis and information to catalyse sustainable development. Ph.D. dissertation. KTH Royal Institute of Technology, Stockholm

Thompson R, Lawrance EL, Roberts LF, Grailey K, Ashrafian H, Maheswaran H, Toledano MB, Darzi A (2023) Ambient temperature and mental health: a systematic review and meta-analysis. Lancet 7(7):e580–e589

Tiu E, Talius E, Patel P et al (2022) Expert-level detection of pathologies from unannotated chest X-ray images via self-supervised learning. Nat Biomed Eng 6:1399–1406

Travisi CM, Camagni R, Nijkamp P (2010) Impacts of urban sprawl and commuting: a modelling study for Italy. J Transp Geogr 18:382–392

United Nations (2023*) World urbanization prospects: the 2018 revision. New York, United Nations, Department of Economic and Social Affairs, Population Division. *Updated version of the original report https://population.un.org/wup/Publications/Files/WUP2018-Report.pdf. Accessed 20 Jan 2024

United Nations Committee for Development Policy, UNCDP (2020) Outcome of the comprehensive review of the LDC criteria. https://www.un.org/development/desa/dpad/wp-content/uploads/sites/45/CDP-2020-Criteria-review-outcome.pdf. Accessed 20 Jan 2024

United Nations Development Programme, UNDP (2017) Leveraging urbanization and governance for growth in Africa: A framework for action. United Nations Development Programme, Addis Ababa

Wan G, Zhang X, Zhao M (2022) Urbanization can help reduce income inequality. npj Urban Sustain 2:1. https://doi.org/10.1038/s42949-021-00040-y

Weigle MC, Balsa-Barreiro J, Chawla NV, Tamas D, Melchior M, Pallante V, Sarker A, Gilliland J (2024) Characterizing social environments in the physical and virtual worlds using digital

data. In: Paus T, Kum H-C (eds) Human behavior in geospatial context., pp. 51–78. The MIT Press, Boston, MA
Wheaton WC, Shishido H (1981) Urban concentration, agglomeration economies, and the level of economic development. Econ Dev Cult Chang 30:17–30
Whyte WH (1980) The social life of small urban spaces. Ed. Project for Public Spaces, INC., New York
Williams DS, Máñez M, Sutherland C, Celliers L, Scheffran J (2019) Vulnerability of informal settlements in the context of rapid urbanization and climate change. Environ Urban 31:157–176
World Bank (2023) How are the income group thresholds updated? https://datahelpdesk.worldbank.org/knowledgebase/articles/378833-how-are-the-income-group-thresholds-determined. Accessed 20 Jan 2024
Xia C, Yeh AG, Zhang A (2020) Analyzing spatial relationships between urban land use intensity and urban vitality at street block level: A case study of five Chinese megacities. Landsc Urban Plan 193:103669. https://doi.org/10.1016/j.landurbplan.2019.103669
Yan Y, Wen H, Zhong S Chen W, Wen Q, Zimmermann R, Liang Y (2023) When urban region profiling meets large language models. https://ar5iv.labs.arxiv.org/html/2310.18340
Zhang P, Kamel Boulos MN (2023) Generative AI in medicine and healthcare: promises, opportunities and challenges. Future Internet 15(9):286. https://doi.org/10.3390/fi15090286
Zhang C, Zhang C, Zhang M, Kweon IS (2023) Text-to-image diffusion models in generative AI: A survey. J Latex Class Files. https://doi.org/10.48550/arXiv.2303.07909

Area-level Measures of the Social Environment: Operationalization, Pitfalls, and Ways Forward

Marco Helbich, Yi Zeng, and Abeed Sarker

Contents

The original version of the chapter has been revised. A correction to this chapter can be found at https://doi.org/10.1007/7854_2024_481

M. Helbich (✉)
Department of Human Geography and Spatial Planning, Faculty of Geosciences, Utrecht University, Utrecht, The Netherlands

Health and Quality of Life in a Green and Sustainable Environment Research Group, Strategic Research and Innovation Program for the Development of MU – Plovdiv, Medical University of Plovdiv, Plovdiv, Bulgaria

Environmental Health Division, Research Institute at Medical University of Plovdiv, Medical University of Plovdiv, Plovdiv, Bulgaria
e-mail: M.Helbich@uu.nl

Y. Zeng
Department of Human Geography and Spatial Planning, Faculty of Geosciences, Utrecht University, Utrecht, The Netherlands

A. Sarker
Emory University School of Medicine, Atlanta, GA, USA

Curr Topics Behav Neurosci (2024) 68: 277–296
https://doi.org/10.1007/7854_2024_464
Published Online: 8 March 2024

Abstract People's mental health is intertwined with the social environment in which they reside. This chapter explores approaches for quantifying the area-level social environment, focusing specifically on socioeconomic deprivation and social fragmentation. We discuss census data and administrative units, egocentric and ecometric approaches, neighborhood audits, social media data, and street view-based assessments. We close the chapter by discussing possible paths forward from associations between social environments and health to establishing causality, including longitudinal research designs and time-series social environmental indices.

Keywords Social environment · Mental health · Socioeconomic deprivation · Social fragmentation · Area-level measures

1 Contextual Social Determinants and Health

Our environment has long been recognized as a significant correlate of people's mental health and well-being (Diez Roux 2001; Macintyre et al. 2002; Tost et al. 2015). In recent decades, it has become evident that the state of people's mental health cannot be exclusively attributed to individual-level factors such as genetic predisposition, lifestyle choices, psychological traits, sex and gender, and age. There is a growing interest in examining area-level determinants that go beyond these individual-level factors when considering mental health (Diez Roux 2001; Oakes et al. 2015). Multiple reviews have suggested that the body of knowledge on individual-level determinants needs to be expanded toward the natural environment (Yang et al. 2021), the built environment (Moore et al. 2018), and the social environment (Kawachi and Subramanian 2018; Sampson et al. 2002), collectively forming both stressors and protective factors vis-à-vis mental health. Understanding these environment–health relations requires robust area-level measures of the neighborhood context (Ivory et al. 2012).

While the natural and built environments have already been discussed in-depth in previous chapters, the overarching aim of this chapter is to examine the area-level social environment, and the way this can be measured. Here, the social neighborhood environment is understood as a multidimensional social and cultural context that shapes people's experiences and health at a residential neighborhood scale. It encompasses various attributes associated with the social dynamics and socioeconomic conditions of neighborhoods (e.g., deprivation), setting the social environment apart from the natural (e.g., green space) and built environment (e.g., urban design). Specifically, we will 1) elaborate on traditional means to capture the area-level social context, 2) we will reflect upon different ways to measure the area-level social environment, and 3) we will discuss some challenges ahead of neighborhood mental-health research and possible ways forward.

An early review (Yen and Syme 1999) emphasized that research on the assessments of the social environment needed to catch up to the physical environment. As a look into the Web of Science today suggests this has changed. We searched for peer-reviewed studies from inception to April 2023 using search terms on "mental health," "social neighborhood," and "neighborhood." With 1,727 hits in total, it is evident that there has been a renewed interest across disciplines, both on a person and area level, resulting in an exponential growth of studies. Much of the literature has dealt with associations between the area-level social environment and multiple mental-health outcomes cross-sectionally (Barnett et al. 2018; Domènech-Abella et al. 2021; Grigoroglou et al. 2019a; Richardson et al. 2015; Roberts et al. 2021; Williams et al. 2020). There is limited and partly conflicting evidence concerning longitudinal associations. A recent meta-analysis of 30 longitudinal studies on adults' mental health reported a possible longitudinal association with the neighborhood social environment (Sui et al. 2022). Of note, the authors stressed, among others, the following limitations of previous longitudinal studies on the area-level social environment. First, the health-influencing geographic context has not been universally defined. Second, the time-varying nature of the social environment has rarely been accounted for by most longitudinal studies. Third, the heterogeneity across the study design and the operationalization of the social neighborhood was considerable.

2 Common Area-level Measures to Characterize Social Environment and their Operationalization

The role of the social environment and health has been discussed extensively (Berkman et al. 2000; Kawachi and Subramanian 2018). Area-level measures of the social environment aim to recognize areas having unfavorable social and economic conditions and are typically employed to assess the associations of the social characteristics of a place (i.e., typically representing the place of residence) and mental health. After introducing ways to operationalize the social environment, we do not aim for a comprehensive review of measures in this section but rather discuss two frequently used ones (i.e., socioeconomic deprivation and social fragmentation). Other measures of the social environment have also been used (e.g., neighborhood crime (Baranyi et al. 2021) or social capital (Mohnen et al. 2011)).

2.1 Statistical Operationalization

An area-based assessment of the social environment is challenging and has been done in various ways (Allik et al. 2020; Lou et al. 2023; Morris and Carstairs 1991). Studies typically use a single variable or multiple variables transformed into a composite score (Lou et al. 2023). A widespread single dimension is, for example, area-level poverty. This approach could be questioned, however, because the social environment comprises numerous dimensions that likely interact with each other and manifest in varying social geographies. Therefore, scholars often assess the area-level social circumstances by means of multidimensional indices condensed into an easy-to-use summary score (Allik et al. 2020). If needed, a specific weight can be assigned to a dimension to underscore its importance relative to the others.

Alternatively, one can use principal component analyses to compute latent variables representing the social environment (Allik et al. 2020). Arguably such a data-driven approach is more objective, but it falls short when it comes to the interpretation of the obtained social environmental components. Furthermore, replicability is often challenging because it requires the researcher to decide how many components should be included; typically, the first component explains a large but perhaps not sufficiently large proportion of the variance of the input data.

Both composite and principal component-based measures also have the advantage of mitigating the risk of multicollinearity, which is common when using multiple single measures. Typically, the individual dimensions of the social environment have moderately high correlations (Lou et al. 2023). Given the trend toward longitudinal analyses, there is no guarantee that the composition of the components does not change over time, which again challenges comparability and temporal cross-comparisons in terms of the interpretation (Pornet et al. 2012). Independent of the statistical approach and whether including a specific dimension is theoretically meaningful, data availability often constrains index development.

2.1.1 Area-level Socioeconomic Deprivation

Area-based indicators of socioeconomic deprivation have been employed to understand geographic disparities in health (Havard et al. 2008; Skapinakis et al. 2005). They are typically approximated by means of a composite score, which permits between-area comparisons. There is no unique way to represent area-level deprivation and what kind of domains are included when compiled into a single index. Two pioneering measures are the Carstairs index (Carstairs and Morris 1990) and the Townsend index (Townsend 1987). The former is compiled based on four census-based variables, including social class, car ownership, overcrowding, and male unemployment. The latter includes unemployment, non-car ownership, non-home ownership, and household overcrowding.

Because the underlying data may not be accessible in every country, cross-country comparisons must be done carefully as only a few standardized, cross-cultural indices exist (for an exception, see (Guillaume et al. 2016)). More common are national deprivation measures (e.g., The Netherlands (Hagedoorn et al. 2020) Switzerland (Panczak et al. 2012), United States (Lou et al. 2023)). For example, the Dutch deprivation index includes the following dimensions: the unemployment rate, reverse-coded standardized median household income, and the share of households with a standardized income below the poverty line. As for the Carstairs and Townsend index, the individual domains were standardized in the Dutch case using *z*-scoring (i.e., measured in terms of standard deviations from the mean) and summed to receive an overall deprivation score per area (Hagedoorn et al. 2020). Higher scores refer to higher levels of deprivation.

Numerous studies have shown that people residing in areas with high levels of socioeconomic deprivation are at risk of experiencing poor mental-health outcomes, including depression (Čermaková et al. 2022; Cohen-Cline et al. 2018; Dowdall et al. 2017; Grigoroglou et al. 2019a; Richardson et al. 2015), anxiety (Chung et al. 2021; Remes et al. 2017; Ventimiglia and Seedat 2019), and stress (Algren et al. 2018; Ribeiro et al. 2018; Whitley et al. 2022). Several mechanisms have been suggested to explain such relations, including that people living in deprived areas may be exposed to more social disorders (e.g., danger, crime, and violence), have poorer access to resources and facilities (e.g., medical services), have greater access to substances of abuse (e.g., alcohol and tobacco), and be exposed to higher levels of other health-threatening environmental stressors (e.g., air and noise pollution) that, in turn, may trigger mental-health problems (Galster 2013; Hajat et al. 2015; Mohai et al. 2009).

2.1.2 Area-level Social Fragmentation

Motivated by Durkheim's work, the fragmentation of social environments has remained of concern to scholars for its possible association with mental health (Congdon 2012; Ivory et al. 2012). In contrast to deprivation, measures of area-level social fragmentation are less often available. The concept of social fragmentation refers to the degree of community social connectedness and integration (or a lack thereof), which has been found to be an essential contextual aspect of the social environment (Grigoroglou et al. 2019b; Ivory et al. 2011; Pearson et al. 2014). Studies have established, for example, that social capital and social cohesion are less common in more fragmented neighborhoods, which may also explain why socially fragmented communities tend to be less health-supportive (Sampson et al. 2002).

We are aware of only a few existing national indices (Bagheri et al. 2020; Hagedoorn et al. 2020; Ivory et al. 2012). Like the deprivation index, routinely collected administrative data are employed for index development. An area-level study in Australia conducted a principal component analysis based on multiple candidate variables (e.g., people living less than a year in the area, single-person households, and unmarried persons) (Bagheri et al. 2020). Empirical evidence of how social fragmentation shapes mental health remains, however, limited, and results are partly contradictory (Grigoroglou et al. 2019b; Ivory et al. 2011; Roberts et al. 2021; Zeng et al. 2023). Its significant association with depression prevalence in the Australian context was partially attenuated after controlling for individual-level socioeconomic status. Similarly, more socially fragmented (and deprived) areas (i.e., a composite measure of the z-scored percentage of residents >18 years who are unmarried, live in a single-person household, and moved to their residential address in the last year) in the Netherlands faced a pronounced suicide risk but, again, the associations were largely attenuated and remained only significant for women after individual-level adjustments were made (Hagedoorn et al. 2020).

3 Assigning the Area-level Social Environment to Individuals

Regardless of which social neighborhood environmental indicator the authors have used, most studies relied on the home location as an exposure receptor point. The operationalization of delineating the health-influencing geographic context can be done in various ways, each having its own benefits and challenges.

3.1 Census Data and Administrative Units

Decades of research relied on census data assigned to administrative units to capture the neighborhood's social environment within which people are nested. Among the reasons for the upsurge of neighborhood research on health is the increasing availability of area-level data (e.g., unemployment rate) assumed to capture people's social environmental exposures. Typically, such area-level data are readily available from national statistical agencies and have full-population coverage.

It is key to bear in mind, however, that the underlying territorial representation using administrative units is not intended to represent the health-influencing geographic context accurately (Helbich 2018; Kirby et al. 2017), and the configuration and number of units may change over time. Another complicating factor is that several spatial granularities exist. In the Netherlands, for example, health research used small-scale 6-digit postal code areas (on average 0.00025 km^2), 100 m raster cells (0.01 km^2) and 4-digit postal code areas (8.3 km^2) to assign people's residential exposures, each unit having a different size and delineation. Cross-national comparisons warrant careful examination because unit sizes vary greatly between countries.

There is still much debate in the geospatial literature criticizing such an operationalization as being meaningful. The modifiable areal unit problem (Openshaw 1981) exemplifies those contextual uncertainties by stressing that conclusions on environmental health associations inferred from a statistical assessment are likely sensitive to the utilized assessment scale and zoning. Additional concerns include, but are not limited to, the following (Flowerdew et al. 2008; Schuurman et al. 2007): 1) Administrative units are not intended to capture health-influencing spatial context meaningfully; 2) each person within a unit gets the same exposures assigned independent of the address location, and 3) those living close to the area boundary are possibly more exposed to the neighboring context than to their own.

3.2 Egocentric Approach

Given the methodological deficits of administrative units and the progress in geocoding residential addresses accurately through geographic information systems

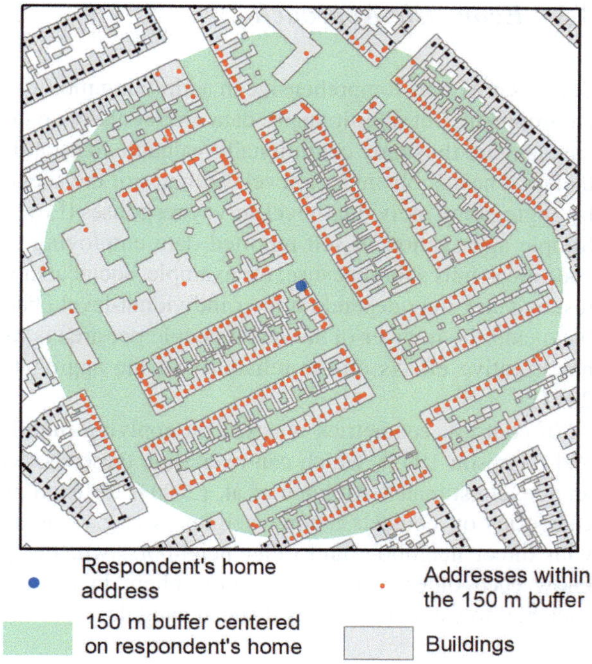

Fig. 1 Egocentric approach to assess the social environment for an address location using a concentric 150 m buffer

coupled with the increasing availability of exact address locations (e.g., in large cohorts or through registries), egocentric approaches are becoming popular (Hagedoorn and Helbich 2021). These approaches superimpose a discrete neighborhood zone (aka a buffer) on a respondent's residential address location. Independent of the buffer shape (i.e., a concentric buffer or one along the street network), scholars need to determine a priori the size of the contextual zone, which ranges from 50 m (Su et al. 2019) to 3 km between studies (Maas et al. 2006). Due to this uncertainty in the geographic scale of the exposure area, it is not surprising that there is much debate on the appropriate buffer size and how this may translate to the estimated associations (Helbich 2018). Nevertheless, egocentric approaches are a step forward but raise concerns about confidentiality and require strict countermeasures for data privacy protection.

In countries where register linkages are feasible (e.g., Denmark and the Netherlands), egocentric approaches allow the computation of indices on the social neighborhood environment at an address level. Figure 1 illustrates an egocentric approach to assess an area-level exposure measure for a specific address location (in blue color). Only those people living at the address locations within the 150-m buffer (colored in red) are included. A similar setting was applied in a Dutch case–control study on suicide risk where full-population register data were first aggregated on an address level, and thereafter, area-level social fragmentation and deprivation indices were determined within the surroundings of each address location (Hagedoorn et al. 2020).

3.3 Ecometric Approach

Surveys are often comprehensive in addressing multiple domains of the social living environment, but questionnaire data are regularly spatially incompatible with other data and are thus only conditionally valuable to establish area-level indices. Nevertheless, one can obtain area-level estimates of the social neighborhood environment by aggregating individual-level survey responses (Leyland and Groenewegen 2020). Simple aggregation of self-reported data employing census geographies or postal code areas falls short because, for example, social area variables, as perceived by the respondents, may depend on their individual-level characteristics (i.e., same-source bias), and the number of surveyed people per area is likely to vary, which possibly has negative effects on the reliability of the aggregated measures (Mohnen et al. 2011).

In contrast, ecometrics is based on applying multilevel modeling (Raudenbush and Sampson 1999), which can overcome these shortcomings to construct reliable area-level estimates. Mohnen et al. (2011), for example, used a series of response categories on contacts among neighbors (e.g., whether respondents have contact with direct neighbors and other neighbors, whether people in the neighborhood know each other, whether the atmosphere in the neighborhood is friendly and sociable) to estimate area-level social capital.

3.4 Neighborhood Audit

While census data or registries capture a few social neighborhood characteristics, they are by no means comprehensive and lack numerous aspects, especially small-scale features shaping the appearance of places. Neighborhood audits offer a more nuanced and complementary view (Kan et al. 2020). Based on predefined rating scales, auditors systematically walk along selected streets and rate the place's appearance regarding safety, littering, pleasantness, etc., using a Likert scale (Van Dillen et al. 2012).

Nonetheless, this approach is not without criticism and has methodological drawbacks. Because such in situ assessments are highly labor-intensive, time-consuming, and thus costly to conduct (Helbich et al. 2021), it is not uncommon that audits are only conducted for a few (randomly selected) streets. It is questionable if such a geographically constrained data collection represents the neighborhood social environment well because relative to the physical environment, the social environment is more difficult to be visually observed and accurately rated by auditors who often lack necessary local neighborhood knowledge. Consequently, linking these neighborhood audits with health data is practically challenging because epidemiological cohorts are usually large in terms of their size.

A perhaps more contemporary way to audit the living environment is through web services (Clarke et al. 2010; Rzotkiewicz et al. 2018). Street view services hosted by Baidu, Google, or crowdsourced photos available via Mapillary let researchers to walk virtually through urban spaces composed of millions of geotagged street-level images (Rundle et al. 2011). Taking Google Street View as an example, these geotagged images have a wide geographic coverage (at least in the Western World). A meta-analysis reported that street view image collections are a reliable and effective data source to assess neighborhoods, although require observer training (Nesse and Airt 2020). While these services, in combination with desktop-based audit tools, are beneficial for collecting place-based neighborhood information (Rzotkiewicz et al. 2018), the manual assessment of many images across larger areas remains tedious and time-consuming (but see below a section on machine learning-based street view assessment).

3.5 *Social Media*

Social media have emerged as sources of data for studying area-level social environments. Estimates suggest that close to five billion people globally used social media (Ruby 2023), and, consequently, social media often provide access to representative population groups (Yang et al. 2023). Unlike surveys and neighborhood audits, however, social media data are not structured. Converting data from social media (e.g., free texts, images, and videos) can be challenging and requires the development and application of advanced data-driven artificial intelligence methods (e.g., natural language processing and machine learning). Substantial advances in data science have been made, and end-to-end data processing pipelines have been proposed to convert social media data into knowledge. Providing an in-depth discussion of all the data science methods that are applied to convert noisy social media data into structured insights is outside the scope of this chapter. Instead, we focus on outlining how target cohorts or population samples are detected from social media and how area-level characterization is performed.

Data from social media are typically collected via application programming interfaces provided by specific social networks such as Twitter (X) and Reddit (Boe 2017; Twitter 2019). The most common data collection method typically uses topic-specific keywords (e.g., medication names). Following data collection, further filtering is often applied to limit the cohort or content to only the most relevant, or to remove potential noise. For example, when chatter about specific health conditions are collected from social media, filtering may be applied to keep only posts that are about personal experiences and not just general chatter about health conditions (Al-Garadi et al. 2021; Sarker et al. 2017).

After the data collection and/or filtering step, geolocation-centric or area-level characterization of data is generally performed using geotags associated with social media posts. Many social media subscribers have their geolocation publicly available, and those can be leveraged to aggregate data/cohort based on geolocation at different levels of granularity (e.g., country, state, postcode). Twitter subscribers, for example, can enable geotagging of their posts. Geolocation information associated with subscriber profiles can be leveraged for area-level aggregation. But only small portions of posts are geotagged. For example, on Twitter, some studies suggest that less than 1% of posts are geotagged (Sloan and Morgan 2015). In the absence of explicitly geotagged information, some researchers have attempted to apply methods for automatically inferring or estimating geolocations based on profile descriptions (e.g., subscriber bios), self-disclosed geolocation information, or other social network specific characteristics (e.g., city-specific forums on the social network Reddit) (Banweer et al. 2018; Zheng et al. 2020). Even though only a small number of posts are geotagged, the large volume of social media data means that the raw volume is still often quite large, particularly in highly populated areas (Sarker et al. 2019). While automatic inference methods, such as those based on natural language processing, may increase the volume of data from specific geolocations, the methods are not 100% accurate and generally introduce errors.

Compared to traditional methods for obtaining area-level measures of social environments, social media present several advantages – social media-based methods are low cost, can be obtained in close to real-time, enable the collection of big data, and may enable the inclusion of hard-to-reach populations. Primary limitations of social media-based approaches include, but are not limited to, skewed representation (younger people are overrepresented on social media) (Dunlap et al. 2023; Pew Research Center 2021), presence of misinformation (Melchior and Oliveira 2021), errors in inference, and methodological challenges in deriving insights from the data.

Research progress in social media mining has overcome or mitigated many of the limitations over the years, enabling better utilization and operationalization of such data. The global subscriber base of social media has led to studies focusing on diverse areas of the world, including countries with complex and large populations where traditional surveying mechanisms are difficult to execute. For example, social media data were shown to reveal differences in the distribution of stress in China (Cui et al. 2022) and the outbreak of COVID-19 in India (Lakamana et al. 2022). Although the high utility of social media for conducting area- or geolocation-specific studies has been demonstrated in numerous publications, these sources of information are still largely underutilized.

3.6 Machine Learning-based Street View Assessment

Urban big data analytics is well suited for assessing automatically street view imageries. Due to progress in computer vision and hardware, it became feasible to evaluate accurately millions of street view scenes within a reasonable amount of time (Biljecki and Ito 2021). The operationalization of these street view-based indices is not trivial. In principle, this methodological innovation obtains street view images (or panoramas) along an area's street network in small distance increments (e.g., every 20 m). These image databases are typically large. For example, a study in Amsterdam used more than 205,000 images (Helbich et al. 2021). Next, these images feed into a (pre-trained) deep learning architecture, typically a convolutional neural network, which has the ability to count either the number of visible objects or segment objects visible in a scene. These deep learning models are trained through large-scale training data. A widely applied training source is the ADE20k semantic segmentation database, including 20,000 annotated images of 150 object types (e.g., cars, trees) (Zhou et al. 2019). Recently, deep learning-based exposure assessments resonated well in the epidemiology community, but the focus was primarily on eye-level green spaces, thought to be associated with mental health. Pilot studies based on observer ratings (e.g., via Amazon Mechanical Turk) began to unlock the perceived and not directly measurable qualities of streetscapes, including safety, littering, or pleasantness (Larkin et al. 2022; Wang et al. 2019). The potential of street view data to assess social environments has not been fully exploited yet, however.

4 Possible Ways Forward

4.1 From Associations to Causation

Despite growing evidence on associations between the area-level social environment and health, whether these associations reflect causal relationships remains under debate (Blair et al. 2014; Jivraj et al. 2020). Previous studies typically face some limitations that hinder causal inferences. First, most studies are cross-sectional, failing to account for the temporal sequence between social environmental exposure and mental-health outcomes (Barnett et al. 2018; Lund et al. 2018; Pearce et al. 2018; Silva et al. 2016). This might be problematic since the social environment–mental health association can be bi-directional. For example, people with better mental-health condition may self-select into residential neighborhoods with better social environment (Barnett et al. 2018; Galster and Hedman 2013; Oakes 2004). Also, mental-health problems, especially externalizing problems (e.g., antisocial behaviors and drug abuse), might lead to more social disorders and impair social cohesion (Salvatore and Grundy 2021), making the neighborhood more socially fragmented (van der Wal et al. 2021). The possibility of a reverse causality bias cannot be ruled out from cross-sectional data, and thus more longitudinal research is warranted for exploring the social environmental causal effects on health.

Second, although longitudinal analyses have methodological advantages toward a more causal understanding, causality remains difficult to infer. Previous longitudinal studies mostly treated the neighborhood social environment as time-invariant by using social environmental data at a single point in time (Lekkas et al. 2017). This operationalization is questionable because the neighborhood environment is shaped by social aspects of the population, which changes dynamically (e.g., due to population turnover) (Pearce et al. 2018; Zeng et al. 2023). Ignoring the potential time-varying nature of the social environment may lead to a misrepresentation of people's actual social environmental exposures over time and further bias the analytical results.

Third, studies typically disregard the correlations among multiple social and physical environmental factors (Sui et al. 2022; Zeng et al. 2023). For example, deprived neighborhoods are often more socially fragmented (Congdon 2011), have lower levels of greenness coverage (Markevych et al. 2017) and more harmful environmental nuisances (e.g., air pollution, noise, and sewage) (Casey et al. 2017; Hajat et al. 2021). Considering the interplay among multiple environments contributes to mitigating the risk of incorrect inferences due to omitted area-level confounding in estimating the health effects of a certain social environmental factor (Zeng et al. 2023).

4.2 Time Series of Indices

The neighborhood social environment is not static, but evolving with structural population shifts driven by changes in the geographic distribution of, for example, resource availability, housing affordability, and employment opportunities (Patias et al. 2022). Nonetheless, the housing stock and affordability, especially in developed countries, generally remain static in a short period of time, and thus people moving in and out of neighborhoods are similar in socioeconomic position, making the neighborhood socioeconomic structure relatively stable (Zwiers et al. 2017). For example, Fig. 2 shows some, but not large, changes in area-level socioeconomic deprivation and social fragmentation scores between 2006 and 2016 in Greater Utrecht, the Netherlands.

Significant transformations in the neighborhood population composition are usually increasingly more noticeable over time (Zwiers et al. 2017). For instance, most regions in England remained stable in the socioeconomic attributes over 10 years (Grigoroglou et al. 2019b; Kontopantelis et al. 2018), but socioeconomic mobility was apparent after extending the study period (Patias et al. 2022). Even in a short period, more remarkable changes in the area-level social environment have emerged in some developing countries undergoing rapid urbanization, which leads to substantial in-flow migration given the increased resources and opportunities. For example, the number of internal migrants in China reached 244 million in 2017 (Mai and Wang 2022). This "floating population" is residentially mobile and has a temporal and weak connection with the local community (Mai and Wang 2022). The influx of migrants may therefore lead to changes in the neighborhood social environment, such as increased neighborhood social fragmentation and decreased social cohesion.

Taken together, three potential avenues are suggested for future research to address the challenges mentioned and support causal inference between the social environment and health. First, longitudinal studies that track individuals' health conditions over time alongside the evolving area-level social environment are crucial. Such designs help address potential reverse causation resulting from

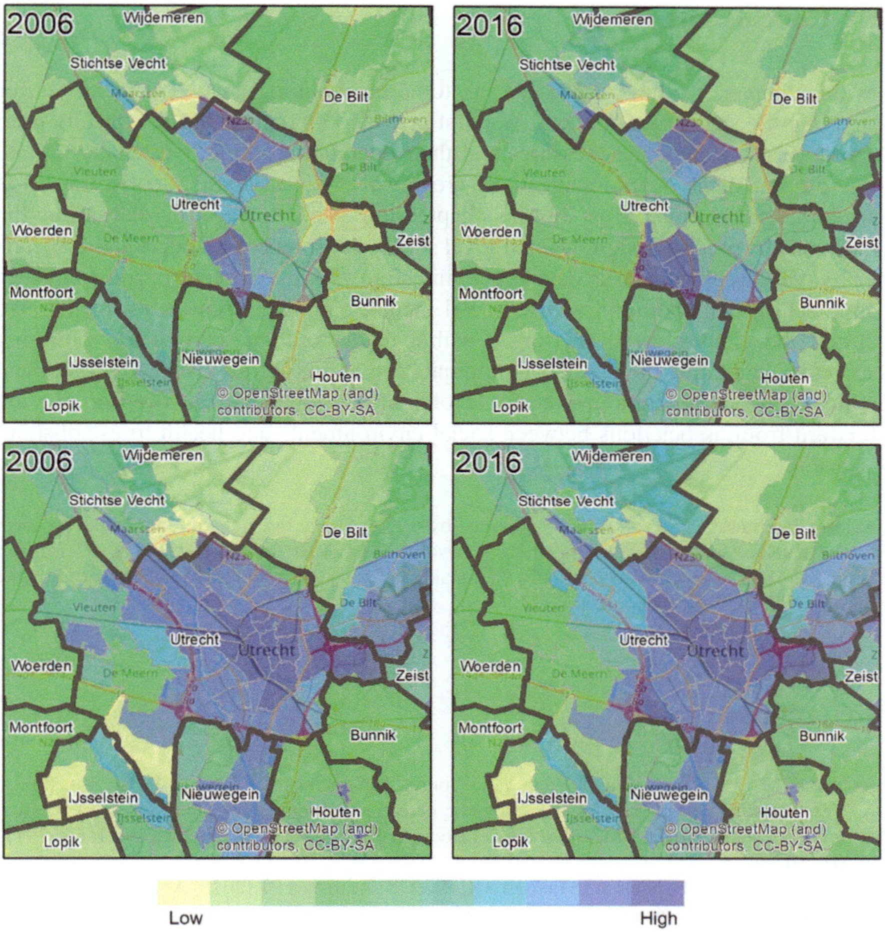

Fig. 2 Changes in area-level measures of the social environment. The top row shows the socio-economic deprivation index for 2006 and 2016 based on the 4 digital postal code areas in Greater Utrecht, the Netherlands, while the bottom row shows the social fragmentation index. In either case, the darker the color, the higher is the deprivation or fragmentation

peoples' choices in residential locations and the possibility of social environmental changes influenced by residents' health conditions. Second, there is a need to use time series of social environmental data to minimize exposure misestimation, especially in studies with extended follow-up and those focused on areas with rapid turnover of residents. Lastly, researchers should consider accounting for the shared variance between social environmental factors and other area-level exposures of the natural and built environment to reduce confounding from these factors on the estimated associations between social environments and health.

5 Conclusions

In this chapter, we have discussed the topic of obtaining area-level measures for studying the social environment. We first provided an outline of the typically used area-level environmental measures and their operationalization, particularly focusing on statistical operationalization, area-level socioeconomic deprivation, and social fragmentation. Most of the chapter has been devoted to discussing the mechanisms used to assign the area-level social environment to individuals. Specifically, we discussed census data and administrative units, egocentric and ecometric approaches, neighborhood audits, social media, and street view assessments. We closed the chapter by discussing possible paths forward, particularly outlining strategies, including the use of longitudinal design, time-series social environmental indices, and incorporation of the role of multiple area-level factors, for moving forward from associations between social environments and health to establishing causality.

Acknowledgments This work was supported by the "Strategic research and innovation program for the development of Medical University – Plovdiv" № BG-RRP-2.004-0007-C01, Establishment of a network of research higher schools, National plan for recovery and resilience, financed by the European Union – NextGenerationEU.

References

Al-Garadi MA, Yang YC, Cai H, Ruan Y, O'Connor K, Graciela GH, Perrone J, Sarker A (2021) Text classification models for the automatic detection of nonmedical prescription medication use from social media. BMC Med Inform Decis Mak 21:1–13. https://doi.org/10.1186/s12911-021-01394-0

Algren MH, Ekholm O, Nielsen L, Ersbøll AK, Bak CK, Andersen PT (2018) Associations between perceived stress, socioeconomic status, and health-risk behaviour in deprived neighbourhoods in Denmark: a cross-sectional study. BMC Public Health 18:1. https://doi.org/10.1186/s12889-018-5170-x

Allik M, Leyland A, Ichihara MYT, Dundas R (2020) Creating small-area deprivation indices: a guide for stages and options. J Epidemiol Community Health 74:20. https://doi.org/10.1136/jech-2019-213255

Bagheri N, Batterham PJ, Salvador-Carulla L, Chen Y, Page A, Calear AL, Congdon P (2020) Development of the Australian neighborhood social fragmentation index and its association with spatial variation in depression across communities. Soc Psychiatry Psychiatr Epidemiol 54: 1189–1198

Banweer K, Graham A, Ripberger J, Cesare N, Nsoesie E, Grant C (2018) Multi-stage collaborative filtering for Tweet geolocation. LocalRec 2018. Proceedings of the 2nd ACM SIGSPATIAL international workshop on recommendations for location-based services and social networks. https://doi.org/10.1145/3282825.3282831

Baranyi G, Di Marco MH, Russ TC, Dibben C, Pearce J (2021) The impact of neighbourhood crime on mental health: a systematic review and meta-analysis. Soc Sci Med 282:114106. https://doi.org/10.1016/j.socscimed.2021.114106

Barnett A, Zhang CJP, Johnston JM, Cerin E (2018) Relationships between the neighborhood environment and depression in older adults: a systematic review and meta-analysis. Int Psychogeriatr 30:1153. https://doi.org/10.1017/S104161021700271X

Berkman LF, Glass T, Brissette I, Seeman TE (2000) From social integration to health: Durkheim in the new millennium. Soc Sci Med 51:843. https://doi.org/10.1016/S0277-9536(00)00065-4

Biljecki F, Ito K (2021) Street view imagery in urban analytics and GIS: a review. Landsc Urban Plan 215:104217

Blair A, Ross NA, Gariepy G, Schmitz N (2014) How do neighborhoods affect depression outcomes? A realist review and a call for the examination of causal pathways. Soc Psychiatry Psychiatr Epidemiol 49:873. https://doi.org/10.1007/s00127-013-0810-z

Boe B (2017) PRAW: The Python Reddit API Wrapper [WWW Document]

Carstairs V, Morris R (1990) Deprivation and health in Scotland. Health Bull (Melb) 48:162. https://doi.org/10.1136/pgmj.69.814.665-a

Casey JA, Morello-Frosch R, Mennitt DJ, Fristrup K, Ogburn EL, James P (2017) Race/ethnicity, socioeconomic status, residential segregation, and spatial variation in noise exposure in the contiguous United States. Environ Health Perspect 125. https://doi.org/10.1289/EHP898

Čermaková P, Andrýsková L, Brázdil M, Marečková K (2022) Socioeconomic deprivation in early life and symptoms of depression and anxiety in young adulthood: mediating role of hippocampal connectivity. Psychol Med 52:2671. https://doi.org/10.1017/S0033291720004754

Chung RYN, Marmot M, Mak JKL, Gordon D, Chan D, Chung GKK, Wong H, Wong SYS (2021) Deprivation is associated with anxiety and stress. A population-based longitudinal household survey among Chinese adults in Hong Kong. J Epidemiol Community Health 75:335. https://doi.org/10.1136/jech-2020-214728

Clarke P, Ailshire J, Melendez R, Bader M, Morenoff J (2010) Using Google earth to conduct a neighborhood audit: reliability of a virtual audit instrument. Health Place 16:1224. https://doi.org/10.1016/j.healthplace.2010.08.007

Cohen-Cline H, Beresford SAA, Barrington WE, Matsueda RL, Wakefield J, Duncan GE (2018) Associations between neighbourhood characteristics and depression: a twin study. J Epidemiol Community Health 302:72. https://doi.org/10.1136/jech-2017-209453

Congdon P (2011) The spatial pattern of suicide in the us in relation to deprivation, fragmentation and rurality. Urban Stud 48:2101. https://doi.org/10.1177/0042098010380961

Congdon P (2012) Assessing the impact of socioeconomic variables on small area variations in suicide outcomes in England. Int J Environ Res Public Health 10:158–177

Cui J, Zhang T, Jaidka K, Pang D, Sherman G, Jakhetiya V, Ungar LH, Guntuku SC (2022) Social media reveals urban-rural differences in stress across China. Proceedings of the International AAAI Conference on Web and Social Media 16:114–124. https://doi.org/10.1609/ICWSM.V16I1.19277

Diez Roux AV (2001) Investigating neighborhood and area effects on health. Am J Public Health 91:1783. https://doi.org/10.2105/AJPH.91.11.1783

Domènech-Abella J, Switsers L, Mundó J, Dierckx E, Dury S, De Donder L (2021) The association between perceived social and physical environment and mental health among older adults: mediating effects of loneliness. Aging Ment Health 25:962. https://doi.org/10.1080/13607863.2020.1727853

Dowdall N, Ward CL, Lund C (2017) The association between neighbourhood-level deprivation and depression: evidence from the south african national income dynamics study. BMC Psychiatry 17:395. https://doi.org/10.1186/s12888-017-1561-2

Dunlap RE, Gallup GH, Gallup AM (2023) The rise of social media. Our World in Data 35:7–39. https://doi.org/10.1080/00139157.1993.9929122

Flowerdew R, Manley DJ, Sabel CE (2008) Neighbourhood effects on health: does it matter where you draw the boundaries? Soc Sci Med 66:1241–1255

Galster GC (2013) The mechanism(s) of neighbourhood effects: theory, evidence, and policy implications. In: van Ham M, Manley D, Bailey N, Simpson L, Maclennan D (eds)

Neighbourhood effects research: new perspectives. Springer, Dordrecht. https://doi.org/10.1007/978-94-007-2309-2_2

Galster G, Hedman L (2013) Measuring neighbourhood effects non-experimentally: how much do alternative methods matter? Hous Stud 28:473. https://doi.org/10.1080/02673037.2013.759544

Grigoroglou C, Munford L, Webb RT, Kapur N, Ashcroft DM, Kontopantelis E (2019a) Prevalence of mental illness in primary care and its association with deprivation and social fragmentation at the small-area level in England. Psychol Med 50:293. https://doi.org/10.1017/S0033291719000023

Grigoroglou C, Munford L, Webb RT, Kapur N, Ashcroft DM, Kontopantelis E (2019b) Spatial distribution and temporal trends in social fragmentation in England, 2001-2011: a national study. BMJ Open 9:e025881. https://doi.org/10.1136/bmjopen-2018-025881

Guillaume E, Pornet C, Dejardin O, Launay L, Lillini R, Vercelli M, Marí-Dell'Olmo M, Fontelo AF, Borrell C, Ribeiro AI, De Pina MF, Mayer A, Delpierre C, Rachet B, Launoy G (2016) Development of a cross-cultural deprivation index in five European countries. J Epidemiol Community Health 70:493. https://doi.org/10.1136/jech-2015-205729

Hagedoorn P, Helbich M (2021) Longitudinal exposure assessments of neighbourhood effects in health research: what can be learned from people's residential histories? Health Place 68: 102543. https://doi.org/10.1016/j.healthplace.2021.102543

Hagedoorn P, Groenewegen PP, Roberts H, Helbich M (2020) Is suicide mortality associated with neighbourhood social fragmentation and deprivation? A Dutch register-based case-control study using individualised neighbourhoods. J Epidemiol Community Health 74:197–202

Hajat A, Hsia C, O'Neill MS (2015) Socioeconomic disparities and air pollution exposure: a global review. Curr Environ Health Rep 2(4):440–450. https://doi.org/10.1007/s40572-015-0069-5

Hajat A, Maclehose RF, Rosofsky A, Walker KD, Clougherty JE (2021) Confounding by socioeconomic status in epidemiological studies of air pollution and health: challenges and opportunities. Environ Health Perspect 129:65001. https://doi.org/10.1289/EHP7980

Havard S, Deguen S, Bodin J, Louis K, Laurent O, Bard D (2008) A small-area index of socioeconomic deprivation to capture health inequalities in France. Soc Sci Med 67:2007. https://doi.org/10.1016/j.socscimed.2008.09.031

Helbich M (2018) Toward dynamic urban environmental exposure assessments in mental health research. Environ Res 161:129–135

Helbich M, Poppe R, Oberski D, Zeylmans van Emmichoven M, Schram R (2021) Can't see the wood for the trees? An assessment of street view-and satellite-derived greenness measures in relation to mental health. Landsc Urban Plan 214:104181

Ivory VC, Collings SC, Blakely T, Dew K (2011) When does neighbourhood matter? Multilevel relationships between neighbourhood social fragmentation and mental health. Soc Sci Med 72: 1993. https://doi.org/10.1016/j.socscimed.2011.04.015

Ivory V, Witten K, Salmond C, Lin EY, You RQ, Blakely T (2012) The New Zealand index of neighbourhood social fragmentation: integrating theory and data. Environ Plan A 44:972. https://doi.org/10.1068/a44303

Jivraj S, Murray ET, Norman P, Nicholas O (2020) The impact of life course exposures to neighbourhood deprivation on health and Well-being: a review of the long-term neighbourhood effects literature. Eur J Pub Health 30:922. https://doi.org/10.1093/EURPUB/CKZ153

Kan HY, Forsyth A, Molinsky J (2020) Measuring the built environment for aging in place: a review of Neighborhood audit tools. J Plan Lit 35:180. https://doi.org/10.1177/0885412220903497

Kawachi I, Subramanian SV (2018) Social epidemiology for the 21st century. Soc Sci Med 196: 240. https://doi.org/10.1016/j.socscimed.2017.10.034

Kirby RS, Delmelle E, Eberth JM (2017) Advances in spatial epidemiology and geographic information systems. Ann Epidemiol 27:1–9

Kontopantelis E, Mamas MA, Van Marwijk H, Ryan AM, Buchan IE, Ashcroft DM, Doran T (2018) Geographical epidemiology of health and overall deprivation in England, its changes and persistence from 2004 to 2015: a longitudinal spatial population study. J Epidemiol Community Health 72:140. https://doi.org/10.1136/jech-2017-209999

Lakamana S, Yang Y-C, Al-Garadi MA, Sarker A (2022) Tracking the COVID-19 outbreak in India through twitter: opportunities for social media based global pandemic surveillance. AMIA Ann Symp Proc 2022:313

Larkin A, Krishna A, Chen L, Amram O, Avery AR, Duncan GE, Hystad P (2022) Measuring and modelling perceptions of the built environment for epidemiological research using crowd-sourcing and image-based deep learning models. J Expo Sci Environ Epidemiol 32:1–8

Lekkas P, Paquet C, Howard NJ, Daniel M (2017) Illuminating the lifecourse of place in the longitudinal study of neighbourhoods and health. Soc Sci Med 177:239–247. https://doi.org/10.1016/J.SOCSCIMED.2016.09.025

Leyland AH, Groenewegen PP (2020) Ecometrics: using MLA to construct contextual variables from individual data. In: Leyl K (ed) Multilevel modelling for public health and health services research. Springer, Cham. https://doi.org/10.1007/978-3-030-34801-4_8

Lou S, Giorgi S, Liu T, Eichstaedt JC, Curtis B (2023) Measuring disadvantage: a systematic comparison of United States small-area disadvantage indices. Health Place 80:102997

Lund C, Brooke-Sumner C, Baingana F, Baron EC, Breuer E, Chandra P, Haushofer J, Herrman H, Jordans M, Kieling C, Medina-Mora ME, Morgan E, Omigbodun O, Tol W, Patel V, Saxena S (2018) Social determinants of mental disorders and the sustainable development goals: a systematic review of reviews. Lancet Psychiatry 5:357. https://doi.org/10.1016/S2215-0366(18)30060-9

Maas J, Verheij RA, Groenewegen PP, De Vries S, Spreeuwenberg P (2006) Green space, urbanity, and health: how strong is the relation? J Epidemiol Community Health 60:587–592

Macintyre S, Ellaway A, Cummins S (2002) Place effects on health: how can we conceptualise, operationalise and measure them? Soc Sci Med 55:125. https://doi.org/10.1016/S0277-9536(01)00214-3

Mai X, Wang J (2022) Situational differences, migratory duration, and social integration of internal migrants in urban China. Cities 125. https://doi.org/10.1016/j.cities.2022.103596

Markevych I, Schoierer J, Hartig T, Chudnovsky A, Hystad P, Dzhambov AM, de Vries S, Triguero-Mas M, Brauer M, Nieuwenhuijsen MJ, Lupp G, Richardson EA, Astell-Burt T, Dimitrova D, Feng X, Sadeh M, Standl M, Heinrich J, Fuertes E (2017) Exploring pathways linking greenspace to health: theoretical and methodological guidance. Environ Res 158:301–317. https://doi.org/10.1016/j.envres.2017.06.028

Melchior C, Oliveira M (2021) Health-related fake news on social media platforms: a systematic literature review. New Media Soc 24:1500–1522. https://doi.org/10.1177/14614448211038762

Mohai P, Pellow D, Roberts JT (2009) Environmental justice. Annu Rev Environ Resour 34:405–430

Mohnen SM, Groenewegen PP, Völker B, Flap H (2011) Neighborhood social capital and individual health. Soc Sci Med 72:660. https://doi.org/10.1016/j.socscimed.2010.12.004

Moore THM, Kesten JM, López-López JA, Ijaz S, McAleenan A, Richards A, Gray S, Savović J, Audrey S (2018) The effects of changes to the built environment on the mental health and Well-being of adults: systematic review. Health Place 53:237. https://doi.org/10.1016/j.healthplace.2018.07.012

Morris R, Carstairs V (1991) Which deprivation? A comparison of selected deprivation indexes. J Public Health (Bangkok) 13. https://doi.org/10.1093/oxfordjournals.pubmed.a042650

Nesse K, Airt L (2020) Google street view as a replacement for in-person street surveys: meta-analysis of findings from evaluations. J Urban Plan Dev 146:4020013

Oakes JM (2004) The (mis)estimation of neighborhood effects: causal inference for a practicable social epidemiology. Soc Sci Med 58. https://doi.org/10.1016/j.socscimed.2003.08.004

Oakes JM, Andrade KE, Biyoow IM, Cowan LT (2015) Twenty years of Neighborhood effect research: an assessment. Curr Epidemiol Rep 2:80. https://doi.org/10.1007/s40471-015-0035-7

Openshaw S (1981) The modifiable areal unit problem. In: Wrigley N, Bennett RJ (eds) Quantitative geography: a British view. Routledge & K. Paul, London, pp 60–69

Panczak R, Galobardes B, Voorpostel M, Spoerri A, Zwahlen M, Egger M (2012) A Swiss neighbourhood index of socioeconomic position: development and association with mortality. J Epidemiol Community Health 66:1129. https://doi.org/10.1136/jech-2011-200699

Patias N, Rowe F, Arribas-Bel D (2022) Trajectories of neighbourhood inequality in Britain: unpacking inter-regional socioeconomic imbalances, 1971–2011. Geogr J 188:150. https://doi.org/10.1111/geoj.12420

Pearce J, Cherrie M, Shortt N, Deary I, Ward Thompson C (2018) Life course of place: a longitudinal study of mental health and place. Trans Inst Br Geogr 43:555. https://doi.org/10.1111/tran.12246

Pearson AL, Ivory V, Breetzke G, Lovasi GS (2014) Are feelings of peace or depression the drivers of the relationship between neighbourhood social fragmentation and mental health in Aotearoa/New Zealand? Health Place 26:1. https://doi.org/10.1016/j.healthplace.2013.11.002

Pew Research Center (2021) Demographics of social media users and adoption in the United States. Pew Research Center [WWW Document]. URL https://www.pewresearch.org/internet/fact-sheet/social-media/ Accessed 27 May 2021

Pornet C, Delpierre C, Dejardin O, Grosclaude P, Launay L, Guittet L, Lang T, Launoy G (2012) Construction of an adaptable European transnational ecological deprivation index: the French version. J Epidemiol Community Health 66:982. https://doi.org/10.1136/jech-2011-200311

Raudenbush SW, Sampson RJ (1999) Ecometrics: toward a science of assessing ecological settings, with application to the systematic social observation of neighborhoods. Sociol Methodol 29:1. https://doi.org/10.1111/0081-1750.00059

Remes O, Wainwright N, Surtees P, Lafortune L, Khaw KT, Brayne C (2017) Sex differences in the association between area deprivation and generalised anxiety disorder: British population study. BMJ Open 7:e013590. https://doi.org/10.1136/bmjopen-2016-013590

Ribeiro AI, Amaro J, Lisi C, Fraga S (2018) Neighborhood socioeconomic deprivation and allostatic load: a scoping review. Int J Environ Res Public Health 15. https://doi.org/10.3390/ijerph15061092

Richardson R, Westley T, Gariépy G, Austin N, Nandi A (2015) Neighborhood socioeconomic conditions and depression: a systematic review and meta-analysis. Soc Psychiatry Psychiatr Epidemiol 50:1641. https://doi.org/10.1007/s00127-015-1092-4

Roberts H, van Lissa C, Helbich M (2021) Perceived neighbourhood characteristics and depressive symptoms: potential mediators and the moderating role of employment status. Soc Sci Med 268:113533. https://doi.org/10.1016/j.socscimed.2020.113533

Ruby D (2023) Social media users in the world – (2023 demographics)

Rundle AG, Bader MDM, Richards CA, Neckerman KM, Teitler JO (2011) Using Google street view to audit neighborhood environments. Am J Prev Med 40:94–100

Rzotkiewicz A, Pearson AL, Dougherty BV, Shortridge A, Wilson N (2018) Systematic review of the use of Google street view in health research: major themes, strengths, weaknesses and possibilities for future research. Health Place 52:240–246

Salvatore MA, Grundy E (2021) Area deprivation, perceived neighbourhood cohesion and mental health at older ages: a cross lagged analysis of UK longitudinal data. Health Place 67:102470. https://doi.org/10.1016/J.HEALTHPLACE.2020.102470

Sampson RJ, Morenoff JD, Gannon-Rowley T (2002) Assessing "neighborhood effects": social processes and new directions in research. Annu Rev Sociol 28:443. https://doi.org/10.1146/annurev.soc.28.110601.141114

Sarker A, Chandrashekar P, Magge A, Cai H, Klein A, Gonzalez G (2017) Discovering cohorts of pregnant women from social media for safety surveillance and analysis. J Med Internet Res 19:e361. https://doi.org/10.2196/jmir.8164

Sarker A, Gonzalez-Hernandez G, Ruan Y, Perrone J (2019) Machine learning and natural language processing for geolocation-centric monitoring and characterization of opioid-related social media chatter. JAMA Netw Open 2:e1914672. https://doi.org/10.1001/jamanetworkopen.2019.14672

Schuurman N, Bell N, Dunn JR, Oliver L (2007) Deprivation indices, population health and geography: an evaluation of the spatial effectiveness of indices at multiple scales. J Urban Health 84:591–603

Silva M, Loureiro A, Cardoso G (2016) Social determinants of mental health: a review of the evidence. European Journal of Psychiatry 30(4):259–292

Skapinakis P, Lewis G, Araya R, Jones K, Williams G (2005) Mental health inequalities in Wales, UK: multi-level investigation of the effect of area deprivation. Br J Psychiatry 186: 417. https://doi.org/10.1192/bjp.186.5.417

Sloan L, Morgan J (2015) Who tweets with their location? Understanding the relationship between demographic characteristics and the use of Geoservices and geotagging on Twitter. PLoS One 10(11):e0142209. https://doi.org/10.1371/journal.pone.0142209

Su JG, Dadvand P, Nieuwenhuijsen MJ, Bartoll X, Jerrett M (2019) Associations of green space metrics with health and behavior outcomes at different buffer sizes and remote sensing sensor resolutions. Environ Int 126:162–170

Sui Y, Ettema D, Helbich M (2022) Longitudinal associations between the neighborhood social, natural, and built environment and mental health: a systematic review with meta-analyses. Health Place 77:102893

Tost H, Champagne FA, Meyer-Lindenberg A (2015) Environmental influence in the brain, human welfare and mental health. Nat Neurosci 18:1421–1431

Townsend P (1987) Deprivation. J Soc Policy 16:125–146

Twitter (2019) API Reference Index [WWW Document]. URL https://developer.twitter.com/en/docs/api-reference-index

van der Wal JM, van Borkulo CD, Deserno MK, Breedvelt JJF, Lees M, Lokman JC, Borsboom D, Denys D, van Holst RJ, Smidt MP, Stronks K, Lucassen PJ, van Weert JCM, Sloot PMA, Bockting CL, Wiers RW (2021) Advancing urban mental health research: from complexity science to actionable targets for intervention. Lancet Psychiatry 8:991. https://doi.org/10.1016/S2215-0366(21)00047-X

Van Dillen SME, de Vries S, Groenewegen PP, Spreeuwenberg P (2012) Greenspace in urban neighbourhoods and residents' health: adding quality to quantity. J Epidemiol Community Health 66:e8–e8

Ventimiglia I, Seedat S (2019) Current evidence on urbanicity and the impact of neighbourhoods on anxiety and stress-related disorders. Curr Opin Psychiatry 32:248. https://doi.org/10.1097/YCO.0000000000000496

Wang R, Yuan Y, Liu Y, Zhang J, Liu P, Lu Y, Yao Y (2019) Using street view data and machine learning to assess how perception of neighborhood safety influences urban residents' mental health. Health Place 59:102186. https://doi.org/10.1016/j.healthplace.2019.102186

Whitley E, Olsen J, Benzeval M (2022) Impact of subjective and objective neighbourhood characteristics and individual socioeconomic position on allostatic load: a cross-sectional analysis of an all-age UK household panel study. Health Place 78:102930. https://doi.org/10.1016/J.HEALTHPLACE.2022.102930

Williams AJ, Maguire K, Morrissey K, Taylor T, Wyatt K (2020) Social cohesion, mental wellbeing and health-related quality of life among a cohort of social housing residents in Cornwall: a cross sectional study. BMC Public Health 20:985. https://doi.org/10.1186/s12889-020-09078-6

Yang BY, Zhao T, Hu LX, Browning MHEM, Heinrich J, Dharmage SC, Jalaludin B, Knibbs LD, Liu XX, Luo YN, James P, Li S, Huang WZ, Chen G, Zeng XW, Hu LW, Yu Y, Dong GH (2021) Greenspace and human health: an umbrella review. The Innovation 2:100164. https://doi.org/10.1016/j.xinn.2021.100164

Yang Y-C, Al-Garadi MA, Love JS, Cooper HLF, Perrone J, Sarker A (2023) Can accurate demographic information about people who use prescription medications nonmedically be derived from twitter? Proc Natl Acad Sci USA 120:e2207391120. https://doi.org/10.1073/PNAS.2207391120

Yen IH, Syme SL (1999) The social environment and health: a discussion of the epidemiologic literature. Annu Rev Public Health 20:287. https://doi.org/10.1146/annurev.publhealth.20.1.287

Zeng Y, Stevens GWJM, Helbich M (2023) Longitudinal associations of neighbourhood environmental exposures with mental health problems during adolescence: findings from the TRAILS study. Environ Int 179:108142. https://doi.org/10.1016/J.ENVINT.2023.108142

Zheng C, Jiang JY, Zhou Y, Young SD, Wang W (2020) Social media user geolocation via hybrid attention. SIGIR 2020. Proceedings of the 43rd International ACM SIGIR Conference on Research and Development in Information Retrieval:1641–1644. https://doi.org/10.1145/3397271.3401329

Zhou B, Zhao H, Puig X, Xiao T, Fidler S, Barriuso A, Torralba A (2019) Semantic understanding of scenes through the ADE20K dataset. Int J Comput Vis 127:302. https://doi.org/10.1007/s11263-018-1140-0

Zwiers M, Kleinhans R, Van Ham M (2017) The path-dependency of low-income neighbourhood trajectories: an approach for analysing neighbourhood change. Appl Spat Anal Policy 10:363. https://doi.org/10.1007/s12061-016-9189-z

Advancing Population Health Through Open Environmental Data Platforms

Mohammad Noaeen, Dany Doiron, Joey Syer, and Jeffrey Brook

Contents

Abstract Data stand as the foundation for studying, evaluating, and addressing the multifaceted challenges within environmental health research. This chapter highlights the contributions of the Canadian Urban Environmental Health Research Consortium (CANUE) in generating and democratizing access to environmental exposure data across Canada. Through a consortium-driven approach, CANUE standardizes a variety of datasets – including air quality, greenness, neighborhood characteristics, and weather and climatic factors – into a centralized, analysis-ready, postal code-indexed database. CANUE's mandate extends beyond data integration, encompassing the design and development of environmental health-related web applications, facilitating the linkage of data to a wide range of health databases

M. Noaeen, J. Syer, and J. Brook (✉)
Dalla Lana School of Public Health, University of Toronto, Toronto, ON, Canada
e-mail: m.noaeen@utoronto.ca; joey.syer@utoronto.ca; jeff.brook@utoronto.ca

D. Doiron
Respiratory Epidemiology and Clinical Research Unit, Research Institute of the McGill University Health Centre, Montréal, QC, Canada

Department of Environmental and Occupational Health, School of Public Health, Université de Montréal, Montréal, QC, Canada
e-mail: dany.doiron@mail.mcgill.ca

Curr Topics Behav Neurosci (2024) 68: 297–324
https://doi.org/10.1007/7854_2024_512
Published Online: 8 August 2024

and sociodemographic data, and providing educational training and events such as webinars, summits, and workshops. The operational and technical aspects of CANUE are explored in this chapter, detailing its human resources, data sources, computational infrastructure, and data management practices. These efforts collectively enhance research capabilities and public awareness, fostering strategic collaboration and generating actionable insights that promote physical and mental health and well-being.

Keywords Data hub · Data linkage · Data sharing · Environmental data · Environmental health · Epidemiology · Equity · FAIR · Population health

1 Introduction

Urban design has long been recognized as a significant factor in shaping physical and mental health and, with >80% of the population residing in cities in Canada (O'Neill 2023), represents an important intervention target. Research has consistently shown connections between chronic diseases and urban environmental factors such as air quality (Duan et al. 2020; Karimi et al. 2020; To et al. 2015), noise exposure (van Kamp and Davies 2013; H. Wang et al. 2020; L. Yang et al. 2024), neighborhood greenness (Browning and Lee 2017; Jennings et al. 2019; Jia et al. 2021), urban heat islands (educhenin 2010; Kovats and Hajat 2008; F. Li et al. 2022), water pollution and soil contamination (Wimalawansa and Wimalawansa 2016), access to healthy food (Caspi et al. 2012; Holsten 2009; Weaver and Fasel 2018), and radiation exposure, such as ultraviolet radiation from sun (Quan et al. 2023). The increasing impact of climate change, including heat waves, flooding, sea level rise, and air pollution, has intensified the urgency to address the health and well-being of those living in cities to help build resilience. Implementation of climate-motivated urban change represents an important opportunity for improving public health, as long as the enacted policies focus on health risks and systemic inequities. To optimize future interventions and policies, it is essential to develop a deeper understanding of the connections between urban design and health, both physical and mental, and how these relationships contribute to, and vary according to, population vulnerability.

The built, physical, and social environments, while distinct, are intricately interconnected and collectively impact public and population health. The built environment, which includes man-made structures like buildings, roads, and parks, influences physical activity and social interactions, thereby shaping the social environment. For example, accessible parks and forests and walkable streets encourage physical activity and social interactions, enhancing both physical and social health (Northridge and Freeman 2011; S. Park et al. 2022b; Reyes et al. 2014; Roemer et al. 2023). The physical environment, encompassing natural elements like air and water quality, impacts both the built environment and health directly. Poor air quality can exacerbate respiratory and cardiovascular conditions, which are further influenced by access to healthcare and greenspaces (built environment) and social support (social environment), also impacting neurodevelopment (Guxens et al. 2022; Salgado et al. 2020; Yuchi et al. 2022) and neurodegenerative diseases (Chen et al. 2017; Zhang et al. 2023). These interconnections highlight the importance of holistic approaches to urban design, emphasizing the need for comprehensive strategies that consider all environmental aspects that have the potential to improve public health and address environmental inequities. Despite recognition of the connections between urban environment and health, however, many municipalities lack appropriate tools and standardized data to help them identify areas for improvements, set targets, and benchmark progress in a consistent manner (Pineo et al. 2018).

The COVID-19 pandemic underscored the importance of addressing urban environmental issues, as research suggests that various factors such as air pollution (Hammad et al. 2023; Hernandez Carballo et al. 2022; Sheppard et al. 2023), and food insecurity (Otten et al. 2023) are associated with infection rates and severe COVID-19 outcomes. For example, exposure to traffic-related air pollution has been linked to higher COVID-19 mortality rates (Fuller et al. 2022). Additionally, studies emphasize that access to parks and recreational spaces enhances mobility behaviors and patterns (Burgoine et al. 2015; Heo et al. 2020; B. Wang et al. 2021), which are critical not only for sustaining mental and physical health during pandemics and outbreaks of infectious diseases (Levinger et al. 2022; A. H. Park et al. 2022a; Tabrizi et al. 2023), but also offer long-term benefits in enhancing community health and well-being (Laddu et al. 2021).

In the contemporary landscape of public health research, the necessity for standardized datasets concerning the physical environment has become increasingly apparent. Making such data widely available is crucial for accelerating research on how environmental factors relate to health outcomes given the type of expertise needed to develop, curate, and distribute such data. To address this need, centralized data platforms play a valuable role in facilitating accessibility for a wide variety of researchers and enhancing the consistency and comparability of environmental health studies (Brook et al. 2021; Child et al. 2022). The Canadian Urban Environmental Health Research Consortium (CANUE) (https://www.canue.ca/) (Brook et al. 2018) serves as a pioneering model in this domain, illustrating the potential of a collaborative approach to data aggregation and resource sharing.

CANUE has increasingly served as a national resource in Canada for exposure-related data across various domains of environmental health and urban science with holdings that are highly relevant to research carried out in the context of population neuroscience. CANUE's primary role has been the identification, collection, processing, and dissemination of data pertinent to the physical, built, and social environments. Three other chapters in this book provide clear evidence of this relevance ("Integrating the physical environment within a population neuroscience perspective", "Leveraging generative AI models in urban science", and "Area-level measures of the social environment: Operationalization, pitfalls, and ways forward"). Thus, this chapter focuses on the role, characteristics, applications, and impacts of CANUE providing an exemplar for creating a common data and methods resource consistent with FAIR data principles of being findable, accessible, interoperable and reusable.

2 CANUE

Since 2016, CANUE has functioned as a key Canadian environmental exposure data hub and also as a bridge between disciplines. CANUE's strategy involves the creation of a common platform where researchers can both access and contribute to a growing repository of health-relevant environmental data. This approach supports a more integrated understanding of environmental impacts on health, enabling researchers to draw robust conclusions and craft targeted interventions. By leveraging collective expertise and resources, CANUE has managed to assemble a diverse array of data, which are indispensable for conducting nuanced environmental health research.

An overview of CANUE is provided in Fig. 1. Its primary roles are shown on the left side of the figure and are discussed below. The activities on the right are critical for maintaining and promoting the consortium, increasing research capacity and sustainability of urban environmental health research in Canada and knowledge translation (KT) activities. Since 2016, CANUE has organized multiple national webinars and supported or led several training initiatives including with student networks. The CANUE website and regular newsletter to members serve as resources for accessing archived information (e.g., webinar recordings, report from the 2023 Environments and Health Summit) and progress updates. HealthyPlan.City, which is CANUE's main KT platform, is highlighted in Sect. 2.2.2.

2.1 Critical Aspects in the Development of CANUE

The origin of CANUE was the *Environments and Health Signature Initiative* (EHSI) by the Canadian Institutes for Health Research (CIHR). In seeking to advance environments and health research in Canada, CIHR foresaw the need for enriching existing health data/research platforms (i.e., cohorts) with environmental exposure data as opposed to starting new cohorts. The specific funding call (approx. $4.2M CAD was available for 5 years) was for the establishment of a pan-Canadian consortium with a focus on "urban form" given a growing CIHR interest in enhancing research on healthy cities. CANUE emerged as the successful applicant putting forward the creation of a data and methods hub with several key features:

Interoperable, analysis-ready data

- Focus on geospatial data and streamline linkage to cohorts and other health data platforms by resolving to six-digit postal codes and assigning environmental

Fig. 1 Overview of CANUE's role in the Canadian environmental health research and knowledge mobilization landscape

exposure variables to all populated postal codes (in urban areas, Canadian postal codes represent one side of a city block)

- Provide exposure metrics that are relevant to the development of non-communicable disease across the life course
- Focus on environmental data to avoid dealing with privacy concerns associated with individual health data
- Establish partnerships with the main Canadian cohorts to accelerate integration of environmental data directly into their platforms, linked at the individual level based upon postal codes

National coverage

- Ensure consistently derived exposure metrics across Canada
- Document regional datasets for future distribution or extrapolating to the national scale in the future

Curate and distribute existing data first

- Acquire data previously developed and/or used in health research
- Ensure originators of data are credited and recognize that sharing data through CANUE is value-added for their work (e.g., increased usage, new collaborations)

A mission statement and a set of core values, as outlined on CANUE's website (https://canue.ca/about/), were established to operationalize the above features. Four principles were identified: (1) open membership, (2) meaningful opportunities for contribution, (3) consensus-based governance, and (4) recognition of contribution.

Despite the volume of research into how exposures related to urban form interact to affect health and how this changes over time, the mechanisms and magnitude of their relevance are still not well understood. Thus, CANUE's mission is to increase knowledge on how these and other related exposures, individually and in combination, affect health and how they can be modified to improve health. In advancing knowledge, CANUE has four main objectives:

- Build a common, harmonized data and methods platform containing environmental exposure metrics that characterize multiple aspects of urban form
- Support research on novel integrated approaches to quantify the combined health effects of multiple harmful and protective characteristics of the physical environment in urbanized areas
- Identify and communicate the results of real-world applications of integrated approaches that provide specific insights into the physical characteristics of Canadian urban/suburban environments that maximize health
- Increase awareness of the importance of human health in urban development by training current and future scientists, professionals, and civic leaders so that they can promote and apply these concepts in their careers

Urban form covers a wide range of issues pertaining to health impacts or benefits such as air and noise pollution, land use, transportation, physical infrastructure, and socioeconomic conditions. Given this breadth, six working groups (WG) drawing from multiple disciplines were established to identify available data and prioritize future data development. These groups, who met frequently at the initiation of CANUE and on an ad hoc basis thereafter, are (1) air pollution, (2) noise pollution, (3) greenness, (4) transportation, (5) climate and extreme weather, and (6) neighborhood factors (e.g., walkability, access to healthy foods, and socioeconomic indicators). In addition to spanning across disciplines, the WGs engaged both a senior and junior researcher as co-leads (to help ensure long-term sustainability) and sought involvement from researchers across Canada. A small budget was directed toward each group for holding workshops (e.g., the Neighhourhood Factors WG met to discuss metrics for characterizing physical activity or walkability and the Greenness WG convened a workshop on the current status of measuring and characterizing greenness) or exploring approaches to develop new data (e.g., the Air Quality WG undertook research on historical emissions data needed for long-term retrospective air quality model runs and on the development of a multipollutant chronic exposure index).

CANUE's community-oriented approach helps ensure a breadth and depth of data that would be unattainable by individual entities working in isolation. The success of this model for collaboration is evidenced by developers of new datasets choosing CANUE for longer term storage and distribution. For instance, the contribution of the Canadian Bikeway Comfort and Safety Classification System (Can-BICS) (Winters et al. 2020) developed a new national dataset on biking infrastructure, a potentially important factor characterizing community capacity for active transportation and physical activity, and CANUE hosts the data for future users. Similarly, the researchers exploring connections between urban interventions and gentrification produced several indices – GENUINE: Gentrification, Urban Interventions, and Equity data (Firth et al. 2021), while the Canadian Optimized Statistical Smoke Exposure Model (CanOSSEM) provides estimates of daily mean $PM_{2.5}$ concentration resulting from biomass smoke across Canada (Paul et al. 2022), both of which are available from the CANUE data portal.

2.1.1 CANUE Data Structure and Technical Requirements

Datasets curated and distributed by CANUE undergo a rigorous assessment for completeness and the documentation necessary for generating metadata. All pertinent information is available through a dedicated data portal (https://www.canuedata.ca/), which provides users with essential details for effective utilization of the data, including data descriptions, quality assessments, data sources, maintenance records, usage conditions, support documentation, variable descriptions, support contact details, and data source contact information. Using Geographic Information System (GIS) software and computer programming, the data are linked to all six-digit DMTI Spatial single link postal code locations in Canada (https://www.dmtispatial.com/our-story/). The single link indicator (SLI) facilitates data linkage by providing the x,y coordinate within postal code polygons that best represents where people live within it. The resulting data are distributed as CSV files to authorized users. The main requirement for use is that the researcher has a license to use DMTI postal code shapefiles. Most Canadian universities are part of the Smart Consortium Agreement with DMTI (https://canadiangis.com/smart-spatial-mapping-academic-research-tools-program.php) and are thus able to freely access and use CANUE data.

Essential Computing, Storage, Software, and Resources

The data operations in CANUE are categorized into three main stages:

1. Data acquisition: Application programming interfaces (APIs) and data portals are either manually or programmatically accessed to download files.
2. Data processing: Programming and GIS software tools are used to transform the downloaded files into structured exposure datasets.
3. Data indexing and dissemination: After processing, data are systematically indexed and metadata are generated to enhance the efficiency of data storage and dissemination.

CANUE's data infrastructure is primarily hosted on the cloud platforms provided by the Digital Research Alliance of Canada (https://docs.alliancecan.ca/wiki/Cloud_resources). The Alliance supports advanced research computing, data management, and software tools for Canadian researchers, making their resources an ideal choice for handling CANUE's extensive data needs. One of the key cloud platforms used is the Arbutus cloud, located at the University of Victoria, BC. CANUE utilizes an Ubuntu server configured as a virtual machine within this environment, providing over 100 TB of storage capacity. This setup helps to meet the data collection, storage, and processing requirements across various temporal (from hourly to annual) and spatial resolutions. Furthermore, CANUE benefits from access to other cloud systems and virtual machines through the Digital Research Alliance, located at various sites across Canada, enhancing our data storage and processing capabilities. For instance, as part of CANUE's data dissemination efforts, the CANUE data portal currently hosts 51 annual and 10 monthly datasets, which together account for approximately 500 GB of environmental exposure data. This database includes around 500 variables across 61 datasets, resulting in approximately 6,100 unique variable-years and variable-months. The ongoing integration of datasets with finer temporal resolutions, such as daily and hourly data, will require increased storage capacity in the future for data acquisition, processing, indexing, and dissemination.

CANUE uses PostgreSQL as its primary database system. The data processing is primarily performed using programming languages such as Python and R, in conjunction with GIS software like ArcGIS Pro and QGIS. CANUE utilizes GoDaddy for web hosting and employs WordPress and Prismic for managing the content on the CANUE website.

CANUE is supported by a dedicated team led by the Principal Investigator and Scientific Director, alongside the Managing Director, who oversee academic funding, human resources, financial management, IT support, and administrative operations. The team also includes key personnel such as Data Director, Data Scientist, Administrative Assistant, Communications Specialist, Web Developer, Geospatial Data Specialist, and Data Curator. Collectively, they ensure CANUE's effectiveness in advancing environmental health research through robust data management and strategic communication.

2.1.2 Current CANUE Data Holdings

CANUE has curated a comprehensive collection of environmental data, including a wide range of variables that serve to study the impact of urban environments on human health. These data are grouped into 4 major themes: *Greenness*, *Neighborhood*, *Air Quality*, and *Weather* (see Table 1).

The data in Table 1 come from a variety of dependable sources, including satellite observations, ground-based monitoring stations, geospatial databases, and government resources such as Statistics Canada (https://www.statcan.gc.ca/en/start). It encompasses neighborhood-related data that provide detailed insights into urban living conditions and weather-related data essential for understanding the climatic impacts on urban areas and their implications for health and lifestyle. Air quality data are gathered from satellite observations, atmospheric models, and national monitoring networks for various air pollutants. These data are subjected to quality assurance processes (often by the data originators) and are standardized to maintain consistency and reliability for research applications. The primary sources for greenness metrics, including tree canopy coverage, are high-resolution satellite platforms such as Landsat (https://www.usgs.gov/landsat-missions) and Modis (https://modis.gsfc.nasa.gov/), as well as Google Earth Engine. These platforms provide detailed

Table 1 The available postal code-level environmental data prepared and made available by CANUE, June 2024

Category	Available postal code-indexed data
Neighborhood	Active living environments; building density; Canadian access to employment; Canadian Bikeway Comfort and Safety (Can-BICS) classification system metrics; Canadian Food Environment Dataset (Can-FED); Canadian Marginalization Index; dwellings needing repair; facility index; gentrification, urban interventions, and equity; green roads; material and social deprivation index; nighttime light; noise; proximity measures – StatsCan; proximity to roads; proximity to water bodies; public transportation; spatial accessibility measures; street connectivity; urban sprawl
Air quality	Nitrogen dioxide (NO_2); fine particulate matter ($PM_{2.5}$); sulfur dioxide (SO_2); wildfire and biomass burning (smoke exposure ($PM_{2.5}$)); ozone (O_3); multi-pollutant index
Greenness	Normalized difference vegetation index (NDVI); greenness; tree canopy coverage
Weather	Climate metrics; land surface temperature; local climate zone; water balance metrics

imagery that is processed to derive vegetation indices and indicators of neighborhood greenness. The tree canopy coverage specifically identifies areas covered by woody vegetation taller than 5 m.

All datasets are harmonized and indexed at postal code levels using GIS software. CANUE ensures each dataset is accompanied by detailed metadata, which includes information such as data sources, variable descriptions, quality assessments, data use condition such as required acknowledgment and citations, and support documentation. Data access is governed by a data sharing agreement, limiting use to academic, research, and educational professional purposes. This ensures that the data are utilized ethically and responsibly and that CANUE complies with the DMTI SMART Consortium Agreement.

The data can also be visualized through the publicly accessible web mapping interface of the data portal. For example, within the weather theme of the web map browser, CANUE offers satellite-based estimates of Land Surface Temperature (LST), which are derived from LandSat 8 satellite data using Google Earth Engine. Users can visualize each of the derived LST variables at each postal code across Canada, select from various data classification schemes, and use a time slider to examine changes in annual estimates.

2.2 Extending the Application of CANUE's Data Infrastructure

2.2.1 Data Extraction Web-Tools

In the context of population neuroscience, the emphasis is on exploiting geospatial patterns to enhance our understanding of health dynamics. This requires a static snapshot of spatial data, but additional insight can be gained through exploration of temporal variations in health responses at specific locations. Such variations necessitate various temporal resolutions of geospatial data. Alongside monthly data snapshots, which CANUE provides for data exhibiting monthly to seasonal variations, more finely time-resolved information is crucial for certain health study designs (e.g., case–control studies of acute response). This highlights the need for tools that facilitate the access to and linkage of temporal geospatial data to distinct study populations. These tools, such as data extraction web apps, support longitudinal analysis and are essential for investigating both long-term trends and short-term acute changes in health.

For instance, in CANUE's daily data extraction tool, users initiate the data retrieval process by uploading a CSV or XLSX file containing columns labeled "subjectid", "postal code", and "date". Users can then select the environmental exposure data they wish to retrieve from the available datasets. The tool allows users to specify the desired averaging intervals for the data they retrieve, with options including daily, 3-day, 7-day averages. This flexibility enables the tool to accommodate various research needs, particularly in studies requiring precise temporal analysis. The tool extracts data for seven consecutive days prior to the specified date, including specific lag days (0 to 6) from case and control dates, and calculates averages over the selected intervals to provide short-term exposure estimates.

This functionality is crucial for researchers who need to align environmental exposure data with specific health events or conditions in their studies. For instance, in birth-related studies, the expanded version of this tool can provide environmental exposure data tailored to each trimester or other significant prenatal periods. Similarly, in studies of acute health events, such as examining the impact of smoke on hospitalizations, researchers can obtain finely tuned data on exposure around the time of the health data collection, enabling them to draw more precise correlations between exposure and health outcomes.

2.2.2 HealthyPlan.City: A Web App to Advance Environmental Equity

Beyond research, CANUE data have value for developing interventions or addressing a range of practical urban planning and public health questions. The combined experience of adverse environmental exposures and limited capacity to obviate such conditions represents a potentially vulnerable population characteristic and a pressing public health need. Environmental Justice is a current issue of concern (Environment and Climate Change Canada 2024) and CANUE research has shown that neighborhoods with higher levels of material deprivation tend to have lower levels of greenness and walkability and higher concentrations of NO_2 (Doiron et al. 2020). These findings underscore the necessity of improving equity among communities at various levels – neighborhood, city, province, and national.

In combining geospatial data developed by CANUE with sociodemographic data, a new publicly available web app called "*HealthyPlan.City*" (Doiron et al. 2024) has been created for mapping inequitable distribution of various environmental exposures by combining CANUE datasets with data from the 2021 Canadian Census. This app was officially launched in November 2023 and is the first tool in Canada to allow evaluation of how multiple environmental indicators vary across communities with different levels of (socioeconomic) vulnerability in 125+ Canadian cities. By mapping the areas where vulnerable populations are exposed to unfavorable environmental conditions, urban planners, public health professionals, and decision makers can prioritize these areas in their investments, and community advocates can ensure environmental investments are being directed to neighborhoods where residents will benefit the most.

HealthyPlan.City includes physical and built environment indicators such as air pollution, proximity to parks, proximity to transit stops and other amenities, and climate-related variables such as heat and flood susceptibility. To identify locations where environmental inequity exists and to address disparities between vulnerable populations and others, the tool also incorporates variables that represent the social environment, including vulnerable populations, such as visible minorities, low-income individuals, immigrants, children, older adults, and individuals living alone, as well as the custom cross-tabulated indicators prepared by Statistics Canada, such as low-income children. This approach (https://healthyplan.city/en/methods) helps pinpoint regions where social vulnerabilities intersect with environmental risks, thereby informing targeted interventions to mitigate inequity.

Nevertheless, it is essential to acknowledge and address the potential risks of identifying areas with high levels of vulnerability and environmental exposure (Millum et al. 2019), which can have potential negative outcomes such as leading to decreased property values or misuse of the data by political or financial groups, and more importantly, spatial stigma (Keene and Padilla 2014). Research indicates that stigmatization can result in breaches of confidentiality and further marginalization of already vulnerable groups (Millum et al. 2019). Vulnerable or at-risk communities often face numerous challenges, lacking access to effective solutions to address their issues. Highlighting their poor environmental and socioeconomic conditions without providing actionable solutions may exacerbate feelings of depression and negatively impact mental health (Koster and van Ommeren 2022). But providing these communities with access to detailed environmental and demographic data can also empower them to advocate for changes they were previously unable to achieve. This democratization of data fosters community-driven efforts to address local environmental issues. We believe that the benefits of making this data available – empowering communities, informing public health initiatives, and driving equitable urban planning – outweigh the risks of stigmatization and other unintended consequences. By doing so, we aim for the HealthyPlan.City app to serve as a valuable tool for promoting environmental equity and community resilience without reinforcing negative stereotypes or stigmatizing vulnerable populations.

To illustrate briefly the app, let us consider the tree canopy coverage, which is critical in interventions against urban heat islands affecting vulnerable populations, and explore equity for two distinct vulnerable populations: children and low-income children. The data presented in Fig. 2 reveal a significant disparity in tree canopy coverage, in Quebec City, particularly affecting low-income children. While Fig. 2a indicates that less than half (46%) of children reside in areas where additional resources targeting tree canopy cover could enhance equity, positioning it as the top-ranked city among similar-sized Canadian cities for tree canopy equity, the situation appears quite different for low-income children. According to Fig. 2b, almost 3 in 4 (72%) children from low-income families reside in areas where more resources would improve equity in Quebec City, ranking it as the least equitable for this combination of factors among the 12 cities of comparable size. This stark contrast identified by the tool clearly highlights the need for targeted efforts to improve canopy cover in neighborhoods inhabited by low-income children to address these equity issues effectively.

Moreover, research has consistently shown that access to quality greenspaces can improve mental health outcomes among youth (Richardson et al. 2017; Vanaken and Danckaerts 2018). For vulnerable populations like low-income children, who are already at higher risk of mental health issues, the lack of sufficient greenspace might exacerbate stress and anxiety. Therefore, improving tree canopy cover in these

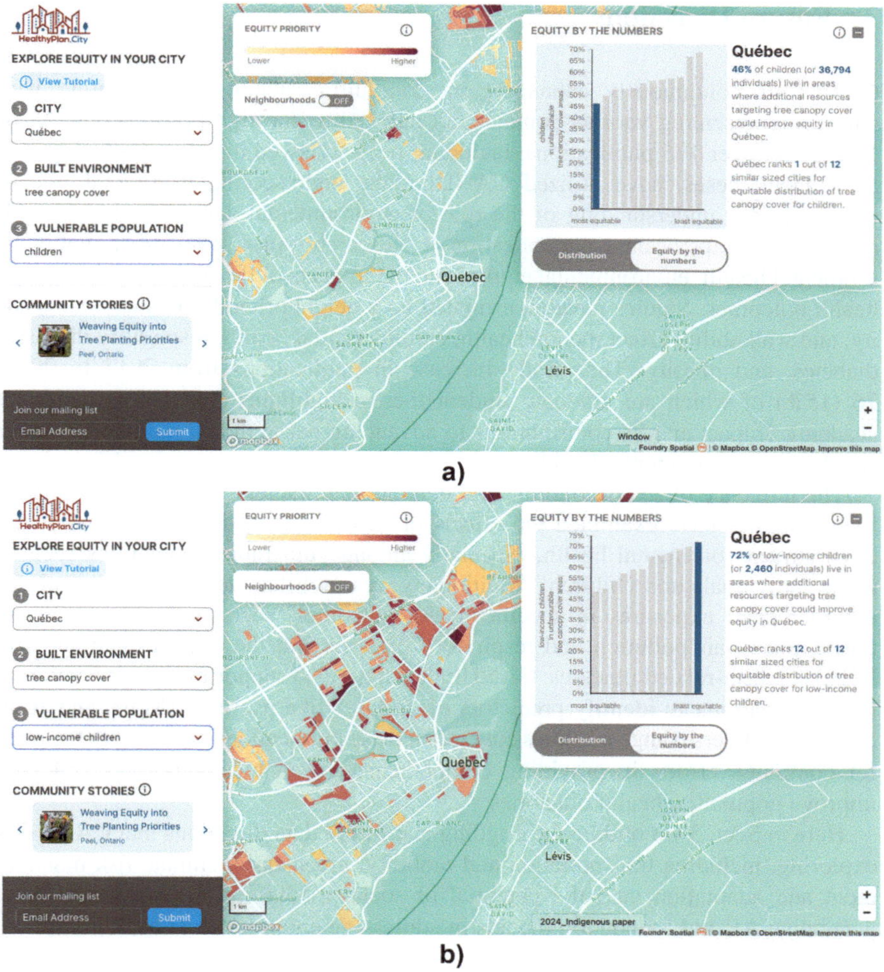

Fig. 2 Exploring equity in tree canopy coverage in Quebec City for two different vulnerable populations, including (**a**) children and (**b**) low-income children

underserved areas could not only promote physical and environmental equity but also serve as a vital intervention for enhancing youth mental health. This dual benefit underscores the importance of integrating urban planning with public health strategies to foster environments that support the well-being of all community members, particularly the most vulnerable. Note, however, that demonstrating the above relationship is only the first step on the way to establishing causal role greenspace may play in this context (see other chapters in this book for the discussion of this issue: "Population neuroscience: Principles and advances" and "Population neuroscience: Understanding concepts of generalizability and transportability and their application to improving the public's health").

3 Future Directions

CANUE is continuously enhancing its data infrastructure by integrating new datasets representing various domains, all indexed to six-digit postal codes. As of June 2024, over 175 publications, including peer-reviewed manuscripts and graduate student theses, have utilized CANUE data, underscoring its pivotal role in advancing our understanding of the role of environmental determinants of health and disease (https://canue.ca/publications/). Research topics cover urban and environmental health, environmental equity, and the impacts of the built environment on health outcomes. Studies have investigated how neighborhood greenness, air quality, and walkability affect mental health, cardiovascular health, respiratory diseases, diabetes, and obesity. The steady increase in publications from 2017 to 2024, 26 (15%) of which are graduate student theses, highlights CANUE's significant academic impact and its contributions to advancing research and cultivating future scholars in environmental health.

The influence of socioeconomic and demographic factors has also been a major focus. Studies have examined the effects of neighborhood deprivation and social environments on mental health, especially among vulnerable populations such as aging individuals and children. Additionally, research has delved into the role of environmental exposures in specific health outcomes, such as the link between traffic-related air pollution and respiratory diseases and the impact of urban heat islands on heat-related illnesses. CANUE data have also supported the application of machine learning to identify predictors of diseases like psoriasis (Muntyanu et al. 2024). The diverse range of topics covered in these publications emphasizes the need for continued efforts in providing comprehensive data to advance research and improve population health outcomes.

Here in Sect. 3 we outline the ongoing and future data directions that CANUE is exploring, including transportation and mobility, greenspace, climate risk management and adaptation, and the potential of social media and crowdsourced data, generative AI, and foundation models in addressing environmental health issues.

3.1 Transportation and Mobility

In the context of today's and future cities, the topic of transportation and mobility, encompassing daily commuting for work, school, and other regular activities, has emerged as a critical factor in shaping the well-being and accessibility of urban communities. Mobility, when addressed effectively, can be a powerful tool (i.e., intervention) in promoting physical and mental health, as well as fostering social and economic inclusion (Cass et al. 2005). For many, especially those new to a city or with limited resources, the lack of accessible and efficient transportation options can be a significant barrier, trapping them in suburban areas with limited mobility (Affordability Action Council 2024). This can have detrimental effects on both physical and mental health, as it restricts access to essential services, employment opportunities, and social networks. Understanding the public health and equity implications and subsequently addressing this challenge is crucial in creating healthier, sustainable cities.

Many types of data relevant to transportation and mobility have the potential to improve understanding of their connections with physical and mental health, equity, and opportunities for intervention. This includes traffic density and flow from municipal and provincial transportation departments, Google's mobility data (https://mobilitydata.org/), and HERE Technologies (https://www.here.com/). These data, which include average vehicle speeds and traffic flow patterns, can also be useful in refining exposure estimates for traffic-related air pollution and noise. Moreover, freight and delivery traffic data from logistics companies and municipal traffic management systems are important for understanding the environmental and health impacts of commercial traffic in urban areas (Bosona 2020; Crainic et al. 2009).

Road infrastructure and conditions data from sources like Google Maps and OpenStreetMap are helpful for assessing the safety and usability of transportation networks, influencing accident rates and commuter stress levels (Keller et al. 2020; Kloog et al. 2018). Public transit data from agencies like the Toronto Transit Commission (https://www.ttc.ca/) and TransLink (https://www.translink.ca/), along with General Transit Feed Specification (https://gtfs.org/) data, provide valuable insights into transit routes and schedules. This information supports the study of health impacts related to public transportation and promotes sustainable, transit-oriented development (Pereira et al. 2017; Higgins and Kanaroglou 2016). Data on parking availability and usage from municipal parking authorities and commercial parking operators can help analyze the impact of parking on urban congestion and emissions (Guo 2013; Shoup 2021). Active transportation data sourced from bike-sharing programs, municipal planning departments, and platforms like Strava Metro (https://metro.strava.com/) can help examine the health benefits of walking and cycling. Increasing active transportation can help reduce vehicle emissions and increase physical activity, improving cardiovascular and overall health (Chapman et al. 2018; Mizdrak et al. 2019; Mueller et al. 2015).

The data from social media, known as social or participatory sensors, can provide valuable insights into various health and environmental topics (Shakeri Hossein Abad et al. 2021, 2022) such as flooding (Rosser et al. 2017) and mobility patterns (Hasan et al. 2013; C. Yang et al. 2019). Platforms like X (formerly known as Twitter), Facebook, and Instagram offer geotagged data points that reveal congestion hotspots and frequently used routes, enabling urban planners to optimize transportation networks and improve travel experiences (Noaeen and Far 2019, 2020). Analyzing these data helps identify issues efficiently, provided the data are reliable. Social media data also help gauge public sentiment and engagement, which is crucial for planning and implementing new mobility initiatives (Ebrahimpour et al. 2020; Noaeen et al. 2017). Public feedback on social platforms can highlight user experiences and preferences, informing adjustments to transit services or the introduction of new transportation solutions.

By integrating mobility data from sources like Mapbox (MapBox. Location intelligence for business 2024), which tracks people's movements and time spent in various locations, researchers and urban planners can gain valuable insights into how people navigate their cities. Aggregated data from mobile phone GPS, such as Mapbox, offer insights into variations in movement patterns across space and changes in mobility behavior, aiding in the analysis of how mobility directly affects health and also alters exposure to environmental hazards (Balzotti et al. 2018; Gao et al. 2021; Kishore et al. 2022). Mapbox Traffic data, drawn from over 700 million monthly active users and updated in real time, enables researchers to analyze mobility patterns, estimate road speeds, and understand the dynamics of urban movement. This rich dataset, which CANUE is integrating into new exposure metrics, allows for examining how transportation accessibility relates to physical health, mental well-being, and social inclusion. Additionally, Mapbox Movement data, which aggregates device density and activity, offers insights into population interactions with their environment, helping to evaluate public health implications of urban planning decisions, monitor economic activity, and assess the effectiveness of health interventions. Understanding how people move through the city can inform the design of more walkable and bikeable infrastructure, encouraging active transportation.

3.2 Urban Greenspaces and Nature-Based Solutions

Urban greenspaces play a critical role in enhancing the quality of life in cities, offering both physical and mental health benefits (Buckley and Brough 2017; Feda et al. 2015; McCormick 2017; Tillmann et al. 2018). Incorporating features like serene walking paths (Sundling and Jakobsson 2023; Xie et al. 2021) and calming water features (Sundling and Jakobsson 2023; Xie et al. 2021) can create a more immersive and restorative environment. Providing ample seating, shaded areas, and amenities that encourage social interaction and relaxation can further enhance the park experience (Dhar and Dash 2022; Pleasant et al. 2013).

It is important to recognize that not all parks and greenspaces are created equal. Some may feature lush tree canopies, while others are predominantly grassy areas. The size of these spaces can also vary, from small pocket parks (Dong et al. 2023) to expansive urban forests. While larger parks may encourage more physical activity and exercise, smaller greenspaces can still offer restorative benefits for the brain (Balai Kerishnan and Maruthaveeran 2021). Even the presence of greenery within buildings and public spaces, such as indoor gardens or living walls, can positively impact mental well-being by providing a calming and rejuvenating environment (Al-Kodmany 2023; Moya et al. 2019).

When it comes to parks and greenspaces, simply having one within a kilometer does not guarantee its accessibility or quality. The distance to the nearest park entrance plays a crucial role in determining actual access to these resources (EnviroAtlas 2015; Rigolon 2016). If walking or cycling to a park is impractical due to obstacles or distance, the availability of mass transit or alternative transportation modes becomes crucial. Understanding how to measure access and distance effectively is key in evaluating the availability of greenspaces for urban residents. Utilizing GIS technology and data analysis can greatly enhance our ability to assess accessibility and distance to parks. By developing effective metrics for measuring access, cities can compare their performance and create city rankings that provide a holistic view of greenspace accessibility. This approach allows for a more nuanced understanding of how park access impacts different populations.

In the context of natural environments, CANUE utilizes land use and land cover data from Natural Resources Canada (https://natural-resources.canada.ca/home), the Canada Centre for Mapping and Earth Observation (https://profils-profiles.science.gc.ca/en/research-centre/canada-centre-mapping-and-earth-observation), and the Copernicus Land Monitoring Service (https://land.copernicus.eu/en). Detailed maps from these sources help assess the availability of forests, parks, wetlands, and urban green spaces. Information on the location, size, and facilities of parks from Parks Canada (https://parks.canada.ca/) and municipal parks departments is also invaluable for studying the health benefits of recreational spaces, promoting physical activity, and reducing the risk of chronic diseases (Parker et al. 2020).

Lastly, high-resolution satellite imagery from Sentinel-2 (https://sentiwiki.copernicus.eu/web/s2-mission), Landsat, and MODIS is used to analyze vegetation density, canopy cover, and accessibility. Data on lakes, rivers, and coastal waters from Fisheries and Oceans Canada (https://www.dfo-mpo.gc.ca/index-eng.html) and Environment and Climate Change Canada (https://www.canada.ca/en/environment-climate-change.html) can also help study the health benefits of blue spaces, which are known to improve mental well-being and encourage physical activity (Bell et al. 2021a, b; Gascon et al. 2015).

3.3 Climate Change Risk Management and Adaptation

Climate change poses multifaceted risks that require attention to effectively manage and mitigate their impacts. Mental health is an area that demands closer attention in regard to climate change. The exacerbation of mental health issues in extreme heat conditions (Cianconi et al. 2020) is a significant factor that often goes unnoticed in traditional vulnerability assessments. Another critical dimension is the biological and health-related vulnerabilities of specific groups, such as seniors. Commonly, seniors are categorized as vulnerable due to factors like physical limitations and complex biological realities that may contribute to their vulnerability, such as impaired thermoregulation, reduced sweating, or circulation problems, often compounded by comorbidities like diabetes (Kirkman et al. 2012; Lu et al. 2024; Samsel et al. 2024). Recognizing such nuances moves us beyond simplistic categorizations and toward a more accurate identification of risk based on a comprehensive understanding of health conditions and vulnerability.

CANUE and particularly HealthyPlan.City, with national urban environmental data, have the potential to assist in understanding climate change challenges. By merging high-resolution data on climatic factors, such as flooding risk and urban heat islands, with comprehensive demographic and socioeconomic details, HealthyPlan.City supports the identification of populations and areas at heightened risk. This approach enables the mapping of risks associated with extreme weather events, alongside the socioeconomic factors that might intensify these risks.

Identifying and mapping climate change-related variables such as heat islands, floods, and smoke from wildfires involve using remote sensing data and various models. By indexing demographic and environmental data to standard geographic indicators (census geographies or postal codes), CANUE strives to achieve detailed, granular mapping. Visualization techniques like heat maps using, for example, quintiles or deciles display temperature variations, flood risks, and wildfire hazards, allowing researchers to identify vulnerable populations accurately and guide targeted interventions effectively.

Current heat vulnerability maps used by CANUE rely primarily on satellite and outdoor temperature data and thus fail to capture the indoor environments where people spend most of their time. Characterizing indoor conditions, such as the availability of air conditioning in homes, schools, and other buildings, is important to understanding vulnerability (Bao et al. 2015; Nayak et al. 2018). Access to air conditioning can significantly mitigate the adverse effects of high outdoor temperatures, particularly for groups like the elderly, young children, and those with preexisting medical conditions (Gronlund 2014). Thus, by incorporating data on air conditioning prevalence, building characteristics, and indoor temperature monitoring, vulnerability assessments can better reflect the actual conditions experienced by individuals and communities (Nayak et al. 2018; Uejio et al. 2011). This information can better inform targeted interventions, such as the establishment of cooling centers, energy assistance programs, or building retrofits, to enhance resilience and protect the most vulnerable populations during heat waves (Gronlund 2014; Hayden et al. 2011). Furthermore, the concept of community cohesion plays a crucial role in vulnerability. In communities with strong social ties, individuals are more likely to receive support during heat waves (Kafeety et al. 2020). Obtaining relevant data on this aspect of climate risk could involve analyzing social media activity to gauge community engagement and support for vulnerable individuals like seniors living alone.

Not all households and facilities have the means or resources to install and operate air conditioning systems effectively. Factors such as socioeconomic status, housing quality, and energy costs can create disparities in access to cooling, exacerbating heat-related vulnerabilities (Hayden et al. 2011). Beyond air conditioning, other indoor conditions can vary substantially from outdoor temperatures due to factors like building materials, insulation, and ventilation (Uejio et al. 2011). Schools, for instance, may experience elevated indoor temperatures even when outdoor conditions are relatively mild, putting students and staff at risk of heat-related illnesses and even compromising learning and academic performance (Bidassey-Manilal et al. 2016; Lala and Hagishima 2023; R. J. Park et al. 2020).

3.4 *Integrating Large Language and Multimodal Foundation Models into the Tools*

Generative multimodal AI models (Wu and Goodman 2018) are designed to create new content, whether it is text, images, audio, or video, by learning patterns from extensive datasets. These models can handle various types of data and generate meaningful outputs across multiple domains.

Large language models (LLMs) (Brown et al. 2020; Minaee et al. 2024) are a specific class of generative AI models focused on text. They analyze a wide array of text data sources – from scientific publications to social media posts – to identify patterns in environmental health, for example. LLMs excel in understanding and generating text, providing thorough analyses of environmental issues.

Multimodal foundation models (C. Li et al. 2024), which include but are not limited to LLMs, are pre-trained on extensive datasets and fine-tuned for specific tasks (Moor et al. 2023). These models are designed to generalize across various domains, making them invaluable for integrating diverse data types such as text, images, audio, and video. This capability enhances the precision of environmental health assessments (Jakubik et al. 2023).

They can leverage data such as satellite imagery and textual weather reports to evaluate the health impacts of climate change. Additionally, foundation models can analyze social media data to identify emerging trends in public sentiment about urban planning and environmental policies, providing actionable feedback for city planning processes. This includes monitoring misinformation during crises, engaging communities in development projects, and generating educational outreach content. Furthermore, these models can tailor public health alerts and create targeted interventions based on socioeconomic data and living conditions. For example, during heat waves, LLMs can identify areas lacking air conditioning through social media discussions, informing mitigation strategies. Similarly, during floods, LLMs can analyze real-time social media data to determine immediate community needs, such as emergency shelters and rescue services.

Designing targeted chatbots is another area where LLMs can make a significant contribution. For instance, to further enhance the user experience on HealthyPlan. City, an interactive chatbot powered by LLMs could be implemented. This chatbot would allow users to ask questions and receive information in real time, making assessments with HealthyPlan.City more accessible and user-friendly. A foundation model leveraging HealthyPlan.City could access additional data such as Google Maps and Street View, aerial images, live traffic videos, official land use plans, municipal policies, etc., and other "undiscovered data", to become a more powerful tool for users. By inputting queries, users could receive personalized insights and recommendations based on the extensive data integrated within the platform, ranging from urban greenspace accessibility to detailed environmental health data.

For a deeper exploration of how LLMs and social media analytics are integrated into urban science and environmental health research, refer to subsequent chapters, "Leveraging generative AI models in urban science" and "Area-level measures of the social environment: Operationalization, pitfalls, and ways forward".

4 Concluding Remarks

The establishment of CANUE has aimed to advance the field of environmental health research by facilitating the integration of disparate data sources from multiple teams of investigators into a cohesive, accessible framework. This consortium-driven platform not only democratizes data access but also enhances the scope of population health research, including linking environmental data directly in major Canadian health data platforms. Through development of the CANUE data portal, HealthyPlan.City, and web-based tools for data extraction, and recent developments in the integration of data in various domains such as mobility, flood risk, and greenspace data, CANUE is at the forefront of addressing urban health and equity challenges. The consortium's efforts in mapping and analyzing greenspace accessibility and its impact on mental health exemplify its comprehensive approach to population health. As CANUE continues to evolve, it remains a pivotal resource in the development of strategies that promote healthier, more sustainable urban environments. This ongoing work underscores the importance of collaborative data utilization in crafting informed, effective population health policies and interventions that are capable of meeting the complex demands of Canada's diverse populations.

Acknowledgments The authors would like to acknowledge Tomás Paus for his motivation in initiating this chapter and for his extremely helpful reviews, which significantly contributed to the development and refinement of this chapter.

References

Affordability Action Council (2024) Rethinking urban mobility: providing more affordable and equitable transportation options. Institute for Research on Public Policy https://irpp.org/research-studies/rethinking-urban-mobility/

Al-Kodmany K (2023) Greenery-covered tall buildings: a review. Buildings 13(9):2362. https://doi.org/10.3390/buildings13092362

Balai Kerishnan P, Maruthaveeran S (2021) Factors contributing to the usage of pocket parks – a review of the evidence. Urban For Urban Green 58:126985. https://doi.org/10.1016/j.ufug.2021.126985

Balzotti C, Bragagnini A, Briani M, Cristiani E (2018) Understanding human mobility flows from aggregated mobile phone data. IFAC-PapersOnLine 51(9):25–30. https://www.sciencedirect.com/science/article/pii/S2405896318307213

Bao J, Li X, Yu C (2015) The construction and validation of the heat vulnerability index, a review. Int J Environ Res Public Health 12(7):7220–7234. https://doi.org/10.3390/ijerph120707220

Bell S, Fleming LE, Grellier J, Kuhlmann F, Nieuwenhuijsen MJ, White MP (2021a) Urban blue spaces: planning and design for water, health and well-being. Routledge. https://play.google.com/store/books/details?id=WYVAEAAAQBAJ

Bell S, Kuhlmann F, White MP (2021b) Urban blue spaces: planning and design for water, health and well-being. Taylor & Francis Group. https://play.google.com/store/books/details?id=lc9zzgEACAAJ

Bidassey-Manilal S, Wright CY, Engelbrecht JC, Albers PN, Garland RM, Matooane M (2016) Students' perceived heat-health symptoms increased with warmer classroom temperatures. Int J Environ Res Public Health 13(6). https://doi.org/10.3390/ijerph13060566

Bosona T (2020) Urban freight last mile logistics – challenges and opportunities to improve sustainability: a literature review. Sustainability 12(21):8769. https://doi.org/10.3390/su12218769

Brook JR, Setton EM, Seed E, Shooshtari M, Doiron D, Awadalla P, Brauer M, Hu H, McGrail K, Stieb D, Subarrao P, Demers P, Manuel D, McLaughlin J, Carlsten C, Azad M, Atkinson S, Burnett R, Lou W et al (2018) The Canadian Urban Environmental Health Research Consortium – a protocol for building a national environmental exposure data platform for integrated analyses of urban form and health. BMC Public Health 18(1):114. https://doi.org/10.1186/s12889-017-5001-5

Brook JR, Doiron D, Setton E, Lakerveld J (2021) Centralizing environmental datasets to support (inter)national chronic disease research: values, challenges, and recommendations. Environ Epidemiol 5(1):e129. https://doi.org/10.1097/EE9.0000000000000129

Brown TB, Mann B, Ryder N, Subbiah M, Kaplan J, Dhariwal P, Neelakantan A, Shyam P, Sastry G, Askell A, Agarwal S, Herbert-Voss A, Krueger G, Henighan T, Child R, Ramesh A, Ziegler DM, Wu J, Winter C et al (2020) Language models are few-shot learners. Adv Neural Inf Process Syst. abs/2005.14165. http://arxiv.org/abs/2005.14165

Browning M, Lee K (2017) Within what distance does "Greenness" best predict physical health? A systematic review of articles with GIS buffer analyses across the lifespan. Int J Environ Res Public Health 14(7):675. https://doi.org/10.3390/ijerph14070675

Buckley RC, Brough P (2017) Economic value of parks via human mental health: an analytical framework. Front Ecol Evol 5. https://doi.org/10.3389/fevo.2017.00016

Burgoine T, Jones AP, Namenek Brouwer RJ, Benjamin Neelon SE (2015) Associations between BMI and home, school and route environmental exposures estimated using GPS and GIS: do we see evidence of selective daily mobility bias in children? Int J Health Geogr 14:8. https://doi.org/10.1186/1476-072X-14-8

Caspi CE, Sorensen G, Subramanian SV, Kawachi I (2012) The local food environment and diet: a systematic review. Health Place 18(5):1172–1187. https://doi.org/10.1016/j.healthplace.2012.05.006

Cass N, Shove E, Urry J (2005) Social exclusion, mobility and access. Sociol Rev 53(3):539–555. https://doi.org/10.1111/j.1467-954X.2005.00565.x

Chapman R, Keall M, Howden-Chapman P, Grams M, Witten K, Randal E, Woodward A (2018) A cost benefit analysis of an active travel intervention with health and carbon emission reduction benefits. Int J Environ Res Public Health 15(5). https://doi.org/10.3390/ijerph15050962

Chen H, Kwong JC, Copes R, Tu K, Villeneuve PJ, van Donkelaar A, Hystad P, Martin RV, Murray BJ, Jessiman B, Wilton AS, Kopp A, Burnett RT (2017) Living near major roads and the incidence of dementia, Parkinson's disease, and multiple sclerosis: a population-based cohort study. Lancet 389(10070):718–726. https://doi.org/10.1016/S0140-6736(16)32399-6

Child AW, Hinds J, Sheneman L, Buerki S (2022) Centralized project-specific metadata platforms: toolkit provides new perspectives on open data management within multi-institution and multidisciplinary research projects. BMC Res Notes 15(1):106. https://doi.org/10.1186/s13104-022-05996-3

Cianconi P, Betrò S, Janiri L (2020) The impact of climate change on mental health: a systematic descriptive review. Front Psych 11:74. https://doi.org/10.3389/fpsyt.2020.00074

Crainic TG, Gendreau M, Potvin J-Y (2009) Intelligent freight-transportation systems: assessment and the contribution of operations research. Transp Res C Emerg Technol 17(6):541–557. https://doi.org/10.1016/j.trc.2008.07.002

Dhar G, Dash SP (2022) A systematic literature review on the impact of open spaces on human physiological and mental Well-being in post-pandemic housing in urban context. ECS Trans 107(1):7723. https://doi.org/10.1149/10701.7723ecst

Doiron D, Setton EM, Shairsingh K, Brauer M, Hystad P, Ross NA, Brook JR (2020) Healthy built environment: spatial patterns and relationships of multiple exposures and deprivation in Toronto, Montreal and Vancouver. Environ Int 143:106003. https://doi.org/10.1016/j.envint.2020.106003

Doiron D, Setton EM, Syer J, Redivo A, McKee A, Noaeen M, Patel P, Booth GL, Brauer M, Fuller D, Kestens Y, Rosella LC, Stieb D, Villeneuve PJ, Brook JR (2024) HealthyPlan.City: a web tool to support urban environmental equity and public health in Canadian communities. J Urban Health Bull N Y Acad Med. https://doi.org/10.1007/s11524-024-00855-x

Dong J, Guo R, Guo F, Guo X, Zhang Z (2023) Pocket parks-a systematic literature review. Environ Res Lett 18(8):083003. https://doi.org/10.1088/1748-9326/ace7e2

Duan R-R, Hao K, Yang T (2020) Air pollution and chronic obstructive pulmonary disease. Chronic Dis Transl Med 6(4):260–269. https://doi.org/10.1016/j.cdtm.2020.05.004

Ebrahimpour Z, Wan W, Velázquez García JL, Cervantes O, Hou L (2020) Analyzing social-geographic human mobility patterns using large-scale social media data. ISPRS Int J Geo Inf 9(2):125. https://doi.org/10.3390/ijgi9020125

educhenin (2010) Heat-waves: risks and responses. https://apps.who.int/iris/bitstream/handle/10665/107552/9789289010948%E2%80%90eng.pdf?sequence=1

EnviroAtlas (2015) Estimated walking distance to a park entrance. {EnviroAtlas}. https://enviroatlas.epa.gov/enviroatlas/DataFactSheets/pdf/ESC/Estimatedwalkingdistancetoaparkentrance.pdf

Environment and Climate Change Canada (2024) Canada taking next steps on the right to a healthy environment and environmental justice and racism. Government of Canada https://www.canada.ca/en/environment-climate-change/news/2024/02/canada-taking-next-steps-on-the-right-to-a-healthy-environment-and-environmental-justice-and-racism.html

Feda DM, Seelbinder A, Baek S, Raja S, Yin L, Roemmich JN (2015) Neighbourhood parks and reduction in stress among adolescents: results from Buffalo, New York. Indoor Built Environ 24(5):631–639. https://doi.org/10.1177/1420326X14535791

Firth CL, Thierry B, Fuller D, Winters M, Kestens Y (2021) Gentrification, urban interventions and equity (GENUINE): a map-based gentrification tool for Canadian metropolitan areas. Health Rep 32(5):15–28. https://doi.org/10.25318/82-003-x202100500002-eng

Fuller R, Landrigan PJ, Balakrishnan K, Bathan G, Bose-O'Reilly S, Brauer M, Caravanos J, Chiles T, Cohen A, Corra L, Cropper M, Ferraro G, Hanna J, Hanrahan D, Hu H, Hunter D, Janata G, Kupka R, Lanphear B et al (2022) Pollution and health: a progress update. Lancet Planet Health 6(6):e535–e547. https://doi.org/10.1016/S2542-5196(22)00090-0

Gao X, Wang H, Liu L (2021) Profiling residents' mobility with grid-aggregated mobile phone trace data using Chengdu as the case. Sustainability 13(24):13713. https://doi.org/10.3390/su132413713

Gascon M, Triguero-Mas M, Martínez D, Dadvand P, Forns J, Plasència A, Nieuwenhuijsen MJ (2015) Mental health benefits of long-term exposure to residential green and blue spaces: a systematic review. Int J Environ Res Public Health 12(4):4354–4379. https://doi.org/10.3390/ijerph120404354

Gronlund CJ (2014) Racial and socioeconomic disparities in heat-related health effects and their mechanisms: a review. Curr Epidemiol Rep 1(3):165–173. https://doi.org/10.1007/s40471-014-0014-4

Guo Z (2013) Does residential parking supply affect household car ownership? The case of New York City. J Transp Geogr 26:18–28. https://doi.org/10.1016/j.jtrangeo.2012.08.006

Guxens M, Lubczynska MJ, Perez-Crespo L, Muetzel RL, El Marroun H, Basagana X, Hoek G, Tiemeier H (2022) Associations of air pollution on the brain in children: a brain imaging study. Res Rep 2022(209):1–61. https://www.ncbi.nlm.nih.gov/pubmed/36106707

Hammad HM, Nauman HMF, Abbas F, Jawad R, Farhad W, Shahid M, Bakhat HF, Farooque AA, Mubeen M, Fahad S, Cerda A (2023) Impacts of COVID-19 pandemic on environment, society, and food security. Environ Sci Pollut Res Int 30(44):99261–99272. https://doi.org/10.1007/s11356-023-25714-1

Hasan S, Zhan X, Ukkusuri SV (2013) Understanding urban human activity and mobility patterns using large-scale location-based data from online social media. Proceedings of the 2nd ACM SIGKDD international workshop on urban computing, pp 1–8. https://doi.org/10.1145/2505821.2505823

Hayden MH, Brenkert-Smith H, Wilhelmi OV (2011) Differential adaptive capacity to extreme heat: a Phoenix, Arizona, case study. Weather Clim Soc 3(4):269–280. https://doi.org/10.1175/WCAS-D-11-00010.1

Heo S, Lim CC, Bell ML (2020) Relationships between local green space and human mobility patterns during COVID-19 for Maryland and California, USA. Sustainability 12(22):9401. https://doi.org/10.3390/su12229401

Hernandez Carballo I, Bakola M, Stuckler D (2022) The impact of air pollution on COVID-19 incidence, severity, and mortality: A systematic review of studies in Europe and North America. Environ Res 215(Pt 1):114155. https://doi.org/10.1016/j.envres.2022.114155

Higgins CD, Kanaroglou PS (2016) A latent class method for classifying and evaluating the performance of station area transit-oriented development in the Toronto region. J Transp Geogr 52:61–72. https://doi.org/10.1016/j.jtrangeo.2016.02.012

Holsten JE (2009) Obesity and the community food environment: a systematic review. Public Health Nutr 12(3):397–405. https://doi.org/10.1017/S1368980008002267

Jakubik J, Roy S, Phillips CE, Fraccaro P, Godwin D, Zadrozny B, Szwarcman D, Gomes C, Nyirjesy G, Edwards B, Kimura D, Simumba N, Chu L, Karthik Mukkavilli S, Lambhate D, Das K, Bangalore R, Oliveira D, Muszynski M, . . . Ramachandran R (2023) Foundation models for generalist geospatial artificial intelligence. In: arXiv [cs.CV]. arXiv. http://arxiv.org/abs/2310.18660

Jennings V, Browning MHE, Rigolon A (2019) Urban green spaces: public health and sustainability in the United States. Springer. https://play.google.com/store/books/details?id=omqPDwAAQBAJ

Jia P, Cao X, Yang H, Dai S, He P, Huang G, Wu T, Wang Y (2021) Green space access in the neighbourhood and childhood obesity. Obes Rev 22(Suppl 1):e13100. https://doi.org/10.1111/obr.13100

Kafeety A, Henderson SB, Lubik A, Kancir J, Kosatsky T, Schwandt M (2020) Social connection as a public health adaptation to extreme heat events. Can J Public Health 111(6):876–879. https://doi.org/10.17269/s41997-020-00309-2

Karimi SM, Maziyaki A, Ahmadian Moghadam S, Jafarkhani M, Zarei H, Moradi-Lakeh M, Pouran H (2020) Continuous exposure to ambient air pollution and chronic diseases: prevalence, burden, and economic costs. Rev Environ Health 35(4):379–399. https://doi.org/10.1515/reveh-2019-0106

Keene DE, Padilla MB (2014) Spatial stigma and health inequality. Crit Public Health 24(4):392–404. https://doi.org/10.1080/09581596.2013.873532

Keller S, Gabriel R, Guth J (2020) Machine learning framework for the estimation of average speed in rural road networks with OpenStreetMap data. ISPRS Int J Geo Inf 9(11):638. https://doi.org/10.3390/ijgi9110638

Kirkman MS, Briscoe VJ, Clark N, Florez H, Haas LB, Halter JB, Huang ES, Korytkowski MT, Munshi MN, Odegard PS, Pratley RE, Swift CS (2012) Diabetes in older adults. Diabetes Care 35(12):2650–2664. https://doi.org/10.2337/dc12-1801

Kishore N, Taylor AR, Jacob PE, Vembar N, Cohen T, Buckee CO, Menzies NA (2022) Evaluating the reliability of mobility metrics from aggregated mobile phone data as proxies for SARS-CoV-2 transmission in the USA: a population-based study. Lancet Digit Health 4(1):e27–e36. https://doi.org/10.1016/S2589-7500(21)00214-4

Kloog I, Kaufman LI, de Hoogh K (2018) Using open street map data in environmental exposure assessment studies: eastern Massachusetts, Bern region, and South Israel as a case study. Int J Environ Res Public Health 15(11). https://doi.org/10.3390/ijerph15112443

Koster HRA, van Ommeren J (2022) Neighbourhood stigma and place-based policies. Econ Policy 38(114):289–339. https://doi.org/10.1093/epolic/eiac039

Kovats RS, Hajat S (2008) Heat stress and public health: a critical review. Annu Rev Public Health 29:41–55. https://doi.org/10.1146/annurev.publhealth.29.020907.090843
Laddu D, Paluch AE, LaMonte MJ (2021) The role of the built environment in promoting movement and physical activity across the lifespan: implications for public health. Prog Cardiovasc Dis 64:33–40. https://doi.org/10.1016/j.pcad.2020.12.009
Lala B, Hagishima A (2023) Impact of escalating heat waves on students' Well-being and overall health: a survey of primary school teachers. Climate 11(6):126. https://www.mdpi.com/2225-1154/11/6/126
Levinger P, Cerin E, Milner C, Hill KD (2022) Older people and nature: the benefits of outdoors, parks and nature in light of COVID-19 and beyond– where to from here? Int J Environ Health Res 32(6):1329–1336. https://doi.org/10.1080/09603123.2021.1879739
Li F, Yigitcanlar T, Nepal M, Thanh KN, Dur F (2022) Understanding urban heat vulnerability assessment methods: a PRISMA review. Energies 15(19):6998. https://doi.org/10.3390/en15196998
Li C, Gan Z, Yang Z, Yang J, Li L, Wang L, Gao J (2024) Multimodal foundation models: from specialists to general-purpose assistants. Found Trends Comput Graph Vis 16(1–2):1–214. https://doi.org/10.1561/0600000110
Lu K, Sheth P, Zhou ZL, Kazari K, Guergachi A, Keshavjee K, Noaeen M, Shakeri Z (2024) Identifying prediabetes in Canadian populations using machine learning. bioRxiv. https://doi.org/10.1101/2024.02.03.24302301
MapBox (2024) Location intelligence for business. Mapbox. https://www.mapbox.com/
McCormick R (2017) Does access to green space impact the mental Well-being of children: a systematic review. J Pediatr Nurs 37:3–7. https://doi.org/10.1016/j.pedn.2017.08.027
Millum J, Campbell M, Luna F, Malekzadeh A, Karim QA (2019) Ethical challenges in global health-related stigma research. BMC Med 17(1):84. https://doi.org/10.1186/s12916-019-1317-6
Minaee S, Mikolov T, Nikzad N, Chenaghlu M, Socher R, Amatriain X, Gao J (2024) Large language models: a survey. In: arXiv [cs.CL]. arXiv. http://arxiv.org/abs/2402.06196
Mizdrak A, Blakely T, Cleghorn CL, Cobiac LJ (2019) Potential of active transport to improve health, reduce healthcare costs, and reduce greenhouse gas emissions: a modelling study. PLoS One 14(7):e0219316. https://doi.org/10.1371/journal.pone.0219316
Moor M, Banerjee O, Abad ZSH, Krumholz HM, Leskovec J, Topol EJ, Rajpurkar P (2023) Foundation models for generalist medical artificial intelligence. Nature 616(7956):259–265. https://doi.org/10.1038/s41586-023-05881-4
Moya TA, van den Dobbelsteen A, Ottelé M, Bluyssen PM (2019) A review of green systems within the indoor environment. Indoor and Built Environment 28(3):298–309. https://doi.org/10.1177/1420326X18783042
Mueller N, Rojas-Rueda D, Cole-Hunter T, de Nazelle A, Dons E, Gerike R, Götschi T, Int Panis L, Kahlmeier S, Nieuwenhuijsen M (2015) Health impact assessment of active transportation: a systematic review. Prev Med 76:103–114. https://doi.org/10.1016/j.ypmed.2015.04.010
Muntyanu A, Milan R, Kaouache M, Ringuet J, Gulliver W, Pivneva I, Royer J, Leroux M, Chen K, Yu Q, Litvinov IV, Griffiths CEM, Ashcroft DM, Rahme E, Netchiporouk E (2024) Tree-based machine learning to identify predictors of psoriasis incidence at the neighborhood level: a populational study from Quebec, Canada. Am J Clin Dermatol 25(3):497–508. https://doi.org/10.1007/s40257-024-00854-3
Nayak SG, Shrestha S, Kinney PL, Ross Z, Sheridan SC, Pantea CI, Hsu WH, Muscatiello N, Hwang SA (2018) Development of a heat vulnerability index for New York State. Public Health 161:127–137. https://doi.org/10.1016/j.puhe.2017.09.006
Noaeen M, Far BH (2019) Social media analysis for traffic management. Proceedings of the 14th international conference on global software engineering, pp 72–73. https://doi.org/10.1109/ICGSE.2019.00012
Noaeen M, Far BH (2020) The efficacy of using social media data for designing traffic management systems. 2020 4th international workshop on crowd-based requirements engineering (CrowdRE), pp 11–17. https://doi.org/10.1109/CrowdRE51214.2020.00009

Noaeen M, Abad ZSH, Far BH (2017) Let's hear it from RETTA: a requirements elicitation tool for TrAffic management systems. 2017 IEEE 25th international requirements engineering conference (RE), pp 450–451. https://doi.org/10.1109/RE.2017.78
Northridge ME, Freeman L (2011) Urban planning and health equity. J Urban Health 88(3): 582–597. https://doi.org/10.1007/s11524-011-9558-5
O'Neill A (2023, 25 September) Canada: degree of urbanization from 1960 to 2022, Urbanization in Canada 2022. Statista. https://www.statista.com/statistics/271208/urbanization-in-canada/#:~:text=Urbanization%20in%20Canada%202022&text=Canada's%20urban%20population%20now%20accounts,percent%20of%20it's%20total%20population
Otten JJ, Averill MM, Spiker ML (2023) Food security and food access during the COVID-19 pandemic: impacts, adaptations, and looking ahead. JPEN J Parenter Enteral Nutr 47(Suppl 1): S11–S15. https://doi.org/10.1002/jpen.2445
Park RJ, Goodman J, Hurwitz M, Smith J (2020) Heat and learning. Am Econ J Econ Pol 12(2): 306–339. https://doi.org/10.1257/pol.20180612
Park AH, Zhong S, Yang H, Jeong J, Lee C (2022a) Impact of COVID-19 on physical activity: a rapid review. J Glob Health 12:05003. https://doi.org/10.7189/jogh.12.05003
Park S, Kim E, Kim G, Kim S, Choi Y, Paek D (2022b) What activities in forests are beneficial for human health? A systematic review. Int J Environ Res Public Health 19(5). https://doi.org/10.3390/ijerph19052692
Parker P, Rollins R, Murray G, Adam C, Cannessa R (2020) Community perceptions of the contributions of parks to sustainability in Canada. In: Leisure and sustainability. Routledge, pp 87–111. https://www.taylorfrancis.com/chapters/edit/10.4324/9780429441615-6/community-perceptions-contributions-parks-sustainability-canada-pete-parker-rick-rollins-grant-murray-chafey-adam-rosaline-cannessa
Paul N, Yao J, McLean KE, Stieb DM, Henderson SB (2022) The Canadian optimized statistical smoke exposure model (CanOSSEM): a machine learning approach to estimate national daily fine particulate matter (PM2.5) exposure. Sci Total Environ 850:157956. https://doi.org/10.1016/j.scitotenv.2022.157956
Pereira RHM, Schwanen T, Banister D (2017) Distributive justice and equity in transportation. Transp Rev 37(2):170–191
Pineo H, Glonti K, Rutter H, Zimmermann N, Wilkinson P, Davies M (2018) Urban health indicator tools of the physical environment: a systematic review. J Urban Health 95(5):613–646. https://doi.org/10.1007/s11524-018-0228-8
Pleasant A, Scanlon MM, Pereira-Leon M (2013) Literature review: environmental design and research on the human health effects of open spaces in urban areas. Hum Ecol Rev 20(1): 36–49. http://www.jstor.org/stable/24707570
Quan Q-L, Yoon K-N, Lee JS, Kim EJ, Lee DH (2023) Impact of ultraviolet radiation on cardiovascular and metabolic disorders: the role of nitric oxide and vitamin D. Photodermatol Photoimmunol Photomed 39(6):573–581. https://doi.org/10.1111/phpp.12914
Reyes M, Páez A, Morency C (2014) Walking accessibility to urban parks by children: a case study of Montreal. Landsc Urban Plan 125:38–47. https://doi.org/10.1016/j.landurbplan.2014.02.002
Richardson EA, Pearce J, Shortt NK, Mitchell R (2017) The role of public and private natural space in children's social, emotional and behavioural development in Scotland: a longitudinal study. Environ Res 158:729–736. https://doi.org/10.1016/j.envres.2017.07.038
Rigolon A (2016) A complex landscape of inequity in access to urban parks: a literature review. Landsc Urban Plan 153:160–169. https://doi.org/10.1016/j.landurbplan.2016.05.017
Roemer J, Komulainen TM, Scholz M, Moatamedi M (2023) The review of 102 design support tools for nature-based solutions applications. Int J Hydrol Sci Technol 15(4):327–348. https://doi.org/10.1504/IJHST.2023.131169
Rosser JF, Leibovici DG, Jackson MJ (2017) Rapid flood inundation mapping using social media, remote sensing and topographic data. Nat Hazards 87(1):103–120. https://doi.org/10.1007/s11069-017-2755-0

Salgado M, Madureira J, Mendes AS, Torres A, Teixeira JP, Oliveira MD (2020) Environmental determinants of population health in urban settings. A systematic review. BMC Public Health 20(1):853. https://doi.org/10.1186/s12889-020-08905-0

Samsel K, Tiwana A, Ali S, Sadeghi A, Guergachi A, Keshavjee K, Noaeen M, Shakeri Z (2024) Predicting depression among Canadians at-risk or living with diabetes using machine learning. bioRxiv. https://doi.org/10.1101/2024.02.03.24302303

Shakeri Hossein Abad Z, Kline A, Sultana M, Noaeen M, Nurmambetova E, Lucini F, Al-Jefri M, Lee J (2021) Digital public health surveillance: a systematic scoping review. NPJ Digit Med 4(1):41. https://doi.org/10.1038/s41746-021-00407-6

Shakeri Hossein Abad Z, Butler GP, Thompson W, Lee J (2022) Crowdsourcing for machine learning in public health surveillance: lessons learned from Amazon mechanical Turk. J Med Internet Res 24(1):e28749. https://doi.org/10.2196/28749

Sheppard N, Carroll M, Gao C, Lane T (2023) Particulate matter air pollution and COVID-19 infection, severity, and mortality: a systematic review and meta-analysis. Sci Total Environ 880: 163272. https://doi.org/10.1016/j.scitotenv.2023.163272

Shoup D (2021) High cost of free parking. Routledge. https://play.google.com/store/books/details?id=KxwfEAAAQBAJ

Sundling C, Jakobsson M (2023) How do urban walking environments impact pedestrians' Experience and psychological health? A systematic review. Sustainability 15(14):10817. https://doi.org/10.3390/su151410817

Tabrizi N, Lak A, Moussavi A, Mohammad Reza Moussavi.A S (2023) Green space and the health of the older adult during pandemics: a narrative review on the experience of COVID-19. Front Public Health 11:1218091. https://doi.org/10.3389/fpubh.2023.1218091

Tillmann S, Tobin D, Avison W, Gilliland J (2018) Mental health benefits of interactions with nature in children and teenagers: a systematic review. J Epidemiol Community Health 72(10): 958–966. https://doi.org/10.1136/jech-2018-210436

To T, Zhu J, Villeneuve PJ, Simatovic J, Feldman L, Gao C, Williams D, Chen H, Weichenthal S, Wall C, Miller AB (2015) Chronic disease prevalence in women and air pollution – a 30-year longitudinal cohort study. Environ Int 80:26–32. https://doi.org/10.1016/j.envint.2015.03.017

Uejio CK, Wilhelmi OV, Golden JS, Mills DM, Gulino SP, Samenow JP (2011) Intra-urban societal vulnerability to extreme heat: the role of heat exposure and the built environment, socioeconomics, and neighborhood stability. Health Place 17(2):498–507. https://doi.org/10.1016/j.healthplace.2010.12.005

van Kamp I, Davies H (2013) Noise and health in vulnerable groups: a review. Noise Health 15(64): 153. https://doi.org/10.4103/1463-1741.112361

Vanaken G-J, Danckaerts M (2018) Impact of green space exposure on children's and adolescents' mental health: a systematic review. Int J Environ Res Public Health 15(12). https://doi.org/10.3390/ijerph15122668

Wang H, Sun D, Wang B, Gao D, Zhou Y, Wang N, Zhu B (2020) Association between noise exposure and diabetes: meta-analysis. Environ Sci Pollut Res Int 27(29):36085–36090. https://doi.org/10.1007/s11356-020-09826-6

Wang B, Xu T, Gao H, Ta N, Chai Y, Wu J (2021) Can daily mobility alleviate green inequality from living and working environments? Landsc Urban Plan 214:104179. https://doi.org/10.1016/j.landurbplan.2021.104179

Weaver LJ, Fasel CB (2018) A systematic review of the literature on the relationships between chronic diseases and food insecurity. Food Nutr Sci 9(5):519–541. https://doi.org/10.4236/fns.2018.95040

Wimalawansa SA, Wimalawansa SJ (2016) Environmentally induced, occupational diseases with emphasis on chronic kidney disease of multifactorial origin affecting tropical countries. Ann Occup Environ Med 28:33. https://doi.org/10.1186/s40557-016-0119-y

Winters M, Zanotto M, Butler G (2020) At-a-glance – the Canadian bikeway comfort and safety (Can-BICS) classification system: a common naming convention for cycling infrastructure, vol

40. Government of Canada – Public Health Agency of Canada, p 288. https://doi.org/10.24095/hpcdp.40.9.04

Wu M, Goodman N (2018) Multimodal generative models for scalable weakly-supervised learning. Adv Neural Inf Proces Syst:31. https://proceedings.neurips.cc/paper_files/paper/2018/hash/1102a326d5f7c9e04fc3c89d0ede88c9-Abstract.html

Xie Q, Lee C, Lu Z, Yuan X (2021) Interactions with artificial water features: a scoping review of health-related outcomes. Landsc Urban Plan 215:104191. https://doi.org/10.1016/j.landurbplan.2021.104191

Yang C, Xiao M, Ding X, Tian W, Zhai Y, Chen J, Liu L, Ye X (2019) Exploring human mobility patterns using geo-tagged social media data at the group level. J Spat Sci 64(2):221–238. https://doi.org/10.1080/14498596.2017.1421487

Yang L, Gutierrez DE, Guthrie OW (2024) Systemic health effects of noise exposure. J Toxicol Environ Health B Crit Rev 27(1):21–54. https://doi.org/10.1080/10937404.2023.2280837

Yuchi W, Brauer M, Czekajlo A, Davies HW, Davis Z, Guhn M, Jarvis I, Jerrett M, Nesbitt L, Oberlander TF, Sbihi H, Su J, van den Bosch M (2022) Neighborhood environmental exposures and incidence of attention deficit/hyperactivity disorder: a population-based cohort study. Environ Int 161:107120. https://doi.org/10.1016/j.envint.2022.107120

Zhang B, Weuve J, Langa KM, D'Souza J, Szpiro A, Faul J, Mendes de Leon C, Gao J, Kaufman JD, Sheppard L, Lee J, Kobayashi LC, Hirth R, Adar SD (2023) Comparison of particulate air pollution from different emission sources and incident dementia in the US. JAMA Intern Med 183(10):1080–1089. https://doi.org/10.1001/jamainternmed.2023.3300

Part V
Brains

Development and Maturation of the Human Brain, from Infancy to Adolescence

Tomáš Paus

Contents

Abstract This chapter describes basic principles and key findings regarding the development and maturation of the human brain, the former referring to the pre-natal and early post-natal periods and the latter concerning childhood and adolescence. In both cases, we focus on brain structure as revealed in vivo with multi-modal magnetic resonance imaging (MRI). We begin with a few numbers about the human brain and its cellular composition and a brief overview of a number of MRI-based metrics used to characterize age-related variations in grey and white matter. We then proceed with synthesizing current knowledge about developmental and maturational changes in the cerebral cortex (its thickness, surface area, and intra-cortical myelination) and the underlying white matter (volume and structural properties). To facilitate biological interpretations of MRI-derived metrics, we introduce the concept of virtual histology. We conclude the chapter with a few notes about future directions in the study of factors shaping the human brain from conception onwards.

This chapter was first published in *The Cambridge Handbook of Cognitive Development*, edited by Olivier Houdé and Grégoire Borst, reprinted courtesy of The Cambridge University Press.

T. Paus (✉)
Departments of Psychiatry and Neuroscience, Faculty of Medicine and Centre Hospitalier Universitaire, University of Montréal, Montreal, QC, Canada
e-mail: tomas.paus@umontreal.ca

Curr Topics Behav Neurosci (2024) 68: 327–348
https://doi.org/10.1007/7854_2024_514
Published Online: 14 August 2024

Keywords Adolescents · Cerebral cortex · Children · Infants · Magnetic resonance imaging · Transcriptomics · White matter

1 General Introduction

1.1 Brain in Numbers

Our capacity to process information depends on the smooth functioning of neuronal networks, which consist of both neuronal and non-neuronal cells. In the cerebral cortex, the two main classes of neurons are pyramidal cells (~70% of all neurons) and interneurons (~30%) (Druga 2009; Sloper et al. 1979). Non-neuronal cells are mostly glia, namely astrocytes (~19% of all glial cells), microglia (~6%), and oligodendrocytes (~75%) (Pelvig et al. 2008), as well as endothelial and mural cells of the cerebral vasculature (Fig. 1). The human brain of 1.5 kg (adult) has eighty-six billion neurons and eighty-five billion non-neuronal cells; sixteen billion neurons (sixty billion non-neuronal cells) are found in the cerebral cortex, while sixty-nine billion neurons (and sixteen billion non-neuronal cells) are in the cerebellum; note the strikingly different ratio between the non-neuronal cells and neurons in the cerebral cortex (3.76 non-neuronal cells to one neuron) and the cerebellum (0.23 non-neuronal cells to one neuron) (Azevedo et al. 2009).

Axons – the cellular extensions of neurons carrying electrical impulses as well as various cargoes (e.g. vesicles, mitochondria) between the cell body and synapse (Paus et al. 2014) – form a network that totals an astounding 176,000 km in length

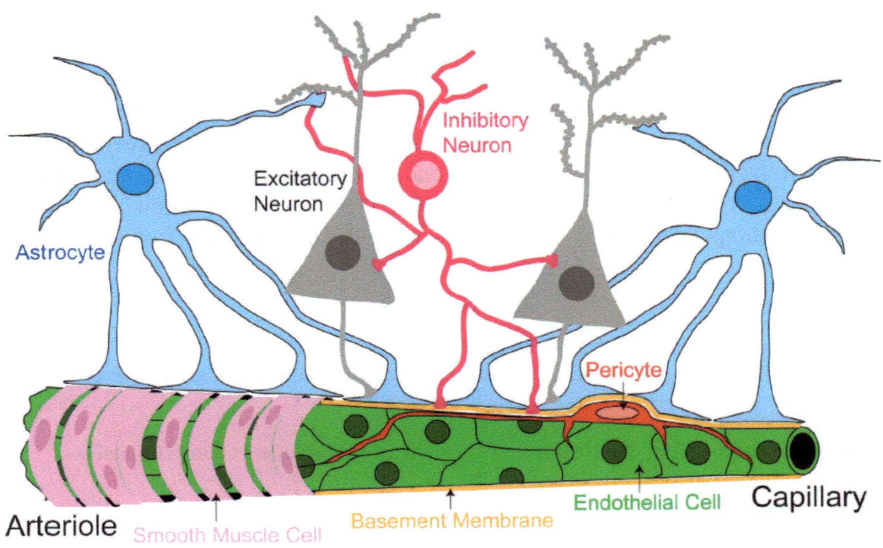

Fig. 1 Neurovascular unit. From Hermann et al. (2015)

(adult) (Marner and Pakkenberg 2003). The length of individual axons varies by four orders of magnitude, from micrometres (interneurons) to centimetres (corticospinal neurons). Schüz and Braitenberg (2002) classified axons into three different groups based on their length and location in the human brain: (1) short (<3 mm) axons within the cortex (intra-cortical horizontal axons); (2) longer (3 to 30 mm) axons in WM adjacent to the cortex (U-fibres); and (3) very long (>30 mm) axons organized in bundles connecting different lobes within a hemisphere (e.g. longitudinal fascicles) and across hemispheres (corpus callosum). They estimated that the majority of axons are located within the short group (i.e. within the cerebral cortex), with only 4% of all axons found in the very long group containing the long-range fibre bundles (Schüz and Braitenberg 2002). Axon diameter is related to its length, and the thickness of the myelin sheath varies with axon diameter (Paus and Toro 2009).

Variations in the cellular composition of the cerebral cortex and those in the relative abundance of thick-or-thin and more-or-less myelinated axons affect MRI signals. We come back to the neurobiology underlying various MRI-based metrics in Sect. 3. But let us now review briefly what can be measured with MRI to capture development and maturations of the human brain.

1.2 Imaging the Brain

Over the past 30+ years, MRI has become the method of choice for studying both structure and function of the human brain across the lifespan. This is due to its versatility (same scanner used for measuring a variety of brain phenotypes[1]), availability (e.g. ~11,000 MRI units in the USA), and the non-invasive nature of MRI technology, namely no exposure to ionizing radiation (not the case in computerized axial tomography [CAT scan] and positron emission tomography [PET scan]). The non-invasiveness makes MRI particularly suitable for research studies of brain development in the general (non-clinical) population. As pointed out above, the majority of MRI scanners come equipped with basic hardware and software allowing one to acquire a wide variety of brain "images" using a particular temporal sequence of radiofrequency pulses, gradients, and read-outs (a "scan") (Moore and Chung 2017) These scans allow one to capture different types of MR signal, which is based on electromagnetic energy emitted by precessing nuclei of hydrogen; Box 1 describes the basic principles of MRI and the main sources of contrast in some of the common types of images acquired in developmental studies, namely T1-weighted and T2-weighted images, diffusion tensor images (DTI), and magnetization transfer (MT) images.

[1] Phenotypes are organism's observable characteristics. Clearly, MRI allows us to observe – in vivo – many characteristics of the human brain that, in the past, could be observed only post mortem.

Box 1 Principles of Magnetic Resonance Imaging (Adapted from Paus 2013)
Nuclei that have an odd number of nucleons (protons and neutrons) possess both a magnetic moment and angular momentum (or spin). In the presence of an external magnetic field, such nuclei "line up" with the main (external) magnetic field and precess (or spin) around their axis at a rate proportional to the strength of the magnetic field, emitting electromagnetic energy as they do. The hydrogen atom contains only a single proton and therefore precesses when exposed to a magnetic field.

In the majority of MR imaging studies, precessing nuclei of hydrogen associated with water and fat are the source of the signal. The signal is generated and measured in the following way. First, the person is exposed to a large static magnetic field (B_0) that aligns hydrogen nuclei along the direction of the applied field. In scanners most commonly used in research studies, the strength of B_0 is 1.5 or 3 Tesla; B_0 is oriented horizontally, pointing from head to toe along the long axis of the cylindrical magnet. Second, a pulse of electromagnetic energy is applied at a specific radiofrequency (RF), with an RF coil placed around the head. This radiofrequency is the same as the precession frequency of the imaged nuclei at a given strength of B_0. The RF pulse rotates the precessing nuclei away from their axes, thus allowing one to measure, with a receiver coil, the time it takes for the nuclei to "relax" back to their original position pointing along B_0. The spatial origin of the signal is determined using subtle, position-related changes in B_0 induced by gradient coils (switching of the gradient coils generates the knocking noise heard during scanning).

The most common pulse sequences used to characterize brain structure are T1- and T2-weighted images, DTI, and MTR. We will review these in turn (for a more detailed overview of pulse sequences, see Roberts and Mikulis (2007).

Contrast in structural **T1- and T2-weighted MR** images is based on local differences in proton density (number of hydrogen nuclei per unit of tissue volume) and on the following two relaxation times: (1) longitudinal relaxation time (T1) and (2) transverse relaxation time (T2). Thus, longitudinal (T1) relaxation time represents an exponential recovery of the total magnetization over time. In a complex manner, it depends on the local structural pattern (lattice: an array of points repeating periodically in three dimensions) assumed by the hydrogen nuclei; in the brain, the more structured the tissue, the shorter T1. Transverse (T2) reflects the local "dephasing" rate for precessing nuclei due to local magnetic-field non-homogeneities, with a concomitant decay of the total MR signal; in the brain, the more structured the tissues, the more rapid the dephasing and hence the shorter the T2. Relaxation times, and therefore tissue "brightness" on T1- and T2-weighted images, depend on a variety of biological and structural properties of the brain tissue. Water content is one of the most important influences on T1 in the brain: the more water there is in a given tissue compartment, the longer T1—and the lower the signal on a T1-weighted image—will be in that compartment. In the adult brain, T1 is the longest in cerebrospinal fluid, intermediate in grey matter, and the shortest in white matter. Lipid content in white matter may also influence T1, through magnetic interactions with hydrogen nuclei of the lipids, which are hydrophobic. Iron content is another important influence, primarily by changing local magnetic non-homogeneities: the higher the iron content, the shorter the T2. Finally, the anatomical arrangement of axons may influence the amount of interstitial water and, in turn, T1 values; more tightly bundled axons would have shorter T1 and therefore appear brighter on T1-weighted images.

The introduction of **diffusion tensor imaging** (DTI) in the mid-1990s opened up new avenues for in vivo studies of white matter microstructure (Le Bihan 1995). This imaging technique allows one to estimate several parameters of water diffusion, such as mean diffusivity (MD) and fractional anisotropy (FA). The latter parameter reflects the degree of water diffusion directionality; voxels containing water that moves predominantly along a single direction have higher FA. In white matter, FA is believed to depend on structural properties of fibre tracts, including the relative alignment of individual axons, how tightly they are packed (which affects the amount of interstitial water; see above), myelin content, and axon diameter.

Finally, **magnetization transfer (MT) imaging** is another MR technique employed in studies of the structural properties of white matter. Contrast in MT images reflects the interaction between free water and water bound to macromolecules (McGowan 1999). The macromolecules of myelin are the dominant source of the MT signal in white matter (Kucharczyk et al. 1994). Post-mortem studies revealed a significant positive correlation between myelin content and MTR (Schmierer et al. 2004; Schmierer et al. 2008).

The fastest way to acquire magnetization transfer data is to use a dual acquisition, with and without a MT saturation pulse. MTR images are calculated as the percent signal change between the two acquisitions (Pike 1996). Mean MTR values, subsequently, can be summed across all white matter voxels constituting, for example, the four lobes of the brain.

Multi-modal imaging of brain structure, which is a combination of multiple "scans" of the person's brain during a single session, is particularly helpful for deriving multiple brain phenotypes and triangulating their neurobiological underpinnings. Figure 2 shows a particular example of such a multi-modal protocol, as employed in one of our population-based studies of the human brain (Bjornholm et al. 2017).

Using these images, one can derive a number of characteristics informative with regard to normal and abnormal brain development. Thus, owing to the high contrast between grey and white matter, T1-weighted images are particularly useful for

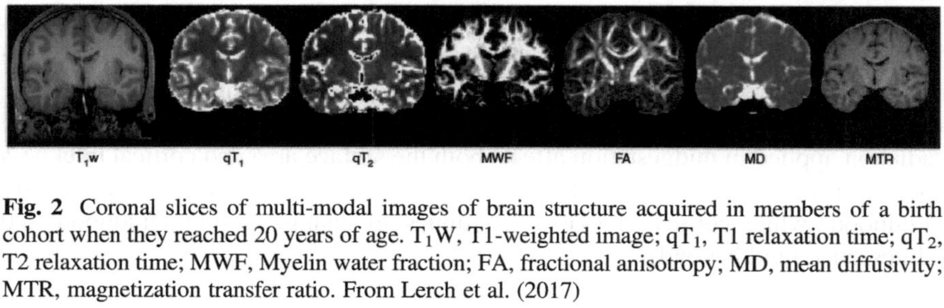

Fig. 2 Coronal slices of multi-modal images of brain structure acquired in members of a birth cohort when they reached 20 years of age. T_1W, T1-weighted image; qT_1, T1 relaxation time; qT_2, T2 relaxation time; MWF, Myelin water fraction; FA, fractional anisotropy; MD, mean diffusivity; MTR, magnetization transfer ratio. From Lerch et al. (2017)

estimating – using automatic image-processing tools – the size of the cerebral cortex and its subdivisions (thickness, surface area) (Fischl and Dale 2000) and volumes of various subcortical structures, such as the hippocampus (Pipitone et al. 2014). Quantitative estimates of T1 and T2 relaxation times provide valuable information about tissue properties of various grey matter (e.g. cerebral cortex) and white matter (e.g. corpus callosum) structures. Images of myelin water fraction (MWF) and magnetic transfer ratio (MTR) capture – in a complementary fashion – *relative* amount of myelin in a given 3D voxel (relative to other types of tissue occupying the same voxel, such as the axon) (Bjornholm et al. 2017). Finally, DTI-derived estimates of mean diffusivity (MD) and fractional anisotropy (FA) reflect constraints imposed on water diffusion by the geometry (and biological properties) of cellular elements that make up a given voxel of imaged tissue. Taking advantage of this feature when modelling spatial patterns of water motion across voxels, the so-called tractography is used commonly to identify bundles of fibres (Jbabdi et al. 2015); note that this approach works well for long-range fibres that constitute, however, only ~4% of all axons (Schüz and Braitenberg 2002). The above MRI-derived metrics provide a toolbox with which one can embark on mapping development and maturation of the human brain in vivo.

2 Development

The growth of the human cerebral cortex begins with the emergence of the ventricular (fifth post-conception week) and subventricular (seventh post-conception week) zones (Zecevic et al. 2005). The radial unit hypothesis provides a framework for global (proliferation) and regional (distribution) expansion of the primate cerebral cortex emanating from the two zones (Fig. 3) (Rakic 1974).

The two most common MRI-derived metrics of the human cerebral cortex, namely surface area and thickness, provide insights into different developmental processes, each with a different timeline. Thus, *cortical surface area* reflects primarily the tangential growth of the cerebral cortex during pre-natal development; the phase of *symmetric division of progenitor cells* in the proliferative zones during the first trimester is particularly important for the tangential growth through additions of ontogenetic columns (Rakic 1988). The subsequent phase of *asymmetric division* continues to increase the number of ontogenetic columns (and thus surface area), but it also begins to contribute to the *thickness of the cerebral cortex* formed by post-mitotic neurons migrating from the proliferative zones to the cortical plate in the inside-out manner (Rakic 1988). Ionizing radiation of the (monkey) foetus during early gestation reduces the surface area (sparing cortical thickness) while the same radiation applied in midgestation affects both the surface area and cortical thickness (Selemon et al. 2013). As we will show below, surface area remains stable after early childhood, while cortical thickness continues to change during childhood and adolescence (and beyond).

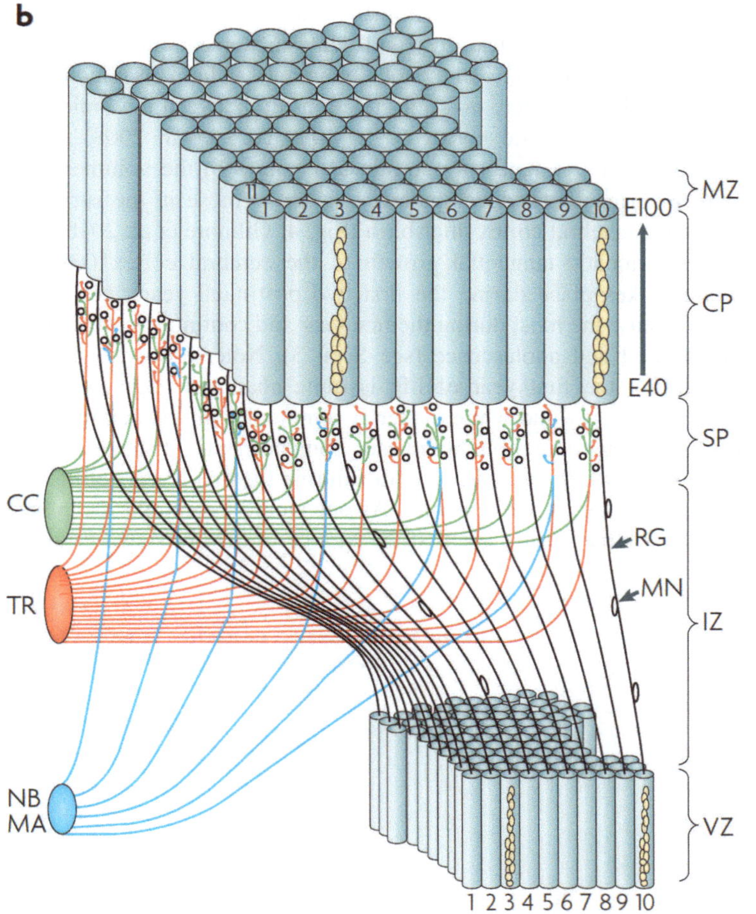

Fig. 3 The radial unit hypothesis. "The model of radial neuronal migration that underlies columnar organization based on (Rakic 1974, 1995). The cohorts of neurons generated in the ventricular zone (VZ) traverse the intermediate zone (IZ) and subplate zone (SP) containing 'waiting' afferents from several sources (cortico-cortical connections (CC), thalamic radiation (TR), nucleus basalis (NB), monoamine subcortical centers (MA)) and finally pass through the earlier generated deep layers before settling in at the interface between the cortical plate (CP) and marginal zone (MZ). The timing of neurogenesis (E40–E100) refers to the embryonic age in the macaque monkey (Bystron et al. 2006; Rakic 1974, 1988). The positional information of the neurons in the VZ and corresponding protomap within the SP and CP is preserved during cortical expansion by transient radial glial scaffolding. Further details can be viewed in the Rakic laboratory animated video of radial migration. RG, radial glia cell; MN, migrating neuron". Legend to Fig. 3 in Rakic (2009)

At birth, the human brain has reached ~420 cm^3 in volume (~36% of adult values) (Knickmeyer et al. 2008). Post-natal growth is particularly rapid during the first 2 years: the brain volume doubles by the end of the first year (855 cm^3; 72% of adult) and increases further to reach 83% (983 cm^3) of adult values by the end of the second year (Knickmeyer et al. 2008). In mammals, cerebral cortex represents the majority of brain tissue; with its cortical grey and white matter combined, the mass of the cerebral cortex constitutes 82% of the total brain mass in humans (Herculano-Houzel 2009). Not surprisingly, there is a tight relationship between the brain volume and surface area of the cerebral cortex ($r^2 \sim 0.7$) (Toro et al. 2008). Gilmore and colleagues provided estimates of age-related changes in the volumes of grey and white matter during foetal development, as well as those in the surface area and thickness of the cerebral cortex after birth (Fig. 4; Gilmore et al. 2018).

The most dramatic tangential growth of the cerebral cortex (i.e. increase in surface area) takes place during the first two post-natal years; cortical thickness appears to increase the most during the first year and continues to change (decrease) during childhood and adolescence (see Sect. 3). Note, however, that estimating thickness during the first year of life is quite challenging given the switch of relaxation time in grey and white matter at some point in this period (possibly due to decreases in water content in both tissue compartments) (Paus et al. 2001).

The robust growth of the cerebral cortex during the first 2 years of life is mirrored in age-related changes in white matter. As one would expect, the steep increases in white matter volume during the first 2 years of life (Matsuzawa et al. 2001) are

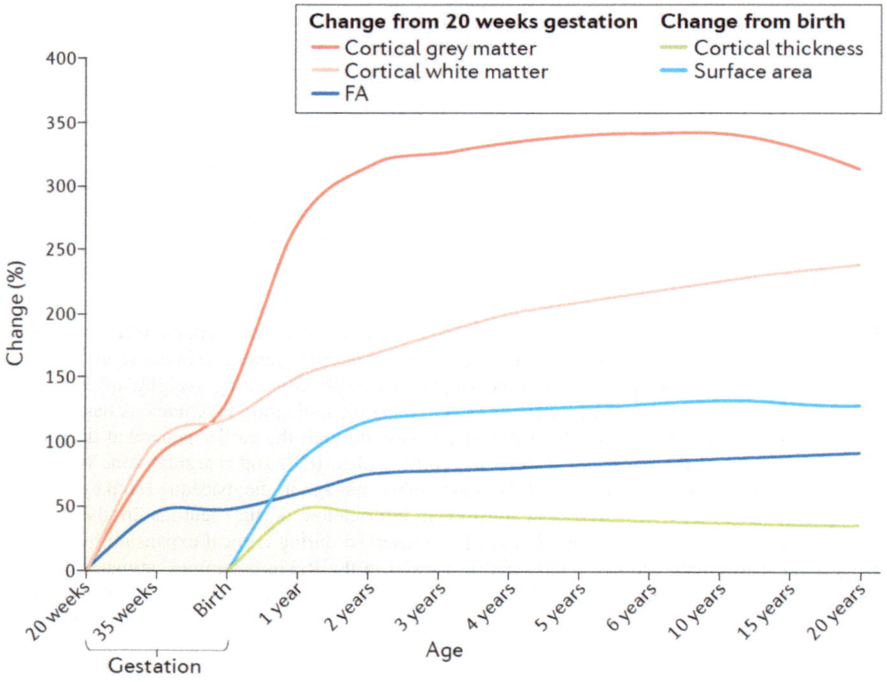

Fig. 4 Estimated trajectories of cortical grey-matter and white-matter volumes during pre-natal development, and cortical surface area and thickness in post-natal development. From Gilmore et al. (2018)

accompanied by changes in structural properties of white matter, as estimated using some of the MRI techniques shown in Fig. 2. As reviewed recently by Lebel and Deoni (2018), most of these (and similar) metrics show a steep change during this period (Fig. 5).

Zooming onto one of the metrics, namely myelin water fraction (MWF), an index of myelin content, we can appreciate the steep increase in this parameter across various compartments of white matter in the infant brain (Fig. 6). The rate of change in MWF appears to peak during the first 2 years of life across all regions examined in this study, with some subtle differences between the four cerebral lobes and virtually no differences across the corpus callosum (genu, body, and splenium) (Dean et al. 2015). Although myelination is likely to be one of the main factors driving age-related changes in structural properties of white matter illustrated in Figs. 5 and 6, the complexity of the underlying neurobiology prevents one to draw such categorical conclusions based on observational data only.

To summarize, the human brain develops at a fast pace during the first three post-conception years. This is reflected in particular in the exponential increase of the surface area of the cerebral cortex and the volume of white matter, the latter tissue compartment including mostly axons either originating or terminating in the cerebral cortex. During the same (post-natal) period, there is also a dramatic change in the structural properties of white matter that, in part, reflect the initial wave of myelination. Thus, to put it simply, *development* of the human brain is completed by 2–3 years of age, but its *maturation* continues.

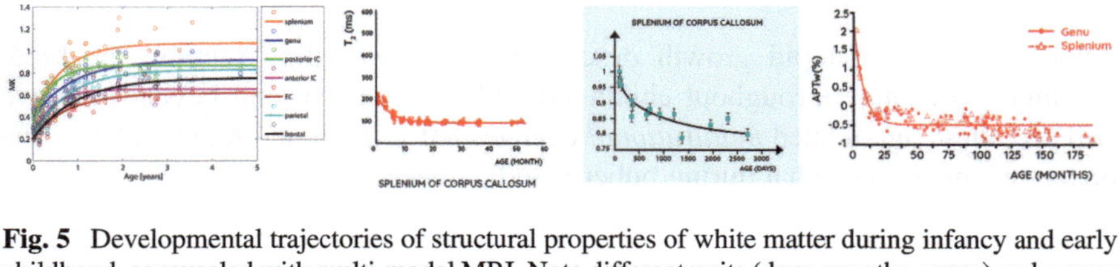

Fig. 5 Developmental trajectories of structural properties of white matter during infancy and early childhood, as revealed with multi-modal MRI. Note different units (days, months, years) and ranges on the X axis of the individual plots. DTI, diffusion tensor imaging; DKI, diffusion kurtosis imaging; MTR, magnetization transfer ratio; APT, amide proton transfer. From Lebel and Deoni (2018)

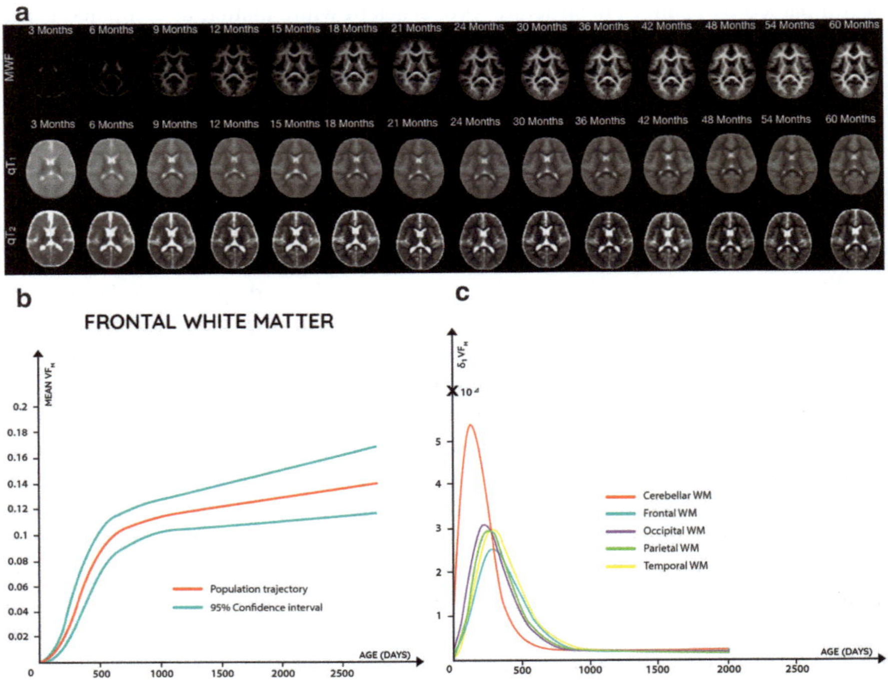

Fig. 6 Myelin water fraction and relaxation times in the developing brain. A. Parametric images of myelin water fraction (MWF) and T1 and T2 relaxation times during infancy and early childhood (from 3 to 60 months); B: Mean values of MWF (denoted as VF_M on the Y axis) as a function of age in white matter of the frontal lobe; C. Curves of the growth rate (denoted as $\delta_1 VF_M$ on the Y axis) in MWF in white matter of the four cerebral lobes and in the cerebellum. WM, white matter. A from Lebel and Deoni; B and C from Dean et al. (2015)

3 Maturation

After the phase of rapid growth described in the preceding section, the brain continues to change throughout childhood and adolescence (and beyond). As we will see, these age-related *maturational* changes are gradual and rather subtle. This picture changes somewhat during puberty and adolescence.[2]

[2] As defined by Sick and Foster, "... puberty refers to the activation of the hypothalamus-pituitary-gonadal axis that culminates in gonadal maturation. Adolescence refers to the maturation of adult social and cognitive behavior" (Sisk and Foster 2004, p. 1040).

3.1 Grey Matter

Figure 7 illustrates the gradual change in whole-brain (absolute) volumes of grey and white matter using cross-sectional data acquired in the NIH-PD project (Brain Development Cooperative Group 2012). On average (based on Table 4 in) (Brain Development Cooperative Group 2012), volumes of grey matter in the brains of pre-pubertal children (4 to 10 years of age) were somewhat larger (by ~10%) while those of white matter were slightly smaller (by ~10%) as compared with the volumes obtained in 17 to 18 years old participants. Contrast this growth rate (~10% over 10 years) with that during the early post-natal development (~100% over 2 years; Fig. 4).

When it comes to age-related changes in grey matter of the cerebral cortex, the initial report by Giedd and colleagues suggested a non-linear trajectory between 4 and 18 years of age, with the cortical volumes increasing to about 10 to 12 years of age (frontal and parietal lobes), and then gradually decreasing (Giedd et al. 1999). But no such non-linear age curves have been observed in subsequent reports of cortical volumes, including the first (cross-sectional) analysis of data acquired in the NIH-PD Study discussed above (Brain Development Cooperative Group 2012). Global and regional volumes of grey matter of the cerebral cortex are the product of surface area and cortical thickness. Based on a combination of cross-sectional and longitudinal data from the NIHD-PD Study (753 scans, 1–3 per scans per participant), cortical thickness decreases between 5 and 20 years of age in a linear fashion (Fig. 8) (Ducharme et al. 2016); this observation has been confirmed in other datasets and re-analyses (Walhovd et al. 2017). One possible explanation of discrepancies between the early and subsequent reports may be the influence of artefacts related to head movement during scanning (more common in younger subjects); non-linear trajectories were observed only if scans with such artefacts were not excluded from the analysis (Ducharme et al. 2016).

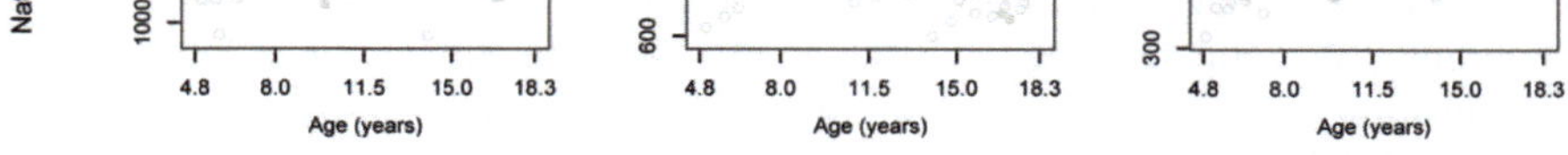

Fig. 7 Total brain volume, as well as whole-brain (absolute) volumes of grey and white matter, derived from a combination of T1-weighted, T2-weighted, and proton density-weighted images obtained in the NIH Pediatric Study. Cross-sectional age curves are indicated in blue (males) and red (females). From the Brain Development Cooperative Group (2012). In all three plots, the Y axis does not start at 0 cm^3; the differences between the youngest (5 years of age) and oldest (18 years of age) do not exceed 10%

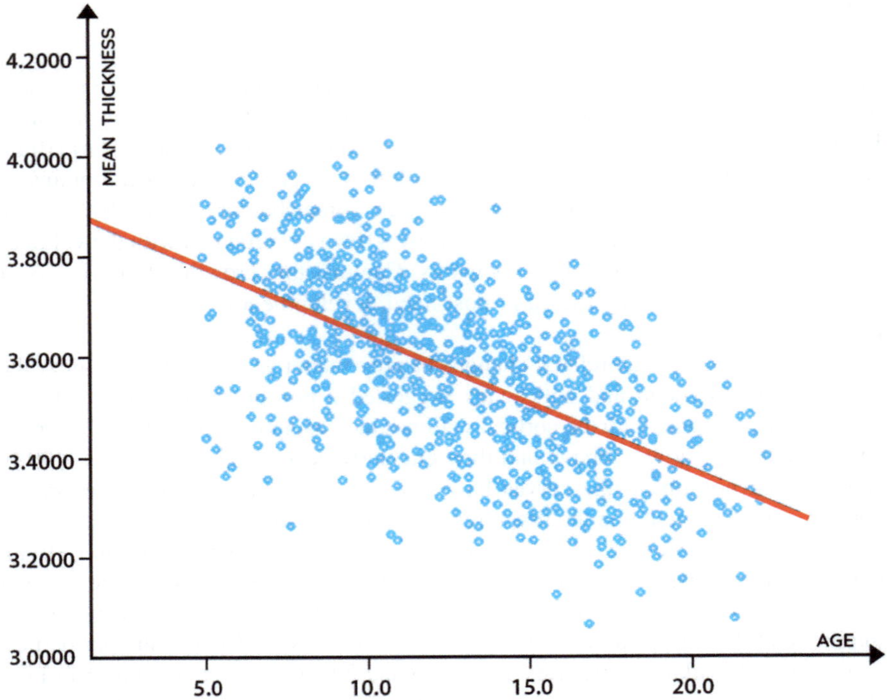

Fig. 8 Decrease in cortical thickness between 10 and 20 years of age. From Ducharme et al. (2016)

Let us now review briefly maturational changes in the adolescent cerebral cortex. Using cross-sectional data from the Saguenay Youth Study (Pausova et al. 2007, 2017), we can see age-related decreases in cortical thickness, which appear to be more pronounced in male vs. female adolescents (Fig. 9, top); note that, in the same individuals, the surface area does not vary with age (Fig. 9, bottom).

3.1.1 Virtual Histology of the Human Cerebral Cortex

Since the initial reports in the late 1990s (Giedd et al. 1999), neurobiological interpretations of age-related decreases in the cortical grey matter have varied from stressing "synaptic pruning" (Giedd et al. 1999) to emphasizing intra-cortical myelination[1,2] (Gogtay et al. 2004; Paus 2005). To address the gap between in vivo (MRI) and ex vivo (histology) observations (Paus 2018), we have recently developed a "virtual histology" approach aimed at estimating cellular contributions to inter-regional variations in various MRI-derived metrics across the human cerebral cortex (Shin et al. 2018). Using this approach, we ask which of the cell types found in the cerebral cortex explains best inter-regional variations of a metric of interest, such as cortical thickness or age-related cortical thinning. We do so by

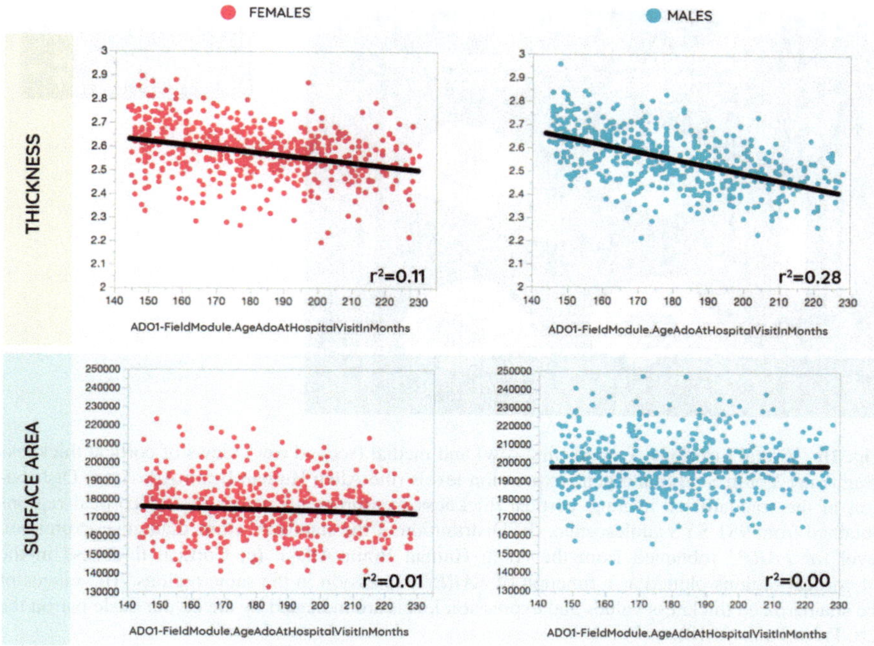

Fig. 9 Cortical thickness and surface area in female ($n = 509$) and male ($n = 479$) adolescents. Data from the Saguenay Youth Study (Pausova et al. 2017)

comparing the profile of a given MRI-derived metric with profiles of cell-specific gene expression across the same cortical regions. To do so, we first extracted gene-expression data from the Allen Human Brain Atlas (Hawrylycz et al. 2012) and estimated the mean values of gene expression for each of 34 cortical regions segmented by FreeSurfer per hemisphere (French and Paus 2015). Using a two-stage procedure, we identified genes with consistent inter-regional profiles of their expression (2,511 out of the total of 20,737 genes passed this filter). From these "consistent" genes, we selected cell type-specific genes as markers of nine types of cells, namely pyramidal neurons, interneurons, astrocytes, microglia, oligodendrocytes, and ependymal, endothelial, and mural (pericytes and vascular smooth muscle cells) cells; these cell-specific marker-genes were identified (in the mouse brain) by others using single-cell transcriptomes (Zeisel et al. 2015). Finally, *across* the 34 "FreeSurfer" regions of the left hemisphere (Desikan-Killiany atlas), we related expression profiles of these cell-specific markers to those in cortical thickness measured with MRI in the same set of regions in 987 adolescents. Figure 10 illustrates this approach by showing inter-regional variations in cortical thickness (left) and expression of *FARP1*, one of 54 gene markers of astrocytes (middle), and their correlation across the 34 cortical regions (right).

In our first study using this approach, we determined that inter-regional profiles of cortical thickness (in the adolescent brain) are related to the expression profiles of

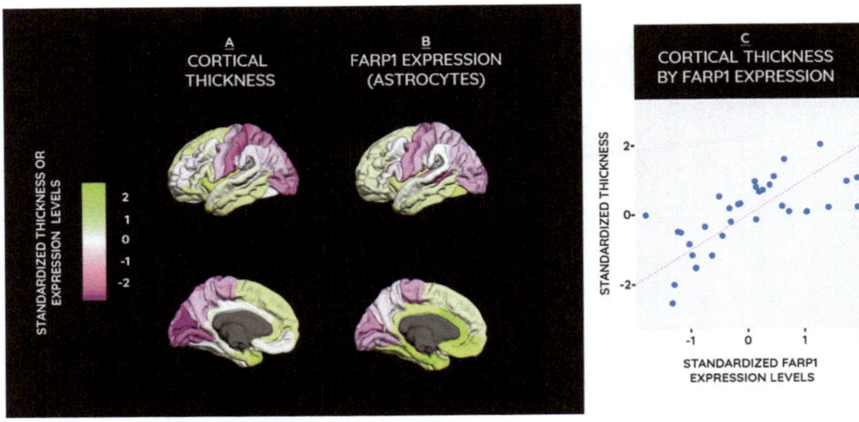

Fig. 10 Virtual histology. Lateral (first row) and medial (second row) views of cortical thickness (Saguenay Youth Study) and gene-expression levels (the Allen Human Brain Atlas). (**a**) Distribution of the standardized average cortical thickness measurements across the 34 cortical regions obtained from 981 SYS adolescents;. (**b**) Distribution of the standardized median gene-expression level for *FARP1* (obtained from the Allen Human Brain Atlas). (**c**) Cortical thickness in the 34 cortical regions plotted as a function of *FARP1* expression in the same regions. The ranges of the standardized thickness values and expression levels are indicated by the colour-scale bar on the left. From Paus (2018)

three cell types, namely astrocytes, microglia, and (CA1) pyramidal cells, but not to any other cell type; altogether, the three cell types explained about 70% of variance in inter-regional differences in cortical thickness (Shin et al. 2018). Furthermore, we observed that the same three cell types also contribute to inter-regional variations in cortical thinning during adolescence; in male adolescents only, we have also observed a contribution of oligodendrocytes to the inter-regional profile of cortical thinning (Shin et al. 2018). In a follow-up study, we used the same approach to identify the cellular substrate of cross-sectional and longitudinal profiles of magnetization transfer ratio (MTR), a scan sensitive to the relative density of macromolecules (including myelin) in the sampled tissue (Henkelman et al. 2001). We found that CA1 pyramidal cells and ependymal cells explain the cross-sectional profiles of MTR in the cerebral cortex, while oligodendrocytes matched closely the longitudinal profiles of the same MRI-derived metric (Patel et al. 2019). Taking these two studies together, we speculate that cortical thickness is dominated by dendritic arborization while cortical thinning during adolescence reflects intra-cortical myelination (which makes grey matter less "grey"), the latter consistent with the results obtained in another study (Whitaker et al. 2016).

To summarize, during childhood and adolescence, the key age-related variations in the human cerebral cortex include those in its thickness and structural properties (rather than its surface area, as seen during infancy). Studies with virtual histology suggest that inter-regional variations and thickness and thinning during adolescence are best explained by parallel variations in pyramidal cells, astrocytes, and microglia. It is likely, but not proven, that these cells drive the dynamics in dendritic arborization. Finally, it is now clear that intra-cortical myelination contributes to the pattern of cortical thinning during adolescence, as well as to the age-related changes in the structural properties captured with multi-modal MRI of the cerebral cortex during this developmental period.

3.2 *White Matter*

As seen in Fig. 7 above, the overall volume of white matter continues to increase during childhood and adolescence (by ~10%). Structural properties of white matter fibre tracts also continue to change during this period; for example, FA in fibre tracts increases by ~20% between 5 and 25 years of age (Fig. 11).

During adolescence, the volume of white matter increases more steeply in males as compared with females (Fig. 12). In male adolescents, this change is at least in part driven by the increasing levels of testosterone; adolescents with a more (vs. less) "efficient" variant of the androgen-receptor gene show a stronger relationship between circulating levels of bioavailable testosterone and white-matter volume (Perrin et al. 2008).

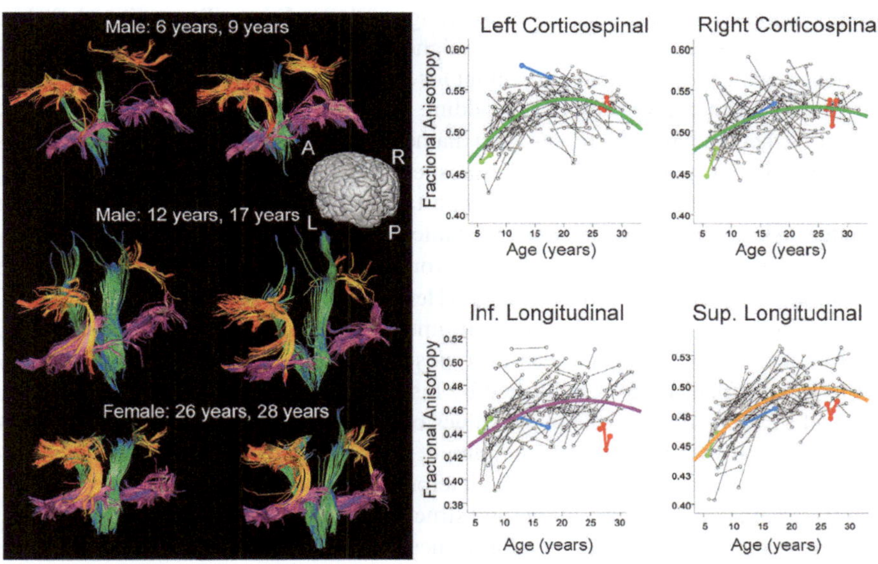

Fig. 11 "Individual tractography results for the superior longitudinal fasciculus (orange), inferior longitudinal fasciculus (magenta), and corticospinal tracts (green) are shown in three representative healthy individuals at different ages. The whole dataset is shown in scatterplots at right, with data from the individuals shown at left identified in colour. The scatter plots show later maturation in the superior longitudinal fasciculus compared to the other regions. Ages of peak FA values were 21 and 23 years for the left and right corticospinal tracts, and 24 and 25 years for the inferior and superior longitudinal fasciculi, respectively". Legend to Fig. 5 in Lebel and Deoni (2018)

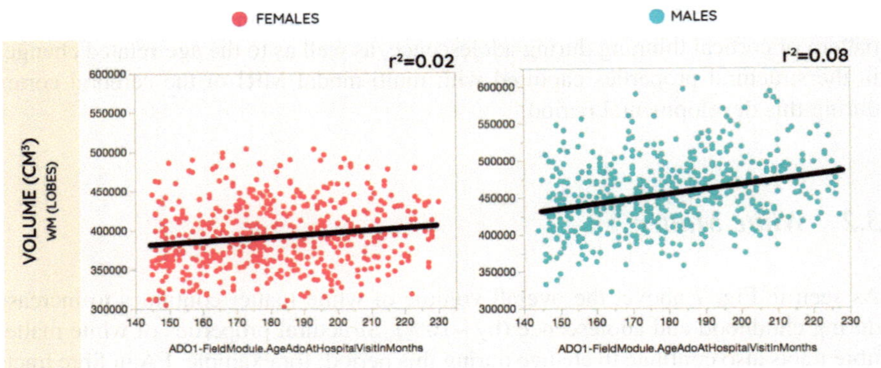

Fig. 12 Volume of white matter (cerebral lobes) in female (n = 509) and male (n = 476) adolescents. Data from the Saguenay Youth Study (Pausova et al. 2017)

Given the parallel *decrease* in MTR values in white matter during male adolescence, we interpreted these findings as reflecting the radial growth of the axon (i.e. an increase in axon diameter) rather than myelination (Perrin et al. 2008). This interpretation was confirmed in a follow-up study in rats where we revealed, with electron microscopy, sex differences in axon diameter (M > F); these sex differences were eliminated by castration (after weaning) of the male rats (Pesaresi et al. 2015).

Axons of larger diameter contain a higher number of microtubules, a key component of axonal cytoskeleton important for axonal transport (Paus et al. 2014). In an in vitro experiment, we showed that adding synthetic testosterone to a tissue culture of sympathetic neurons increases axon diameter and – most importantly – modulates anterograde transport of vesicles (Pesaresi et al. 2015).

Axon diameter appears to be an important contributor to MR signals in white matter. As compared with adjacent white matter, fibre tracts containing large-diameter (long) (Paus and Toro 2009) axons, such as the corticospinal tract, have lower intensity on T1-weighted images (Herve et al. 2009; Pangelinan et al. 2016), as well as longer T1 and T2 relaxation times and lower MTR (Herve et al. 2011). Across the corpus callosum (anterior to posterior), variations in the relative density of large and small fibres (established by histology) are tracked almost perfectly by T1 and T2 relaxation times and the values of myelin water fraction (MWF) but less so by FA and MD (Fig. 13) (Bjornholm et al. 2017), thus emphasizing the importance of multi-modal imaging for the interpretations of MRI data.

To summarize, both the volume and structural properties of white matter continue to change during childhood and adolescence. Testosterone exerts a strong influence on the axon during male adolescence, influencing both its structure (radial growth) and function (axonal transport).

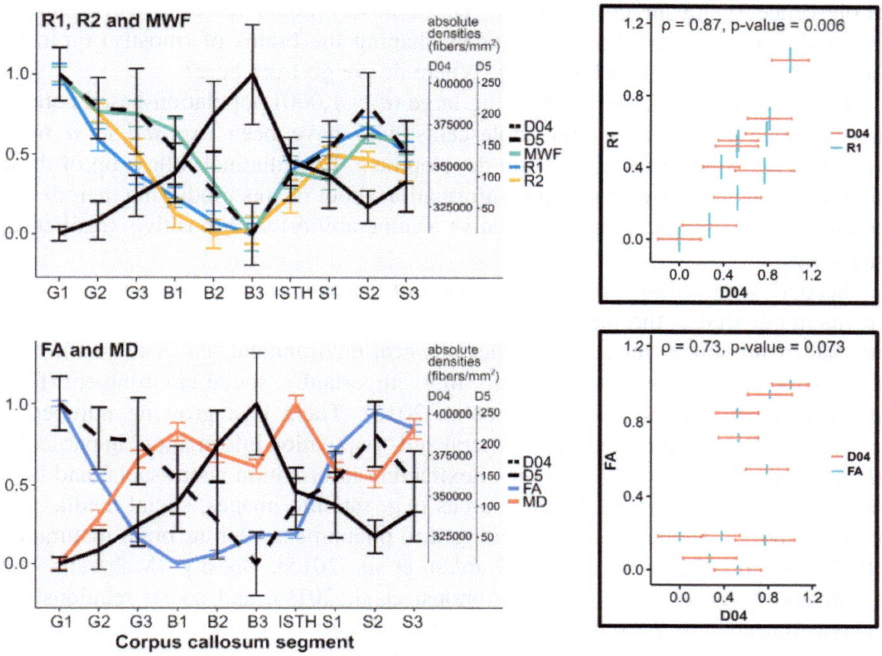

Fig. 13 Multi-modal imaging of the human corpus callosum: a comparison with histology. Left: Anterior-posterior profiles of callosal mid-sagittal histology (black; histological data are from (Aboitiz et al. 1992)) and MRI measures (colour) with 95% confidence intervals. All values were scaled to 0–1. Original absolute densities for "small" (D04 for d > 0.4 μm) and "large" (D5 for d > 5 μm) axons are shown on the right side of the plots. Abbreviations: R1 and R2 (relaxation rate 1 and 2), MWF (myelin water fraction), FA (fractional anisotropy), and MD (mean diffusivity). Right: Spearman correlations between R1 (top) and FA (bottom) and the density of "small" axons. G1–G3, segments of the genu; B1–B3, segments of the body; ISTH, isthmus; S1–S3, segments of the splenium. From Paus (2018) modified from Bjornholm et al. (2017)

4 Future Directions

In 1989, the Child Psychiatry Brain of the National Institute of Mental Health (USA) initiated the first systematic MRI-based study of brain development during childhood and adolescence (Giedd et al. 2015). Over the last 30 years, other groups have established a number of population-based cohorts, with sample sizes exceeding 1,000 individuals, including Saguenay Youth Study (Pausova et al. 2007). IMAGEN (Schumann et al. 2010), Generation R (White et al. 2013), the Philadelphia Neurodevelopmental Cohort (Satterthwaite et al. 2014), and the Pediatric Imaging, Neurocognition, and Genetics (PING) Study (Jernigan et al. 2016). These efforts continue, with the latest addition being the Adolescent Brain Cognitive Development (ABCD) Study that has recruited over 11,000 children (9 to 10 years old) to participate in longitudinal assessments of their brains and behaviour throughout adolescence (Garavan et al. 2018). These observational datasets will continue generating new knowledge about forces shaping the brains of (mostly) typically developing children and adolescents. Where do we go from here?

First, I would argue for establishing large ($n > 1{,}000$) population-based cohorts oversampled for children and adolescents who have been *exposed to a well-recognized risk* vis-à-vis early brain development. Longitudinal follow-up of these individuals would provide valuable information about factors modifying their developmental trajectories, both in negative (vulnerability) and positive (resilience) manner.

Second, we need to enhance assessment of the *environment* surrounding participants in our studies throughout their lives. The developing human being is under a myriad of influences related to his/her physical environment (e.g. air, trees), built environment (e.g. transportation), and, most importantly, social environment (family, peers, neighbours) (Ruiz Jdel et al. 2016). There is a growing number of examples whereby data science in general and "population informatics" in particular (Kum et al. 2014), show the power of extracting information about social and built environment from various digital sources (e.g. satellite images, social media, patterns of mobile-phone use), and relating it to phenomena such as brain maturation (Parker et al. 2017), wellbeing (Kardan et al. 2015), obesity (Maharana and Okanyene Nsoesie 2018), health (Abnousi et al. 2019), and social relationships (David-Barrett et al. 2016; Gruzd and Haythornthwaite 2013). Twitter-based studies of social behaviour include a cyclic nature of coordinated social activity (Morales et al. 2017) or public attention and temporal patterns of tweets on specific social issues (Peng et al. 2017). To start, such information – mapped in space and time – can be considered at an aggregate level (e.g. neighbourohood, census tract) and related to the individual member of a cohort through a geospatial information system (Paus 2016).

Third, there is a great desire to uncover *causal* influences shaping the developing brain. But observational studies do not allow for causal inferences to be made, although the use of Mendelian randomization (Davey-Smith and Hemani 2014) has become a method of choice in this context. Some would argue that the ultimate test of a hypothesis is an experiment. In studies of human development, interventions represent unique opportunities for examining causal effects of a variety of influences tested by randomizing individuals (or groups of individuals) into "experimental" and "control" arms and measuring outcomes before and after the intervention (see Houdé et al. 2000 for a seminal example using brain imaging in a reasoning task, before and after training). Randomized control trials and quasi-experimental designs have been used, for example, to evaluate effects of breasfeeding on cognitive development (Kramer et al. 2008) or effectiveness of various psycho-social interventions on mental health in children and adolescents (Sandler et al. 2014). Incorporating state-of-the-art assessment of brain and behaviour in future interventions has the potential to provide insights relevant for understanding brain development as well as mechanisms underlying the success (or failure) of a given intervention.

References

Abnousi F, Krumholz HM, Rumsfeld JS (2019) Social determinants of health in the digital age: determining the source code for nurture. JAMA 321(3):247–248

Aboitiz F, Scheibel AB, Fisher RS et al (1992) Fiber composition of the human corpus callosum. Brain Res 598(1–2):143–153

Azevedo FA, Carvalho LR, Grinberg LT et al (2009) Equal numbers of neuronal and nonneuronal cells make the human brain an isometrically scaled-up primate brain. J Comp Neurol 513:532–541

Bjornholm L, Nikkinen J, Kiviniemi V et al (2017) Structural properties of the human corpus callosum: multimodal assessment and sex differences. Neuroimage 152:108–118

Brain Development Cooperative Group (2012) Total and regional brain volumes in a population-based normative sample from 4 to 18 years: the NIH MRI study of normal brain development. Cereb Cortex 22:1–12

Bystron I, Blakemore C, Rakic P, Molnar Z (2006) The first neurons of the human cerebral cortex. Nat Neurosci 9:880–886

Davey-Smith G, Hemani G (2014) Mendelian randomization: genetic anchors for causal inference in epidemiological studies. Hum Mol Genet 23:R89–R98

David-Barrett T, Kertesz J, Rotkirch A et al (2016) Communication with family and friends across the life course. PLoS One 11:e0165687

Dean DC 3rd, O'Muircheartaigh J, Dirks H et al (2015) Characterizing longitudinal white matter development during early childhood. Brain Struct Funct 220:1921–1933

Druga R (2009) Neocortical inhibitory system. Folia Biol 55:201–217

Ducharme S, Albaugh MD, Nguyen TV et al (2016) Trajectories of cortical thickness maturation in normal brain development – the importance of quality control procedures. Neuroimage 125:267

Fischl B, Dale AM (2000) Measuring the thickness of the human cerebral cortex from magnetic resonance images. Proc Natl Acad Sci U S A 97:11050–11055

French L, Paus TA (2015) FreeSurfer view of the cortical transcriptome generated from the Allen human brain atlas. Front Neurosci 9:323

Garavan H, Bartsch H, Conway K et al (2018) Recruiting the ABCD sample: design considerations and procedures. Dev Cogn Neurosci 32:16–22

Giedd JN, Blumenthal J, Jeffries NO et al (1999) Brain development during childhood and adolescence: a longitudinal MRI study. Nat Neurosci 1999(2):861–863

Giedd JN, Raznahan A, Alexander-Bloch A, Schmitt E, Gogtay N, Rapoport; J. L. (2015) Child psychiatry branch of the National Institute of Mental Health longitudinal structural magnetic resonance imaging study of human brain development. Neuropsychopharmacology 40:43–49

Gilmore JH, Knickmeyer RC, Gao W (2018) Imaging structural and functional brain development in early childhood. Nat Rev Neurosci 19:123–137

Gogtay N, Giedd JN, Lusk L et al (2004) Dynamic mapping of human cortical development during childhood through early adulthood. Proc Natl Acad Sci U S A 101:8174–8179

Gruzd A, Haythornthwaite C (2013) Enabling community through social media. J Med Internet Res 15:e248

Hawrylycz MJ, Lein ES, Guillozet-Bongaarts AL et al (2012) An anatomically comprehensive atlas of the adult human brain transcriptome. Nature 489:391–399

Henkelman RM, Stanisz GJ, Graham SJ (2001) Magnetization transfer in MRI: a review. NMR Biomed 2001(14):57–64

Herculano-Houzel S (2009) The human brain in numbers: a linearly scaled-up primate brain. Front Hum Neurosci 3:31

Hermann DM, Buga AM, Popa-Wagner A (2015) Neurovascular remodeling in the aged ischemic brain. J Neural Transm (Vienna) Suppl 1:S25–S33

Herve PY, Leonard G, Perron M et al (2009) Handedness, motor skills and maturation of the corticospinal tract in the adolescent brain. Hum Brain Mapp 30:3151–3162

Herve PY, Cox EF, Lotfipour AK et al (2011) Structural properties of the corticospinal tract in the human brain: a magnetic resonance imaging study at 7 Tesla. Brain Struct Funct 216:255–262
Houdé O et al (2000) Shifting from the perceptual brain to the logical brain: the neural impact of cognitive inhibition training. J Cogn Neurosci 12:721–728
Jbabdi S, Sotiropoulos SN, Haber SN, Van-Essen DC, Behrens TE (2015) Measuring macroscopic brain connections in vivo. Nat Neurosci 18:1546–1555
Jernigan TL, Brown TT, Hagler DJ Jr et al (2016) The pediatric imaging, neurocognition, and genetics (PING) data repository. Neuroimage 124:1149–1154
Kardan O, Gozdyra P, Misic B et al (2015) Neighborhood greenspace and health in a large urban center. Sci Rep 5:11610
Knickmeyer RC, Gouttard S, Kang C et al (2008) A structural MRI study of human brain development from birth to 2 years. J Neurosci 28:12176–12182
Kramer MS, Aboud F, Mironova E et al (2008) Breastfeeding and child cognitive development: new evidence from a large randomized trial. Arch Gen Psychiatry 65:578–584
Kucharczyk W, Macdonald PM, Stanisz GJ, Henkelman RM (1994) Relaxivity and magnetization transfer of white matter lipids at MR imaging: importance of cerebrosides and pH. Radiology 192:521–529
Kum HC, Krishnamurthy A, Machanavajjhala A, Ahalt S et al (2014) Social genome: putting big data to work for population informatics. IEEE computer special outlook issue, pp 56–63
Le Bihan D (1995) Molecular diffusion, tissue microdynamics and microstructure. NMR Biomed 8: 375–386
Lebel C, Deoni S (2018) The development of brain white matter microstructure. Neuroimage 182: 207–218
Lerch JP, Van der Kouwe AJ, Raznahan A et al (2017) Studying neuroanatomy using MRI. Nat Neurosci 20:314–326
Maharana A, Okanyene Nsoesie E (2018) Use of deep learning to examine the association of the built environment with prevalence of neighborhood adult obesity. JAMA Netw Open 1:e181535
Marner L, Pakkenberg B (2003) Total length of nerve fibers in prefrontal and global white matter of chronic schizophrenics. J Psychiatr Res 37:539–547
Matsuzawa J, Matsui M, Konishi T et al (2001) Age-related volumetric changes of brain gray and white matter in healthy infants and children. Cereb Cortex 11:335–342
McGowan JC (1999) The physical basis of magnetization transfer imaging. Neurology 53:S3–S7
Moore MM, Chung T (2017) Review of key concepts in magnetic resonance physics. Pediatr Radiol 47:497–506
Morales AJ, Vavilala V, Benito RM, Bar-Yam Y (2017) Global patterns of synchronization in human communications. J R Soc Interface 14
Pangelinan MM, Leonard G, Perron M et al (2016) Puberty and testosterone shape the corticospinal tract during male adolescence. Brain Struct Funct 221:1083–1094
Parker N, Wong AP, Leonard G et al (2017) Income inequality, gene expression, and brain maturation during adolescence. Sci Rep 7:7397
Patel Y, Shin J, Gowland PA, Pausova Z, Paus T (2019) Maturation of the human cerebral cortex during adolescence: myelin or dendritic arbor? Cereb Cortex 29:3351–3362
Paus T (2005) Mapping brain maturation and cognitive development during adolescence. Trends Cogn Sci 9:60–68
Paus T (2013) Population neuroscience. Springer, Berlin
Paus T (2016) Population neuroscience. Handb Clin Neurol 138:17–37
Paus T (2018) Imaging microstructure in the living human brain: a viewpoint. Neuroimage 182:3–7
Paus T, Toro R (2009) Could sex differences in white matter be explained by g ratio? Front Neuroanat 3:14
Paus T, Collins DL, Evans AC, Leonard G, Pike B, Zijdenbos A (2001) Maturation of white matter in the human brain: a review of magnetic resonance studies. Brain Res Bull 54:255–266
Paus T, Pesaresi M, French L (2014) White matter as a transport system. Neuroscience 276:117–125

Pausova Z, Paus T, Abrahamowicz M et al (2007) Genes, maternal smoking, and the offspring brain and body during adolescence: design of the Saguenay youth study. Hum Brain Mapp 28:502–518

Pausova Z, Paus T, Abrahamowicz M et al (2017) Cohort profile: the Saguenay youth study (SYS). Int J Epidemiol 46:e19

Pelvig DP, Pakkenberg H, Stark AK, Pakkenberg B (2008) Neocortical glial cell numbers in human brains. Neurobiol Aging 29:1754–1762

Peng TQ, Sun G, Wu Y (2017) Interplay between public attention and public emotion toward multiple social issues on Twitter. PLoS One 12:e0167896

Perrin JS, Herve PY, Leonard G et al (2008) Growth of white matter in the adolescent brain: role of testosterone and androgen receptor. J Neurosci 28:9519–9524

Pesaresi M, Soon-Shiong R, French L et al (2015) Axon diameter and axonal transport: In vivo and in vitro effects of androgens. Neuroimage 115:191–201

Pike GB (1996) Pulsed magnetization transfer contrast in gradient echo imaging: a two-pool analytic description of signal response. Magn Reson Med 36:95–103

Pipitone J, Park MT, Winterburn J et al (2014) Multi-atlas segmentation of the whole hippocampus and subfields using multiple automatically generated templates. Neuroimage 2014(101): 494–512

Rakic P (1974) Neurons in rhesus monkey visual cortex: systematic relation between time of origin and eventual disposition. Science 183:425–427

Rakic P (1988) Specification of cerebral cortical areas. Science 241:170–176

Rakic P (1995) A small step for the cell, a giant leap for mankind: a hypothesis of neocortical expansion during evolution. Trends Neurosci 18:383–388

Rakic P (2009) Evolution of the neocortex: a perspective from developmental biology. Nat Rev Neurosci 10:724–735

Roberts TP, Mikulis D (2007) Neuro MR: principles. J Magn Reson Imaging 26:823–837

Ruiz Jdel C, Quackenboss JJ, Tulve NS (2016) Contributions of a child's built, natural, and social environments to their general cognitive ability: a systematic scoping review. PLoS One 11: e0147741

Sandler I, Wolchik SA, Cruden G et al (2014) Overview of meta-analyses of the prevention of mental health, substance use, and conduct problems. Annu Rev Clin Psychol 10:243–273

Satterthwaite TD, Elliott MA, Ruparel K et al (2014) Neuroimaging of the Philadelphia neurodevelopmental cohort. Neuroimage 86:544–553

Schmierer K, Scaravilli F, Altmann DR, Barker GJ, Miller DH (2004) Magnetization transfer ratio and myelin in postmortem multiple sclerosis brain. Ann Neurol 56:407–415

Schmierer K, Wheeler-Kingshott CA, Tozer DJ et al (2008) Quantitative magnetic resonance of postmortem multiple sclerosis brain before and after fixation. Magn Reson Med 59:268–277

Schumann G, Loth E, Banaschewski T et al (2010) The IMAGEN study: reinforcement-related behaviour in normal brain function and psychopathology. Mol Psychiatry 2010(15):1128–1139

Schüz A, Braitenberg V (2002) The human cortical white matter: quantitative aspects of cortico-cortical long-range connectivity. Taylor & Francis, London

Selemon LD, Ceritoglu C, Ratnanather JT et al (2013) Distinct abnormalities of the primate prefrontal cortex caused by ionizing radiation in early or midgestation. J Comp Neurol 521: 1040–1053

Shin J, French L, Xu T et al (2018) Cell-specific gene-expression profiles and cortical thickness in the human brain. Cereb Cortex 28:3267–3277

Sisk CL, Foster DL (2004) The neural basis of puberty and adolescence. Nat Neurosci 7:1040–1047

Sloper JJ, Hiorns RW, Powell TP (1979) A qualitative and quantitative electron microscopic study of the neurons in the primate motor and somatic sensory cortices. Philos Trans R Soc Lond B Biol Sci 285:141–171

Toro R, Perron M, Pike B et al (2008) Brain size and folding of the human cerebral cortex. Cereb Cortex 18:2352–2357

Walhovd KB, Fjell AM, Giedd J, Dale AM, Brown TT (2017) Through thick and thin: a need to reconcile contradictory results on trajectories in human cortical development. Cereb Cortex 27: 1472–1481
Whitaker KJ, Vertes PE, Romero-Garcia R et al (2016) Adolescence is associated with genomically patterned consolidation of the hubs of the human brain connectome. Proc Natl Acad Sci U S A 113:9105–9110
White T, El Marroun H, Nijs I et al (2013) Pediatric population-based neuroimaging and the generation R study: the intersection of developmental neuroscience and epidemiology. Eur J Epidemiol 28:99–111
Zecevic N, Chen Y, Filipovic R (2005) Contributions of cortical subventricular zone to the development of the human cerebral cortex. J Comp Neurol 491:109–122
Zeisel A, Munoz-Manchado AB, Codeluppi S et al (2015) Brain structure. Cell types in the mouse cortex and hippocampus revealed by single-cell RNA-seq. Science 347:1138–1142

Aging Brain from a Lifespan Perspective

Anders Martin Fjell

Contents

Abstract Research during the last two decades has shown that the brain undergoes continuous changes throughout life, with substantial heterogeneity in age trajectories between regions. Especially, temporal and prefrontal cortices show large changes, and these correlate modestly with changes in the corresponding cognitive abilities such as episodic memory and executive function. Changes seen in normal aging overlap with changes seen in neurodegenerative conditions such as Alzheimer's disease; differences between what reflects normal aging vs. a disease-related change are often blurry. This calls for a dimensional view on cognitive decline in aging, where clear-cut distinctions between normality and pathology cannot be always drawn. Although much progress has been made in describing typical patterns of age-related changes in the brain, identifying risk and protective factors, and mapping cognitive correlates, there are still limits to our knowledge that should be addressed by future research. We need more longitudinal studies following the same participants over longer time intervals with cognitive testing and brain imaging, and an increased focus on the representativeness vs. selection bias in neuroimaging research of aging.

A. M. Fjell (✉)
Department of Psychology, Center for Lifespan Changes in Brain and Cognition, University of Oslo, Oslo, Norway
e-mail: andersmf@psykologi.uio.no

Curr Topics Behav Neurosci (2024) 68: 349–370
https://doi.org/10.1007/7854_2024_476
Published Online: 26 May 2024

Keywords Alzheimer's disease · Brain aging · Hippocampus · Magnetic resonnance imaging

1 Why Should We Care about Brain Aging?

Brain disorders are the leading cause of disability, the second leading cause of death worldwide (Group 2017), and their burden is increasing with the aging of the population (Collaborators 2019). This situation makes the task of understanding brain changes in aging critical. Dementias, such as those caused by Alzheimer's disease and Parkinson's disease, are the second most prevalent non-communicable neurological diseases and the most prevalent among those aged 65 years and above (Bazargan-Hejazi et al. 2020). By far, the greatest risk factor for dementias in the population is age, and it is difficult to conceive that we will ever understand such neurodegenerative conditions without understanding why they are so intimately connected to processes and changes unfolding in the aging brain. To date, it is also recognized that various types of dementia are related to a multitude of risk factors and that they have many different causes and display heterogeneous trajectories of disease progression (Devi and Scheltens 2018; Birkenbihl et al. 2021). This makes a dimensional view on dementias and other forms of age-related cognitive decline necessary, which means that the distinction between disease and what is often called "normal aging" becomes blurred. Hence, it is not always possible to distinguish which changes – in the aging brain – are normal and which are not, as more than 30% of cognitively normal older adults will eventually show "pathological" levels of the major Alzheimer's disease neuropathological marker amyloid-β (Fjell et al. 2014a).

This often-blurry distinction between what is regarded as normal and what is regarded as a pathological change in the aging brain also reflects another important principle of brain aging: there are large individual differences between persons at any age, also in aging. The difference in cognitive function between two 80-year-olds may be very large, even to the extent that one functions as an average 30-year-old while a same-age peer functions at a very low level. Importantly, the individual differences are great at any age, and it is not obvious to which extent variations in function in older age are due to differences in the *rate of aging*, and to which extent they rather reflect stable differences in function established early in life (Walhovd et al. 2023). This is an important point to have in mind in studies of brain causes and correlates of high and low cognitive function in aging; differences in older age do not necessarily index differences in *aging* per se.

In this chapter, we will start by discussing why age-related changes in brain structure are relevant to our understanding of cognitive function in aging. We will focus on morphometric measures, such as regional volumes and cortical thickness, but also touch upon other relevant magnetic resonance imaging (MRI) methods, such as diffusion tensor imaging and measures of tissue signal intensity and contrast. Next, we will give an overview of typical lifespan trajectories for different brain regions, focusing on the heterogeneity among the regions and between different metrics. Of particular interest is to understand the structural patterns of brain aging, and how they can be understood. During the discussion of these topics, we will also include studies of individual differences in aging, and some of the potential causes for these. Finally, we will discuss prospects of future research and suggest important questions and avenues that need to be addressed to bring our understanding of brain aging in a lifespan perspective further.

2 Are Structural Brain Measures Relevant for Understanding Cognitive Function in Aging?

At any point in life, there are large individual differences in various measures of brain structure such as regional volumes, cortical thickness, and surface area. At the same time, there are marked differences between older and younger adults in these measures. The first question to ask is: Does it matter? In samples of healthy and young participants, cross-sectional relationships between structural measures and scores on specific cognitive tests are relatively weak and often unstable (Van Petten 2004; Marek et al. 2022). In contrast, gross volumetric measures, such as total brain size or total gray matter volume, show robust although modest relationships with general cognitive abilities. This is especially true for cognitive tests with high loadings on the *g-factor* (Deary et al. 2021; Deary et al. 2022). Several studies have also found associations between measures of general cognitive function and regional (apparent) cortical thickness[1] (Walhovd et al. 2017), surface area, and volume in developmental samples (Shaw et al. 2006; Ostby et al. 2012; Tamnes et al. 2018), across the lifespan (Walhovd et al. 2016a; Cox et al. 2019; Walhovd et al. 2022) and in aging (Karama et al. 2014; Ritchie et al. 2016; Cox et al. 2021)). There is no one-to-one relationship between MRI-derived brain volumetric measures and underlying neurobiological features (see chapter on "Population Neuroscience: Principles and Advances" in this book), but the simplest explanation for these relationships may be that larger brains likely have more neurons. Still, this is not a reasonable explanation for the volumetric differences between younger and older participants, because the loss of neurons in aging is likely limited (Hof and Morrison 2004), and volumetric reductions must be due to other factors, such as reduced dendritic complexity (Fjell et al. 2014a). Most likely, multiple factors contribute to the observed age-related morphometric brain differences, of which several may be relevant to explain age decline in cognitive function.

Even though gross brain measures show robust correlations with scores on cognitive tests with high g-loadings, there are still regional differences regarding in which regions such relationships tend to be strongest. For instance, it has been

[1] MRI-derived measures of cortical development are merely our best approximation; hence, the term "apparent cortical thickness" may be preferable.

shown that regions that are high expanding in development, with protracted age-trajectories during childhood and adolescence, are also high expanding in brain evolution. As a proxy for evolutionary expansion, MRI-derived regional cortical area has been compared between humans and other primates, and the regions differing the most are often regarded as most expanded in human evolution (Hill et al. 2010; Fjell et al. 2015c). Further, an intriguing hypothesis is that cortical regions that are high expanding in development and evolution also show the strongest correlations with intellectual function in humans. In line with this, it has been shown that regions in which cortical area correlates with the capacity for visuospatial reasoning in humans are generally high expanded in both development and evolution (Fjell et al. 2015c). Such results can be interpreted within the radial unit model (Rakic 1988; Rakic 2009). This model provides a common framework for interpretation of the findings in the context of evolution and prenatal development, linking brain structure in adulthood and aging to early development (see more below). Interestingly, when studied using functional MRI, it is clear that the high-expanding cortices shares functional connections with different brain networks in a context-dependent manner (Sneve et al. 2019). More specifically, the high-expanding cortex shows high internetwork functional connectivity, and it is recruited flexibly over many different cognitive tasks. This capacity of high-expanding cortex to connect flexibly with various specialized brain networks as a function of the demands of a cognitive task suggests that these cortical regions support supra-modal cognition. In accordance with an evolutionary-developmental view, this ability of high-expanding regions to modulate flexibly functional connections as a function of cognitive demands develops gradually through childhood and into young adulthood. This is relevant because, as will be discussed later in this chapter, the so-called *evo-devo* view provides a framework for interpreting anatomical heterogeneity in aging and dementia-related brain decline.

In sum, associations between cross-sectional measures of brain structure and cognitive functions in healthy adults appear to be strongest for gross morphometric measures and cognitive tests with high g-loadings (Deary et al. 2022), which means that lifespan changes in the same brain measures may potentially be relevant targets for understanding age-related changes in cognitive functions. This will be discussed in more detail below.

Comparing structural measures between normal controls and groups with different risk factors or neurological conditions often shows large effect sizes. This is especially true for age-related neurodegenerative conditions, such as Alzheimer's disease. For instance, volumes of the medial temporal lobe differ markedly between patients with Alzheimer's disease and age-matched controls, often yielding Cohen's $D > 1$ (Fjell et al. 2014b; Zhang et al. 2023). Similarly, gross morphometric differences are found between groups of individuals with different risk factors associated with dementia (Livingston et al. 2020), such as smoking and hypertension (Nobis et al. 2019), high body mass index (Opel et al. 2021), hearing loss (Rigters et al. 2017), diabetes (Shang et al. 2023), high alcohol consumption (Topiwala et al. 2022), physical inactivity (Hamer et al. 2018), high levels of air pollution (Hedges et al. 2020), and late-life depression (Geerlings and Gerritsen 2017). Thus, measures of brain structure can be used to detect late-life neurodegenerative conditions and associated with multiple risk factors for dementia, some of which are considered modifiable (Livingston et al. 2020). This is often interpreted to suggest that brain aging and dementia risk can be substantially affected by lifestyle-related factors. Caution must be used, however, when interpreting correlational results from observational studies, because it is highly unclear what the causal structure of the relationships is, there are substantial covariation between different risk factors that are challenging to tease apart, and targeted intervention studies have so far not yielded convincing evidence to back up the inferences about causality from observational studies (see chapter on "Population Neuroscience: Understanding Concepts of Generalizability and Transportability and their Application to Improving the Public's Health" in this book). Large multi-domain intervention studies such as the Finnish FINGER study, targeting multiple lifestyle factors simultaneously, including diet, exercise, cognitive training, and vascular risk monitoring, have shown positive effects on cognitive functions in older adults (Ngandu et al. 2015). Such controlled interventions are very important as they are able to assess causality and potential for improved brain health and cognitive function by non-pharmacological means. Still, effect sizes are so small that it appears unlikely that these types of lifestyle-focused interventions will have a large effect on rates of cognitive decline and dementia at a societal level. Hence, although it may be (potentially) possible to influence brain aging and reduce dementia risk through some of these factors, the scientific evidence that this will be effective is still lacking.

With the exception of valuable interventions such as FINGER, it is important to keep in mind that most of the findings referred above do not inform us about aging per se, as they reflect cross-sectional associations. Cross-sectional relationships may be established early in life as lifelong stable associations, which means that they do not necessarily arise due to actual changes associated with aging (Walhovd et al. 2023). As aging – in contrast to age – implies change, detection of longitudinal relationships between the brain change and the cognitive change is often more relevant than cross-sectional brain–cognition associations with regard to understanding the consequences and cognitive correlates of brain aging (Raz and Lindenberger 2011).

This difference between what can be inferred from cross-sectional versus longitudinal relationships is evident from current research on *BrainAge*. The concepts of *BrainAge* and *BrainAge Gap* have recently gained popularity (Cole 2017), as they may be related to several conditions and symptoms associated with older age (Elliott et al. 2021; Wagenmakers et al. 2023). In essence, the approach is that a machine learning model is trained to predict chronological age from MRI scans, yielding normative age values. Individuals can then be compared with the normative, age-expected values according to the model, and the difference between an individual's chronological age and BrainAge can be calculated: this is the BrainAge Gap. Importantly, however, when BrainAge and BrainAge Gap are based on cross-sectional observations, deviations from normative age values do not necessarily imply differences in the pace of aging itself. Rather, stable, individual differences, which can be manifest throughout adult life, and likely even in development, may account for the major share of the BrainAge Gap. This was exactly what was found in the first large longitudinal study of BrainAge. The study showed that deviation from age-expected brain volumes – the BrainAge Gap – did not reflect individual differences in brain change when the same participants were measured over time (Vidal-Pineiro et al. 2021). Rather, BrainAge Gap was related to stable individual differences in brain structure that probably originated in early development. Evidence for this was that BrainAge Gap was related to inter-individual differences in birth weight, in addition to polygenic scores of brain age. Hence, BrainAge Gap did not reflect differences in rate of aging per se in these data, and the results suggest that differences in BrainAge between groups of participants should not be taken to suggest that there are differences in brain *aging* without further evidence. The point of this discussion of the BrainAge metric is to illustrate that differences between individuals in older age cannot automatically be explained by potential individual differences in brain *aging*. Hence, longitudinal observations are needed to address this issue (Raz and Lindenberger 2011).

Longitudinal studies following participants over various time intervals with repeated MRI scans have been able to detail actual changes in the brain during aging, such as volume loss signifying brain atrophy. These studies have shown that although increased rates of atrophy and accompanying cognitive decline are seen in neurodegenerative conditions, this pattern also characterizes normal aging (Fraser et al. 2015), albeit of a lesser magnitude (Fjell et al. 2014a). For example, atrophy in the medial temporal lobe, which is highly elevated in Alzheimer's disease, can be seen in cognitively normal older adults even with a very low risk of Alzheimer's disease, based on genetic, biomarker, cognitive, and clinical screening (Fjell et al. 2013a). Still, age-related changes in the brain of these low-risk older participants correlate with reductions in memory scores observed over time. Thus, although they are not related to Alzheimer's disease, they predicted degree of age-related cognitive decline. This suggests that changes in brain structure during aging, even when not related to neurodegenerative disease, may still be associated with an amplified age-related reduction in cognitive function. A later study from the large, multinational European Lifebrain consortium (Walhovd et al. 2018) found that relationships between hippocampal atrophy and memory decline only in carriers of the AD risk allele APOE ϵ4, suggesting that although such relationships are seen in low-risk populations, they may also be sensitive to very early pathological changes in cognitively normal older adults (Gorbach et al. 2020).

To conclude, although stable relationships between specific cognitive functions and regional measures of brain structure are not found often in young and healthy populations, these relationships are stronger for more general cognitive functions with higher g-loadings and between groups of patients vs. controls. Further, such measures appear to be associated with established risk factors for cognitive decline. Most importantly, longitudinal investigations of brain change vs. cognitive change are of great value to understand the potential role for cognitive decline played by changes in brain structure during aging. Hence, understanding these changes in aging is relevant to our understanding of normal age-related changes in cognitive abilities, as well as for the important question of why some individuals show better functioning in older age than others.

3 How Does the Brain Change with Age?

Very large cross-sectional mega-analyses of MRI brain scans show that there are substantial differences in brain structure between older and younger participants (Bethlehem et al. 2022). These age-related differences are sometimes even larger than those seen between patients with Alzheimer's disease and same-age older adults (Fjell et al. 2010a). Importantly, correlations between neuroanatomical volumes and age are found throughout most of adult life and are not confined to old age only. Rather, brain change is a continuous (temporal) process characterizing adult life even in healthy populations.

Age-related changes in brain structure may best be regarded as the sum of several underlying positive and negative processes, which may even operate at the same time (Fjell et al. 2013b). This creates age trajectories of continuous, nonlinear, and sometimes even non-monotonous change across adult life, which differ substantially across different neuroanatomical regions. For example, while the hippocampus shows relative stability in volume across young adulthood and middle age, before accelerated reductions are seen on a group lever from about 60 years (Fraser et al. 2015), volume of the cerebral cortex shows a much more steady, monotonous, and almost linear decline in the same age range (Fjell et al. 2015b). Volume of white matter deviates from both hippocampus and cortical volume, showing increases through young adulthood and into middle age, before volume loss becomes evident. The brain stem stands out from most other structures, showing no or very modest age-related differences (Fjell et al. 2013b).

The above review and conclusions are to a major extent based on cross-sectional observations, as it has been impossible to follow the same individuals over decades, not the least due to development and replacement of MRI scanners, hardware, and software/sequences. Still, as referred to above, longitudinal studies have shown structural changes over time in the same individuals, also in normal aging (Fjell et al. 2010b). We have reviewed systematically research on changes in brain structure in aging and its cognitive correlates (Fjell and Walhovd 2010). Although a large amount of research has been conducted since then, the three main conclusions drawn at that point have been confirmed by later studies. First, as briefly described above, normal aging is characterized by volumetric loss of brain tissue and expansion of the ventricular system. The pattern of change is, however, highly heterogeneous across structures, regions, and metrics, from linear to accelerating, decelerating, and non-monotonous age trajectories. If a study includes a sufficiently wide age range and a reasonable sample size, e.g., $n > \sim 100$, these basic age trajectories are usually found. Second, volumetric changes are not driven by neuronal loss. Shrinkage of neurons, reductions of synaptic spines, and lower numbers of synapses are more likely candidates to account for the volumetric reductions of gray matter (Terry et al. 1987; Peters et al. 1998; Pakkenberg et al. 2003; Freeman et al. 2008). Finally, reductions in cognitive abilities with age are often mediated by changes in brain structure. The magnitude of the relationships between brain change and cognitive change will depend on several factors, including how healthy the sample is. Usually, the relationships are of modest strength in cohorts of cognitively normal older adults. In the following, we will discuss in more detail the first and the last of these principles, and not dive much further into the neurobiological underpinnings of the macrostructural brain changes in aging.

First, although the pattern of brain change in aging is more or less global, affecting all cortical regions, the strength of the age relationships and rates of age-related change is highly heterogeneous across regions. Frontal and temporal cortices show the most dramatic changes as mentioned above, and such a fronto-temporal pattern of atrophy is replicated across multiple longitudinal studies (Raz et al. 2005; Fjell et al. 2014b). Also medial parietal regions, including precuneus, retrosplenial, and posterior cingulate cortices, are highly age-sensitive. Thus, although age-related changes in brain structure are pervasive, the pattern of change is by no means random.

Several models or theories are developed to account for the typical pattern of age-related change across the brain, i.e., why different brain regions change at different pace at different ages. One such account is the *retrogenesis* hypothesis, or the "*last in-first out*" principle, according to which late-maturing brain regions show the largest aging-related changes. As prefrontal regions show protracted development and are vulnerable to effects of aging, this theory has common prediction with the classic *frontal lobe theory of aging*. The age-sensitive late-maturing regions tend to have more complex cortical architecture than earlier-maturing, less age-sensitive regions, they take longer to develop, follow more complex developmental trajectories (Giedd et al. 2008; Shaw et al. 2008), and are more vulnerable to the negative effects of normal aging. Although the prefrontal cortex shows late maturation and accelerated age changes, the medial temporal cortex appears to break the pattern predicted from the retrogenesis principle. This region undergoes substantial changes in normal aging and AD alike but shows relatively modest maturation after mid childhood (Shaw et al. 2008; van Soelen et al. 2012; Tamnes et al. 2013; Krogsrud et al. 2014). The "last in-first out" model, although it has some merit, fits only partially the empirical results.

Similarly, as pointed out above, regions of high cortical expansion in humans compared with other primates (Chaplin et al. 2013) seem to show especially protracted developmental trajectories, both structurally (Hill et al. 2010; Amlien et al. 2016) and functionally (Sneve, Grydeland et al. 2019). Regions of high cortical expansion in humans also tend to support both specific cognitive functions (Friederici 2023) and are related to individual differences in general cognitive abilities (Fjell et al. 2015c). This supports a theory that human-specific cognitive adaptations are correlated with enlargement of the neocortex (Sherwood et al. 2008; Friederici 2023) and the "wiring" of the human brain (Ardesch et al. 2019; Pang et al. 2022). Further, as described above, these parts of the cerebral cortex support supra-modal cognition and show especially high flexibility in their ability to change pattern of functional connectivity as a function of cognitive demands. That is, how strongly these cortices communicate with other regions, and which regions they communicate with, can vary flexibly with the cognitive task. Interestingly, these cortical regions, which are high expanded in humans compared with other primates, undergo especially rapid decline during normal aging, very clearly evident in lateral cortical regions, such as the lateral temporal cortex, inferior regions of the parietal cortex, and most of the lateral prefrontal cortex (Fjell et al. 2015a). On average, regions high expanded in humans compared with other primates in one study showed annual 0.53% volume reductions in aging, while low-expanding regions showed only 0.23% reduction (Fjell et al. 2014a). Expansion of the medial temporal lobe is low, however, while again showing a high degree of vulnerability in both normal aging and Alzheimer's disease (Fjell et al. 2015c). This represents a deviation from the pattern of high-expanding cortices being vulnerable in aging and disease. It has been suggested that the higher degree of plasticity in the human medial temporal lobe is more reminiscent of other non-human mammals, compared with the less plastic cortex in the rest of the human brain (Walhovd et al. 2016b). Hence, although the theory can account for major aspects of the typical anatomical pattern of normal aging, there are discrepancies between the predictions from this theory and accumulated evidence detailing robust patterns of brain change in aging.

If we instead apply a functional network approach to characterize brain aging, one typically finds that regions in the so-called default mode network (DMN) (Lustig et al. 2003; Andrews-Hanna et al. 2007) show especially large structural changes in aging. DMN typically increases activity and connectivity during rest or "task-free" conditions in an fMRI experiment, interpreted as a sign of attention being directed away from external stimuli. The DMN comprises multiple regions, including the medial components (i.e., parahippocampal cortex, retrosplenial and posterior cingulate cortices, and dorsal, anterior, and ventral medial-prefrontal cortices) and the lateral components (i.e., temporal pole, lateral temporal cortex, temporo-parietal junction, and the inferior parietal lobule) (Andrews-Hanna et al. 2010). DMN is involved in various aspects of cognitive processes such as episodic memory processing (Buckner and Carroll 2007). Episodic memory performance is known to decline substantially for most people even in normal aging (Nyberg et al. 2012). It is therefore interesting that the mean level of atrophy in aging is substantially higher within DMN regions, both medial and lateral DMN, than the average of the rest of the cerebral cortex (Fjell et al. 2014a). Such longitudinal changes in DMN structure and function correlate, although modestly, with reductions in cognitive performance, including episodic memory (Rodrigue and Raz 2004; Murphy et al. 2010; Nyberg et al. 2012; Persson et al. 2012; Gorbach et al. 2020).

Age-related changes in DMN represent a good fit with age-expected structural and functional brain changes. One speculation is that this is due to high metabolic demands of these regions, which may make them vulnerable to a spectrum of different processes which eventually leads to increased atrophy and vulnerability to age-related degenerative disease (Buckner et al. 2005). Oxidative stress is one such mechanism, but there are almost certainly multiple aspects of brain aging and many neural processes which in sum are responsible for the pattern of brain changes typically seen in normal aging. For instance, although there is a certain regional overlap between DMN and accumulation of amyloid plaques, a study testing directly the anatomical overlap between cortical hypometabolism and amyloid plaque distribution found that metabolic reductions in Alzheimer's disease were linked to global amyloid burden and that regional fibrillar amyloid deposition showed little to no association with regional hypometabolism (Altmann et al. 2015). Such a view of brain networks as vulnerable to many types of pathology was supported by a study identifying a common brain network that linked development, aging, and vulnerability to disease (Douaud et al. 2014). This network was derived from healthy individuals across the lifespan but showed spatial overlap with both schizophrenia and Alzheimer's disease.

Hence, research converges on a view that the pattern of lifespan changes in brain structure is not random but rather shows different degrees of overlap with numerous other factors and conditions. These patterns of change are partly influenced by genes. For example, cortical regions with overlapping genetic architecture (Chen et al. 2011) show correlated developmental and adult-age trajectories of change and vice versa for regions with low genetic overlap (Fjell et al. 2015b). Thus, associations between genes and regional variations in cortical thickness in middle age can be traced to regional differences in the rate of neurodevelopmental change, and extrapolated to adult aging-related cortical thinning. Such results suggest that genetic factors contribute to the spatial patterns of changes in the cerebral cortex throughout the lifespan (Brouwer et al. 2017), including in aging. Further work along these lines suggests that development of cortical regionalization is a continuous process from the embryonic stage throughout life and that cortical regions that developed together in childhood change together even in aging (Fjell et al. 2019). The same fundamental principle of lifespan continuity and change has been shown to apply also to longitudinal changes in subcortical structures, as revealed by twin-based genetic correlations and single-nucleotide polymorphisms-based analyses of genetic covariance (Fjell et al. 2021).

4 Changes in White Matter During Aging

As described above, white matter – mainly comprised of myelinated axons – deviates from most gray-matter structures in that its volume increases through young adulthood and into middle age. Similar trajectories are found when studying the microstructure of the white matter by use of diffusion tensor imaging (DTI). DTI measures restrictions to free water diffusion in the tissue, and several relevant metrics can be derived from a DTI scan.

One of the most popular measures in the context of aging is fractional anisotropy (FA), which indexes the directionality of diffusion. As – for instance – myelin hinders free diffusion, FA is generally high in tissue with parallel and well-myelinated fibers. In areas with less myelin or crossing fibers, FA tends to be lower. In interpreting DTI results, an important caveat must be stressed: water diffusion in brain tissue is affected by several different factors and processes, which preclude any simple interpretation about the neurobiological correlates of DTI-derived metrics (Figley et al. 2021). Although early interpretations suggested that high diffusion along the principal axis, i.e., axial diffusion, reflected axonal integrity, and increased diffusion perpendicular to the principal axis, i.e., radial diffusion (RD), reflected reduced myelination, further investigations soon came to the conclusion that there is no one-to-one relationship between changes in different diffusion metrics and brain tissue (Zatorre et al. 2012). For instance, DTI can be used to delineate main white matter tracts even in the infant brain (Dubois et al. 2014), which contains next to no myelin. This clearly shows that myelin is not the only, or main, driver of directional diffusion.

Still, although the neurobiological underpinnings of the white matter changes in aging that can be detected by different DTI-derived metrics are not always clear, DTI has been and continues to be a useful tool in the study of structural properties of white matter and its role in cognitive function. FA increases during development and young adulthood (Krogsrud et al. 2016; Lebel et al. 2019; Reynolds et al. 2019), before reductions are seen in higher age (Westlye et al. 2010). Another interesting measure is mean diffusion (MD). MD follows the opposite trajectory of FA, with reductions during development and young adulthood, and sharper increases from middle age (Zhu et al. 2024). Although most aging and lifespan studies have been cross-sectional, longitudinal studies have confirmed the main conclusions (Sexton et al. 2014; Schilling et al. 2022).

FA and MD have become very useful measures to study lifespan changes in the brain and have yielded much new knowledge about the role of age-related changes in white matter in cognitive development and age-related cognitive decline (Carmichael and Lockhart 2012; Bennett and Madden 2014; Goddings et al. 2021). Especially, changes in FA, MD, and RD have been shown to relate to slower speed of processing and dysexecutive problems often seen in older age (Bucur et al. 2008; Madden et al. 2012; Fjell et al. 2017). This suggests that the structural properties of white matter are important for optimal cognitive function and that preservation of white matter is associated with less age-related cognitive decline.

Other measures of white matter characteristics have also been used to study aging. One approach has been to use the ratio of T1w/T2w image intensities to eliminate the MR-related image intensity bias and thereby enhance the contrast to noise ratio for myelin. Although it is clear that the T1w/T2w contrast reflects multiple aspects of brain microstructure besides myelin, it has proven to be an interesting metric for studying white matter in a lifespan and aging context (Sui et al. 2022). The reasoning is that the T1w/T2w may provide a non-invasive, in vivo approximation of myelin content in the tissue (Glasser and Van Essen 2011). Later studies have supported the view that the T1w/T2w ratio yields better information than T1w scans alone, provides indirect information that partly reflects myelin content, and thus may be a valuable tool for understanding the neurobiology of myelin-related conditions (Pokosova et al. 2023). For example, one paper studied age differences in T1w/T2w contrast over an age range from 8 to 85 years. Waves of maturation and aging-related effects of this putative myelin-sensitive metric were identified, consistent with regionally heterogeneous waves of intracortical myelinogenesis and age-related demyelination (Grydeland et al. 2019).

Furthermore, it has been shown that the contrast between the signal in cortical gray matter and the adjacent white matter, so-called gray-white contrast, is reduced over time and that this reduction happens faster in regions where the T1/T2 ratio is lower to begin with (Vidal-Pineiro et al. 2016). The rate of blurring of the contrast was especially high in temporal and medial frontal regions and was primarily driven by intensity reductions of the white matter, which could reflect that lightly myelinated areas are more sensitive to aging. This further connects the above discussed models of brain aging based on cortical expansion – the *evo-devo model* – as myelin content and hence brain plasticity is assumed to be lower in high-expanding cortices (Glasser and Van Essen 2011). Accordingly, a study found that low-expanding regions had higher (apparent) intracortical myelin values than the high-expanding regions based on T1w/T2w contrast mapping, connecting myelin to aging and cortical expansion in development and putative evolution (Hill et al. 2010; Amlien et al. 2016). Hence, although affected by several other neurobiological features besides myelin, these different contrast-based MRI metrics have been shown to be useful tools in mapping microstructural changes and age-related differences in the brain's white matter.

Interestingly, although mostly applied to white matter, DTI can also be used to study tissue characteristics of gray matter. Since the fiber architecture in gray matter is not as unidirectional as in white matter, and each MRI voxel will contain multiple small fibers crossing in many directions, MD is often seen as a better measure than FA when applied to gray matter structures. For example, the hippocampus shows clearly distinguishable lifespan trajectories for microstructure (mean diffusion) versus macrostructure (volume) (Langnes et al. 2020). Both classes of measures showed, however, substantial age effect after mid-life, with increases in mean diffusion and parallel reductions in volume, while the age changes in development and earlier adulthood were more different. Recently, it was shown that diffusion measured from the cortical gray matter was related to amyloid pathology in a group of Alzheimer's disease patients (Spotorno et al. 2023). In both studies, the gray matter microstructure was related to memory performance. Hence, microstructural changes in gray matter seem to represent important aspects of brain aging, both in cognitively normal older adults and in patients with a neurodegenerative condition.

5 Implications for Cognitive Functions

As described above, age-related changes in brain structure that characterize normal aging are not totally benign, as the degree of age-related change seems to correlate with the degree of cognitive decline and explain at least some proportion of the individual differences in cognitive changes in aging. Cross-sectional studies have reported that more than 25% of the differences between young and old participants in selected cognitive functions can be explained by group differences in structural brain characteristics (Fjell and Walhovd 2010). Such estimates are likely inflated, however, and when testing correlations between brain change and cognitive change, effects sizes are modest, with explained variance usually not exceeding 10% in samples of cognitively healthy adults (Fjell et al. 2014a). There are several studies detecting correlations between changes in cognitive function and changes in relevant brain structures supporting that specific function, e.g., that changes in medial temporal lobe and other episodic memory-related regions are related to decline in memory function. An example of this is reported associations between the rate of hippocampal atrophy and episodic memory decline, with reported effect sizes of 10–15% of the variation in memory change explained by differences in the hippocampal atrophy rate (Gorbach et al. 2020; Lovden et al. 2023). The relatively large rates of atrophy typically seen in the prefrontal cortex fit well with classic theories of cognitive aging focusing on the vulnerability of the frontal lobes (Rabbitt et al. 2001) and decline in performance on executive cognitive tests (Head et al. 2009). Overall, as described above, temporal and frontal cortical regions show comparable rates of atrophy and are among the brain areas showing the largest changes in normal aging (Fjell et al. 2014a).

Nevertheless, there is no simple one-to-one relationship between regional brain change and change in the cognitive functions typically associated with each specific region. Several more recent studies have detected patterns of correlated change across brain regions and related these to specific cognitive changes. Anatomical clusters of change identified in development showed distinguishable lifespan trajectories and specific relationships to cognitive functions. Especially, cognitive tests with high g-loadings, i.e., working memory and general cognitive abilities, showed relationships to surface area clusters, while more specific tests, i.e., episodic memory and executive functions, were associated with cortical thickness clusters (Fjell et al. 2019). This was expected based on the literature review in the beginning of the chapter. One longitudinal study of older adults found high mean correlation between volumetric change across cortical regions (mean $r = 0.81$), which was much higher than the cross-sectional correlations ($r = 0.35$) (Cox et al. 2021). A main, cortex-wide dimension accounted for 66% of the variance in change, while a fronto-temporal and an occipito-parietal dimension, orthogonal to the general dimension, was also identified. The general dimension further correlated with reductions in general cognitive ability, visuospatial ability, processing speed, and memory, with correlations around $r = 0.4$ (Cox et al. 2021). Another study identified distinct patterns of atrophy in striatal versus hippocampal-cortical systems, with the latter being related to memory decline (Nyberg et al. 2023).

These studies find that regional aging-related brain changes cluster in meaningful systems, which are associated with changes in cognitive functions relying on these systems. Thus, although gross brain measures are related to gross cognitive measures in cross-sectional studies, longitudinal studies of change–change relationships tend to find anatomically meaningful associations with age-related cognitive changes. In groups of patients with degenerative conditions, such as Alzheimer's disease, these relationships are expected to be substantially stronger. This is in accordance with a classical "neuropsychological hypothesis" that normal brain structure supports normal cognition and that correlations between the two classes of measures will tend to increase as a function of brain decline (Van Petten 2004). Hence, robust relationships between specific cognitive functions and specific brain measures may not be found in young and in cross-sectional datasets, but structural brain changes may still be sensitive markers or causes of cognitive decline in normal aging as well as in age-related neurodegenerative conditions.

6 Concluding Remarks: The Way Forward

While two decades ago, the common opinion was that the brain was relatively stable and preserved throughout young adulthood and middle age and that changes mostly characterized old age, research has now established that the brain changes continuously across life. Sampling of participants covering wide age ranges made it possible to use statistical methods to show that age trajectories are often not linear, and for some parts of the brain, such as the white matter, even non-monotonous. Mapping the heterogeneity in age trajectories between structures and regions of the brain was another important achievement of the past two decades, and we now have a good overview about the differences in vulnerability across the brain. One important avenue forward will be to integrate this knowledge about common aging patterns with an individual-differences perspective. That is, can longitudinal mapping of brain scans identify subgroups of participants with different patterns of brain change, and will such subpopulations show different cognitive trajectories? There are recent studies suggesting that this can be a fruitful approach, but how different people are in their age trajectories is yet to be established. To address this and other important questions about lifespan changes in brain and cognition, longitudinal studies spanning several years are necessary. Not the least, it is important to acknowledge that differences between older adults in brain health and cognitive function do not necessarily imply that they change differently over time. This issue cannot be overcome by larger samples alone, but requires an increased focus on collection of longitudinal data. Finally, an inherent limitation in most, possibly all, large-scale MRI studies of aging is the issue of select sampling into the study and then non-random attrition over time. These issues are very hard to tackle because people who volunteer to participate in comprehensive studies, which includes MRIs of the brain, are on average different from those who do not sign up. This selection bias not only reduced generalizability of the results of the research but also affects the relationships between the variables under study. For instance, a biomarker of brain health will tend to be more strongly related to cognitive decline in a sample of patients than a well-screened sample of cognitive controls. Hence, conclusions will tend to vary as a function of the composition of the sample. This is an issue that should be addressed better in future research, by population-based iterative and selective sampling to reduce bias, as well as the use of statistical methods to assess and attempt to compensate for these biases. More population-representative longitudinal samples spanning longer time periods require large efforts and investments, but have the potential to really move the field of lifespan and aging research forward.

References

Altmann A, Ng B, Landau SM, Jagust WJ, Greicius MD, I. Alzheimer's Disease Neuroimaging (2015) Regional brain hypometabolism is unrelated to regional amyloid plaque burden. Brain 138(Pt 12):3734–3746

Amlien IK, Fjell AM, Tamnes CK, Grydeland H, Krogsrud SK, Chaplin TA, Rosa MGP, Walhovd KB (2016) Organizing principles of human cortical development—thickness and area from 4 to 30 years: insights from comparative primate neuroanatomy. Cereb Cortex 26(1):257–267

Andrews-Hanna JR, Snyder AZ, Vincent JL, Lustig C, Head D, Raichle ME, Buckner RL (2007) Disruption of large-scale brain systems in advanced aging. Neuron 56(5):924–935

Andrews-Hanna JR, Reidler JS, Sepulcre J, Poulin R, Buckner RL (2010) Functional-anatomic fractionation of the brain's default network. Neuron 65(4):550–562

Ardesch DJ, Scholtens LH, Li L, Preuss TM, Rilling JK, van den Heuvel MP (2019) Evolutionary expansion of connectivity between multimodal association areas in the human brain compared with chimpanzees. Proc Natl Acad Sci USA 116(14):7101–7106

Bazargan-Hejazi S, Dehghan K, Edwards C, Mohammadi N, Attar S, Sahraian MA, Eskandarieh S (2020) The health burden of non-communicable neurological disorders in the USA between 1990 and 2017. Brain Commun 2(2):fcaa097

Bennett IJ, Madden DJ (2014) Disconnected aging: cerebral white matter integrity and age-related differences in cognition. Neuroscience 276:187–205

Bethlehem RAI, Seidlitz J, White SR, Vogel JW, Anderson KM, Adamson C, Adler S, Alexopoulos GS, Anagnostou E, Areces-Gonzalez A, Astle DE, Auyeung B, Ayub M, Bae J, Ball G, Baron-Cohen S, Beare R, Bedford SA, Benegal V, Beyer F, Blangero J, Cabez MB, Boardman JP, Borzage M, Bosch-Bayard JF, Bourke N, Calhoun VD, Chakravarty MM, Chen C, Chertavian C, Chetelat G, Chong YS, Cole JH, Corvin A, Costantino M, Courchesne E, Crivello F, Cropley VL, Crosbie J, Crossley N, Delarue M, Delorme R, Desrivieres S, Devenyi GA, Di Biase MA, Dolan R, Donald KA, Donohoe G, Dunlop K, Edwards AD, Elison JT, Ellis CT, Elman JA, Eyler L, Fair DA, Feczko E, Fletcher PC, Fonagy P, Franz CE, Galan-Garcia L, Gholipour A, Giedd J, Gilmore JH, Glahn DC, Goodyer IM, Grant PE, Groenewold NA, Gunning FM, Gur RE, Gur RC, Hammill CF, Hansson O, Hedden T, Heinz A, Henson RN, Heuer K, Hoare J, Holla B, Holmes AJ, Holt R, Huang H, Im K, Ipser J, Jack CR Jr, Jackowski AP, Jia T, Johnson KA, Jones PB, Jones DT, Kahn RS, Karlsson H, Karlsson L, Kawashima R, Kelley EA, Kern S, Kim KW, Kitzbichler MG, Kremen WS, Lalonde F, Landeau B, Lee S, Lerch J, Lewis JD, Li J, Liao W, Liston C, Lombardo MV, Lv J, Lynch C, Mallard TT, Marcelis M, Markello RD, Mathias SR, Mazoyer B, McGuire P, Meaney MJ, Mechelli A, Medic N, Misic B, Morgan SE, Mothersill D, Nigg J, Ong MQW, Ortinau C, Ossenkoppele R, Ouyang M, Palaniyappan L, Paly L, Pan PM, Pantelis C, Park MM, Paus T, Pausova Z, Paz-Linares D, Binette AP, Pierce K, Qian X, Qiu J, Qiu A, Raznahan A, Rittman T, Rodrigue A, Rollins CK, Romero-Garcia R, Ronan L, Rosenberg MD, Rowitch DH, Salum GA, Satterthwaite TD, Schaare HL, Schachar RJ, Schultz AP, Schumann G, Scholl M, Sharp D, Shinohara RT, Skoog I, Smyser CD, Sperling RA, Stein DJ, Stolicyn A, Suckling J, Sullivan G, Taki Y, Thyreau B, Toro R, Traut N, Tsvetanov KA, Turk-Browne NB, Tuulari JJ, Tzourio C, Vachon-Presseau E, Valdes-Sosa MJ, Valdes-Sosa PA, Valk SL, van Amelsvoort T, Vandekar SN, Vasung L, Victoria LW, Villeneuve S, Villringer A, Vertes PE, Wagstyl K, Wang YS, Warfield SK, Warrier V, Westman E, Westwater ML, Whalley HC, Witte AV, Yang N, Yeo B, Yun H, Zalesky A, Zar HJ, Zettergren A, Zhou JH, Ziauddeen H, Zugman A, Zuo XN,

Aibl BR, I. Alzheimer's Disease Neuroimaging, I. Alzheimer's Disease Repository Without Borders, C. Team, C. A. N. Cam, Ccnp, Cobre, cVeda, E. D. B. A. W. Group, P. Developing Human Connectome, FinnBrain, S. Harvard Aging Brain, Imagen, Kne, A. Mayo Clinic Study of, Nspn, Pond, P.-A. R. Group, Vetsa, Bullmore ET, Alexander-Bloch AF (2022) Brain charts for the human lifespan. Nature 604(7906):525–533

Birkenbihl C, Salimi Y, Frohlich H, I. Japanese Alzheimer's Disease Neuroimaging, I. Alzheimer's Disease Neuroimaging (2021) Unraveling the heterogeneity in Alzheimer's disease progression across multiple cohorts and the implications for data-driven disease modeling. Alzheimers Dement 18(2):251–261

Brouwer RM, Panizzon MS, Glahn DC, Hibar DP, Hua X, Jahanshad N, Abramovic L, de Zubicaray GI, Franz CE, Hansell NK, Hickie IB, Koenis MMG, Martin NG, Mather KA, McMahon KL, Schnack HG, Strike LT, Swagerman SC, Thalamuthu A, Wen W, Gilmore JH, Gogtay N, Kahn RS, Sachdev PS, Wright MJ, Boomsma DI, Kremen WS, Thompson PM, Hulshoff Pol HE (2017) Genetic influences on individual differences in longitudinal changes in global and subcortical brain volumes: results of the ENIGMA plasticity working group. Hum Brain Mapp 38(9):4444–4458

Buckner RL, Carroll DC (2007) Self-projection and the brain. Trends Cogn Sci 11(2):49–57

Buckner RL, Snyder AZ, Shannon BJ, LaRossa G, Sachs R, Fotenos AF, Sheline YI, Klunk WE, Mathis CA, Morris JC, Mintun MA (2005) Molecular, structural, and functional characterization of Alzheimer's disease: evidence for a relationship between default activity, amyloid, and memory. J Neurosci 25(34):7709–7717

Bucur B, Madden DJ, Spaniol J, Provenzale JM, Cabeza R, White LE, Huettel SA (2008) Age-related slowing of memory retrieval: contributions of perceptual speed and cerebral white matter integrity. Neurobiol Aging 29(7):1070–1079

Carmichael O, Lockhart S (2012) The role of diffusion tensor imaging in the study of cognitive aging. Curr Top Behav Neurosci 11:289–320

Chaplin TA, Yu HH, Soares JG, Gattass R, Rosa MG (2013) A conserved pattern of differential expansion of cortical areas in simian primates. J Neurosci 33(38):15120–15125

Chen CH, Panizzon MS, Eyler LT, Jernigan TL, Thompson W, Fennema-Notestine C, Jak AJ, Neale MC, Franz CE, Hamza S, Lyons MJ, Grant MD, Fischl B, Seidman LJ, Tsuang MT, Kremen WS, Dale AM (2011) Genetic influences on cortical regionalization in the human brain. Neuron 72(4):537–544

Cole JH (2017) Neuroimaging-derived brain-age: an ageing biomarker? Aging (Albany NY) 9(8): 1861–1862

Collaborators GBDN (2019) Global, regional, and national burden of neurological disorders, 1990-2016: a systematic analysis for the global burden of disease study 2016. Lancet Neurol 18(5): 459–480

Cox SR, Ritchie SJ, Fawns-Ritchie C, Tucker-Drob EM, Deary IJ (2019) Structural brain imaging correlates of general intelligence in UK biobank. Intelligence 76:101376

Cox SR, Harris MA, Ritchie SJ, Buchanan CR, Valdes Hernandez MC, Corley J, Taylor AM, Madole JW, Harris SE, Whalley HC, McIntosh AM, Russ TC, Bastin ME, Wardlaw JM, Deary IJ, Tucker-Drob EM (2021) Three major dimensions of human brain cortical ageing in relation to cognitive decline across the eighth decade of life. Mol Psychiatry 26(6):2651–2662

Deary IJ, Hill WD, Gale CR (2021) Intelligence, health and death. Nat Hum Behav 5(4):416–430

Deary IJ, Cox SR, Hill WD (2022) Genetic variation, brain, and intelligence differences. Mol Psychiatry 27(1):335–353

Devi G, Scheltens P (2018) Heterogeneity of Alzheimer's disease: consequence for drug trials? Alzheimers Res Ther 10(1):122

Douaud G, Groves AR, Tamnes CK, Westlye LT, Duff EP, Engvig A, Walhovd KB, James A, Gass A, Monsch AU, Matthews PM, Fjell AM, Smith SM, Johansen-Berg H (2014) A common brain network links development, aging, and vulnerability to disease. Proc Natl Acad Sci USA 111(49):17648–17653

Dubois J, Dehaene-Lambertz G, Kulikova S, Poupon C, Huppi PS, Hertz-Pannier L (2014) The early development of brain white matter: a review of imaging studies in fetuses, newborns and infants. Neuroscience 276:48–71

Elliott ML, Belsky DW, Knodt AR, Ireland D, Melzer TR, Poulton R, Ramrakha S, Caspi A, Moffitt TE, Hariri AR (2021) Brain-age in midlife is associated with accelerated biological aging and cognitive decline in a longitudinal birth cohort. Mol Psychiatry 26(8):3829–3838

Figley CR, Uddin MN, Wong K, Kornelsen J, Puig J, Figley TD (2021) Potential pitfalls of using fractional anisotropy, axial diffusivity, and radial diffusivity as biomarkers of cerebral white matter microstructure. Front Neurosci 15:799576

Fjell AM, Walhovd KB (2010) Structural brain changes in aging: courses, causes and cognitive consequences. Rev Neurosci 21(3):187–221

Fjell AM, Amlien IK, Westlye LT, Stenset V, Fladby T, Skinningsrud A, Eilsertsen DE, Bjornerud A, Walhovd KB (2010a) CSF biomarker pathology correlates with a medial temporo-parietal network affected by very mild to moderate Alzheimer's disease but not a fronto-striatal network affected by healthy aging. NeuroImage 49(2):1820–1830

Fjell AM, Westlye LT, Espeseth T, Reinvang I, Dale AM, Holland D, Walhovd KB (2010b) Cortical gray matter atrophy in healthy aging cannot be explained by undetected incipient cognitive disorders: a comment on Burgmans et al. (2009). Neuropsychology 24(2):258–263. discussion 264–266

Fjell AM, McEvoy L, Holland D, Dale AM, Walhovd KB, I. Alzheimer's Disease Neuroimaging (2013a) Brain changes in older adults at very low risk for Alzheimer's disease. J Neurosci 33(19):8237–8242

Fjell AM, Westlye LT, Grydeland H, Amlien I, Espeseth T, Reinvang I, Raz N, Holland D, Dale AM, Walhovd KB, I. Alzheimer Disease Neuroimaging (2013b) Critical ages in the life course of the adult brain: nonlinear subcortical aging. Neurobiol Aging 34(10):2239–2247

Fjell AM, McEvoy L, Holland D, Dale AM, Walhovd KB, I. Alzheimer's Disease Neuroimaging (2014a) What is normal in normal aging? Effects of aging, amyloid and Alzheimer's disease on the cerebral cortex and the hippocampus. Prog Neurobiol 117:20–40

Fjell AM, Westlye LT, Grydeland H, Amlien I, Espeseth T, Reinvang I, Raz N, Dale AM, Walhovd KB, I. Alzheimer Disease Neuroimaging (2014b) Accelerating cortical thinning: unique to dementia or universal in aging? Cereb Cortex 24(4):919–934

Fjell AM, Amlien IK, Sneve MH, Grydeland H, Tamnes CK, Chaplin TA, Rosa MG, Walhovd KB (2015a) The roots of Alzheimer's disease: are high-expanding cortical areas preferentially targeted? Dagger. Cereb Cortex 25(9):2556–2565

Fjell AM, Grydeland H, Krogsrud SK, Amlien I, Rohani DA, Ferschmann L, Storsve AB, Tamnes CK, Sala-Llonch R, Due-Tonnessen P, Bjornerud A, Solsnes AE, Haberg AK, Skranes J, Bartsch H, Chen CH, Thompson WK, Panizzon MS, Kremen WS, Dale AM, Walhovd KB (2015b) Development and aging of cortical thickness correspond to genetic organization patterns. Proc Natl Acad Sci USA 112(50):15462–15467

Fjell AM, Westlye LT, Amlien I, Tamnes CK, Grydeland H, Engvig A, Espeseth T, Reinvang I, Lundervold AJ, Lundervold A, Walhovd KB (2015c) High-expanding cortical regions in human development and evolution are related to higher intellectual abilities. Cereb Cortex 25(1):26–34

Fjell AM, Sneve MH, Grydeland H, Storsve AB, Walhovd KB (2017) The disconnected brain and executive function decline in aging. Cereb Cortex 27(3):2303–2317

Fjell AM, Chen CH, Sederevicius D, Sneve MH, Grydeland H, Krogsrud SK, Amlien I, Ferschmann L, Ness H, Folvik L, Beck D, Mowinckel AM, Tamnes CK, Westerhausen R, Haberg AK, Dale AM, Walhovd KB (2019) Continuity and discontinuity in human cortical development and change from embryonic stages to old age. Cereb Cortex 29(9):3879–3890

Fjell AM, Grydeland H, Wang Y, Amlien IK, Bartres-Faz D, Brandmaier AM, Duzel S, Elman J, Franz CE, Haberg AK, Kietzmann TC, Kievit RA, Kremen WS, Krogsrud SK, Kuhn S, Lindenberger U, Macia D, Mowinckel AM, Nyberg L, Panizzon MS, Sole-Padulles C, Sorensen O, Westerhausen R, Walhovd KB (2021) The genetic organization of longitudinal subcortical volumetric change is stable throughout the lifespan. elife 10

Fraser MA, Shaw ME, Cherbuin N (2015) A systematic review and meta-analysis of longitudinal hippocampal atrophy in healthy human ageing. NeuroImage 112:364–374

Freeman SH, Kandel R, Cruz L, Rozkalne A, Newell K, Frosch MP, Hedley-Whyte ET, Locascio JJ, Lipsitz LA, Hyman BT (2008) Preservation of neuronal number despite age-related cortical brain atrophy in elderly subjects without Alzheimer disease. J Neuropathol Exp Neurol 67(12): 1205–1212

Friederici AD (2023) Evolutionary neuroanatomical expansion of Broca's region serving a human-specific function. Trends Neurosci 46(10):786–796

Geerlings MI, Gerritsen L (2017) Late-life depression, hippocampal volumes, and hypothalamic-pituitary-adrenal Axis regulation: a systematic review and meta-analysis. Biol Psychiatry 82(5): 339–350

Giedd JN, Lenroot RK, Shaw P, Lalonde F, Celano M, White S, Tossell J, Addington A, Gogtay N (2008) Trajectories of anatomic brain development as a phenotype. Novartis Found Symp 289: 101–112. discussion 112–108, 193–105

Glasser MF, Van Essen DC (2011) Mapping human cortical areas in vivo based on myelin content as revealed by T1- and T2-weighted MRI. J Neurosci 31(32):11597–11616

Goddings AL, Roalf D, Lebel C, Tamnes CK (2021) Development of white matter microstructure and executive functions during childhood and adolescence: a review of diffusion MRI studies. Dev Cogn Neurosci 51:101008

Gorbach T, Pudas S, Bartres-Faz D, Brandmaier AM, Duzel S, Henson RN, Idland AV, Lindenberger U, Macia Bros D, Mowinckel AM, Sole-Padulles C, Sorensen O, Walhovd KB, Watne LO, Westerhausen R, Fjell AM, Nyberg L (2020) Longitudinal association between hippocampus atrophy and episodic-memory decline in non-demented APOE epsilon4 carriers. Alzheimers Dement (Amst) 12(1):e12110

Group, G. B. D. N. D. C (2017) Global, regional, and national burden of neurological disorders during 1990-2015: a systematic analysis for the global burden of disease study 2015. Lancet Neurol 16(11):877–897

Grydeland H, Vertes PE, Vasa F, Romero-Garcia R, Whitaker K, Alexander-Bloch AF, Bjornerud A, Patel AX, Sederevicius D, Tamnes CK, Westlye LT, White SR, Walhovd KB, Fjell AM, Bullmore ET (2019) Waves of maturation and senescence in micro-structural MRI markers of human cortical myelination over the lifespan. Cereb Cortex 29(3):1369–1381

Hamer M, Sharma N, Batty GD (2018) Association of objectively measured physical activity with brain structure: UK biobank study. J Intern Med 284(4):439–443

Head D, Kennedy KM, Rodrigue KM, Raz N (2009) Age differences in perseveration: cognitive and neuroanatomical mediators of performance on the Wisconsin card sorting test. Neuropsychologia 47(4):1200–1203

Hedges DW, Erickson LD, Gale SD, Anderson JE, Brown BL (2020) Association between exposure to air pollution and thalamus volume in adults: a cross-sectional study. PLoS One 15(3):e0230829

Hill J, Inder T, Neil J, Dierker D, Harwell J, Van Essen D (2010) Similar patterns of cortical expansion during human development and evolution. Proc Natl Acad Sci USA 107(29): 13135–13140

Hof PR, Morrison JH (2004) The aging brain: morphomolecular senescence of cortical circuits. Trends Neurosci 27(10):607–613

Karama S, Bastin ME, Murray C, Royle NA, Penke L, Munoz Maniega S, Gow AJ, Corley J, Mdel CVH, Lewis JD, Rousseau ME, Lepage C, Fonov V, Collins DL, Booth T, Rioux P, Sherif T, Adalat R, Starr JM, Evans AC, Wardlaw JM, Deary IJ (2014) Childhood cognitive ability accounts for associations between cognitive ability and brain cortical thickness in old age. Mol Psychiatry 19(5):555–559

Krogsrud SK, Tamnes CK, Fjell AM, Amlien I, Grydeland H, Sulutvedt U, Due-Tonnessen P, Bjornerud A, Solsnes AE, Haberg AK, Skrane J, Walhovd KB (2014) Development of hippocampal subfield volumes from 4 to 22 years. Hum Brain Mapp 35(11):5646–5657

Krogsrud SK, Fjell AM, Tamnes CK, Grydeland H, Mork L, Due-Tonnessen P, Bjornerud A, Sampaio-Baptista C, Andersson J, Johansen-Berg H, Walhovd KB (2016) Changes in white matter microstructure in the developing brain—a longitudinal diffusion tensor imaging study of children from 4 to 11years of age. NeuroImage 124(Pt A):473–486

Langnes E, Sneve MH, Sederevicius D, Amlien IK, Walhovd KB, Fjell AM (2020) Anterior and posterior hippocampus macro- and microstructure across the lifespan in relation to memory-a longitudinal study. Hippocampus 30(7):678–692

Lebel C, Treit S, Beaulieu C (2019) A review of diffusion MRI of typical white matter development from early childhood to young adulthood. NMR Biomed 32(4):e3778

Livingston G, Huntley J, Sommerlad A, Ames D, Ballard C, Banerjee S, Brayne C, Burns A, Cohen-Mansfield J, Cooper C, Costafreda SG, Dias A, Fox N, Gitlin LN, Howard R, Kales HC, Kivimaki M, Larson EB, Ogunniyi A, Orgeta V, Ritchie K, Rockwood K, Sampson EL, Samus Q, Schneider LS, Selbaek G, Teri L, Mukadam N (2020) Dementia prevention, intervention, and care: 2020 report of the lancet commission. Lancet 396(10248):413–446

Lovden M, Pagin A, Bartres-Faz D, Boraxbekk CJ, Brandmaier AM, Demnitz N, Drevon CA, Ebmeier KP, Fjell AM, Ghisletta P, Gorbach T, Lindenberger U, Plachti A, Walhovd KB, Nyberg L (2023) No moderating influence of education on the association between changes in hippocampus volume and memory performance in aging. Aging Brain 4:100082

Lustig C, Snyder AZ, Bhakta M, O'Brien KC, McAvoy M, Raichle ME, Morris JC, Buckner RL (2003) Functional deactivations: change with age and dementia of the Alzheimer type. Proc Natl Acad Sci USA 100(24):14504–14509

Madden DJ, Bennett IJ, Burzynska A, Potter GG, Chen NK, Song AW (2012) Diffusion tensor imaging of cerebral white matter integrity in cognitive aging. Biochim Biophys Acta 1822(3): 386–400

Marek S, Tervo-Clemmens B, Calabro FJ, Montez DF, Kay BP, Hatoum AS, Donohue MR, Foran W, Miller RL, Hendrickson TJ, Malone SM, Kandala S, Feczko E, Miranda-Dominguez-O, Graham AM, Earl EA, Perrone AJ, Cordova M, Doyle O, Moore LA, Conan GM, Uriarte J, Snider K, Lynch BJ, Wilgenbusch JC, Pengo T, Tam A, Chen J, Newbold DJ, Zheng A, Seider NA, Van AN, Metoki A, Chauvin RJ, Laumann TO, Greene DJ, Petersen SE, Garavan H, Thompson WK, Nichols TE, Yeo BTT, Barch DM, Luna B, Fair DA, Dosenbach NUF (2022) Publisher correction: reproducible brain-wide association studies require thousands of individuals. Nature 605(7911):E11

Murphy EA, Holland D, Donohue M, McEvoy LK, Hagler DJ Jr, Dale AM, Brewer JB, I. Alzheimer's Disease Neuroimaging (2010) Six-month atrophy in MTL structures is associated with subsequent memory decline in elderly controls. NeuroImage 53(4):1310–1317

Ngandu T, Lehtisalo J, Solomon A, Levalahti E, Ahtiluoto S, Antikainen R, Backman L, Hanninen T, Jula A, Laatikainen T, Lindstrom J, Mangialasche F, Paajanen T, Pajala S, Peltonen M, Rauramaa R, Stigsdotter-Neely A, Strandberg T, Tuomilehto J, Soininen H, Kivipelto M (2015) A 2 year multidomain intervention of diet, exercise, cognitive training, and vascular risk monitoring versus control to prevent cognitive decline in at-risk elderly people (FINGER): a randomised controlled trial. Lancet 385(9984):2255–2263

Nobis L, Manohar SG, Smith SM, Alfaro-Almagro F, Jenkinson M, Mackay CE, Husain M (2019) Hippocampal volume across age: nomograms derived from over 19,700 people in UK biobank. Neuroimage Clin 23:101904

Nyberg L, Lovden M, Riklund K, Lindenberger U, Backman L (2012) Memory aging and brain maintenance. Trends Cogn Sci 16(5):292–305

Nyberg L, Andersson M, Lundquist A, Baare WFC, Bartres-Faz D, Bertram L, Boraxbekk CJ, Brandmaier AM, Demnitz N, Drevon CA, Duezel S, Ebmeier KP, Ghisletta P, Henson R, Jensen DEA, Kievit RA, Knights E, Kuhn S, Lindenberger U, Plachti A, Pudas S, Roe JM, Madsen KS, Sole-Padulles C, Sommerer Y, Suri S, Zsoldos E, Fjell AM, Walhovd KB (2023) Individual differences in brain aging: heterogeneity in cortico-hippocampal but not caudate atrophy rates. Cereb Cortex 33(9):5075–5081

Opel N, Thalamuthu A, Milaneschi Y, Grotegerd D, Flint C, Leenings R, Goltermann J, Richter M, Hahn T, Woditsch G, Berger K, Hermesdorf M, McIntosh A, Whalley HC, Harris MA, MacMaster FP, Walter H, Veer IM, Frodl T, Carballedo A, Krug A, Nenadic I, Kircher T, Aleman A, Groenewold NA, Stein DJ, Soares JC, Zunta-Soares GB, Mwangi B, Wu MJ, Walter M, Li M, Harrison BJ, Davey CG, Cullen KR, Klimes-Dougan B, Mueller BA, Samann PG, Penninx B, Nawijn L, Veltman DJ, Aftanas L, Brak IV, Filimonova EA, Osipov EA, Reneman L, Schrantee A, Grabe HJ, Van der Auwera S, Wittfeld K, Hosten N, Volzke H, Sim K, Gotlib IH, Sacchet MD, Lagopoulos J, Hatton SN, Hickie I, Pozzi E, Thompson PM, Jahanshad N, Schmaal L, Baune BT, Dannlowski U (2021) Brain structural abnormalities in obesity: relation to age, genetic risk, and common psychiatric disorders: evidence through univariate and multivariate mega-analysis including 6420 participants from the ENIGMA MDD working group. Mol Psychiatry 26(9):4839–4852

Ostby Y, Tamnes CK, Fjell AM, Walhovd KB (2012) Dissociating memory processes in the developing brain: the role of hippocampal volume and cortical thickness in recall after minutes versus days. Cereb Cortex 22(2):381–390

Pakkenberg B, Pelvig D, Marner L, Bundgaard MJ, Gundersen HJ, Nyengaard JR, Regeur L (2003) Aging and the human neocortex. Exp Gerontol 38(1–2):95–99

Pang JC, Rilling JK, Roberts JA, van den Heuvel MP, Cocchi L (2022) Evolutionary shaping of human brain dynamics. elife 11:e80627

Persson J, Pudas S, Lind J, Kauppi K, Nilsson LG, Nyberg L (2012) Longitudinal structure-function correlates in elderly reveal MTL dysfunction with cognitive decline. Cereb Cortex 22(10): 2297–2304

Peters A, Morrison JH, Rosene DL, Hyman BT (1998) Feature article: are neurons lost from the primate cerebral cortex during normal aging? Cereb Cortex 8(4):295–300

Pokosova P, Kala D, Sanda J, Jezdik P, Prysiazhniuk Y, Faridova A, Jahodova A, Belohlavkova A, Kalina A, Holubova Z, Jurasek B, Kyncl M, Otahal J (2023) Magnetic resonance imaging techniques for indirect assessment of myelin content in the brain using standard T1w and T2w MRI sequences and postprocessing analysis. Physiol Res 72(S5):S573–S585

Rabbitt P, Lowe C, Shilling V (2001) Frontal tests and models for cognitive ageing. Eur J Cogn Psychol 13:5–28

Rakic P (1988) Specification of cerebral cortical areas. Science 241(4862):170–176

Rakic P (2009) Evolution of the neocortex: a perspective from developmental biology. Nat Rev Neurosci 10(10):724–735

Raz N, Lindenberger U (2011) Only time will tell: cross-sectional studies offer no solution to the age-brain-cognition triangle: comment on Salthouse (2011). Psychol Bull 137(5):790–795

Raz N, Lindenberger U, Rodrigue KM, Kennedy KM, Head D, Williamson A, Dahle C, Gerstorf D, Acker JD (2005) Regional brain changes in aging healthy adults: general trends, individual differences and modifiers. Cereb Cortex 15(11):1676–1689

Reynolds JE, Grohs MN, Dewey D, Lebel C (2019) Global and regional white matter development in early childhood. NeuroImage 196:49–58

Rigters SC, Bos D, Metselaar M, Roshchupkin GV, Baatenburg de Jong RJ, Ikram MA, Vernooij MW, Goedegebure A (2017) Hearing impairment is associated with smaller brain volume in aging. Front Aging Neurosci 9:2

Ritchie SJ, Tucker-Drob EM, Starr JM, Deary IJ (2016) Do cognitive and physical functions age in concert from age 70 to 76? Evidence from the Lothian birth cohort 1936. Span J Psychol 19:E90

Rodrigue KM, Raz N (2004) Shrinkage of the entorhinal cortex over five years predicts memory performance in healthy adults. J Neurosci 24(4):956–963

Schilling KG, Archer D, Yeh FC, Rheault F, Cai LY, Hansen C, Yang Q, Ramdass K, Shafer AT, Resnick SM, Pechman KR, Gifford KA, Hohman TJ, Jefferson A, Anderson AW, Kang H, Landman BA (2022) Aging and white matter microstructure and macrostructure: a longitudinal multi-site diffusion MRI study of 1218 participants. Brain Struct Funct 227(6):2111–2125

Sexton CE, Walhovd KB, Storsve AB, Tamnes CK, Westlye LT, Johansen-Berg H, Fjell AM (2014) Accelerated changes in white matter microstructure during aging: a longitudinal diffusion tensor imaging study. J Neurosci 34(46):15425–15436

Shang X, Hill E, Liu J, Zhu Z, Ge Z, Wang W, He M (2023) Association of type 1 diabetes and age at diagnosis of type 2 diabetes with brain volume and risk of dementia in the UK biobank: a prospective cohort study of community-dwelling participants. Diabet Med 40(2):e14966

Shaw P, Greenstein D, Lerch J, Clasen L, Lenroot R, Gogtay N, Evans A, Rapoport J, Giedd J (2006) Intellectual ability and cortical development in children and adolescents. Nature 440(7084):676–679

Shaw P, Kabani NJ, Lerch JP, Eckstrand K, Lenroot R, Gogtay N, Greenstein D, Clasen L, Evans A, Rapoport JL, Giedd JN, Wise SP (2008) Neurodevelopmental trajectories of the human cerebral cortex. J Neurosci 28(14):3586–3594

Sherwood CC, Subiaul F, Zawidzki TW (2008) A natural history of the human mind: tracing evolutionary changes in brain and cognition. J Anat 212(4):426–454

Sneve MH, Grydeland H, Rosa MGP, Paus T, Chaplin T, Walhovd K, Fjell AM (2019) High-expanding regions in primate cortical brain evolution support Supramodal cognitive flexibility. Cereb Cortex 29(9):3891–3901

Spotorno N, Strandberg O, Vis G, Stomrud E, Nilsson M, Hansson O (2023) Measures of cortical microstructure are linked to amyloid pathology in Alzheimer's disease. Brain 146(4):1602–1614

Sui YV, Masurkar AV, Rusinek H, Reisberg B, Lazar M (2022) Cortical myelin profile variations in healthy aging brain: a T1w/T2w ratio study. NeuroImage 264:119743

Tamnes CK, Walhovd KB, Dale AM, Ostby Y, Grydeland H, Richardson G, Westlye LT, Roddey JC, Hagler DJ Jr, Due-Tonnessen P, Holland D, Fjell AM, I. Alzheimer's Disease Neuroimaging (2013) Brain development and aging: overlapping and unique patterns of change. NeuroImage 68:63–74

Tamnes CK, Overbye K, Ferschmann L, Fjell AM, Walhovd KB, Blakemore SJ, Dumontheil I (2018) Social perspective taking is associated with self-reported prosocial behavior and regional cortical thickness across adolescence. Dev Psychol 54(9):1745–1757

Terry RD, DeTeresa R, Hansen LA (1987) Neocortical cell counts in normal human adult aging. Ann Neurol 21(6):530–539

Topiwala A, Ebmeier KP, Maullin-Sapey T, Nichols TE (2022) Alcohol consumption and MRI markers of brain structure and function: cohort study of 25,378 UK biobank participants. Neuroimage Clin 35:103066

Van Petten C (2004) Relationship between hippocampal volume and memory ability in healthy individuals across the lifespan: review and meta-analysis. Neuropsychologia 42(10):1394–1413

van Soelen IL, Brouwer RM, Peper JS, van Leeuwen M, Koenis MM, van Beijsterveldt TC, Swagerman SC, Kahn RS, Hulshoff Pol HE, Boomsma DI (2012) Brain SCALE: brain structure and cognition: an adolescent longitudinal twin study into the genetic etiology of individual differences. Twin Res Hum Genet 15(3):453–467

Vidal-Pineiro D, Walhovd KB, Storsve AB, Grydeland H, Rohani DA, Fjell AM (2016) Accelerated longitudinal gray/white matter contrast decline in aging in lightly myelinated cortical regions. Hum Brain Mapp 37(10):3669–3684

Vidal-Pineiro D, Wang Y, Krogsrud SK, Amlien IK, Baare WFC, Bartres-Faz D, Bertram L, Brandmaier AM, Drevon CA, Duzel S, Ebmeier K, Henson RN, Junque C, Kievit RA, Kuhn S, Leonardsen E, Lindenberger U, Madsen KS, Magnussen F, Mowinckel AM, Nyberg L, Roe JM, Segura B, Smith SM, Sorensen O, Suri S, Westerhausen R, Zalesky A, Zsoldos E, Walhovd KB, Fjell A (2021) Individual variations in 'brain age' relate to early-life factors more than to longitudinal brain change. elife 10:e69995

Wagenmakers MJ, Oudega ML, Klaus F, Wing D, Orav G, Han LKM, Binnewies J, Beekman ATF, Veltman DJ, Rhebergen D, van Exel E, Eyler LT, Dols A (2023) Brain age of patients with severe late-life depression referred for electroconvulsive therapy. J Affect Disord 330:1–6

Walhovd KB, Krogsrud SK, Amlien IK, Bartsch H, Bjornerud A, Due-Tonnessen P, Grydeland H, Hagler DJ Jr, Haberg AK, Kremen WS, Ferschmann L, Nyberg L, Panizzon MS, Rohani DA,

Skranes J, Storsve AB, Solsnes AE, Tamnes CK, Thompson WK, Reuter C, Dale AM, Fjell AM (2016a) Neurodevelopmental origins of lifespan changes in brain and cognition. Proc Natl Acad Sci USA 113(33):9357–9362

Walhovd KB, Westerhausen R, de Lange AM, Brathen AC, Grydeland H, Engvig A, Fjell AM (2016b) Premises of plasticity - and the loneliness of the medial temporal lobe. NeuroImage 131:48–54

Walhovd KB, Fjell AM, Giedd J, Dale AM, Brown TT (2017) Through thick and thin: a need to reconcile contradictory results on trajectories in human cortical development. Cereb Cortex 27(2):1472–1481

Walhovd KB, Fjell AM, Westerhausen R, Nyberg L, Ebmeier KP, Lindenberger U, Bartres-Faz D, Baare WFC, Siebner HR, Henson R, Drevon CA, Stromstad Knudsen GP, Ljosne IB, Penninx B, Ghisletta P, Rogeberg O, Tyler L, Bertram L, Lifebrain C (2018) Healthy minds 0-100 years: optimising the use of European brain imaging cohorts ("Lifebrain"). Eur Psychiatry 50:47–56

Walhovd KB, Nyberg L, Lindenberger U, Amlien IK, Sorensen O, Wang Y, Mowinckel AM, Kievit RA, Ebmeier KP, Bartres-Faz D, Kuhn S, Boraxbekk CJ, Ghisletta P, Madsen KS, Baare WFC, Zsoldos E, Magnussen F, Vidal-Pineiro D, Penninx B, Fjell AM (2022) Brain aging differs with cognitive ability regardless of education. Sci Rep 12(1):13886

Walhovd KB, Lovden M, Fjell AM (2023) Timing of lifespan influences on brain and cognition. Trends Cogn Sci 27(10):901–915

Westlye LT, Walhovd KB, Dale AM, Bjornerud A, Due-Tonnessen P, Engvig A, Grydeland H, Tamnes CK, Ostby Y, Fjell AM (2010) Life-span changes of the human brain white matter: diffusion tensor imaging (DTI) and volumetry. Cereb Cortex 20(9):2055–2068

Zatorre RJ, Fields RD, Johansen-Berg H (2012) Plasticity in gray and white: neuroimaging changes in brain structure during learning. Nat Neurosci 15(4):528–536

Zhang J, Xie L, Cheng C, Liu Y, Zhang X, Wang H, Hu J, Yu H, Xu J (2023) Hippocampal subfield volumes in mild cognitive impairment and alzheimer's disease: a systematic review and meta-analysis. Brain Imaging Behav 17(6):778–793

Zhu AH, Nir TM, Javid S, Villalon-Reina JE, Rodrigue AL, Strike LT, de Zubicaray GI, McMahon KL, Wright MJ, Medland SE, Blangero JE, Glahn DC, Kochunov P, Haberg AK, Thompson PM, Jahanshad N (2024) Lifespan reference curves for harmonizing multi-site regional brain white matter metrics from diffusion MRI. bioRxiv. https://doi.org/10.1101/2024.02.22.581646

Large-Scale Neuroimaging of Mental Illness

Christopher R. K. Ching, Melody J. Y. Kang, and Paul M. Thompson

Contents

Abstract Neuroimaging has provided important insights into the brain variations related to mental illness. Inconsistencies in prior studies, however, call for methods that lead to more replicable and generalizable brain markers that can reliably predict illness severity, treatment course, and prognosis. A paradigm shift is underway with large-scale international research teams actively pooling data and resources to drive consensus findings and test emerging methods aimed at achieving the goals of

---

C. R. K. Ching (✉), M. J. Y. Kang, and P. M. Thompson
Imaging Genetics Center, Mark and Mary Stevens Neuroimaging & Informatics Institute, Keck School of Medicine, University of Southern California, Marina del Rey, CA, USA

Curr Topics Behav Neurosci (2024) 68: 371–398
https://doi.org/10.1007/7854_2024_462
Published Online: 31 March 2024

precision psychiatry. In parallel with large-scale psychiatric genomics studies, international consortia combining neuroimaging data are mapping the transdiagnostic brain signatures of mental illness on an unprecedented scale. This chapter discusses the major challenges, recent findings, and a roadmap for developing better neuroimaging-based tools and markers for mental illness.

Keywords Biological psychiatry · Machine learning · Mental illness · Neuroimaging · Neuroimaging genetics · Transdiagnostic studies

1 Introduction

Variations in brain structure and function shape an individual's experience of the world (and vice versa), and thus their experience of mental health and illness. Brain structure and function are influenced by a multitude of genetic and environmental factors that determine each individual's dynamic neurobiological fingerprint and their risk for mental illness. In this chapter we will discuss 1) key challenges facing brain research, 2) recent large-scale neuroimaging efforts aimed at driving consensus findings and boosting the generalizability of brain markers for mental illness, and 3) a path forward to discover more powerful translational brain markers.

Some notes regarding terminology used in this chapter: There is an important distinction to be made between *mental health* and *mental illness*. Here, mental health will refer to an individual's mental well-being – including emotional, psychological, and social interactions. Positive mental health can include the ability to cope with stress and daily challenges while maintaining meaningful social relationships, professional commitments, and emotional stability. This sense of well-being or resilience in daily functioning can be experienced on a continuum and is not simply the absence of mental illness but an overall mental balance. Mental illness or psychiatric disorders are thoughts and behaviors that cause significant distress and impairment in daily life, often requiring prolonged treatment and support. Mental illnesses are formally categorized by the Diagnostic and Statistical Manual of Mental Disorders (DSM) and the International Classification of Diseases (ICD) and grouped together based on symptom profiles. While mental health and mental illness are sometimes used interchangeably, the chapter will focus on efforts to map the underlying biology of mental illness.

Brain states versus traits. Measurements of the brain's structure and, to some extent, consistent spatial patterns of brain activity are thought to represent relatively stable traits of the brain over time. This contrasts with brain states, which are more dynamic, moment to moment fluctuations in brain activity that change depending on environmental stimuli, thoughts, or emotions. Much of the chapter will be devoted to large-scale consortium studies using structural brain magnetic resonance imaging (MRI) with a particular focus on cortical and subcortical morphometry, white matter microstructure, and structural connectivity. The term *biomarker* will refer throughout to biological measures that have strong predictive power for illness processes, whereas *markers* will denote correlates of illness processes still in need of further scientific replication and validation.

2 Key Challenges

The aim of neuroimaging research in psychiatry is to establish a link between measures of brain variations and different processes associated with mental illness, thus providing a better understanding of the factors that influence the onset, progression, and treatment of mental illness. Non-invasive, in vivo neuroimaging techniques, such as MRI, have revealed important insights into variations in brain structure and function underlying mental illness. Unlike self-reported symptoms, neuroimaging brain trait and state measures serve as endophenotypes that, in turn, may be closely related to the underlying biology of the illness and may offer a more tractable genetic architecture than symptom profiles (Glahn et al. 2007; Le and Stein 2019). Neuroimaging research, like many other biomedical fields, has suffered from inconsistent findings across studies. Replication and generalizability have been challenged by factors such as diagnostic heterogeneity, relatively high costs of data collection, variable analysis methods, underpowered/underrepresented study samples, a reliance on cross-sectional study design, and a poor understanding of the causal links between biological scales (Fig. 1).

2.1 Diagnostic Heterogeneity

To date, most candidate markers of mental illness – including brain measures and genetic risk variants – exhibit small-to-moderate effect sizes, especially for predicting group differences between patients and controls (Paulus and Thompson 2019). No validated biomarkers (neuroimaging or otherwise) exist for any psychiatric disorder to aid diagnostics, predict course of illness, or guide personalized treatment. A fundamental research challenge involves the way clinicians and researchers draw the line between those with and without a mental illness. The aforementioned DSM and ICD are the most common diagnostic systems, providing reliable assessments of the type, severity, and duration of symptoms that meet consensus thresholds for illness. The refinement of these behavior-based assessment tools has revolutionized the diagnosis and treatment of psychiatric disorders and has led to highly effective treatments (Clarke and Kuhl 2014). Even so, limitations of these diagnostic criteria and their lack of scientific validity have contributed to a period of stagnation in the development of new treatments and clinically actionable biomarkers (Grinker 2021). Patients with the same diagnosis can experience vastly different symptom profiles, comorbid health conditions, and environmental risk factors related to distinct underlying neurobiological processes. This diagnostic-

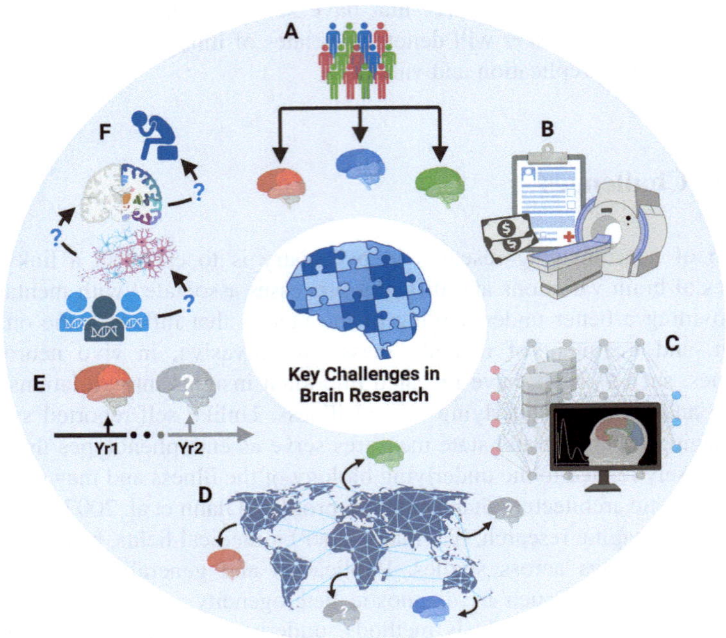

Fig. 1 Key challenges facing brain imaging research on mental illness. (**a**) Diagnostic heterogeneity – where patients with the same psychiatric disorder can present with different symptom profiles, comorbidities, and polygenic risk markers makes it difficult to map consensus underlying brain signatures. (**b**) High cost of neuroimaging research can limit study design and create disparities in access to research tools and data. (**c**) Vast array of neuroimaging processing analysis methods makes it challenging to compare results from one study to the next. (**d**) Small and underrepresented samples contribute to inconsistent results and narrow generalizability of study findings. (**e**) Limited longitudinal data impede modeling of chronic, episodic, and progressive illness markers. (**f**) Limited understanding of the explanatory links between different neurobiological scales from symptoms > brain structure/function > cells/circuits > molecular and genetic mechanisms. Created with help from BioRender.com

level heterogeneity is reflected in highly variable responses to medication, illness course, and long-term outcomes (Zimmerman et al. 2015). Ultimately, the challenge of linking noisy diagnostic labels to consistent underlying biological signals has proved challenging.

2.2 *Cost, Access, and the Need for Larger Longitudinal Samples*

Mounting evidence points to the need for large study samples to improve replication and generalizability of neuroimaging study findings (Grasby et al. 2020; Marek et al.

2022; Liu et al. 2023). Small, underpowered samples can contribute to false positives or overestimates of effect sizes across studies. The high cost of MRI data collection means that most existing neuroimaging studies have evaluated fewer than 100 participants, often focusing on a single (cross-sectional) visit. In an effort to limit interindividual heterogeneity, researchers will often focus recruitment on participants with particular symptom attributes, or "pure forms" of a disorder, which may limit the generalizability of findings to the wider, more heterogeneous population. Furthermore, relatively few psychiatric neuroimaging studies have followed individuals for more than 1 year (Abé et al. 2022; Horga et al. 2014; Giedd and Rapoport 2010), neglecting the chronic, episodic, and progressive nature of many mental illnesses.

As is the unfortunate case for most biomedical research, many neuroimaging studies of mental illness lack ancestral diversity of the wider population (Sharma and Palaniappan 2021). Ambitious (i.e., burdensome) study design may limit participation given that those suffering from mental illness often lack the support needed to participate (Woodall et al. 2010). Access to expensive research tools like an MRI scanner has also created disparities between researchers and institutions with the established infrastructure and financial backing and those without these resources. Markers and treatments derived from studies with limited sample diversity risk failed replication (Palk et al. 2020). Sex differences are also important factors to consider when modeling brain variations related to illness (Salminen et al. 2022). While not all mental illnesses show sex-related disparities in prevalence, sex-related factors (e.g., symptom patterns, prescribed treatments) can contribute to differential illness trajectories.

2.3 *Variability in the Processing and Analysis Techniques Across Studies*

The emergence and popularity of neuroimaging as a research tool can be attributed to modern advances in both scanning and computing technology. Researchers have access to an overwhelming selection of software methodologies to extract a wide array of metrics from brain images. Yet, the diversity of neuroimaging processing and analysis protocols – including their varied implementation and quality control procedures – further complicates the interpretation of findings from one study to the next (Botvinik-Nezer et al. 2020; Carp 2012). Studies purporting to measure the same brain features (e.g., hippocampal volume) may use different techniques (e.g., manual segmentation, automated segmentation, subfield parcellation). Traditional literature-based meta-analyses often combine effect sizes from studies using heterogeneous processing/analysis methods and may suffer from publication bias or quality of data/analysis reporting.

2.4 Building Links Across Scales of Measurement

The search for biomarkers of mental illness has also been challenged by a lack of explanatory connections between neurobiological scales. Links between changes in macro-scale neuroimaging signatures and underlying micro-scale variations in neural circuits, cellular/synaptic functioning, neurotransmitters, and genetic variation remain unclear. In the Alzheimer's disease (AD) research field, neuroimaging has helped shift a once behaviorally defined illness to one rooted in more clinically actionable biomarkers (Veitch et al. 2022). For example, the A/T/N system is a biologically driven classification scheme that quantifies an individual's A: β-amyloid, T: tau, and N: neurodegeneration or neural injury from cerebrospinal fluid (CSF), positron emission tomography (PET), and structural MRI. A/T/N brain trait measures may provide an understanding of a patient's illness processes before cognitive/behavioral deficits become pronounced, when treatments are most likely to be effective at slowing, stopping, or someday reversing progression (Dubois et al. 2014; Jack Jr et al. 2017). These efforts are changing the way we diagnose and identify those at risk for developing AD, helping to guide the design of more biologically informed treatment trials (van Dyck et al. 2023). While clear differences exist between AD etiology and the uncertain causal mechanisms of mental illness, these efforts offer a potential road map for psychiatric research. More recently, international team science initiatives such as the Psychiatric Genomics Consortium, the ENIGMA Consortium, and others have made strides to pool data and resources to address some of these core challenges, aiming to boost study power, replication, and generalizability of mental illness markers.

3 Large-Scale Consortium Science to Improve Study Replication and Generalizability

Prior small-scale efforts to identify polymorphisms and candidate genes for mental illness often failed to replicate in large independent samples (Sullivan 2007). The Psychiatric Genomics Consortium (PGC) is a global effort to discover replicable genetic risk loci for mental illnesses through coordinated, large-scale projects (Sullivan et al. 2018). PGC studies have now revealed hundreds of common genetic variants reliably associated with psychiatric disorders and mapped overlapping and distinct genetic markers across disorders (Gandal et al. 2018; Lam et al. 2019; Mattheisen et al. 2022; Hindley et al. 2022).

In a parallel effort with the PGC, the Enhancing NeuroImaging Genetics through Meta-Analysis (ENIGMA) Consortium launched a project in 2009 to identify both common and rare genetic variants associated with brain structure and function using MRI (Thompson et al. 2020). With the majority of common genetic variants explaining less than 1% of the total variation in any given brain measure, ENIGMA implemented standardized processing and analysis methods to increase study power.

By applying standard protocols, often based on widely adopted tools from the field (e.g., FreeSurfer brain segmentation software (Fischl et al. 2002)), ENIGMA was able to boost statistical power to 99.92% for detecting genetic loci accounting for at least 1% of the variance in a brain region (Stein et al. 2012). The first ENIGMA studies revealed novel genetic variants associated with overall brain volume and morphometry of regions such as the hippocampus (Stein et al. 2012; Hibar et al. 2016). Ongoing collaborative efforts between ENIGMA and other large-scale international consortia, such as the Cohorts for Heart and Aging Research (CHARGE) Consortium (Psaty et al. 2009), IMAGEN (Schumann et al. 2010), and others, continue to validate earlier markers and discover new genetic variants that influence brain structure (Hibar et al. 2017; Satizabal et al. 2019; Grasby et al. 2020), function (Smit et al. 2018), and brain change over time (Brouwer et al. 2022).

The ENIGMA Consortium now includes a growing effort from over 50 working groups focused on the development of standardized neuroimaging and genomics methods, with focused projects on a range of brain conditions, including 26 working groups devoted to the study of mental illness as well as healthy brain variations over the lifespan. Each working group has brought together collaborators from around the world to form the largest standardized neuroimaging datasets in their respective fields and have published the largest studies of their kind. ENIGMA studies now include over 100,000 study participants through partnerships with over 2,400 scientists spanning 340 institutions and 45 countries.

A key advantage to the large-scale consortium approach over more traditional meta-analyses is the implementation of standardized processing and analysis protocols. Whereas traditional neuroimaging meta-analyses combine published effect sizes from studies using variable analysis methods, the consortium approach can empower both decentralized meta- and pooled mega-analyses using standard brain measures from participating cohorts. In theory, data standardization leads to more unbiased investigations of brain measures and can overcome potential publication bias while boosting statistical power by including cohorts that may have been underpowered to detect subtle effects on their own. Furthermore, the coordinated use of standard processing and quality control allows for the direct comparison of data across ENIGMA working groups and the mapping of overlapping and distinct brain variations across traditional diagnostic groups. Initial ENIGMA working group studies often take the form of prospective meta-analyses, where groups process, quality inspect, and then fit harmonized statistical models independently before combining effect size estimates meta-analytically. This process has the benefit of avoiding potential obstacles of sharing primary data sources across institutions, while still boosting analysis power over the traditional, literature-based meta-analysis approach. Mega-analyses – now the norm across ENIGMA studies – pool primary data, allowing for greater data analysis flexibility, which has led to more complex analyses including the application of more advanced machine learning methods that would be difficult to implement via meta-analysis.

It is important to emphasize that large-scale consortium efforts complement and do not replace well-designed, small-scale neuroimaging studies. Multisite consortium studies often require a level of consensus and cooperation between collaborators that requires more pragmatic study design. Smaller studies can often collect detailed phenotype information and apply more complex analysis strategies that are not feasible in larger multisite efforts. Pooling international data to build more ecologically valid study samples better powered to capture generalizable effects is an important goal. In practice, it is important to control for the potential introduction of confounders and artifacts (e.g., greater movement in patients compared with controls (Yao et al. 2017)) as well as site effects (e.g., inclusion/exclusion sample criteria, scanner differences). The modeling of site effects poses a significant analysis challenge, especially in the context of machine learning where algorithms have shown a proclivity to learning the differences across study sites as opposed to the biological signals of interest (Bayer et al. 2022). Here, studies with focused translational targets (e.g., neuroimaging guided optimization of brain stimulation to improve a particular symptom domain) could help isolate potential sources of clinically relevant heterogeneity (Abi-Dargham et al. 2023).

4 Recent Findings from Large-Scale Brain Mapping of Mental Illness

Early neuroimaging (computerized axial tomography) studies showing larger ventricles in patients with schizophrenia (Johnstone et al. 1976) ushered in a new era of psychiatric research, helping to shift perceptions toward an underlying biological basis for psychiatric disorders. While other medical disciplines have moved toward more biologically driven diagnostic and treatment systems, psychiatry remains in the early phases of biomarker discovery and validation. The ENIGMA psychiatric working groups represent the largest coordinated neuroimaging studies to date and have revealed consistent but subtle patterns of variations in the structural properties of gray (Fig. 2) and white matter and in functional features across disorders (Thompson et al. 2022). Machine learning studies making use of large-scale, standardized neuroimaging data have shown modest patient versus control classification performance (Gleichgerrcht et al. 2021; Bruin et al. 2023; Nunes et al. 2020), where the merging of multimodal structural, diffusion, and functional MRI data may help boost predictive power (Rashid and Calhoun 2020).

ENIGMA working groups have also shown that illness-related brain alterations can be influenced by complex factors such as medication (van Erp et al. 2018; Schmaal et al. 2020), comorbid health risk (e.g., body mass index and insomnia) (McWhinney et al. 2022; Leerssen et al. 2020), environmental risk (e.g., childhood trauma (Tozzi et al. 2020)), substance use (Mackey et al. 2019; Navarri et al. 2022; Grace et al. 2021), sex differences (Jahanshad and Thompson 2017), as well as polygenic risk (Opel et al. 2021). For example, large-scale analyses of patients with bipolar disorder show that lithium treatment is associated with thicker cortex, larger subcortical volumes, and a potential normalization (compared with controls) of white matter properties, whereas anticonvulsant mood stabilizers and antipsychotic medications are associated with the opposite patterns when compared with those not receiving these medications (Ching et al. 2022). Such studies show the importance of modeling the complex array of clinical and demographic factors that also influence brain features to understand their role in mental illness. ENIGMA working groups have also shown that combining existing neuroimaging samples from around the world allows for more formal testing of the generalizability of brain markers beyond those from primarily European ancestry (Koshiyama et al. 2022; Palk et al. 2020).

Traditionally, neuroimaging studies of mental illness have focused on single disorders. While these patient-versus-control comparisons provide an important benchmark for understanding associations between psychiatric diagnoses and particular variations in brain structure and function, recent efforts to compare neuroimaging measures across disorders and derive multivariate, individual-level predictors of illness have shown promise. Cross-disorder neuroimaging studies aim to map the shared and distinct brain alterations spanning traditional, behaviorally defined diagnoses. In combination with data-driven techniques, these data provide an opportunity to derive new transdiagnostic clusters of patients based on shared (ideally clinically relevant) biology. Recent ENIGMA cross-disorder analyses show overlapping cortical profiles with larger deviations from controls in schizophrenia, followed by bipolar disorder and major depression (Fig. 2). The shared psychiatric brain signatures appear to be distinct from those associated with "normative" aging (Opel et al. 2020; Cheon et al. 2022). In the largest study of its kind, the ENIGMA OCD, ADHD, and autism spectrum Working Groups combined individual-level brain measures across 151 research sites including 2,271 participants with ADHD, 1,777 with ASD, 2,323 with OCD, and 5,827 healthy controls and found subtle but distinct differences across all three groups (i.e., no shared structural variations) (Boedhoe et al. 2020).

Efforts to parse heterogeneous patient groups into more homogeneous subgroups have been a goal of precision psychiatry for decades. With recent computational advancements and the availability of larger transdiagnostic data samples, supervised/unsupervised machine learning models are now being trained and applied to create illness predictions based on complex multivariate brain and other biometric data (Marquand et al. 2016). Clustering and finite mixture modeling are popular approaches, but most studies have used psychometrics or clinical assessments rather than neuroimaging markers. Normative modeling is a complementary method that quantifies an individual's deviation from a normal/healthy distribution, providing individualized risk scores. These models are widely known from height/weight growth charts to monitor childhood development and can be used to guide clinical interventions when individuals deviate from normal trajectories (Marquand et al. 2019). Recent normative modeling studies from larger transdiagnostic samples have shown that individual-level brain deviations from healthy norms appear to be highly heterogeneous within disorders. While some transdiagnostic networks appear to be shared across disorders, the heterogeneity of brain alterations within disorders highlights the limitations of group averages or *the average patient* to inform individual-level predictions, the ultimate goal of precision psychiatry (Segal et al. 2023; Wolfers et al. 2018).

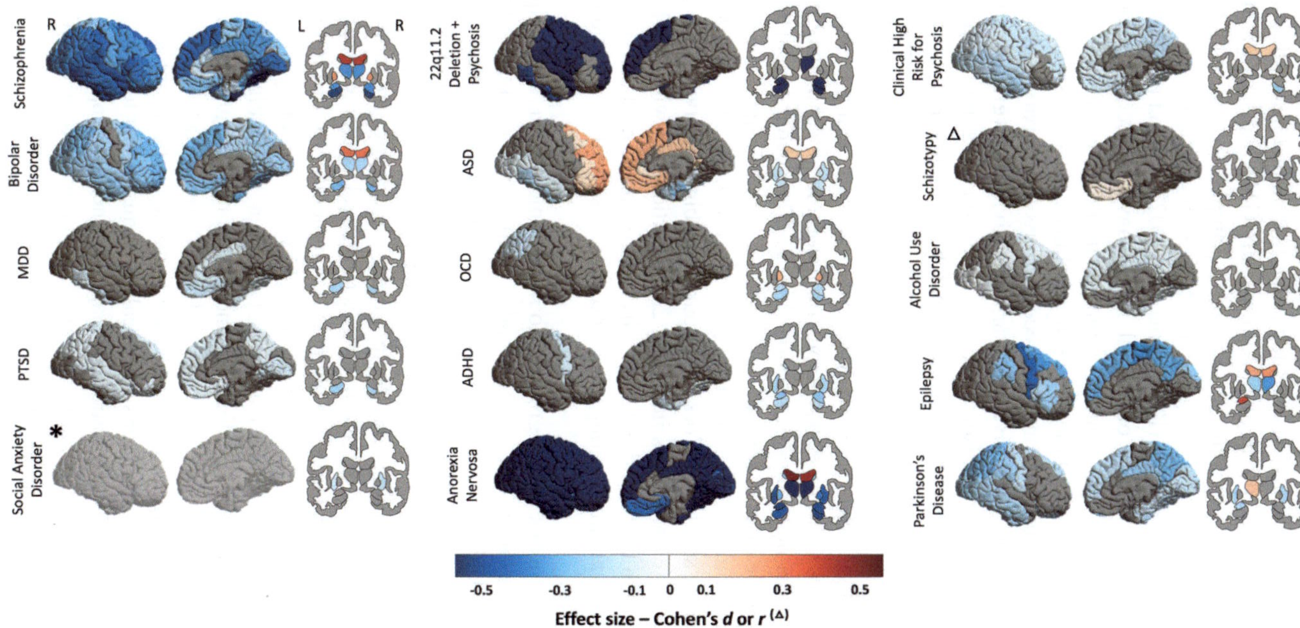

Fig. 2 ENIGMA cortical thickness and subcortical volume findings. Patient-versus-control cortical thickness (unless otherwise specified) and subcortical volume results are displayed for either averaged left and right hemisphere analyses (✦) or hemisphere specific analyses using Cohen's *d* or *r* effect size as provided in the original publication. Findings from: ENIGMA Schizophrenia (van Erp et al. 2018, 2016); Bipolar Disorder (Hibar et al. 2018, 2016); Major Depressive Disorder (MDD) (Schmaal et al. 2017, 2016); ✦Posttraumatic Stress Disorder (PTSD cortical volume) (Wang et al. 2021; Logue et al. 2018); ✦Social Anxiety Disorder (cortical thickness findings not yet published*, lateral ventricle data unavailable) (Groenewold et al. 2023); ✦22q11.2 Deletion Syndrome (deletion carriers with a history of psychosis versus those without) (Sun et al. 2020; Ching et al. 2020), Autism Spectrum Disorder (ASD) (van Rooij et al. 2018); Obsessive Compulsive Disorder (OCD) (Boedhoe et al. 2018, 2017); Attention Deficit Hyperactivity Disorder (ADHD) (Hoogman et al. 2019, 2017); ✦Anorexia Nervosa (Walton et al. 2022); ✦Clinical High Risk for Psychosis (ENIGMA Clinical High Risk for Psychosis Working Group et al. 2021); Schizotypy (correlation between cortical thickness and symptom severity, *r* values reported) (Kirschner et al. 2022), ✦Epilepsy (Whelan et al. 2018), and ✦Parkinson's Disease (Laansma et al. 2021)

The ENIGMA lifespan, brain age, centile brain projects, and others are building normative models from the largest, standardized MRI datasets (Han et al. 2021; Dima et al. 2022; Frangou et al. 2022; Bethlehem et al. 2022), with several publicly available tools now available to compute predicted brain-age difference (https://photon-ai.com/enigma_brainage) and normative deviation scores (https://centilebrain.org; http://www.brainchart.io/).

5 Bridging the Gap between Neuroimaging and Underlying Neurobiology and Genetics

Important advances have been made linking genetic variants to psychiatric diagnoses as well as mapping brain states/traits to illness and clinical dimensions. Many studies lack, however, mechanistic explanations that span biological levels. For example, smaller hippocampal volume is a common neuroimaging finding associated with a range of mental illnesses (Fig. 2), but the mechanisms driving these macro-scale brain alterations remain unclear. Smaller hippocampal volumes in major depression may be driven by similar underlying processes as those in bipolar disorder, but are likely distinct from hippocampal atrophy in Alzheimer's disease – a neurodegenerative disorder that appears genetically and etiologically distinct from most psychiatric disorders (Brainstorm Consortium et al. 2018). Researchers often provide hypotheses about potential links, speculating on whether macro-scale brain alterations are a (proximal) cause or consequence of the illness, often including hypotheses about potential circuits, cells, molecules, and genetic risk factors to contextualize their findings. Empirical support is limited and a better understanding of the causal relationship between biological scales could prove essential to parsing heterogeneity within and between disorders that could help guide new, actionable treatment targets. Here we summarize some recent large-scale efforts underway attempting to bridge these gaps.

5.1 Genetic Influences on Brain Traits and Mental Illness

Large-scale genomic studies show a significant overlap of risk variants across psychiatric disorders (Cross-Disorder Group of the Psychiatric Genomics Consortium et al. 2013, 2019; Bulik-Sullivan et al. 2015). Less is known, however, about the degree to which neuroimaging brain traits across disorders are related to these genetic risk factors. A recent study found significant correlations between brain signatures from the ENIGMA bipolar, major depression, schizophrenia, OCD, ADHD, and autism spectrum Working Group studies (24,360 patients and 37,425 controls) that also mirrored the underlying genetic correlations for the same disorders based on genome-wide association studies (Radonjić et al. 2021) (Fig. 3).

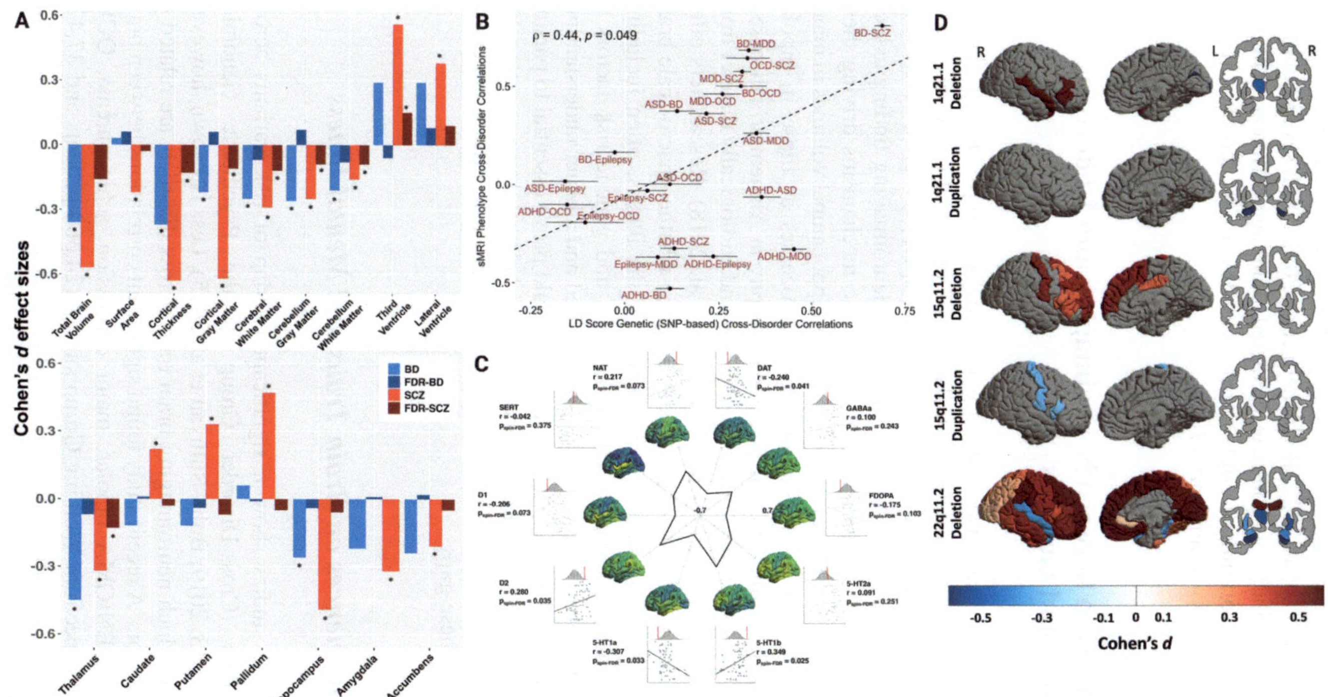

Fig. 3 Bridging the gap between neuroimaging markers of mental illness and underlying neurobiology. (**a**) Cohen's *d* effect sizes comparing global brain and subcortical volume measures in patients with bipolar disorder (BD), schizophrenia (SCZ), and unaffected first-degree relatives (FDRs). Result plots recreated from de Zwarte et al. (2019). (**b**) Correlations in transdiagnostic brain signatures across disorders mirror underlying correlated genetic risk architecture. Pairwise correlations between published ENIGMA (patient vs. control) cortical thickness, surface area, and subcortical volume effects plotted on the y-axis, and genetic correlations (LD-score regression) from independent genome-wide association studies on the x-axis. Dotted line shows best-fit regression line, including Spearman's rho (ρ) and p-value from Radonjić et al. (2021). (**c**) Shared cortical alterations across published ENIGMA bipolar, schizophrenia, major depression,

The ENIGMA Genetics Working Group has also identified a growing list of over 500 genome-wide significant variants influencing the structure of the human brain. Overlap between genetic markers associated with brain traits and genetic variants related to mental illness could help identify the brain regions critical for the development of mental illness. In the largest GWAS of cortical structure, variants associated with cortical surface area were found to overlap those for cognitive function and risk for depression, ADHD, insomnia, and Parkinson's disease (Grasby et al. 2020). The largest GWAS of longitudinal brain change found that novel variants associated with brain growth/atrophy overlapped variants associated with cognitive functioning, height, body mass index, as well as risk for smoking, insomnia, depression, and schizophrenia (Brouwer et al. 2022).

5.2 *From Neuroimaging Variation to Gene Expression*

Several recent large-scale studies have begun to dissect gene expression and illness-related MRI signatures across disorders (see Chapter "Population Neuroscience: Principles and Advances"). By combining maps of cortical illness-related variations from over 12,000 patients and 15,000 controls from the ENIGMA bipolar, schizophrenia, major depression, ADHD, OCD, ADHD, and autism spectrum groups, along with expression profiles from the Allen Human Brain Atlas (Shen et al. 2012), researchers mapped a shared pattern of cell-specific gene expression associated with group differences in cortical thickness, between patients and controls. Genes involved in neurodevelopment and synaptic plasticity were found to act as potential drivers of spatial variations in the thickness differences across disorders (Writing Committee for the Attention-Deficit/Hyperactivity Disorder et al. 2021). Measures of cortical surface area were associated with interregional profiles of prenatal gene expression, where regions with the largest group differences in surface

Fig. 3 (continued) ADHD, OCD, and autism spectrum disorder studies show spatial correlations with in vivo neurotransmitter topography for both serotonin and dopamine receptors and transporters. Neurotransmitter topographic maps are plotted on brain surfaces, scatter plots show spatial correlations of neurotransmitter maps with transdiagnostic brain alterations and histograms represent distributions of correlation coefficients after spin tests and *r*-values (red bars). FDOPA 18 F fluorodopa, D1 and D2 dopamine receptors, dopamine transporter (DAT), noradrenaline transporter (NAT), 5-HT1a/5-HT1b/5-HT2a serotonin receptors, serotonin transporter (SERT), γ-Aminobutyric acid type A receptor (GABAa). Figure from Park et al. (2022). (**d**) Significant copy number variant (CNV) carrier versus control cortical thickness and subcortical volume differences mapped using Cohen's *d* effect size from published ENIGMA CNV Working Group studies. Cortical thickness and subcortical volume results are displayed for either averaged left and right hemisphere analyses (✦) or hemisphere specific analyses. Findings from ENIGMA ✦1q21.1 deletion and duplication (Sønderby et al. 2021; Kumar et al. 2023), 15q11.2 deletion and duplication (Writing Committee for the ENIGMA-CNV Working Group et al. 2020), and ✦22q11.2 deletion (Sun et al. 2020; Ching et al. 2020)

area also showed greater prenatal expression of proliferative cells (intermediate progenitor cells, radial glia) and lower expression of differentiated cells (excitatory neurons, endothelial, and mural cells) during the first trimester (Patel et al. 2022). Studies such as these provide a framework to generate new, testable hypotheses about the potential genetic factors that contribute to the mental illness-related alterations observed from large-scale neuroimaging studies.

5.3 Studying the Family Tree as a Window into Genetic Risk for Mental Illness

Many psychiatric disorders show strong heritability. By studying brain variations in unaffected, first-degree relatives of those diagnosed with a mental illness, researchers can learn how familial risk contributes to resilience and subclinical markers of mental illness. The ENIGMA Relatives Working Group has mapped brain variations across individuals with bipolar disorder and schizophrenia, including their unaffected relatives. In over 6,008 individuals from cohorts around the world, relatives of schizophrenia patients showed lower general intelligence, smaller total brain volumes, and larger ventricle volumes compared with controls (Fig. 3). Relatives of bipolar participants had significantly larger intracranial volumes, showing that despite a shared genetic liability, unaffected relatives from genetically related mental illnesses may show differential patterns of structural brain variation (de Zwarte et al. 2019, 2022).

5.4 Copy Number Variants as a Genetics-First Framework to Study Complex Mental Illness

Copy number variants (CNVs) are structural genetic variants that include, among others, deletions or duplications of portions of the genome. Some (recurrent) CNVs are associated with a range of medical conditions including higher (vs. non-carriers) risk for the development of mental illness, where roughly 10% of children with a neurodevelopmental disorder have a CNV. For example, those carrying the 22q11.2 deletion have 20- to 25-fold higher (vs. non-carriers) risk for developing a schizophrenia spectrum disorder as well as greater risk for developmental disorders such as intellectual disability, ADHD, anxiety, and autism spectrum (Fiksinski et al. 2018). Carriers of the reciprocal 22q11.2 duplication appear to have a different risk profile, including a lower risk for schizophrenia (Marshall et al. 2017). Compared with idiopathic mental illness, CNVs offer a more genetically homogeneous framework to study how highly penetrant variants disrupt neurobiological pathways and contribute to the development of mental illness. Recent large-scale studies pooling participants with a variety of rare CNVs show cortical, subcortical, white matter, and functional brain differences (carriers vs. non-carriers) and significant overlap with brain signatures from large-scale studies of idiopathic mental illness (Sønderby et al. 2022; Ching et al. 2020; Moreau et al. 2021; Sun et al. 2020; Moreau et al. 2020; Jalbrzikowski et al. 2022) (Figs. 2 and 3). Compared with the polygenicity and more subtle brain effects of idiopathic psychiatric disorders, CNVs offer a powerful genetics-first approach to study psychiatric pathophysiology.

5.5 Gene-Environmental Influences on Brain Traits and Mental Illness

Environmental factors including perinatal, childhood environment, pollutants, climate conditions, and more (i.e., the exposome) play a significant role in the development of mental illness (Uher and Zwicker 2017; Lakhani et al. 2019). Twin studies have helped to estimate the genetic and environmental risk, but a large meta-analysis showed that estimates of the neuroimaging variance explained by environmental factors are inconsistent across studies and likely depend on variable analysis methodology (Tooley et al. 2021). For example, socioeconomic status (SES) during childhood development is a strong predictor of well-being later in life. SES impacts the pace of childhood brain development, influencing the maturation of functional networks critical to mental health (e.g., stress may accelerate brain maturation, whereas cognitive enrichment may decelerate maturation (Tooley et al. 2021)). Large-scale efforts to model environmental effects on the brain and mental illness include the environMENTAL project (Schumann et al. 2023) and ENIGMA's Environment Working Group, each of which use geocoding to link neuroimaging and clinical data to geographical data on pollution, toxins, availability of healthcare, urban density, climate, and available data on epidemics.

5.6 Tools to Bridge the Gap

A range of neuroscience resources exist to help generate new testable hypotheses regarding the potential relationships between micro-level mechanisms and macroscale variations in brain structure and function. Resources such as the UK Biobank (large-scale neuroimaging, genetics and deep phenotyping (Sudlow et al. 2015)), Human Connectome Project (high-quality structural/functional connectivity (Van Essen et al. 2013)), Big Brain (human brain histology (Amunts et al. 2013)), and the Allen Human Brain Atlas (post mortem brain structure/gene expression (Shen et al. 2012)) to name a few, provide access to a wide array of data types from genes to circuits to brain variation and behavior.

The ENIGMA Toolbox is an example of an open platform providing multiscale integration between postmortem microstructure, gene expression data, and published neuroimaging effect-size maps from the ENIGMA Consortium (Larivière et al.

2021). Several studies have used ENIGMA cortical thickness findings to investigate structural covariance in networks across psychiatric disorders including bipolar, depression, schizophrenia, autism spectrum, ADHD, and OCD (Fig. 3). Patterns of overlapping illness-related brain variations were observed across bipolar, depression, and schizophrenia and were linked to shared (spatial) gradients across multiple levels of hierarchical cortical organization – including cortical microstructure and cortex-wide variations in serotonin and dopamine receptor topography (Park et al. 2022; Hettwer et al. 2022). In a separate study, brain maps derived from 21,000 patients and 26,000 controls across 13 neurodevelopmental, neurological, and psychiatric disorders were used to relate shared and unique cortical alterations across disorders to multiple micro-architectural measures, including gene expression, neurotransmitter density, metabolism, myelination, and connectivity measures (Hansen et al. 2022). Transdiagnostic cortical variations could be predicted by local attributes, such as neurotransmitter receptor profiles, which highlight a possible link between molecular vulnerability and macro-scale connectivity in shaping disorder-specific cortical abnormalities as well as cross-disorder similarities.

6 Continuing Challenges and the Path Forward

6.1 Moving beyond Patient vs. Control

As discussed, the majority of mental illness neuroimaging research has been focused on group comparisons between those with and without a psychiatric diagnosis. While these studies have made essential contributions to our understanding of average brain alterations associated with many conditions, most well-powered studies show small-moderate effect sizes, including a high degree of overlap between patients and controls, and significant variability across individuals with the same diagnosis.

Alternative research frameworks such as the Research Domain Criteria (RDoC) from the US National Institute of Mental Health (Insel 2014), European Roadmap for Mental Health Research (ROAMER) (Haro et al. 2014), the Hierarchical Taxonomy of Psychopathology (Kotov et al. 2017), as well as other clinical staging models (Uhlhaas et al. 2023) attempt to avoid the challenging task of relating heterogeneous diagnostic labels to underlying neurobiology. These alternatives attempt to embrace known diagnostic heterogeneity, creating an opportunity for a bottom-up approach where transdiagnostic biological signals guide the discovery of new diagnostic criteria and biomarker targets (Insel et al. 2010). Growing evidence supports a general psychopathology factor (i.e., *p*-factor) that captures overall illness burden across disorders and has been shown to be heritable, predictive of certain functional outcomes, and associated with biological measures such as brain imaging (Patel et al. 2021; Martel et al. 2017). Significant challenges to these transdiagnostic construct approaches remain. Prospective, large-scale, ancestrally diverse, and deeply phenotyped cohorts with standardized behavioral and biospecimen batteries will be necessary to power such methods. Population-based cohorts such as the UK

Biobank (adults only) and the Adolescent Brain Cognitive Development (ABCD) (children aged 9–10 years at their baseline visit) lack large samples of individuals with a serious mental illness. Ongoing transdiagnostic studies from the ENIGMA Consortium currently rely on existing, independently collected cohorts, making use of highly variable neuropsychological and behavioral batteries across study sites. Prospective efforts would benefit from standardized symptom severity measures across individuals as well as the inclusion of those in prodromal or subthreshold stages of illness.

6.2 Taking the Long View

Psychiatric neuroimaging continues to rely heavily on cross-sectional measures or a biological snapshot in time. Given the chronic, episodic, and progressive nature of many mental illnesses, longitudinal modeling is desperately needed to assess biomarker performance in predicting illness severity, best treatment course, and prognosis. The cost of adequately powered longitudinal studies runs counter to many existing academic research funding mechanisms, which tend to support shorter duration studies (<5 years) with limited budgets to accomplish deep phenotyping in large cohorts.

Population health registries, especially those from Nordic countries (Laugesen et al. 2021; Schmidt et al. 2019), as well as recent ambitious investments such as the UK Biobank (Sudlow et al. 2015), ABCD (Casey et al. 2018), and the Healthy Brain Network (Alexander et al. 2017) are changing the scale and scope of biomedical research. The return on investment for similar "moon-shot" efforts focused on larger-scale, longitudinal mental-illness data could radically shift the discovery of more actionable mechanisms. An untapped source of discovery could be found through the expansion of clinical health care data and electronic medical records. Millions of patients are seen every day by medical providers around the world, representing a potential source of much needed longitudinal and prognostic mental health information. Privacy and cost concerns persist, but efforts such as the All of Us Study (Barr et al. 2022), Million Veterans (Kimbrel et al. 2022), and others that access large healthcare network data show the promise of using (existing) big-data sources to advance mental-illness discoveries (Lakhani et al. 2019).

6.3 Better Representation = Better Science = Better Translational Markers

Global consortium projects such as ENIGMA (Yan et al. 2019; Koshiyama et al. 2022), the PGC (Lam et al. 2019), and others (Bigdeli et al. 2020) are making important efforts to improve the ancestral diversity in large-scale studies of mental illness. These studies are beginning to show how potential mental illness risk markers generalize to groups beyond European ancestry. In addition to ancestry, future research would greatly benefit from the inclusion of underrepresented groups such as children and those at high risk or in prodromal stages of the illness. Most mental illnesses have an onset between 12–25 years of age and growing evidence points to critical developmental stages that play an important role in psychopathology. These critical periods also create potential windows of interventional opportunity (Paus et al. 2008; Uhlhaas et al. 2023). Most psychiatric research, including neuroimaging, has been focused, however, on adults already diagnosed with a disorder. With more mental illnesses found to exhibit prodromal stages, an increased focus on markers that precede illness onset could lead to more actionable therapeutic strategies for developmental periods when such interventions are most likely to be effective. The success of recent initiatives focusing on clinical high risk for and prodromal stages of psychosis (e.g., The North American Prodrome Longitudinal Study and The Personalized Prognostic Tools for Early Psychosis Management Consortium) provides a road map for the wider psychiatric research community, showing that integration of psychometric, biospecimen, and neuroimaging measures can lead to potential translational predictions for those at highest risk for developing mental illness (ENIGMA Clinical High Risk for Psychosis Working Group et al. 2021; Cannon et al. 2016).

6.4 The Promises of AI

Recent advances in artificial intelligence (AI) including techniques such as machine learning have promised to change the way we diagnose and treat all types of illness. Recent AI models are able to detect subtle multivariate patterns hidden in biomedical data and significant advances have been made in certain fields including protein folding (Jumper et al. 2021) and some imaging-based diagnostics (Oren et al. 2020). There remain several key challenges limiting the performance of AI models in psychiatric neuroimaging.

AI models often require very large samples of training and testing data, and few studies have been adequately powered or applied rigorous external validation to evaluate model generalizability and performance. Given the relatively subtle variations in brain structure and function across mental illnesses, it remains to be seen whether large enough sample sizes (with adequate phenotyping) exist to train and test such models (Marek et al. 2022). AI tools are progressing rapidly and techniques such as deep neural networks, transfer learning, generative models, and the use of multimodal and time-varying predictors are just some of the advancements helping to address current challenges (Nielsen et al. 2020; Koppe et al. 2021; Dhamala et al. 2023). ENIGMA Consortium studies represent some of the largest samples of standardized mental illness neuroimaging data, providing real-world, multisite tests of the latest AI methods to detect illness effects (Sun et al. 2022; Bayer et al. 2022; Radua et al. 2020). The latest machine learning methods are particularly susceptible to site effects (e.g., scanner differences, exclusion criteria), and methods to ensure site-invariance and cope with "domain shift" are needed to improve performance and generalizability (Dinsdale et al. 2021). Lastly, most AI models provide little explanatory information for their predictions (Farahani et al. 2022), and more explainable models are needed to guarantee that such tools will be as transparent, usable, reliable, and as trustworthy as possible.

7 Conclusions

Advances in neuroimaging have vastly improved our understanding of the brain variations associated with mental illness. International consortia are successfully pooling data and resources to map the subtle and generalizable brain signatures of mental illness. We are just beginning to appreciate the complex genetic and environmental factors at play. A paradigm shift toward big data is underway, but significant investments are needed to build the individual-level predictive biomarkers of illness, treatment, and prognosis needed to improve the lives of patients. The latest AI tools are only as powerful as the data used to train them. Future prospective, large-scale, longitudinal, diverse ancestry, transdiagnostic cohorts (including healthy, at risk, and ill individuals), including coordinated deep phenotyping across biological scales are needed to test whether precision psychiatry is within reach.

Acknowledgments We would like to thank the thousands of study participants and their families, the many scientists and colleagues from around the world collaborating in large-scale team science initiatives, and the entire team at the Imaging Genetics Center, part of the Mark and Mary Stevens Neuroimaging and Informatics Institute, Keck School of Medicine of USC.

References

Abé C, Ching CRK, Liberg B, Lebedev AV, Agartz I, Akudjedu TN, Alda M et al (2022) Longitudinal structural brain changes in bipolar disorder: a Multicenter neuroimaging study of 1232 individuals by the ENIGMA Bipolar Disorder Working Group. Biol Psychiatry 91(6): 582–592

Abi-Dargham A, Moeller SJ, Ali F, DeLorenzo C, Domschke K, Horga G, Jutla A et al (2023) Candidate biomarkers in psychiatric disorders: state of the field. World Psychiatry 22(2): 236–262

Alexander LM, Escalera J, Ai L, Andreotti C, Febre K, Mangone A, Vega-Potler N et al (2017) An open resource for transdiagnostic research in Pediatric mental health and learning disorders. Scientific Data 4(December):170181

Amunts K, Lepage C, Borgeat L, Mohlberg H, Dickscheid T, Rousseau M-É, Bludau S et al (2013) BigBrain: an ultrahigh-resolution 3D human brain model. Science 340(6139):1472–1475

Barr PB, Bigdeli TB, Meyers JL (2022) Prevalence, comorbidity, and sociodemographic correlates of psychiatric disorders reported in the all of us research program. JAMA Psychiatry 79(6): 622–628

Bayer JMM, Thompson PM, Ching CRK, Liu M, Chen A, Panzenhagen AC, Jahanshad N, Marquand A, Schmaal L, Sämann PG (2022) Site effects how-to and when: an overview of retrospective techniques to accommodate site effects in multi-site neuroimaging analyses. Front Neurol 13(October):923988

Bethlehem RAI, Seidlitz J, White SR, Vogel JW, Anderson KM, Adamson C, Adler S et al (2022) Brain charts for the human lifespan. Nature 604(7906):525–533

Bigdeli TB, Genovese G, Georgakopoulos P, Meyers JL, Peterson RE, Iyegbe CO, Medeiros H et al (2020) Contributions of common genetic variants to risk of schizophrenia among individuals of African and Latino ancestry. Mol Psychiatry 25(10):2455–2467

Boedhoe PSW, Schmaal L, Abe Y, Ameis SH, Arnold PD, Batistuzzo MC, Benedetti F et al (2017) Distinct subcortical volume alterations in Pediatric and adult OCD: a worldwide meta- and mega-analysis. Am J Psychiatry 174(1):60–69

Boedhoe PSW, Schmaal L, Abe Y, Alonso P, Ameis SH, Anticevic A, Arnold PD et al (2018) Cortical abnormalities associated with Pediatric and adult obsessive-compulsive disorder: findings from the ENIGMA obsessive-compulsive disorder working group. Am J Psychiatry 175(5):453–462

Boedhoe PSW, van Rooij D, Hoogman M, Twisk JWR, Schmaal L, Abe Y, Alonso P et al (2020) Subcortical brain volume, regional cortical thickness, and cortical surface area across disorders: findings from the ENIGMA ADHD, ASD, and OCD working groups. Am J Psychiatry 177(9): 834–843

Botvinik-Nezer R, Holzmeister F, Camerer CF, Dreber A, Huber J, Johannesson M, Kirchler M et al (2020) Variability in the analysis of a single neuroimaging dataset by many teams. Nature 582(7810):84–88

Brainstorm Consortium, Anttila V, Bulik-Sullivan B, Finucane HK, Walters RK, Bras J, Duncan L et al (2018) Analysis of shared heritability in common disorders of the brain. Science 360(6395)

Brouwer RM, Klein M, Grasby KL, Schnack HG, Jahanshad N, Teeuw J, Thomopoulos SI et al (2022) Genetic variants associated with longitudinal changes in brain structure across the lifespan. Nat Neurosci 25(4):421–432

Bruin WB, Abe Y, Alonso P, Anticevic A, Backhausen LL, Balachander S, Bargallo N et al (2023) The functional connectome in obsessive-compulsive disorder: resting-state mega-analysis and machine learning classification for the ENIGMA-OCD consortium. Mol Psychiatry., May 28: 4307–4319

Bulik-Sullivan B, Finucane HK, Anttila V, Gusev A, Day FR, Loh PR, ReproGen Consortium et al (2015) An atlas of genetic correlations across human diseases and traits. Nat Genet 47 (11):1236–1241

Cannon TD, Changhong Y, Addington J, Bearden CE, Cadenhead KS, Cornblatt BA, Heinssen R et al (2016) An individualized risk calculator for research in prodromal psychosis. Am J Psychiatry 173(10):980–988

Carp J (2012) On the plurality of (methodological) worlds: estimating the analytic flexibility of FMRI experiments. Front Neurosci 6(October):149

Casey BJ, Cannonier T, Conley MI, Cohen AO, Barch DM, Heitzeg MM, Soules ME et al (2018) The adolescent brain cognitive development (ABCD) study: imaging acquisition across 21 sites. Dev Cogn Neurosci 32(August):43–54

Cheon E-J, Bearden CE, Sun D, Ching CRK, Andreassen OA, Schmaal L, Veltman DJ et al (2022) Cross disorder comparisons of brain structure in schizophrenia, bipolar disorder, major depressive disorder, and 22q11.2 deletion syndrome: a review of ENIGMA findings. Psychiatry Clin Neurosci 76(5):140–161

Ching CRK, Gutman BA, Sun D, Reina JV, Ragothaman A, Isaev D, Zavaliangos-Petropulu A et al (2020) Mapping subcortical brain alterations in 22q11.2 deletion syndrome: effects of deletion size and convergence with idiopathic neuropsychiatric illness. Am J Psychiatry 177(7):589–600

Ching CRK, Hibar DP, Gurholt TP, Nunes A, Thomopoulos SI, Abé C, Agartz I et al (2022) What we learn about bipolar disorder from large-scale neuroimaging: findings and future directions from the ENIGMA bipolar disorder working group. Hum Brain Mapp 43(1):56–82

Clarke DE, Kuhl EA (2014) DSM-5 cross-cutting symptom measures: a step towards the future of psychiatric care? World Psychiatry 13(3):314–316

Cross-Disorder Group of the Psychiatric Genomics Consortium, Lee SH, Ripke S, Neale BM, Faraone SV, Purcell SM, Perlis RH et al (2013) Genetic relationship between five psychiatric disorders estimated from genome-wide SNPs. Nat Genet 45(9):984–994

Cross-Disorder Group of the Psychiatric Genomics Consortium. Electronic address: plee0@mgh.harvard.edu and Cross-Disorder Group of the Psychiatric Genomics Consortium (2019) Genomic Relationships, Novel Loci, and Pleiotropic Mechanisms across Eight Psychiatric Disorders. Cell 179(7):1469–1482.e11

de Zwarte SMC, Brouwer RM, Agartz I, Alda M, Aleman A, Alpert KI, Bearden CE et al (2019) The association between familial risk and brain abnormalities is disease specific: an ENIGMA-relatives study of schizophrenia and bipolar disorder. Biol Psychiatry 86(7):545–556

de Zwarte SMC, Brouwer RM, Agartz I, Alda M, Alonso-Lana S, Bearden CE, Bertolino A et al (2022) Intelligence, educational attainment, and brain structure in those at familial high-risk for schizophrenia or bipolar disorder. Hum Brain Mapp 43(1):414–430

Dhamala E, Thomas Yeo BT, Holmes AJ (2023) One size does not fit all: methodological considerations for brain-based predictive Modeling in psychiatry. Biol Psychiatry 93(8): 717–728

Dima D, Modabbernia A, Papachristou E, Doucet GE, Agartz I, Aghajani M, Akudjedu TN et al (2022) Subcortical volumes across the lifespan: data from 18,605 healthy individuals aged 3-90 years. Hum Brain Mapp 43(1):452–469

Dinsdale NK, Jenkinson M, Namburete AIL (2021) Deep learning-based unlearning of dataset bias for MRI harmonisation and confound removal. NeuroImage 228(March):117689

Dubois B, Feldman HH, Jacova C, Hampel H, Molinuevo JL, Blennow K, DeKosky ST et al (2014) Advancing research diagnostic criteria for Alzheimer's disease: the IWG-2 criteria. Lancet Neurol 13(6):614–629

ENIGMA Clinical High Risk for Psychosis Working Group, Jalbrzikowski M, Hayes RA, Wood SJ, Nordholm D, Zhou JH, Fusar-Poli P et al (2021) Association of Structural Magnetic Resonance Imaging Measures with Psychosis Onset in individuals at clinical high risk for developing psychosis: an ENIGMA working group mega-analysis. JAMA Psychiatry 78(7): 753–766

Farahani FV, Fiok K, Lahijanian B, Karwowski W, Douglas PK (2022) Explainable AI: a review of applications to neuroimaging data. Front Neurosci 16(December):906290

Fiksinski AM, Schneider M, Murphy CM, Armando M, Vicari S, Canyelles JM, Gothelf D et al (2018) Understanding the Pediatric psychiatric phenotype of 22q11.2 deletion syndrome. Am J Med Genet A 176(10):2182–2191

Fischl B, Salat DH, Busa E, Albert M, Dieterich M, Haselgrove C, van der Kouwe A et al (2002) Whole brain segmentation: automated Labeling of neuroanatomical structures in the human brain. Neuron 33(3):341–355

Frangou S, Modabbernia A, Williams SCR, Papachristou E, Doucet GE, Agartz I, Aghajani M et al (2022) Cortical thickness across the lifespan: data from 17,075 healthy individuals aged 3-90 years. Hum Brain Mapp 43(1):431–451

Gandal MJ, Haney JR, Parikshak NN, Leppa V, Ramaswami G, Hartl C, Schork AJ et al (2018) Shared molecular neuropathology across major psychiatric disorders parallels polygenic overlap. Science 359(6376):693–697

Giedd JN, Rapoport JL (2010) Structural MRI of Pediatric brain development: what have we learned and where are we going? Neuron 67(5):728–734

Glahn DC, Thompson PM, Blangero J (2007) Neuroimaging endophenotypes: strategies for finding genes influencing brain structure and function. Hum Brain Mapp 28(6):488–501

Gleichgerrcht E, Munsell BC, Alhusaini S, Alvim MKM, Bargalló N, Bender B, Bernasconi A et al (2021) Artificial intelligence for classification of temporal lobe epilepsy with ROI-level MRI data: a worldwide ENIGMA-epilepsy study. NeuroImage Clinical 31(July):102765

Grace S, Rossetti MG, Allen N, Batalla A, Bellani M, Brambilla P, Chye Y et al (2021) Sex differences in the neuroanatomy of alcohol dependence: hippocampus and amygdala subregions in a sample of 966 people from the ENIGMA addiction working group. Transl Psychiatry 11(1): 156

Grasby KL, Jahanshad N, Painter JN, Colodro-Conde L, Bralten J, Hibar DP, Lind PA et al (2020) The genetic architecture of the human cerebral cortex. Science 367(6484)

Grinker RR (2021) Nobody's Normal: how culture created the stigma of mental illness. W. W. Norton & Company

Groenewold NA, Bas-Hoogendam JM, Amod AR, Laansma MA, Van Velzen LS, Aghajani M, Hilbert K et al (2023) Volume of subcortical brain regions in social anxiety disorder: mega-analytic results from 37 samples in the ENIGMA-anxiety working group. Mol Psychiatry 28(3): 1079–1089

Han LKM, Dinga R, Hahn T, Ching CRK, Eyler LT, Aftanas L, Aghajani M et al (2021) Brain aging in major depressive disorder: results from the ENIGMA major depressive disorder working group. Mol Psychiatry 26(9):5124–5139

Hansen JY, Shafiei G, Vogel JW, Smart K, Bearden CE, Hoogman M, Franke B et al (2022) Local molecular and global Connectomic contributions to cross-disorder cortical abnormalities. Nat Commun 13(1):4682

Haro JM, Ayuso-Mateos JL, Bitter I, Demotes-Mainard J, Leboyer M, Lewis SW, Linszen D et al (2014) ROAMER: roadmap for mental Health Research in Europe. Int J Methods Psychiatr Res 23(Suppl 1):1–14

Hettwer MD, Larivière S, Park BY, van den Heuvel OA, Schmaal L, Andreassen OA, Ching CRK et al (2022) Coordinated cortical thickness alterations across six neurodevelopmental and psychiatric disorders. Nat Commun 13(1):6851

Hibar DP, Westlye LT, van Erp TGM, Rasmussen J, Leonardo CD, Faskowitz J, Haukvik UK et al (2016) Subcortical volumetric abnormalities in bipolar disorder. Mol Psychiatry 21(12): 1710–1716

Hibar DP, Adams HHH, Jahanshad N, Chauhan G, Stein JL, Hofer E, Renteria ME et al (2017) Novel genetic loci associated with hippocampal volume. Nat Commun 8(January):13624

Hibar DP, Westlye LT, Doan NT, Jahanshad N, Cheung JW, Ching CRK, Versace A et al (2018) Cortical abnormalities in bipolar disorder: an MRI analysis of 6503 individuals from the ENIGMA bipolar disorder working group. Mol Psychiatry 23(4):932–942

Hindley G, Frei O, Shadrin AA, Cheng W, O'Connell KS, Icick R, Parker N et al (2022) Charting the landscape of genetic overlap between mental disorders and related traits beyond genetic correlation. Am J Psychiatry 179(11):833–843

Hoogman M, Bralten J, Hibar DP, Mennes M, Zwiers MP, Schweren LSJ, van Hulzen KJE et al (2017) Subcortical brain volume differences in participants with attention deficit hyperactivity disorder in children and adults: a cross-sectional mega-analysis. Lancet Psychiatry 4(4): 310–319

Hoogman M, Muetzel R, Guimaraes JP, Shumskaya E, Mennes M, Zwiers MP, Jahanshad N et al (2019) Brain imaging of the cortex in ADHD: a coordinated analysis of large-scale clinical and population-based samples. Am J Psychiatry 176(7):531–542

Horga G, Kaur T, Peterson BS (2014) Annual research review: current limitations and future directions in MRI studies of child- and adult-onset developmental psychopathologies. J Child Psychol Psychiatry 55(6):659–680

Insel TR (2014) The NIMH research domain criteria (RDoC) project: precision medicine for psychiatry. Am J Psychiatry 171(4):395–397

Insel T, Cuthbert B, Garvey M, Heinssen R, Pine DS, Quinn K, Sanislow C, Wang P (2010) Research domain criteria (RDoC): toward a new classification framework for research on mental disorders. Am J Psychiatry 167(7):748–751

Jack CR Jr, Wiste HJ, Weigand SD, Therneau TM, Lowe VJ, Knopman DS, Gunter JL et al (2017) Defining imaging biomarker cut points for brain aging and Alzheimer's disease. Alzheimers Dement 13(3):205–216

Jahanshad N, Thompson PM (2017) Multimodal neuroimaging of male and female brain structure in health and disease across the life span. J Neurosci Res 95(1–2):371–379

Jalbrzikowski M, Lin A, Vajdi A, Grigoryan V, Kushan L, Ching CRK, Schleifer C et al (2022) Longitudinal trajectories of cortical development in 22q11.2 copy number variants and typically developing controls. Mol Psychiatry 27(10):4181–4190

Johnstone EC, Crow TJ, Frith CD, Husband J, Kreel L (1976) Cerebral ventricular size and cognitive impairment in chronic schizophrenia. Lancet 2(7992):924–926

Jumper J, Evans R, Pritzel A, Green T, Figurnov M, Ronneberger O, Tunyasuvunakool K et al (2021) Highly accurate protein structure prediction with AlphaFold. Nature 596(7873):583–589

Kimbrel NA, Ashley-Koch AE, Qin XJ, Lindquist JH, Garrett ME, Dennis MF, Hair LP et al (2022) A genome-wide association study of suicide attempts in the million veterans program identifies evidence of Pan-ancestry and ancestry-specific risk loci. Mol Psychiatry 27(4):2264–2272

Kirschner M, Hodzic-Santor B, Antoniades M, Nenadic I, Kircher T, Krug A, Meller T et al (2022) Cortical and subcortical neuroanatomical signatures of Schizotypy in 3004 individuals assessed in a worldwide ENIGMA study. Mol Psychiatry 27(2):1167–1176

Koppe G, Meyer-Lindenberg A, Durstewitz D (2021) Deep learning for small and big data in psychiatry. Neuropsychopharmacology 46(1):176–190

Koshiyama D, Miura K, Nemoto K, Okada N, Matsumoto J, Fukunaga M, Hashimoto R (2022) Neuroimaging studies within cognitive genetics collaborative research organization aiming to replicate and extend works of ENIGMA. Hum Brain Mapp 43(1):182–193

Kotov R, Krueger RF, Watson D, Achenbach TM, Althoff RR, Michael Bagby R, Brown TA et al (2017) The hierarchical taxonomy of psychopathology (HiTOP): a dimensional alternative to traditional Nosologies. J Abnorm Psychol 126(4):454–477

Kumar K, Modenato C, Moreau C, Ching CRK, Harvey A, Martin-Brevet S, Huguet G et al (2023) Subcortical brain alterations in carriers of genomic copy number variants. Am J Psychiatry., July, appiajp20220304 80(9):685–698

Laansma MA, Bright JK, Al-Bachari S, Anderson TJ, Ard T, Assogna F, Baquero KA et al (2021) International Multicenter analysis of brain structure across clinical stages of Parkinson's disease. Mov Disord 36(11):2583–2594

Lakhani CM, Tierney BT, Manrai AK, Yang J, Visscher PM, Patel CJ (2019) Repurposing large health insurance claims data to estimate genetic and environmental contributions in 560 phenotypes. Nat Genet 51(2):327–334

Lam M, Chen C-Y, Li Z, Martin AR, Bryois J, Ma X, Gaspar H et al (2019) Comparative genetic architectures of schizophrenia in east Asian and European populations. Nat Genet 51(12): 1670–1678

Larivière S, Paquola C, Park B-Y, Royer J, Wang Y, Benkarim O, de Wael RV et al (2021) The ENIGMA toolbox: multiscale neural contextualization of multisite neuroimaging datasets. Nat Methods 18(7):698–700

Laugesen K, Ludvigsson JF, Schmidt M, Gissler M, Valdimarsdottir UA, Lunde A, Sørensen HT (2021) Nordic health registry-based research: a review of health care systems and key registries. Clin Epidemiol 13(July):533–554

Le BD, Stein JL (2019) Mapping causal pathways from genetics to neuropsychiatric disorders using genome-wide imaging genetics: current status and future directions. Psychiatry Clin Neurosci 73(7):357–369

Leerssen J, Blanken TF, Pozzi E, Jahanshad N, Aftanas L, Andreassen OA, Baune BT et al (2020) Brain structural correlates of insomnia severity in 1053 individuals with major depressive disorder: results from the ENIGMA MDD working group. Transl Psychiatry 10(1):425

Liu S, Abdellaoui A, Verweij KJH, van Wingen GA (2023) Replicable brain-phenotype associations require large-scale neuroimaging data. Nat Hum Behav 7(8):1344–1356

Logue MW, van Rooij SJH, Dennis EL, Davis SL, Hayes JP, Stevens JS, Densmore M et al (2018) Smaller hippocampal volume in posttraumatic stress disorder: a multisite ENIGMA-PGC study: subcortical volumetry results from posttraumatic stress disorder consortia. Biol Psychiatry 83(3):244–253

Mackey S, Allgaier N, Chaarani B, Spechler P, Orr C, Bunn J, Allen NB et al (2019) Mega-analysis of Gray matter volume in substance dependence: general and substance-specific regional effects. Am J Psychiatry 176(2):119–128

Marek S, Tervo-Clemmens B, Calabro FJ, Montez DF, Kay BP, Hatoum AS, Donohue MR et al (2022) Reproducible brain-wide association studies require thousands of individuals. Nature 603(7902):654–660

Marquand AF, Wolfers T, Mennes M, Buitelaar J, Beckmann CF (2016) Beyond lumping and splitting: a review of computational approaches for stratifying psychiatric disorders. Biol Psychiatry Cogn Neurosci Neuroimaging 1(5):433–447

Marquand AF, Kia SM, Zabihi M, Wolfers T, Buitelaar JK, Beckmann CF (2019) Conceptualizing mental disorders as deviations from normative functioning. Mol Psychiatry 24(10):1415–1424

Marshall CR, Howrigan DP, Merico D, Thiruvahindrapuram B, Wenting W, Greer DS, Antaki D et al (2017) Contribution of copy number variants to schizophrenia from a genome-wide study of 41,321 subjects. Nat Genet 49(1):27–35

Martel MM, Pan PM, Hoffmann MS, Gadelha A, do Rosário MC, Mari JJ, Manfro GG et al (2017) A general psychopathology factor (P factor) in children: structural model analysis and external validation through familial risk and child global executive function. J Abnorm Psychol 126(1): 137–148

Mattheisen M, Grove J, Als TD, Martin J, Voloudakis G, Meier S, Demontis D et al (2022) Identification of shared and differentiating genetic architecture for autism Spectrum disorder, attention-deficit hyperactivity disorder and case subgroups. Nat Genet 54(10):1470–1478

McWhinney SR, Brosch K, Calhoun VD, Crespo-Facorro B, Crossley NA, Dannlowski U, Dickie E et al (2022) Obesity and brain structure in schizophrenia - ENIGMA study in 3021 individuals. Mol Psychiatry 27(9):3731–3737

Moreau CA, Urchs SGW, Kuldeep K, Orban P, Schramm C, Dumas G, Labbe A et al (2020) Mutations associated with neuropsychiatric conditions delineate functional brain connectivity dimensions contributing to autism and schizophrenia. Nat Commun 11(1):5272

Moreau CA, Ching CR, Kumar K, Jacquemont S, Bearden CE (2021) Structural and functional brain alterations revealed by neuroimaging in CNV carriers. Curr Opin Genet Dev 68 (June):88–98

Navarri X, Afzali MH, Lavoie J, Sinha R, Stein DJ, Momenan R, Veltman DJ et al (2022) How do substance use disorders compare to other psychiatric conditions on structural brain abnormalities? A cross-disorder meta-analytic comparison using the ENIGMA consortium findings. Hum Brain Mapp 43(1):399–413

Nielsen AN, Barch DM, Petersen SE, Schlaggar BL, Greene DJ (2020) Machine learning with neuroimaging: evaluating its applications in psychiatry. Biol Psychiatry Cogn Neurosci Neuroimaging 5(8):791–798

Nunes A, Schnack HG, Ching CRK, Agartz I, Akudjedu TN, Alda M, Alnæs D et al (2020) Using structural MRI to identify bipolar disorders - 13 site machine learning study in 3020 individuals from the ENIGMA bipolar disorders working group. Mol Psychiatry 25(9):2130–2143

Opel N, Goltermann J, Hermesdorf M, Berger K, Baune BT, Dannlowski U (2020) Cross-disorder analysis of brain structural abnormalities in six major psychiatric disorders: a secondary analysis of mega- and meta-analytical findings from the ENIGMA consortium. Biol Psychiatry 88(9): 678–686

Opel N, Thalamuthu A, Milaneschi Y, Grotegerd D, Flint C, Leenings R, Goltermann J et al (2021) Brain structural abnormalities in obesity: relation to age, genetic risk, and common psychiatric disorders : evidence through univariate and multivariate mega-analysis including 6420 participants from the ENIGMA MDD working group. Mol Psychiatry 26(9):4839–4852

Oren O, Gersh BJ, Bhatt DL (2020) Artificial intelligence in medical imaging: switching from radiographic pathological data to clinically meaningful endpoints. The Lancet Digit Health 2(9): e486–e488

Palk A, Illes J, Thompson PM, Stein DJ (2020) Ethical issues in global neuroimaging genetics collaborations. NeuroImage 221(November):117208

Park B-Y, Kebets V, Larivière S, Hettwer MD, Paquola C, van Rooij D, Buitelaar J et al (2022) Multiscale neural gradients reflect transdiagnostic effects of major psychiatric conditions on cortical morphology. Commun Biol 5(1):1024

Patel Y, Parker N, Salum GA, Pausova Z, Paus T (2021) General psychopathology, cognition, and the cerebral cortex in 10-year-old children: insights from the adolescent brain cognitive development study. Front Hum Neurosci 15:781554

Patel Y, Shin J, Abé C, Agartz I, Alloza C, Alnæs D, Ambrogi S et al (2022) Virtual ontogeny of cortical growth preceding mental illness. Biol Psychiatry 92(4):299–313

Paulus MP, Thompson WK (2019) The challenges and opportunities of small effects: the new Normal in academic psychiatry. JAMA Psychiatry 76(4):353–354

Paus T, Keshavan M, Giedd JN (2008) Why do many psychiatric disorders emerge during adolescence? Nat Rev Neurosci 9(12):947–957

Psaty BM, O'Donnell CJ, Gudnason V, Lunetta KL, Folsom AR, Rotter JI, Uitterlinden AG et al (2009) Cohorts for heart and aging research in genomic epidemiology (CHARGE) consortium: Design of Prospective Meta-Analyses of genome-wide association studies from 5 cohorts. Circ Cardiovasc Genet 2(1):73–80

Radonjić NV, Hess JL, Rovira P, Andreassen O, Buitelaar JK, Ching CRK, Franke B et al (2021) Structural brain imaging studies offer clues about the effects of the shared genetic Etiology among neuropsychiatric disorders. Mol Psychiatry 26(6):2101–2110

Radua J, Vieta E, Shinohara R, Kochunov P, Quidé Y, Green MJ, Weickert CS et al (2020) Increased power by harmonizing structural MRI site differences with the ComBat batch adjustment method in ENIGMA. NeuroImage 218(September):116956

Rashid B, Calhoun V (2020) Towards a brain-based Predictome of mental illness. Hum Brain Mapp 41(12):3468–3535

Salminen LE, Tubi MA, Bright J, Thomopoulos SI, Wieand A, Thompson PM (2022) Sex is a defining feature of neuroimaging phenotypes in major brain disorders. Hum Brain Mapp 43(1): 500–542

Satizabal CL, Adams HHH, Hibar DP, White CC, Knol MJ, Stein JL, Scholz M et al (2019) Genetic architecture of subcortical brain structures in 38,851 individuals. Nat Genet 51(11):1624–1636

Schmaal L, Veltman DJ, van Erp TGM, Sämann PG, Frodl T, Jahanshad N, Loehrer E et al (2016) Subcortical brain alterations in major depressive disorder: findings from the ENIGMA major depressive disorder working group. Mol Psychiatry 21(6):806–812

Schmaal L, Hibar DP, Sämann PG, Hall GB, Baune BT, Jahanshad N, Cheung JW et al (2017) Cortical abnormalities in adults and adolescents with major depression based on brain scans from 20 cohorts worldwide in the ENIGMA major depressive disorder working group. Mol Psychiatry 22(6):900–909

Schmaal L, Pozzi E, Ho TC, van Velzen LS, Veer IM, Opel N, Van Someren EJW et al (2020) ENIGMA MDD: seven years of global neuroimaging studies of major depression through worldwide data sharing. Transl Psychiatry 10(1):172

Schmidt M, Schmidt SAJ, Adelborg K, Sundbøll J, Laugesen K, Ehrenstein V, Sørensen HT (2019) The Danish health care system and epidemiological research: from health care contacts to database records. Clin Epidemiol 11(July):563–591

Schumann G, Loth E, Banaschewski T, Barbot A, Barker G, Büchel C, Conrod PJ et al (2010) The IMAGEN study: reinforcement-related behaviour in Normal brain function and psychopathology. Mol Psychiatry 15(12):1128–1139

Schumann G, Andreassen OA, Banaschewski T, Calhoun VD, Clinton N, Desrivieres S, Brandlistuen RE et al (2023) Addressing global environmental challenges to mental health using population neuroscience: a review. JAMA Psychiatry , August 80(10):1066–1074

Segal A, Parkes L, Aquino K, Kia SM, Wolfers T, Franke B, Hoogman M et al (2023) Regional, circuit and network heterogeneity of brain abnormalities in psychiatric disorders. Nat Neurosci 26:1613–1629. August

Sharma A, Palaniappan L (2021) Improving diversity in medical research. Nat Rev Dis Primers 7(1):74

Shen EH, Overly CC, Jones AR (2012) The Allen human brain atlas: comprehensive gene expression mapping of the human brain. Trends Neurosci 35(12):711–714

Smit DJA, Wright MJ, Meyers JL, Martin NG, Ho YYW, Malone SM, Zhang J et al (2018) Genome-wide association analysis links multiple psychiatric liability genes to oscillatory brain activity. Hum Brain Mapp 39(11):4183–4195

Sønderby IE, van der Meer D, Moreau C, Tobias Kaufmann G, Walters B, Ellegaard M, Abdellaoui A et al (2021) 1q21.1 distal copy number variants are associated with cerebral and cognitive alterations in humans. Transl Psychiatry 11(1):182

Sønderby IE, Ching CRK, Thomopoulos SI, van der Meer D, Sun D, Villalon-Reina JE, Agartz I et al (2022) Effects of copy number variations on brain structure and risk for psychiatric illness: large-scale studies from the ENIGMA working groups on CNVs. Hum Brain Mapp 43(1): 300–328

Stein JL, Medland SE, Vasquez AA, Hibar DP, Senstad RE, Winkler AM, Toro R et al (2012) Identification of common variants associated with human hippocampal and intracranial volumes. Nat Genet 44(5):552–561

Sudlow C, Gallacher J, Allen N, Beral V, Burton P, Danesh J, Downey P et al (2015) UK biobank: an open access resource for identifying the causes of a wide range of complex diseases of middle and old age. PLoS Med 12(3):e1001779

Sullivan PF (2007) Spurious genetic associations. Biol Psychiatry 61(10):1121–1126

Sullivan PF, Agrawal A, Bulik CM, Andreassen OA, Børglum AD, Breen G, Cichon S et al (2018) Psychiatric genomics: an update and an agenda. Am J Psychiatry 175(1):15–27

Sun D, Ching CRK, Lin A, Forsyth JK, Kushan L, Vajdi A, Jalbrzikowski M et al (2020) Large-scale mapping of cortical alterations in 22q11.2 deletion syndrome: convergence with idiopathic psychosis and effects of deletion size. Mol Psychiatry 25(8):1822–1834

Sun D, Rakesh G, Haswell CC, Mark Logue C, Baird L, O'Leary EN, Cotton AS et al (2022) A comparison of methods to harmonize cortical thickness measurements across scanners and sites. NeuroImage 261(November):119509

Thompson PM, Jahanshad N, Ching CRK, Salminen LE, Thomopoulos SI, Bright J, Baune BT et al (2020) ENIGMA and global neuroscience: a decade of large-scale studies of the brain in health and disease across more than 40 countries. Transl Psychiatry 10(1):100

Thompson PM, Jahanshad N, Schmaal L, Turner JA, Winkler AM, Thomopoulos SI, Egan GF, Kochunov P (2022) The enhancing NeuroImaging genetics through meta-analysis consortium: 10 years of global collaborations in human brain mapping. Hum Brain Mapp 43(1):15–22

Tooley UA, Bassett DS, Mackey AP (2021) Environmental influences on the pace of brain development. Nat Rev Neurosci 22(6):372–384

Tozzi L, Garczarek L, Janowitz D, Stein DJ, Wittfeld K, Dobrowolny H, Lagopoulos J et al (2020) Interactive impact of childhood maltreatment, depression, and age on cortical brain structure: mega-analytic findings from a large multi-site cohort. Psychol Med 50(6):1020–1031

Uher R, Zwicker A (2017) Etiology in psychiatry: embracing the reality of poly-gene-environmental causation of mental illness. World Psychiatry 16(2):121–129

Uhlhaas PJ, Davey CG, Mehta UM, Shah J, Torous J, Allen NB, Avenevoli S et al (2023) Towards a youth mental health paradigm: a perspective and roadmap. Mol Psychiatry 28:3171–3181, August

van Dyck CH, Swanson CJ, Aisen P, Bateman RJ, Chen C, Gee M, Kanekiyo M et al (2023) Lecanemab in early Alzheimer's disease. N Engl J Med 388(1):9–21

van Erp TGM, Hibar DP, Rasmussen JM, Glahn DC, Pearlson GD, Andreassen OA, Agartz I et al (2016) Subcortical brain volume abnormalities in 2028 individuals with schizophrenia and 2540 healthy controls via the ENIGMA consortium. Mol Psychiatry 21(4):547–553

van Erp TGM, Walton E, Hibar DP, Schmaal L, Jiang W, Glahn DC, Pearlson GD et al (2018) Cortical brain abnormalities in 4474 individuals with schizophrenia and 5098 control subjects via the enhancing neuro imaging genetics through meta analysis (ENIGMA) consortium. Biol Psychiatry 84(9):644–654

Van Essen DC, Smith SM, Barch DM, Behrens TEJ, Yacoub E, Ugurbil K, WU-Minn HCP Consortium (2013) The WU-Minn human connectome project: an overview. NeuroImage 80 (October):62–79
van Rooij D, Anagnostou E, Arango C, Auzias G, Behrmann M, Busatto GF, Calderoni S et al (2018) Cortical and subcortical brain morphometry differences between patients with autism Spectrum disorder and healthy individuals across the lifespan: results from the ENIGMA ASD working group. Am J Psychiatry 175(4):359–369
Veitch DP, Weiner MW, Aisen PS, Beckett LA, DeCarli C, Green RC, Harvey D et al (2022) Using the Alzheimer's Disease Neuroimaging Initiative to improve early detection, diagnosis, and treatment of Alzheimer's disease. Alzheimers Dement 18(4):824–857
Walton E, Bernardoni F, Batury V-L, Bahnsen K, Larivière S, Abbate-Daga G, Andres-Perpiña S et al (2022) Brain structure in acutely underweight and partially weight-restored individuals with anorexia nervosa: a coordinated analysis by the ENIGMA eating disorders working group. Biol Psychiatry 92(9):730–738
Wang X, Xie H, Chen T, Cotton AS, Salminen LE, Logue MW, Clarke-Rubright EK et al (2021) Cortical volume abnormalities in posttraumatic stress disorder: an ENIGMA-psychiatric genomics consortium PTSD workgroup mega-analysis. Mol Psychiatry 26(8):4331–4343
Whelan CD, Altmann A, Botía JA, Jahanshad N, Hibar DP, Absil J, Alhusaini S et al (2018) Structural brain abnormalities in the common epilepsies assessed in a worldwide ENIGMA study. Brain J Neurol 141(2):391–408
Wolfers T, Doan NT, Kaufmann T, Alnæs D, Moberget T, Agartz I, Buitelaar JK et al (2018) Mapping the heterogeneous phenotype of schizophrenia and bipolar disorder using normative models. JAMA Psychiatry 75(11):1146–1155
Woodall A, Morgan C, Sloan C, Howard L (2010) Barriers to participation in mental Health Research: are there specific gender, ethnicity and age related barriers? BMC Psychiatry 10 (December):103
Writing Committee for the Attention-Deficit/Hyperactivity Disorder, Autism Spectrum Disorder, Bipolar Disorder, Major Depressive Disorder, Obsessive-Compulsive Disorder, and Schizophrenia ENIGMA Working Groups, Patel Y et al (2021) Virtual histology of cortical thickness and shared neurobiology in 6 psychiatric disorders. JAMA Psychiatry 78(1):47–63
Writing Committee for the ENIGMA-CNV Working Group, van der Meer D, Sønderby IE, Tobias Kaufmann G, Walters B, Abdellaoui A, Ames D et al (2020) Association of Copy Number Variation of the 15q11.2 BP1-BP2 region with cortical and subcortical morphology and cognition. JAMA Psychiatry 77(4):420–430
Yan C-G, Chen X, Li L, Castellanos FX, Bai T-J, Bo Q-J, Cao J et al (2019) Reduced default mode network functional connectivity in patients with recurrent major depressive disorder. Proc Natl Acad Sci USA 116(18):9078–9083
Yao N, Winkler AM, Barrett J, Book GA, Beetham T, Horseman R, Leach O et al (2017) Inferring pathobiology from structural MRI in schizophrenia and bipolar disorder: Modeling head motion and neuroanatomical specificity. Hum Brain Mapp 38(8):3757–3770
Zimmerman M, Ellison W, Young D, Chelminski I, Dalrymple K (2015) How many different ways do patients meet the diagnostic criteria for major depressive disorder? Compr Psychiatry 56 (January):29–34

Correction to: Area-level Measures of the Social Environment: Operationalization, Pitfalls, and Ways Forward

Marco Helbich, Yi Zeng, and Abeed Sarker

Correction to: Chapter "Area-level Measures of the Social Environment: Operationalization, Pitfalls, and Ways Forward", Curr Topics Behav Neurosci, https://doi.org/10.1007/7854_2024_464

The original version of the chapter was inadvertently published without the following corrections. The chapter has now been corrected.

Correction:

The additional affiliations of the author Dr. Marco Helbich was missed to add in the published chapter. This has been now included with the below additional affiliation details:

M. Helbich (✉)

1 Department of Human Geography and Spatial Planning, Faculty of Geosciences, Utrecht University, Utrecht, The Netherlands

2 Health and Quality of Life in a Green and Sustainable Environment Research Group, Strategic Research and Innovation Program for the Development of MU – Plovdiv, Medical University of Plovdiv, Plovdiv, Bulgaria

3 Environmental Health Division, Research Institute at Medical University of Plovdiv, Medical University of Plovdiv, Plovdiv, Bulgaria

The updated online version of this chapter can be found at https://doi.org/10.1007/7854_2024_464

Curr Topics Behav Neurosci (2024) 68: C1
https://doi.org/10.1007/7854_2024_481
Published Online: 13 June 2024

MIX
Papier aus verantwortungsvollen Quellen
Paper from responsible sources
FSC® C105338

If you have any concerns about our products,
you can contact us on
ProductSafety@springernature.com

In case Publisher is established outside the EU,
the EU authorized representative is:
Springer Nature Customer Service Center GmbH
Europaplatz 3, 69115 Heidelberg, Germany

Printed by Libri Plureos GmbH
in Hamburg, Germany